High-Pressure Studies of Crystalline Materials

High-Pressure Studies of Crystalline Materials

Selected Articles Published by MDPI

MDPI • Basel • Beijing • Wuhan • Barcelona • Belgrade

MDPI

Editorial Office
MDPI
St. Alban-Anlage 66
Basel, Switzerland

This is a reprint of articles published online by the open access publisher MDPI from 2017 to 2018 (available at: www.mdpi.com).

For citation purposes, cite each article independently as indicated on the article page online and as indicated below:

LastName, A.A.; LastName, B.B.; LastName, C.C. Article Title. *Journal Name* **Year**, *Article Number, Page Range*.

ISBN 978-3-03897-131-3 (Pbk)
ISBN 978-3-03897-132-0 (PDF)

Contents

Preface to "High-Pressure Studies of Crystalline Materials"

This book contains a selection of research articles focused on High-Pressure Studies of Crystalline Materials that have been published in Crystals. It includes a collection of articles describing and discussing recent results of high-pressure studies on structural, mechanical, vibrational, and electronic properties of a broad variety of materials. In particular, the articles of the book have been selected with the aim of giving a special emphasis to phase transitions and their effects on different properties and to novel phenomena, but other issues are not excluded. Both experimental and theoretical contributions have been included in the book. The selected works are of relevance for a wide range of disciplines from solid state physics, to chemistry and materials sciences, engineering, and earth and planetary sciences among others.

Daniel Errandonea
Universidad de Valencia
Spain

Article

Evolution of Interatomic and Intermolecular Interactions and Polymorphism of Melamine at High Pressure

Hannah Shelton [1,2] ⓘ, **Przemyslaw Dera** [1,2,*] and **Sergey Tkachev** [3] ⓘ

[1] Department of Geology & Geophysics, School of Ocean and Earth Science and Technology, University of Hawaii at Mānoa, 1680 East West Road, POST Bldg., Honolulu, HI 96822, USA; sheltonh@hawaii.edu

[2] Hawaii Institute of Geophysics and Planetology, School of Ocean and Earth Science and Technology, University of Hawaii at Mānoa, 1680 East West Road, POST Bldg., Honolulu, HI 96822, USA

[3] Center for Advanced Radiation Sources, University of Chicago, Argonne National Laboratory, 9700 S. Cass Ave., Bldg. 434, Argonne, IL 60439, USA; tkachev@cars.uchicago.edu

* Correspondence: pdera@hawaii.edu

Received: 1 June 2018; Accepted: 21 June 2018; Published: 27 June 2018

Abstract: Melamine ($C_3H_6N_6$; 1,3,5-triazine-2,4,6-triamine) is an aromatic substituted *s*-triazine, with carbon and nitrogen atoms forming the ring body, and amino groups bonded to each carbon. Melamine is widely used to produce laminate products, adhesives, and flame retardants, but is also similar chemically and structurally to many energetic materials, including TATB (2,4,6-triamino-1,3,5-trinitrobenzene) and RDX (1,3,5-trinitroperhydro-1,3,5-triazine). Additionally, melamine may be a precursor in the synthesis of superhard carbon-nitrides, such as β-C_3N_4. In the crystalline state melamine forms corrugated sheets of individual molecules, which are stacked on top of one another, and linked by intra- and inter-plane N-H hydrogen bonds. Several previous high-pressure X-ray diffraction and Raman spectroscopy studies have claimed that melamine undergoes two or more phase transformations below 25 GPa. Our results show no indication of previously reported low pressure polymorphism up to approximately 30 GPa. High-pressure crystal structure refinements demonstrate that the individual molecular units of melamine are remarkably rigid, and their geometry changes very little despite volume decrease by almost a factor of two at 30 GPa and major re-arrangements of the intermolecular interactions, as seen through the Hirshfeld surface analysis. A symmetry change from monoclinic to triclinic, indicated by both dramatic changes in diffraction pattern, as well as discontinuities in the vibration mode behavior, was observed above approximately 36 GPa in helium and 30 GPa in neon pressure media. Examination of the hydrogen bonding behavior in melamine's structure will allow its improved utilization as a chemical feedstock and analog for related energetic compounds.

Keywords: melamine; *s*-triazine; hydrogen bond

---

1. Introduction

The six-membered aromatic ring of *s*-triazine, particularly when substituted with nitro (-NO_2) groups, is a structural motif often found among molecular energetic materials [1]. Melamine, ($C_3N_6H_6$, 1,3,5-triazine-2,4,6-triamine) is a simple and very stable heterocyclic aromatic molecule based on the *s*-triazine ring. At ambient conditions melamine forms a crystalline molecular solid with monoclinic symmetry, with space group $P2_1/a$, in which intermolecular interactions are dominated by N-H···N hydrogen bonds. At ambient conditions individual melamine molecules are linked to others via eight N-H hydrogen bonds, balanced between donor and acceptor roles. The superstructure of melamine forms corrugated sheets of individual melamine molecules, where kinked planes of molecules are stacked on top of one another. When used as a salt or mixed with resins, melamine is an effective fire retardant, in part due to the release of flame-smothering nitrogen gas when burned [2]. When combined with formaldehyde, melamine forms a very durable thermosetting plastic used in a broad variety of kitchenware and household goods [3]. Despite its stability and flame-retardant properties, melamine is also very closely related, both structurally and chemically, to the widely used molecular explosives RDX (hexahydro-1,3,5-trinitro-1,3,5-triazine), TATB (2,4,6-triamino-1,3,5-trinitrobenzene), and to 2,4,6-trinitro-1,3,5-triazine, a hypothetical new explosive [4] which has not yet been successfully synthesized [5,6]. As a curiously stable cousin of these explosives, melamine is a worthwhile target of investigation in the search for new energetic materials with enhanced safety and stability while maintaining sufficient explosive potential [7].

The primary motivation for studying the high-pressure behavior of melamine ultimately stems from its intermediate position between energetic species and ultra-hard materials; for instance, melamine may be a functional precursor for synthesis of a hypothetical β-C_3N_4 phase, with a β-Si_3N_4 structure, predicted to be a super-hard material [8]. Melamine appears to be a good reagent for the high-pressure, high-temperature solid state reactions where carbon nitrides are formed; however, the transformations it exhibits are rather complex, and the products strongly depend on the conditions of pyrolysis and presence of catalysts. Montigaud et al. [9] demonstrated that pyrolysis of melamine at 2.5 GPa and 800 °C in presence of hydrazine leads to formation of a bi-dimensional honeycomb-type structure close to those expected for the theoretical graphitic-like g-C_3N_4. Similar syntheses successfully producing g-C_3N_4 and closely related byproducts utilizing pure melamine have also been achieved [8,10]. Synthesis of C-N nanotubes was also observed in high-pressure catalytic pyrolysis of melamine with NaN_3-Fe-Ni at 35 MPa and 650 °C [11]. Interestingly, another recent high-pressure pyrolysis experiments at 5 GPa and 800 °C showed formation of a different molecular crystalline monoclinic solid [12].

Given the complexity of coupled high temperature and pressure pyrolysis, a good starting point for understanding the reaction potential of melamine is through a thorough examination of the compression mechanism. One of the first high-pressure studies of melamine to 4.0 GPa utilized infrared spectroscopy [13], and while noting number of important band frequency shifts, it did not report any discontinuous changes in the compression behavior. In contrast, a Raman study to 8.7 GPa in alcohol pressure medium reported discontinuities in the pressure-dependence of some modes, possibly indicating two phase transitions at 2 and 6 GPa [14]. Indeed, a later synchrotron energy dispersive (EDX) powder diffraction study to 14.7 GPa in alcohol pressure medium [15] seemed to confirm the notion of the two low pressure phase transitions, where a new triclinic phase was suggested to form at 1.3 GPa, and transform to an orthorhombic polymorph above 8.2 GPa. The resolution and sensitivity of the EDX experiments, however, were not sufficient to constrain the structures of the new phases, as less than ten indexable diffraction peaks were observed. Recently, Liu et al. [16] carried out a Raman spectroscopy experiment with melamine powder without pressure medium to 25 GPa, but did not confirm any of the earlier reported discontinuities. A soft mode behavior was observed for one of the N-H vibrations, and N-H vibration peaks gradually disappeared above 10 GPa. A detailed Raman, far-IR, and angular-dispersive X-ray diffraction study conducted by Galley et al. reported three potential phase transitions of powdered melamine with KBr as a pressure-transmitting medium, two of which were observed previously [15,17], with an additional high-pressure phase forming at 16 GPa. As with previous spectroscopic studies, it was observed that the modes associated with N-H stretching, particularly at 3116 and 3325 cm^{-1}, broaden significantly with pressure. This study also documented irreversible amorphization of the sample, attributed to weakening of the internal bonds, which could foster polymerization of the sample.

In each of these previous experiments, the effects of non-hydrostatic stress, hydrogen-bond interactions between melamine and polar solvent-based pressure media, or inter-grain interactions exist. In order to re-examine the details of compression behavior of melamine and reconcile the different results of previous studies we have conducted a combined X-ray diffraction and Raman spectroscopy investigation using high quality single crystal samples in quasihydrostatic noble gas pressure transmitting media.

2. Materials and Methods

The high-pressure X-ray diffraction and Raman spectroscopy experiments took place over several experimental sessions between 2010 and 2015, where experiments were conducted at the Advanced Photon Source (APS) at Argonne National Laboratory. Data were collected at the 13-IDD and 16-IDB beamlines of GeoSoilEnviro-CARS (GSECARS) and the High Pressure Collaborative Access Team (HPCAT), with gas-loading of diamond anvil cells and Raman measurements conducted at GSECARS [18]. Prior to each experiment, melamine powder of >99% purity was recrystallized from saturated aqueous solution, and crystals approximately $10 \times 30 \times 30$ μm in diameter were loaded into symmetric type diamond anvil cell equipped with conical Boehler-Almax type anvils and backing plates [19]. These crystals were accompanied by chips of ruby as a pressure calibrant, with helium (2010) or neon (2015) as a pressure transmitting medium. Diffraction data were collected with Kickpatrick–Baez mirror-focused monochromatic X-rays ($\lambda = 0.3344$ at GSECARS and 0.4066 Å at HPCAT), focused onto the sample crystals. At each pressure point, the sample was rotated during X-ray exposure around the ω-axis of the instrument, with diffraction images collected at three detector positions, each perpendicular to the X-ray beam, and differing by 7 cm detector translation. For the 2015 experiment, the internal pressure was changed without removing the cell from the instrument, using a gas-driven membrane device with a Druck PACE 5000 electronic valve controller. Pressure was measured by the ruby fluorescence method [20] during pressurization and after a complete diffraction data set was collected.

Diffraction images collected during each experiment were analyzed using the ATREX IDL software package [21], where peak intensities were corrected for Lorenz and polarization effects. Peaks were indexed using cell_now [22], while the orientation matrix was refined and unit cell parameters were obtained using RSV [21]. Integrated peak intensities were used for the structure refinement. Refinements were conducted using SHELXL [23] with isotropic atomic displacement parameters and full site occupancies for all atoms. Amino hydrogens were constrained using the riding hydrogen model of AFIX 94, preventing the H-N-H angle from varying from 120°, but allowing some variance in the N-H bond distance and no limitations on the rotation angle of each amino group. Due to limitations on accurately reporting donor-hydrogen distances determined by X-ray diffraction, donor-acceptor distances are primarily used to gauge intermolecular interactions. No additional constraints or restraints were utilized to fix bond distances or the character of the central triazine ring. Equation of state results, such as bulk modulus and its pressure derivatives (K and K′), as well as linear compressibilities were determined with EOSFit7-GUI [24], fitting experimental unit cell values with a third-order Vinet equation of state [25,26]. Additional analysis, including hydrogen bond geometries and Hirshfeld surface generation, was conducted using Olex^2 [27] and CrystalExplorer [28], respectively.

For Raman experiments, a separate symmetric Princeton-type diamond anvil cell was prepared. We used low-fluorescence, modified brilliant cut diamond anvils with 0.3 mm culets, mounted on WC backing plates. The gasket preparation technique was identical to the one used for the X-ray diffraction experiments. One good optical quality, single crystal of melamine, approximately $15 \times 50 \times 50$ μm, and two small ruby spheres were loaded in neon pressure medium. Raman spectra were collected over a total spectral range of 250–3700 relative cm^{-1} with a holographic 1800 gr/mm grating, utilizing the green line of a 200 mW Ar$^+$ laser ($\lambda = 514.532$ nm) for excitation, and a Horiba Triax 550 Spectrograph, equipped with a liquid nitrogen-cooled Princeton Instruments CCD detector, at the GSECARS Raman Spectroscopy Lab. Sample Raman spectra were collected every 3–5 GPa from 1 GPa up to 33.1(1) GPa in three separate spectral ranges of 250–1200 relative cm^{-1}, 1400–1900 relative cm^{-1}, and 3100–3650 relative cm^{-1}. Raman data were processed in Spectragryph 1.2.8 [29], where individual adaptive baseline subtractions were applied to each spectrum to better compare between pressure steps.

3. Results

3.1. X-ray Diffraction

After initial compression and gas-loading with helium to 0.96(5) GPa, the unit cell parameters of melamine were $a = 10.410(6)$ Å, $b = 7.465(2)$ Å, $c = 7.086(6)$ Å, and $\beta = 113.89(4)°$ in the space group setting of P2$_1$/a. Multiple data sets, between 2010 and 2015, provide nineteen pressure points where the compressional effects on the structure of melamine was observed. All nineteen data points were used for equation of state fitting, whereas seventeen were of sufficient data quality to permit hydrogen bonding geometry analysis. The crystal structures obtained during the compression pathway show little deviation in the overall molecular geometry with pressure. At 0.96(5) GPa, each individual melamine unit is connected to its neighbors via eight N-H···N hydrogen bonds, as is the case at ambient pressure, forming layers of melamine molecules, approximately parallel to (010), that are hydrogen bonded within, and also linked to neighboring layers (Figure 1a). Additionally, in neighboring layers, there are pairs of melamine molecules whose central rings lie within parallel planes and overlap. Neighboring pairs of molecules are offset by 33.8(9) degrees, forming the corrugated structural motif. With each pressure step up to 36.2(1) GPa, full structure refinement was conducted, initially based on the structure of melamine reported from ambient pressure neutron diffraction experiments [30]. Selected data collection and refinement information from initial and final pressure steps for this structure are listed in Table 1. The evolution of unit cell parameters with pressure is listed in Table 2. Equation of state and linearized axial compressibilities are depicted in Figures 4 and 5, with the accompanying F-f plot depicted in Figure A1. Additional information, including fractional atomic coordinates, symmetry operators, and selected bond lengths and angles are listed in Appendix A Tables A1–A4. On compression beyond 36.2(1) GPa in helium or 31.8(1) GPa in neon, a reversible phase transition to a twinned triclinic polymorph (where $a = 6.08(1)$ Å, $b = 7.267(2)$ Å, $c = 7.82(1)$ Å, and $\alpha = 78.25(4)°$, $\beta = 80.1(2)°$, $\gamma = 80.48(4)°$ at 38.9(1) GPa) was seen, with pronounced changes in the diffraction pattern, as shown in Figure 2. This transition was reversible, however, further compression past approximately 45 GPa in helium resulted in irreversible amorphization, accompanied by loss of diffraction signal and easily identifiable change of the color and opacity of the crystal (Figure 3).

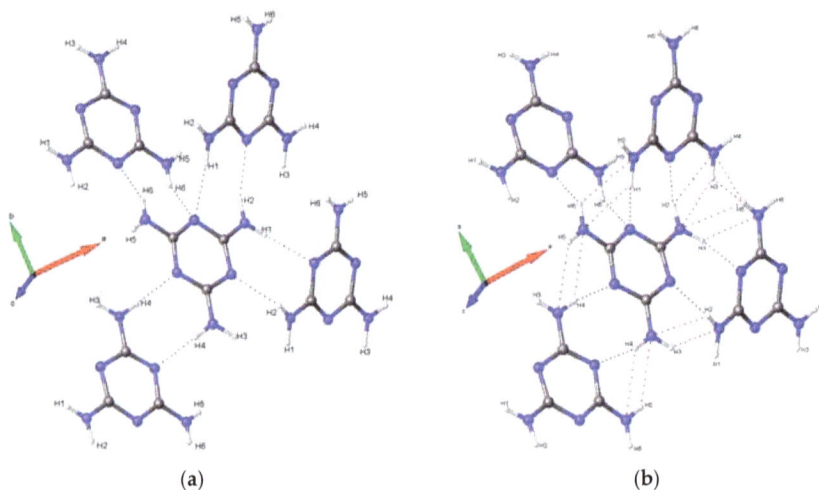

(a)   (b)

Figure 1. Hydrogen bonding network of melamine at (**a**) low (0.96(5) GPa) and (**b**) high (36.21(5) GPa) pressures. Potential hydrogen bonding interactions between anti-parallel amino N-H groups induced by pressure are displayed in red. Bonds behind and in front of the plane of view are omitted for clarity.

Table 1. Selected data collection and refinement details from lowest and highest pressure points. The opening angle of the diamond anvil cell in the 2015_10 experiment was larger by 20° compared to the 2010_1 experiment, producing higher number of observed peaks.

2010_1 (0.96(5) GPa)	Parameters	2015_10 (36.21(5) GPa)	Parameters
No. of reflections collected	453	No. of reflections collected	654
No. of independent reflections	186	No. of independent reflections	221
R_{int}	0.1272	R_{int}	0.0944
$R[F^2 > 4\sigma(F^2)]$	0.0611	$R[F^2 > 4\sigma(F^2)]$	0.0700
$wR(F^2)$	0.1396	$wR(F^2)$	0.1652
Goodness-of-fit	1.114	Goodness-of-fit	1.231
No. of parameters refined	40	No. of parameters refined	40
No. of restraints used	0	No. of restraints used	0

Table 2. Unit cell parameters as a function of pressure. 2010 data were collected in Ne, 2015 data in He.

Data Set	P (GPa)	a (Å)	b (Å)	c (Å)	β (°)	V (Å^3)
2010	10^{-4}	10.606(1)	7.495(1)	7.295(2)	112.26(2)	536.7(2)
2010	0.96(5)	10.410(6)	7.465(2)	7.086(6)	113.90(4)	503.5(6)
2010	2.9(1)	10.148(6)	7.389(2)	6.954(6)	115.93(4)	468.9(5)
2015	3.5(1)	10.099(1)	7.372(1)	6.866(1)	116.47(1)	457.6(6)
2010	4.7(1)	9.962(6)	7.333(2)	6.821(6)	117.14(4)	443.4(5)
2015	6.3(1)	9.906(1)	7.306(1)	6.737(6)	117.62(3)	432.0(4)
2010	7.7(1)	9.807(4)	7.274(1)	6.705(4)	118.07(3)	422.0(4)
2015	9.0(1)	9.740(1)	7.250(1)	6.623(7)	118.47(4)	411.1(5)
2010	10.5(1)	9.678(6)	7.235(1)	6.616(5)	118.34(4)	407.8(4)
2015	12.8(1)	9.572(1)	7.191(1)	6.481(6)	119.21(3)	389.4(4)
2010	16.8(1)	9.446(6)	7.151(2)	6.350(5)	119.89(4)	371.9(4)
2015	17.1(1)	9.404(1)	7.136(1)	6.359(6)	119.86(4)	370.1(4)
2010	19.8(1)	9.366(7)	7.122(7)	6.252(6)	120.47(5)	359.4(5)
2015	22.0(1)	9.246(1)	7.083(1)	6.287(7)	120.00(4)	356.6(4)
2010	23.5(1)	9.254(4)	7.103(4)	6.279(4)	120.14(3)	356.9(3)
2015	27.1(1)	9.219(1)	7.039(1)	6.195(7)	120.44(5)	346.6(4)
2015	30.8(1)	9.057(2)	7.013(1)	6.128(7)	120.57(5)	335.1(4)
2015	34.3(1)	9.006(1)	6.990(1)	6.081(7)	120.67(4)	329.3(4)
2015	36.2(1)	8.965(1)	6.966(1)	6.075(7)	120.48(5)	326.9(4)

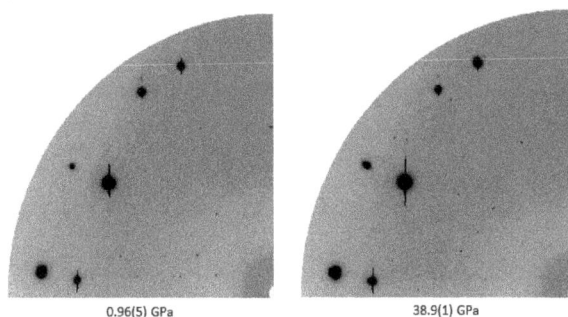

0.96(5) GPa   38.9(1) GPa

Figure 2. Diffraction patterns at the central detector position of low- and high-pressure phases of melamine, displaying the change in pattern before and after the reversible structural phase transition above 36 GPa in helium.

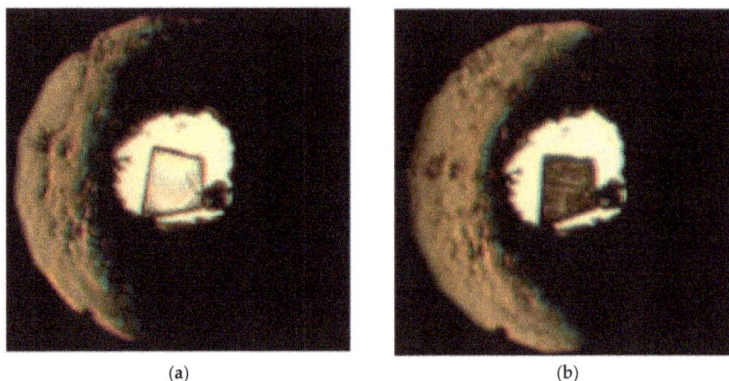

(a)   (b)

Figure 3. Visible changes in a melamine crystal before (**a**) and after (**b**) the irreversible high-pressure amorphization at 45 GPa in helium.

3.2. Equation of State and Bond Compressibility

Previously, Ma et al. [15] documented the P-V equation of state of melamine, but only to approximately 15 GPa. Although that study describes a transformation to a triclinic structure below 2 GPa, there exist similarities between that study and the current results. In both cases, there is a precipitous drop in the unit cell volume below 5 GPa. However, in the present study, there is no evidence of a phase transformation. Rather, there is a smooth and continuous reduction in the unit cell volume that is well described by the Vinet equation of state [25,26].

$$P = 3K_0 \frac{(1 - f_V)}{f_V^2} e^{(\frac{3}{2}(K' - 1)(1 - f_V))}$$

where $f_V = \left(\frac{V}{V_0}\right)^{\frac{1}{3}}$, V_0 is the initial cell volume, K_0 is the isothermal bulk modulus, and K' is its derivative when the pressure is equal to zero. The output of this equation of state, with V_0 fixed to the ambiently determined value of 536.7(2) Å^3, is K_0 = 12.9(8) GPa, and K' = 7.4(3). The linear axial compressibilities (defined as $\beta_{10} = 1/3K_{10}$) [31] were determined using a linearized version of the Vinet equation. The increase of the unit cell's β angle was well-described by a three-parameter exponential rise-to-maximum function. The progression of each of these values with pressure can be seen in Figures 4 and 5. By 36.2(1) GPa the unit cell volume has experienced a 40 percent collapse, driven primarily by the shortening of the *a* and *c* axes. As each axis shrinks the β angle opens, reflecting the shift of molecules with respect to one another, rising to a predicted maximum of 120.7(1) degrees at 40 GPa.

When comparing the bulk moduli other common six-membered ring molecules such as benzene (5.5 GPa) or aniline (5.44 GPa), as well as extended structures with ring motifs such as graphite (33.8 GPa), the compressibility of melamine is closer to the former [32–34]. This can be readily explained by the type of dominating intermolecular interactions and arrangement of molecules with respect to each other, where graphite is held more rigidly in covalently-bonded planar sheets, while individual benzene or aniline molecules are not covalently bonded to their neighbors. Both polymorphs of aniline also form extended stacked layers held together by hydrogen bonding, where its singular amino group participates in N-H···N and N-H···π hydrogen bonds [33,35]. Individual aniline units do not experience significant structural or energetic modification in pressure, yet its hydrogen bonds decrease in length to the point of destabilization [35]. Benzene does not share hydrogen atoms below a theorized point of metallization [32,36], and π-π interactions act as the driving intermolecular interaction with increasing pressure. For solid benzene-III, a recent theoretical study [36] suggested that at 50 GPa, despite an almost two-fold reduction in unit cell volume, the intramolecular bond lengths stay basically unchanged. A similar situation appears to occur with melamine: while the intramolecular deformation is minimal, intermolecular hydrogen bonds accommodate pressure changes within the crystal without departing from the original space group and basic structure.

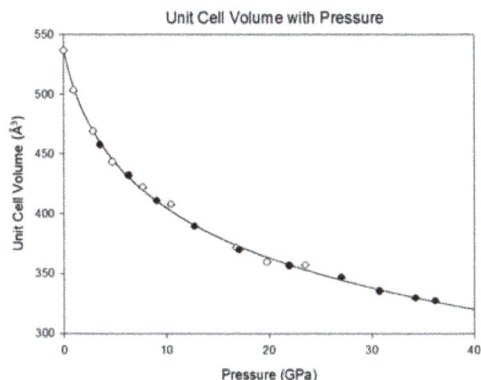

Figure 4. Evolution of unit cell volume of melamine with pressure. Data points were fitted with a third-order Vinet equation of state, $V_0 = 536.7(2)$ Å^3, $K_0 = 12.9(8)$ GPa, and $K' = 7.4(3)$. Open circles indicate 2010 data collected in Ne, filled circles indicate 2015 data collected in He.

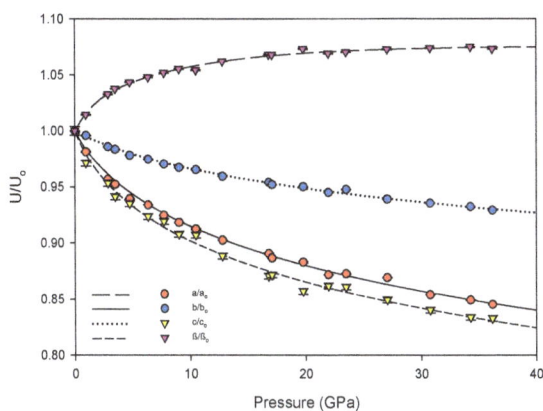

Figure 5. Evolution of normalized unit cell parameters of melamine with pressure. Linear axial compressibilities are $\beta_{a0} = 6.54(5) \times 10^{-3}$ GPa^{-1}, $\beta_{b0} = 1.89(1) \times 10^{-3}$ GPa^{-1}, and $\beta_{c0} = 9.0(1) \times 10^{-3}$ GPa^{-1}.

In comparison to energetic materials, the bulk modulus of melamine is analogous to or slightly less than β-HMX (12.4 GPa), α-RDX (13.9 GPa), and TATB (16.2 GPa) [37]. Interestingly, these values were obtained using powder X-ray diffraction techniques in methanol-ethanol-water, argon, and hexane pressure media, respectively [38–40]. The use of powders, as well as non-hydrostatic and non-inert pressure media likely introduces similar uncertainties and irregularities as those encountered with prior high-pressure experiments of melamine. Significant variations in compressibility for these compounds, including elastic constants and phase transformation behavior, have been shown to greatly depend on the hydrostatic character of the pressure media [41,42]. Hydrogen bonding has also been demonstrated to be the driving interaction in crystalline networks of RDX and TATB, where intermolecular hydrogen bonding networks could be disrupted by participating polar-solvent pressure media or inter-grain boundaries [43]; when compressional hydrostaticity is ensured, highly energetic molecules like TATB have been shown to remain crystalline past 100 GPa [44].

For inorganic substances and minerals, a very common notion for understanding a compression mechanism is the Rigid Unit Mode model, in which it is assumed that each subunit (i.e., tetrahedra or octahedra) is very stiff compared to the framework in which it resides [45]. As a result, rotation of whole units is preferable to alteration of bond lengths within a unit. Although it is an organic molecule, within this frame of reference melamine subunits (primarily the aromatic *s*-triazine ring and amine nitrogen) can be considered as the inflexible subunit, and are relatively unchanged with pressure. In contrast, intermolecular hydrogen bonds greatly compress and shift position with pressure, and individual NH_2 units have some ability to rotate, in order to accommodate pressure changes and avoid repulsive H-H interactions. This is evident from the relative lack of change in carbon-nitrogen bond lengths with substantial increases in pressure, as shown in Figure 6; for both ring and amine carbon-nitrogen bonds, the bond length decrease is less than about 0.05 Å, while the donor-acceptor lengths of hydrogen bonds decrease by nearly 0.5 Å over the same compression path.

The compressional behavior of melamine from the perspective of an individual molecular unit can also be visualized through Hirshfeld surfaces. This method of crystal analysis condenses properties such as interatomic angles and distances, crystal packing schemes, and intermolecular interactions into models that can be easily and qualitatively interpreted, yet are derived from quantitative analysis [46,47]. Hirshfeld surfaces are differentiated from other molecular surface representations such as electron density maps or van der Waals surfaces by accounting for both a molecule and its proximity to its nearest neighbors, making it well-suited for the analysis of molecular crystals. Two surfaces, the shape index and normalized contact distance (d_{norm}), are particularly useful for describing the packing of molecular crystals such as melamine.

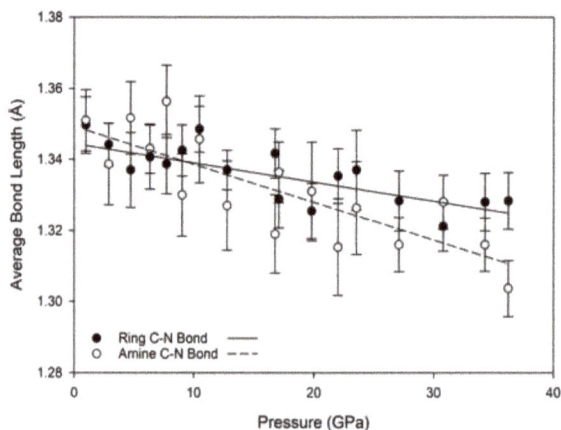

Figure 6. Average carbon-nitrogen bond lengths for ring and amine components of melamine. Each follows a loosely linear decrease in length with pressure.

The shape index is a Hirshfeld surface that identifies concave or convex areas of a molecule's surface based on charge density. Red areas indicate concave areas, whereas blue indicates convex. For the purpose of examining compressional behavior, any change in the intramolecular geometry is distinguishable by changes in color. For melamine, the relative lack of change between low and high pressure is apparent (Figure 7), mirroring the small changes in covalent bond lengths and overall inflexibility of the aromatic component with pressure.

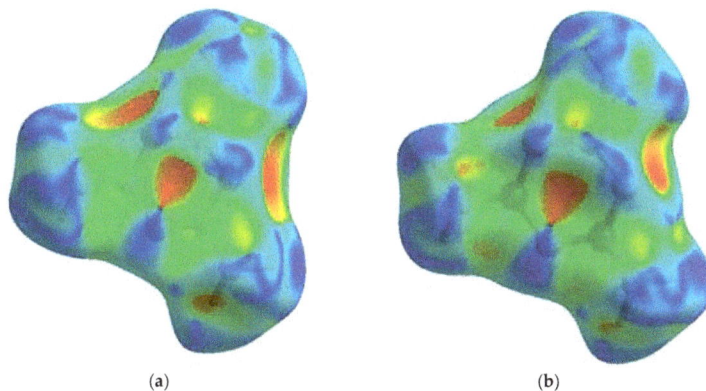

Figure 7. Hirshfeld surfaces displaying the shape index of melamine at low (0.96(5) GPa), (**a**) and high (36.21(5) GPa), (**b**) pressures. Concave areas, shown in red, correspond to areas where a complimentary neighboring molecule may interact.

Hirshfeld surfaces of the normalized contact distances tell the other half of the story, and are shown in Figure 8; this parameter describes the internal (d_i) and external (d_e) contact distances of the Hirshfeld surface to the nearest atomic nucleus, normalized by the van der Waals (vdW) radii of the atoms involved. The result is a surface where intermolecular contacts longer than the sum of the atoms' vdW radii are displayed in blue, and contacts shorter than the vdW radii are displayed in red. At low pressures, this highlights the points of contact for the N-H$\cdots$N hydrogen bonds, where the bond contracts the intermolecular distance. As pressure increases, contact points with distances shorter than the vdW radii appear on the previously non-interacting amine nitrogen and atoms of the central ring.

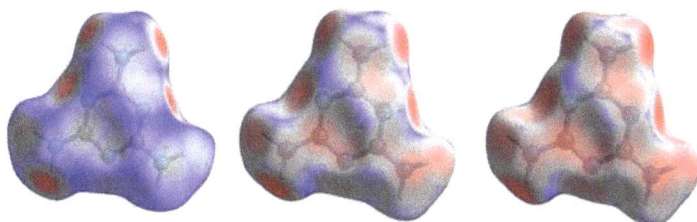

Figure 8. Hirshfeld surfaces displaying d_{norm} at 0.96(5), 17.10(5), and 36.21(5) GPa. As pressure increases, intermolecular contacts are induced, primarily as new hydrogen bonds.

3.3. Hydrogen Bonding Behavior with Pressure

At ambient and low-pressure conditions, individual melamine molecules are linked with neighbors through pairs of complementary hydrogen bonds connecting amino hydrogens exclusively to ring-based nitrogen atoms. In previous ambient-pressure X-ray and neutron diffraction studies, it was observed that of the six symmetry independent hydrogen atoms in NH_2 groups, only four strongly participate in hydrogen bonding [30]. The two remaining hydrogen atoms are subject to hindrances that prevent strong hydrogen bonding interactions, and are denoted in this study as H3 and H5. In the case of H5, the hydrogen atom is in close contact with another H5 on a neighboring molecule, and repulsive interaction occurs as distance decreases. For H3, steric hindrance prevents it from being sufficiently close to a ring nitrogen acceptor atom, allowing only weak interaction with a NH_2 group on a neighboring molecule. Each ring nitrogen atom also acts as a hydrogen bond acceptor for a total of four bonds per ring, with one nitrogen (denoted here as N4) acting as acceptor for two bonds (Figure 1a).

The hydrogen bonds in melamine can also be distinguished by whether they link molecules within a given corrugated plane or between them; for instance, the decrease in donor-acceptor distance for intra-plane molecules decreases smoothly as a function of pressure akin to the behavior of the unit cell parameters, while the inter-plane behavior is less consistent (Figures 9 and 10). This is likely caused by the larger influence of the intra-plane bonding, as there are more hydrogen bonds within a layer, as well as small amounts of rotation and torsion to accommodate the increased intra-layer bonding and repulsive interactions between close-contact hydrogens. Ultimately, the intermolecular hydrogen bonds are capable of significant shortening with response to pressure, without any significant changes in the pattern of the original hydrogen bonds. Compression through 36.2(1) GPa decreases donor-acceptor distances substantially, which increases the covalent character of a bond and increases its strength [48].

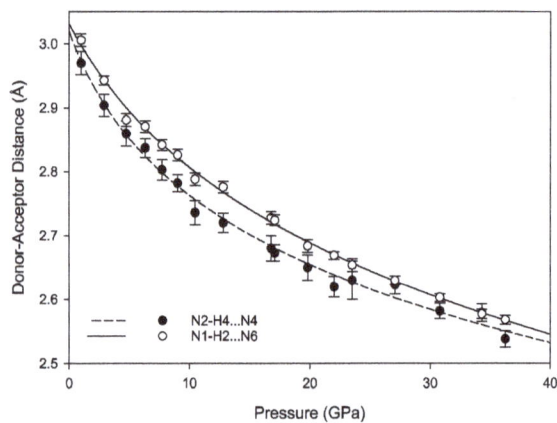

Figure 9. Intra-plane hydrogen bond D⋯A distance as a function of pressure, fitted with a third-order polynomial function.

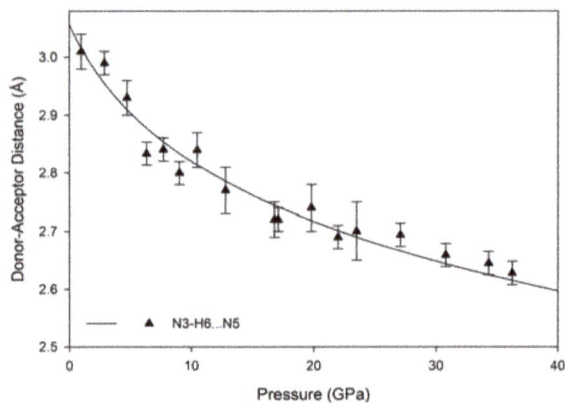

Figure 10. Inter-plane hydrogen bond D⋯A distance as a function of pressure, fitted with a third-order polynomial function.

Notably, none of the original hydrogen bonds present at ambient pressures are broken below the phase transition pressure; instead, new hydrogen bonds between amino hydrogens and ring nitrogens form by 9.0(1) GPa and persist until at least 36.2(1) GPa (Figure 11). As pressure increases further, amino groups are pushed into close contact with one another, and interactions between oppositely aligned N-H atoms occur as each hydrogen is brought closer to an opposing nitrogen's lone pair of electrons. These interactions, although primarily electrostatic in character, are stabilized by their anti-parallel orientation to one another, with pressure overriding unfavorably shallow D-H-A angles, steric crowding, and repulsive interactions. At 36.2(1) GPa, sufficiently short bond distances indicate bi- and tri-furcation of these bonds, as shown in Figure 1b, by the previously un-bonded H3 and H5 atoms.

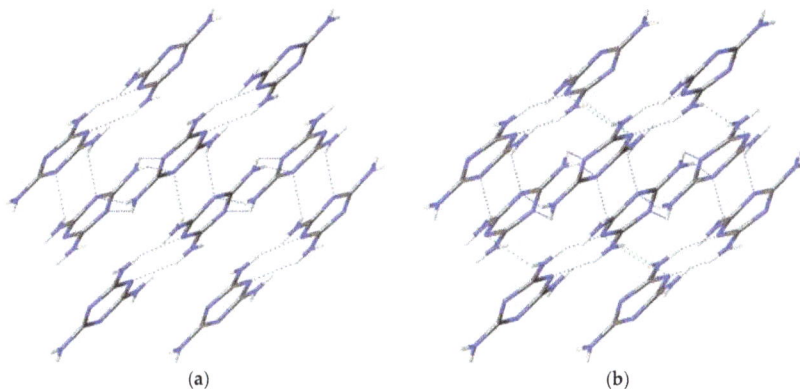

(a)   (b)

Figure 11. Hydrogen bonding network of melamine at (**a**) low (0.96(5) GPa) and (**b**) high (36.21(5) GPa) pressures, as viewed down <010>, displaying the layered structure. New hydrogen bonds to the central ring have been highlighted in green. Some bonds, including weak amino N-H interactions, have been omitted for clarity.

The new pressure-induced hydrogen bonds are not sufficiently strong to create a stable high-pressure configuration, as the transformation between monoclinic and triclinic phases is reversible. This is However, the new pressure-induced hydrogen bonds with ring nitrogen atoms may foster changes in the intermolecular interactions of π electrons in the high pressure phase; at ambient conditions, inter-layer hydrogen bonds link molecules where the centroid planes of the rings are parallel to one another, resulting in a skewed parallel-displaced arrangement where an electron-rich ring nitrogen is roughly aligned with a moderately electron-deficient ring center. At high pressures, the new hydrogen bonds also link molecules whose planes are at an offset to one another, introducing an interaction between these rings not experienced at lower pressures. Furthermore, compression reduces the distance and shift between the ring centroids (depicted in Figure A2), as well as reducing the angle between offset pairs of molecules (Figure 12), increasing the likelihood of extended interactions between multiple pairs of molecules. Although the term "π-π stacking" does not correctly describe the contact between neighboring melamine molecules [49,50], the distances and angles between ring centroids are within limits for attractive electrostatic π-σ interactions at both ambient and high pressures [49,51,52]. However, the N-H⋯N hydrogen bonds ultimately direct the supramolecular changes in the melamine crystal; this provides exceptional stability when compared to un-substituted *s*-triazine, which does not have the ability to act as a hydrogen bond donor [49,53,54].

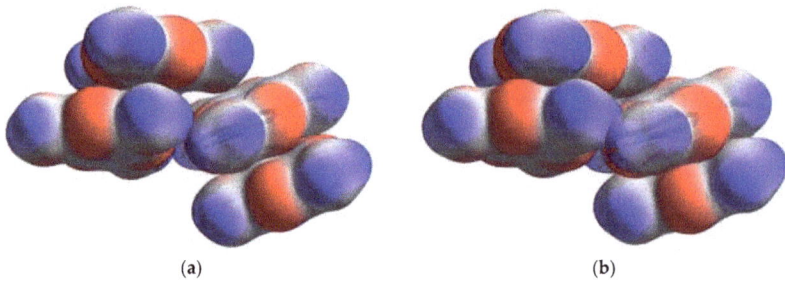

Figure 12. Electrostatic potential surfaces of neighboring melamine molecules at 6.33(5) (**a**) and 36.21(5) (**b**) GPa, scaled from −0.0928 to 0.1388 au. Regions in red are electron-rich, while regions in blue are electron deficient. With pressure, distance between melamine ring centroids oriented in parallel planes decreases, and the offset angle to the next pair of molecules decreases.

3.4. Raman Spectroscopy

In general, Raman spectroscopy is an excellent tool for detecting pressure-induced structural phase transitions in solids, which usually manifest themselves as discontinuous changes in vibration mode behavior. For instance, symmetry lowering related to displacive phase transitions typically results in splitting of Raman peaks. The case of molecular crystals, however, is often more complicated than simple inorganic solids. The starting crystal symmetry is often lower, and the number of Raman modes can be very significant. At the same time, there are more types of competing interatomic and intermolecular interactions (e.g., hydrogen bonds, van der Waals forces, electrostatic interactions, charge transfer), which affect the vibration force constants. At high pressure the balance between these various interactions changes and may cause discontinuities in the Raman mode behavior unrelated to first-order structural phase transitions. An example of such was found in benzene, for which Raman experiments [55] described the existence of phase transitions between the II-III and III-IV phases at about 4 and 11 GPa. However, later studies combining both IR spectroscopy and powder X-ray diffraction cast doubt on tose proposed transitions, as the observed discontinuities and changes in vibrational modes did not correspond to symmetry-altering first-order structural changes [32].

The ambient pressure Raman spectrum of melamine was first quantitatively interpreted in terms of mode assignment by Schneider and Schrader [56]. The assignment of the collective ring vibration modes can be made by analogy to the unsubstituted parent-molecule of *s*-triazine [57]. There are also several recent Raman studies of solid salts of melamine [58,59] that are useful in interpretation of individual vibration modes, such as those from hydrogen bonds. In the spectral range covered by our experiments, the Raman spectrum of melamine can be divided into three regions: the 250–1200 relative cm^{-1} is the collective ring vibration mode region, from 1400–1900 cm^{-1} is the C-NH$_2$ vibration mode range, and the 3000–3700 cm^{-1} range is the N-H vibration mode range. This last region proves to be the most informative; at ambient pressure there are four distinct Raman peaks, at 3128, 3333, 3420, and 3471 relative cm^{-1} [16]. The two peaks at lower wavenumbers are very broad and quite asymmetric, whereas the peaks at higher wavenumbers are sharp and symmetric, as seen in Figure 13.

The non-uniformity of the hydrogen bonding interactions can also be seen in the Raman spectra. The sharp Raman peaks of 3420 and 3471 cm^{-1} at close to ambient pressure conditions can be associated with the non-hydrogen bonded N-H vibrations, such as from H3 and H5. The broad features, reminiscent of the O-H peak shapes in other hydrogen-bonded crystals, e.g., solid H$_2$O, correspond to the two groups of hydrogen-bonded NH$_2$. With increasing pressure, as intermolecular distances are reduced and hydrogens are forced into closer vicinity of nitrogen atoms, these peaks broaden and fade. As these previously non-interacting atoms are forced into hydrogen bonding interactions, the signal for each is muddied until the point of transition.

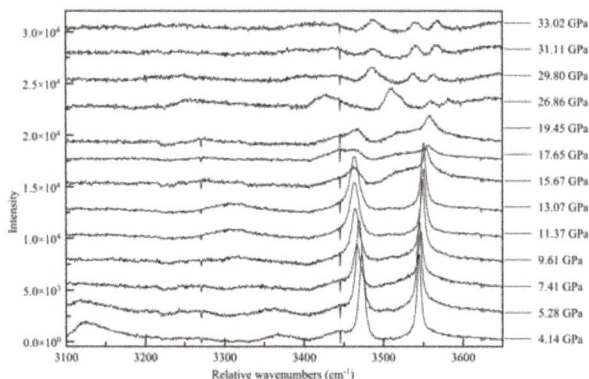

Figure 13. Raman spectra of melamine in N-H vibration region, vertical offset indicates sequential pressure steps in GPa, as shown on the right vertical axis. Intensity is arbitrary with vertical stacking offset.

The changes in the Raman spectra accompanying the monoclinic-to-triclinic phase transition, observed in diffraction data at approximately 30 GPa in Ne and 38 GPa in He, are quite pronounced in all three spectral ranges, with the appearance of new spectral features often occurring slightly before the observed transition pressure. In the ring breathing mode range the high wave number component of the 750 cm^{-1} peak splits into a doublet, as does the 1100 cm^{-1} peak, shown in Figure 14. This is indicative of a change in the interaction between inter-layer molecules, potentially between newly crystallographically and energetically inequivalent ring systems after the phase transition [60–62]. In the C-N vibration range (Figure 15) a whole new family of peaks appear before the point of transition between 1500 and 1600 relative cm^{-1}, with new peaks forming at approximately 1700 and 17900 rel cm^{-1} above approximately 30 GPa. This corroborates a change in the electronic state beginning at 26.86(5) GPa, culminating in a change in symmetry. Similarly, in the N-H range two new high wavenumber peaks appear between 3500 and 3600 cm^{-1} at approximately 30 GPa, and strengthen with increasing pressure.

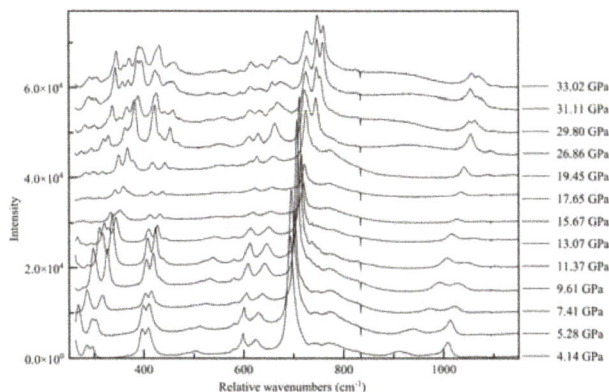

Figure 14. Raman spectra of melamine in the ring vibration-mode range, vertical offset indicates pressure steps in GPa. Splitting of the 750 cm^{-1} and 1100 cm^{-1} peaks is observed at 29.80(5) GPa in Ne.

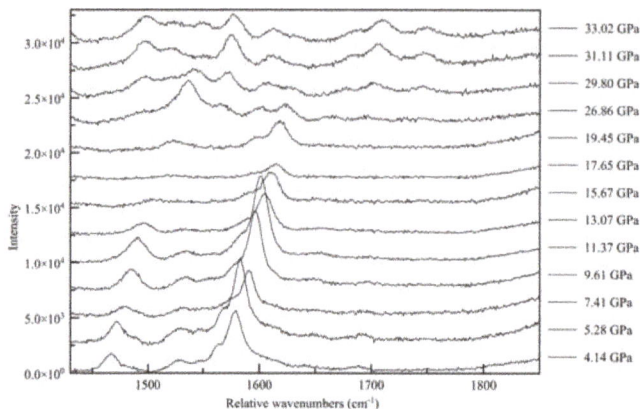

Figure 15. Raman spectra of melamine in the C-N vibration mode range, vertical offset indicates pressure steps in GPa. Appearance of new peaks at approximately 1700 and 17900 rel cm^{-1} are seen above approximately 30 GPa.

4. Discussion and Conclusions

The compressional behavior of melamine closely resembles that of other hydrogen-bonded molecular solids, where individual molecular subunits and internal covalent bonds are not significantly altered with pressure [33,35,54,63,64]. Instead, hydrogen bonds can readily accommodate the increase in pressure, becoming more stable with decreasing donor-acceptor distances and linearization of donor-hydrogen-acceptor angles up to a critical pressure [48,65]. Melamine is notable for its extensive network of hydrogen bonds when compared to other substituted aromatics [54,63], which allows it to remain structurally stable over the same pressure range where many other molecular crystals irreversibly amorphize or decompose [44]. Additionally, this increase in hydrogen bonding stability (and by extension, covalent character) seems to overrule or direct other steric and electronic interactions that would otherwise exhibit more control over compressional behavior and reactivity. This is especially apparent when comparing against non-hydrogen bonded molecular crystals; for instance, although *s*-triazine shares the same fundamental aromatic ring system as melamine, its behavior is controlled by π interactions, resulting in susceptibility to electronic modification with pressure and increased reactivity [53,65,66], ultimately becoming irreversibly amorphous at 15.2 GPa. Similarly, benzene amorphizes into extended polyaromatic compounds at 23 GPa. Crystallinity is also generally limited to lower pressure ranges in molecular solids with limited or weak hydrogen bonding character; pyrimidine amorphizes at 22.4 GPa [67], pyrrole at 14.3 GPa [68], and pyridine at 22 GPa [69]. Each of these amorphization transitions is irreversible, indicating significant changes in the covalent bonding environment, often through ring-opening and polymerization [67].

In molecular crystals with extended hydrogen bonding networks (often with "rosette-type" hydrogen bonding geometry) such as melamine and its related adducts, as well as TATB, the pressure–stability relationship appears to be more complex. In a 1:2 melamine-boric acid adduct, a reversible amorphization transition was reported at 18 GPa, indicating a loss of periodicity but not chemical alteration [70]. For a 1:1 cyanuric acid-melamine adduct, an irreversible phase transition at 4.9 GPa to a quenchable crystalline phase of lower symmetry was reported [71]. This was accompanied by a pronounced change in the crystal's opacity, not unlike what is seen in the irreversible amorphization of melamine. Surprisingly, TATB in hydrostatic helium was found to be stable up to at least 150 GPa, after experiencing several subtle, reversible phase transformations that produced visible color changes in the crystal [44].

In melamine, the phase transformations to the triclinic polymorph and amorphous phase occur at roughly double the pressure than other molecular solids, at upwards of 36 GPa in helium. The reversible transformation to the triclinic structure results from a rearrangement of the hydrogen bonding network in melamine, where electronic and electrostatic interactions not seen at ambient conditions are kept stable at high pressure but revert upon decompression. Upon further compression, the crystal can no longer accommodate these forced interactions, and an irreversible transition occurs to release the imposed stress on the structure. The large pressure stability field of melamine, combined with the interesting behavior of melamine adducts with pressure, indicate some potential for using melamine-based substituted triazines or adducts to design functional crystals held together by extensive pressure-stabilized hydrogen bonding networks.

Author Contributions: H.S. and P.D. performed the high-pressure X-ray diffraction experiments; P.D. and S.T. performed the high-pressure Raman spectroscopy experiments; H.S. and P.D. analyzed the X-ray and Raman data; S.T. assisted with sample preparation, all authors wrote the paper.

Funding: This study was funded by the Carnegie-DOE Alliance Center (CDAC) under cooperative agreement DE FC52-08NA28554. Development of the GSE_ADA software used for data analysis is supported by NSF grant EAR1440005. H.S. was supported by the U.S. Department of Energy and through CDAC. This project was performed at GeoSoilEnviroCARS (GSECARS Sector 13), and the High Pressure Collaborative Access Team (HPCAT Sector 16) of the Advanced Photon Source at Argonne National Laboratory. GeoSoilEnviroCARS is supported by the National Science Foundation—Earth Sciences (EAR-0622171) and Department of Energy—Geosciences (DE-FG02-94ER14466). Use of the COMPRES-GSECARS gas loading system was supported by COMPRES under NSF Cooperative Agreement EAR-1606856, by GSECARS through NSF grant EAR-1634415, and DOE grant DE-FG02-94ER14466. HPCAT operations are supported by DOE-NNSA under Award No. DE-NA0001974 and DOE-BES under Award No. DE-FG02-99ER45775, with partial instrumentation funding by NSF. Use of the Advanced Photon Source was supported by the U. S. Department of Energy, Office of Science, Office of Basic Energy Sciences, under Contract No. DE-AC02-06CH11357.

Conflicts of Interest: The authors declare no conflict of interest.

Appendix A

Table A1. Selected bond lengths and angles for melamine at 36.21(5) GPa (2015_10). Amino C-N-H angles are set to 120° by the AFIX hydrogen bond command.

Bonded Atoms	Length (Å)	Atoms in Angle	Angle (°)
C(1)-N(1)	1.26(2)	N(1)-C(1)-N(6)	121.5(6)
C(1)-N(6)	1.345(17)	N(1)-C(1)-N(5)	118.1(12)
C(1)-N(5)	1.361(8)	N(6)-C(1)-N(5)	120.4(14)
C(2)-N(2)	1.302(14)	N(2)-C(2)-N(6)	120.1(6)
C(2)-N(6)	1.316(17)	N(2)-C(2)-N(4)	115.8(11)
C(2)-N(4)	1.345(8)	N(6)-C(2)-N(4)	124.1(10)
C(3)-N(5)	1.30(2)	N(5)-C(3)-N(4)	125.8(8)
C(3)-N(4)	1.317(15)	N(5)-C(3)-N(3)	117.3(10)
C(3)-N(3)	1.335(10)	N(4)-C(3)-N(3)	116.9(14)

Table A2. Atomic coordinates ($\times 10^4$) and equivalent isotropic displacement parameters ($\text{Å}^2 \times 10^3$) for carbon and nitrogen atoms melamine at 36.21(5) GPa (2015_10). U_{eq} is defined as one third of the trace of the orthogonalized Uij tensor.

Atom	X	Y	Z	U_{eq}
C(1)	1771(16)	6196(8)	290(40)	11(1)
C(2)	1307(16)	4725(7)	3230(40)	9(1)
C(3)	634(16)	7829(8)	2330(40)	11(1)
N(1)	2234(14)	6180(7)	1340(40)	10(1)
N(2)	1280(12)	3193(7)	4440(30)	7(1)
N(3)	252(13)	9514(7)	2970(30)	10(1)
N(4)	586(15)	6305(7)	3570(40)	11(1)
N(5)	1098(13)	7850(7)	610(30)	10(1)
N(6)	2005(14)	4661(7)	1780(40)	11(1)

Table A3. Definitions of symmetry operators.

Atom	Symmetry Operators		
1	x,	y,	z − 1
2	−x + 1/2,	y − 1/2,	−z
3	−x + 1/2,	y + 1/2,	−z
4	x,	y − 1,	z
5	−x + 1/2,	y − 1/2,	−z + 1
6	−x,	−y + 1,	−z + 1
7	x	Y + 1	z
8	−x	−y + 2	−z
9	−x	−y + 2	−z + 1

Table A4. Hydrogen bond D-H··A lengths and angles at 36.21(5) GPa (2015_10). Hydrogen bonds linking to neighboring triazine ring nitrogen atoms are displayed in bold.

D-H··A Symmetry	d(D-H)	d(H··A)	<DHA	d(D··A)
N1-H1···N5_$2	0.935	1.792	155.72	2.672(8)
N1-H2···N6_$3	0.935	1.821	134.85	2.568(7)
N2-H3···N4_$5	0.865	2.16	126.5	2.763(12)
N2-H3···N6_$5	0.865	2.688	117.55	3.182(16)
N3-H6···N5_$8	0.886	1.89	139.51	2.628(18)
N2-H4···N4_$6	0.865	1.703	161.53	2.538(13)
N1-H1···N2_$1	0.935	2.481	120.73	3.067(18)
N1-H1···N3_$2	0.935	2.379	134.58	3.108(10)
N1-H2···N2_$3	0.935	2.477	127.01	3.131(14)
N2-H3···N1_$2	0.865	2.49	131.52	3.131(14)
N2-H3···N3_$4	0.865	2.152	122.47	2.717(10)
N2-H4···N3_$6	0.865	2.561	129.4	3.182(12)
N3-H6···N1_$3	0.886	2.546	121.95	3.108(10)
N3-H6···N2_$7	0.886	2.158	120.4	2.717(10)
N3-H5···N2_$6	0.886	2.565	127.35	3.182(12)
N3-H5···N3_$9	0.886	1.96	159.3	2.81(3)

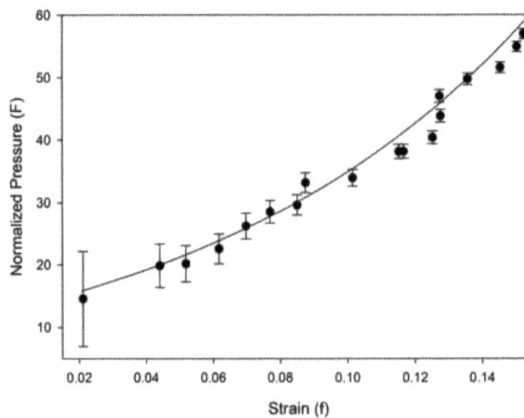

Figure A1. F-f plot for data plotted with a third-order Vinet equation of state.

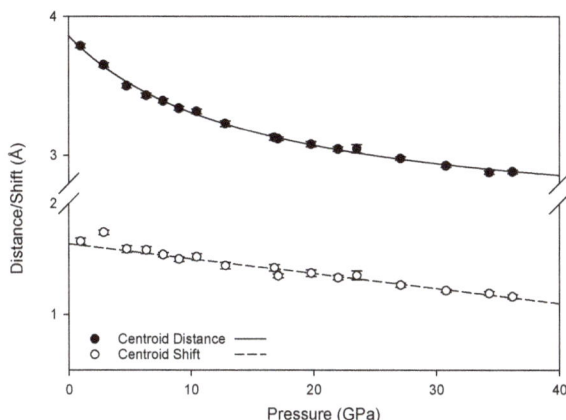

Figure A2. Ring centroid-centroid distance and shift as a function of pressure. Centroid distance is the straight line distance between the centers of the rings, shift indicates the in-plane separation between the centers of ring planes.

References

1. Hiskey, M.A.; Chavez, D.E.; Naud, D.L. *Insensitive High-Nitrogen Compounds*; NTIS No: DE-2001-776133; Los Alamos National Lab.: Los Alamos, NM, USA, 2001.
2. Xu, D.; Lu, H.; Huang, Q.; Deng, B.; Li, L. Flame-retardant effect and mechanism of melamine phosphate on silicone thermoplastic elastomer. *RSC Adv.* **2018**, *8*, 5034–5041. [CrossRef]
3. Liu, A.Y.; Cohen, M.L. Prediction of new low compressibility solids. *Science* **1989**, *245*, 841–842. [CrossRef] [PubMed]
4. Li, J. An ab initio theoretical study of 2,4,6-trinitro-1,3,5-triazine, 3,6-dinitro-1,2,4,5-tetrazine, and 2,5,8-trinitro-tri-s-triazine. *Propellants Explos. Pyrotech.* **2008**, *33*, 443–447. [CrossRef]
5. Coburn, M.D.; Hayden, H.H.; Coon, C.L.; Mitchell, A.R. Synthesis of poly (S, S-dimethylsulfilimino) heterocycles. *Synthesis* **1986**, *6*, 490–492. [CrossRef]
6. Hartman, G.D.; Schwering, J.E.; Hartman, R.D. Dimethyl sulfide ditriflate: A new reagent for the conversion of amino heterocycles to iminosulfuranes. *Tetrahedron Lett.* **1983**, *24*, 1011–1014. [CrossRef]
7. Hejny, C.; Minkov, V.S. High-pressure crystallography of periodic and aperiodic crystals. *IUCrJ* **2015**, *2*, 218–229. [CrossRef] [PubMed]
8. Ma, H.A.; Jia, X.P.; Chen, L.X.; Zhu, P.W.; Guo, W.L.; Guo, X.B.; Wang, Y.D.; Li, S.Q.; Zou, G.T.; Zhang, G.; et al. High-pressure pyrolysis study of $C_3N_6H_6$: A route to preparing bulk C_3N_4. *J. Phys. Condens. Matter* **2002**, *14*, 11269–11273. [CrossRef]
9. Montigaud, H.; Tanguy, B.; Demazeau, G.; Alves, I.; Courjault, S. C_3N_4: Dream or reality? Solvothermal synthesis as macroscopic samples of the C_3N_4 graphitic form. *J. Mater. Sci.* **2000**, *35*, 2547–2552. [CrossRef]
10. Yao, L.D.; Li, F.Y.; Li, J.X.; Jin, C.Q.; Yu, R.C. Study of the products of melamine ($C_3N_6H_6$) treated at high pressure and high temperature. *Phys. Status Solidi* **2005**, *202*, 2679–2685. [CrossRef]
11. Wu, X.; Tao, Y.; Lu, Y.; Dong, L.; Hu, Z. High-pressure pyrolysis of melamine route to nitrogen-doped conical hollow and bamboo-like carbon nanotubes. *Diam. Relat. Mater.* **2006**, *16*, 164–170. [CrossRef]
12. Yu, D.L.; He, J.L.; Liu, Z.Y.; Xu, B.; Li, D.C.; Tian, Y.J. Phase transformation of melamine at high pressure and temperature. *J. Mater. Sci.* **2008**, *43*, 689–695. [CrossRef]
13. Hamann, S.D.; Linton, M. The influence of pressure on the infrared spectra of hydrogen bonded solids. III compounds with N-H ... X bonds. *Aust. J. Chem.* **1976**, *29*, 1641–1647. [CrossRef]
14. Zhao, Y.; Zhang, Z.; Cui, Q.; Liu, Z.; Li, D.; Zou, G. High pressure raman spectra of melamine ($C_3H_6N_6$) and pressure induced transition. *High Press. Res.* **1990**, *3*, 233–235.
15. Ma, H.A.; Jia, X.; Cui, Q.L.; Pan, Y.W.; Zhu, P.W.; Liu, B.B.; Liu, H.J.; Wang, X.C.; Liu, J.; Zou, G.T. Crystal structures of $C_3N_6H_6$ under high pressure. *Chem. Phys. Lett.* **2003**, *368*, 668–672. [CrossRef]

16. Liu, X.R.; Zinin, P.V.; Ming, L.C.; Acosta, T.; Sharma, S.K.; Misra, A.K.; Hong, S.M. Raman spectroscopy of melamine at high pressures. *J. Phys. Conf. Ser.* **2010**, *215*, 012045. [CrossRef]

17. Pravica, M.; Kim, E.; Tkatchev, S.; Chow, P.; Xiao, Y. High-pressure studies of melamine. *High Press. Res.* **2010**, *30*, 65–71. [CrossRef]

18. Rivers, M.L.; Prakapenka, V.B.; Kubo, A.; Pullins, C.; Hall, C.M.; Jacobsen, S.D. The compres/gsecars gas loading system for diamond anvil cells at the advanced photon source. *High Press. Res.* **2008**, *28*, 273–292. [CrossRef]

19. Boehler, R.; De Hantsetters, K. New anvil designs in diamond-cells. *High Press. Res.* **2004**, *24*, 391–396. [CrossRef]

20. Mao, H.K.; Xu, J.; Bell, P.M. Calibration of the ruby pressure gauge to 800 kbar under quasi-hydrostatic conditions. *J. Geophys. Res. Solid Earth* **1986**, *91*, 4673–4676. [CrossRef]

21. Dera, P.; Zhuravlev, K.; Prakapenka, V.; Rivers, M.L.; Finkelstein, G.J.; Grubor-Urosevic, O.; Tschauner, O.; Clark, S.M.; Downs, R.T. High pressure single-crystal micro x-ray diffraction analysis with gse_ada/rsv software. *High Press. Res.* **2013**, *33*, 466–484. [CrossRef]

22. Sheldrick, G.M. *Cell_Now*, Version 2008/4; Georg-August-Universität Göttingen: Göttingen, Germany, 2008.

23. Sheldrick, G. A short history of shelx. *Acta Crystallogr. Sect. A* **2008**, *64*, 112–122. [CrossRef] [PubMed]

24. Gonzalez-Platas, J.; Alvaro, M.; Nestola, F.; Angel, R. EosFit7-GUI: A new graphical user interface for equation of state calculations, analyses and teaching. *J. Appl. Crystallogr.* **2016**, *49*, 1377–1382. [CrossRef]

25. Vinet, P.; Smith, J.R.; Ferrante, J.; Rose, J.H. Temperature effects on the universal equation of state of solids. *Phys. Rev. B* **1987**, *35*, 1945–1953. [CrossRef]

26. Jeanloz, R. Universal equation of state. *Phys. Rev. B* **1988**, *38*, 805–807. [CrossRef]

27. Dolomanov, O.V.; Bourhis, L.J.; Gildea, R.J.; Howard, J.A.K.; Puschmann, H. Olex2: A complete structure solution, refinement and analysis program. *J. Appl. Crystallogr.* **2009**, *42*, 339–341. [CrossRef]

28. Wolff, S.K.; Grimwood, D.J.; McKinnon, J.J.; Turner, M.J.; Jayatilaka, D.; Spackman, M.A. *Crystalexplorer (Version 3.1)*; University of Western Australia: Crawley, Australia, 2012.

29. Menges, F. *Spectragryph—Optical Spectroscopy Software, Version 1.2.8*; Spectroscopy Ninja: Oberstdorf, Germany, 2018.

30. Varghese, J.N.; O'Connell, A.M.; Maslen, E.N. The x-ray and neutron crystal structure of 2,4,6-triamino-1,3,5-triazine (melamine). *Acta Cryst.* **1977**, *B33*, 2102–2108. [CrossRef]

31. Hazen, R.M.; Downs, R.T. High-temperature and high-pressure crystal chemistry. In *Reviews in Mineralogy and Geochemistry*; Hazen, R.M., Downs, R.T., Eds.; Mineralogical Society of America: Washington, DC, USA, 2000; Volume 41.

32. Ciabini, L.; Gorelli, F.A.; Santoro, M.; Bini, R.; Schettino, V.; Mezouar, M. High-pressure and high-temperature equation of state and phase diagram of solid benzene. *Phys. Rev. B* **2005**, *72*, 094108. [CrossRef]

33. Nobrega, M.M.; Temperini, M.L.A.; Bini, R. Probing the chemical stability of aniline under high pressure. *J. Phys. Chem. C* **2017**, *121*, 7495–7501. [CrossRef]

34. Hanfland, M.; Brister, K.; Syassen, K. Graphite under pressure: Equation of state and first-order raman modes. *Phys. Rev. B* **1989**, *39*, 12598–12603. [CrossRef]

35. Funnell, N.P.; Dawson, A.; Marshall, W.G.; Parsons, S. Destabilisation of hydrogen bonding and the phase stability of aniline at high pressure. *CrystEngComm* **2013**, *15*, 1047–1060. [CrossRef]

36. Wen, X.D.; Hoffmann, R.; Ashcroft, N.W. Benzene under high pressure: A story of molecular crystals transforming to saturated networks, with a possible intermediate metallic phase. *J. Am. Chem. Soc.* **2011**, *133*, 9023–9035. [CrossRef] [PubMed]

37. Peiris, S.M.; Piermarini, G.J. *Static Compression of Energetic Materials*; Springer: Berlin, Germany, 2008; pp. 108–110.

38. Yoo, C.S.; Cynn, H.; Howard, W.M.; Holmes, N. *Equations of State of Unreacted High Explosives at High Pressures 11th International Detonation Symposium, Snowmass Village, CO, USA, 31 August–4 September 1998*; Lawrence Livermore National Laboratory: Snowmass Village, CO, USA, 1998.

39. Stevens, L.L.; Velisavljevic, N.; Hooks, D.E.; Dattelbaum, D.M. Hydrostatic compression curve for triamino-trinitrobenzene determined to 13.0 GPa with powder x-ray diffraction. *Propellants Explos. Pyrotech.* **2008**, *33*, 286–295. [CrossRef]

40. Gump, J.C.; Peiris, S.M. Isothermal equations of state of beta octahydro-1,3,5,7-tetranitro-1,3,5,7-tetrazocine at high temperatures. *J. Appl. Phys.* **2005**, *97*, 053513. [CrossRef]

41. Plisson, T.; Pineau, N.; Weck, G.; Bruneton, E.; Guignot, N.; Loubeyre, P. Equation of state of 1,3,5-triamino-2,4,6-trinitrobenzene up to 66 gpa. *J. Appl. Phys.* **2017**, *122*, 235901. [CrossRef]

42. Millar, D.I.A. *Energetic Materials at Extreme Conditions*; Springer: Berlin/Heidelberg, Germany, 2012.

43. Guo, F.; Zhang, H.; Hu, H.-Q.; Cheng, X.-L. Effects of hydrogen bonds on solid state tatb, rdx, and datb under high pressures. *Chin. Phys. B* **2014**, *23*, 046501. [CrossRef]

44. Davidson, A.J.; Dias, R.P.; Dattelbaum, D.M.; Yoo, C.-S. "Stubborn" triaminotrinitrobenzene: Unusually high chemical stability of a molecular solid to 150 GPa. *J. Chem. Phys.* **2011**, *135*, 174507. [CrossRef] [PubMed]

45. Dove, M.T.; Trachenko, K.O.; Tucker, M.G.; Keen, D.A. Rigid unit modes in framework structures: Theory, experiment and applications. *Rev. Mineral. Geochem.* **2000**, *39*, 1–33. [CrossRef]

46. Spackman, M.A.; Jayatilaka, D. Hirshfeld surface analysis. *CrystEngComm* **2009**, *11*, 19–32. [CrossRef]

47. McKinnon, J.J.; Jayatilaka, D.; Spackman, M.A. Towards quantitative analysis of intermolecular interactions with hirshfeld surfaces. *Chem. Commun.* **2007**, *37*, 3814–3816. [CrossRef]

48. Gilli, G.; Gilli, P. *The Nature of the Hydrogen Bond: Outline of a Comprehensive Hydrogen Bond Theory*; Oxford University Press: Oxford, UK, 2009; pp. 36–61.

49. Mooibroek, T.J.; Gamez, P. The s-triazine ring, a remarkable unit to generate supramolecular interactions. *Inorg. Chim. Acta* **2007**, *360*, 381–404. [CrossRef]

50. Martinez, C.R.; Iverson, B.L. Rethinking the term "pi-stacking". *Chem. Sci.* **2012**, *3*, 2191–2201. [CrossRef]

51. Mishra, B.K.; Arey, J.S.; Sathyamurthy, N. Stacking and spreading interaction in n-heteroaromatic systems. *J. Phys. Chem. A* **2010**, *114*, 9606–9616. [CrossRef] [PubMed]

52. Hunter, C.A.; Sanders, J.K.M. The nature of π-π interactions. *J. Am. Chem. Soc.* **1990**, *112*, 5525–5534. [CrossRef]

53. Fanetti, S.; Citroni, M.; Bini, R. Tuning the aromaticity of s-triazine in the crystal phase by pressure. *J. Phys. Chem. C* **2014**, *118*, 13764–13768. [CrossRef]

54. Fanetti, S.; Citroni, M.; Dziubek, K.; Nobrega, M.M.; Bini, R. The role of h-bond in the high-pressure chemistry of model molecules. *J. Phys. Condens. Matter* **2018**, *30*, 094001. [CrossRef] [PubMed]

55. Thiery, M.M.; Kobashi, K.; Spain, I.L. Raman spectra of solid benzene under high pressure. *Solid State Commun.* **1985**, *54*, 95–97. [CrossRef]

56. Schneider, J.R.; Schrader, B. Measurement and calculation of the infrared and raman active molecular and lattice vibrations of the crystalline melamine (1,3,5-triamino-s-triazine). *J. Mol. Struct.* **1975**, *29*, 1–14. [CrossRef]

57. Larkin, P.J.; Makowski, M.P.; Colthup, N.B. The form of the normal modes of s-triazine: Infrared and raman spectral analysis and ab initio force field calculations. *Spectrochim. Acta A* **1999**, *55*, 1011–1020. [CrossRef]

58. Marchewka, M.K. Infrared and raman spectra of the new melaminium salt: 2,4,6-triamino-1,3,5-triazin-1-ium hydrogenphthalate. *Mater. Lett.* **2004**, *58*, 843–848. [CrossRef]

59. Marchewka, M.K. Infrared and raman spectra of melaminium chloride hemihydrate. *Mater. Sci. Eng.* **2002**, *B95*, 214–221. [CrossRef]

60. Goncharov, A.; Gregoryanz, E. Chapter 8—Solid nitrogen at extreme conditions of high pressure and temperature. In *Chemistry at Extreme Conditions*; Manaa, M.R., Ed.; Elsevier: Amsterdam, The Netherlands, 2005; pp. 241–267.

61. Song, Y.; Hemley, R.J.; Mao, H.-K.; Herschbach, D.R. Chapter 6—Nitrogen-containing molecular systems at high pressures and temperature. In *Chemistry at Extreme Conditions*; Manaa, M.R., Ed.; Elsevier: Amsterdam, The Netherlands, 2005; pp. 189–222.

62. Li, W.; Huang, X.; Bao, K.; Zhao, Z.; Huang, Y.; Wang, L.; Wu, G.; Zhou, B.; Duan, D.; Li, F.; et al. A novel high-density phase and amorphization of nitrogen-rich 1h-tetrazole (ch2n4) under high pressure. *Sci. Rep.* **2017**, *7*, 39249. [CrossRef] [PubMed]

63. Shishkina, S.V.; Konovalova, I.S.; Shishkin, O.V.; Boyko, A.N. Acceptor properties of amino groups in aminobenzene crystals: Study from the energetic viewpoint. *CrystEngComm* **2017**, *19*, 6274–6288. [CrossRef]

64. Szatyłowicz, H. Structural aspects of the intermolecular hydrogen bond strength: H-bonded complexes of aniline, phenol and pyridine derivatives. *J. Phys. Org. Chem.* **2008**, *21*, 897–914. [CrossRef]

65. Citroni, M.; Fanetti, S.; Bazzicalupi, C.; Dziubek, K.; Pagliai, M.; Nobrega, M.M.; Mezouar, M.; Bini, R. Structural and electronic competing mechanisms in the formation of amorphous carbon nitride by compressing s-triazine. *J. Phys. Org. Chem. C* **2015**, *119*, 28560–28569. [CrossRef]

66. Citroni, M.; Fanetti, S.; Bini, R. Pressure and laser-induced reactivity in crystalline s-triazine. *J. Phys. Org. Chem. C* **2014**, *118*, 10284–10290. [CrossRef]

67. Li, S.; Li, Q.; Xiong, L.; Li, X.; Li, W.; Cui, W.; Liu, R.; Liu, J.; Yang, K.; Liu, B.; et al. Effect of pressure on heterocyclic compounds: Pyrimidine and s-triazine. *J. Phys. Org. Chem.* **2014**, *141*, 114902. [CrossRef] [PubMed]

68. Li, W.; Duan, D.; Huang, X.; Jin, X.; Yang, X.; Li, S.; Jiang, S.; Huang, Y.; Li, F.; Cui, Q.; et al. Pressure-induced diversity of π-stacking motifs and amorphous polymerization in pyrrole. *J. Phys. Org. Chem. C* **2014**, *118*, 12420–12427. [CrossRef]

69. Zhuravlev, K.K.; Traikov, K.; Dong, Z.; Xie, S.; Song, Y.; Liu, Z. Raman and infrared spectroscopy of pyridine under high pressure. *Phys. Rev. B* **2010**, *82*, 064116. [CrossRef]

70. Wang, K.; Duan, D.; Wang, R.; Lin, A.; Cui, Q.; Liu, B.; Cui, T.; Zou, B.; Zhang, X.; Hu, J.; et al. Stability of hydrogen-bonded supramolecular architecture under high pressure conditions: Pressure-induced amorphization in melamine–boric acid adduct. *Langmuir* **2009**, *25*, 4787–4791. [CrossRef] [PubMed]

71. Wang, K.; Duan, D.; Wang, R.; Liu, D.; Tang, L.; Cui, T.; Liu, B.; Cui, Q.; Liu, J.; Zou, B.; et al. Pressure-induced phase transition in hydrogen-bonded supramolecular adduct formed by cyanuric acid and melamine. *J. Phys. Org. Chem. B* **2009**, *113*, 14719–14724. [CrossRef] [PubMed]

crystals

MDPI

Article

The Behavior of the Deformation Vibration of NH₃ in Semi-Organic Crystals under High Pressure Studied by Raman Spectroscopy

André Luís de Oliveira Cavaignac [1,2,*] , Rilleands Alves Soares [1], Ricardo Jorge Cruz Lima [1], Pedro de Freitas Façanha Filhoo [1] and Paulo de Tarso Cavalcante Freire [3]

[1] Centro de Ciências Sociais, Saúde e Tecnologia, Universidade Federal do Maranhão, Imperatriz-MA 65900-410, Brazil; rilleandsalvessoares@gmail.com (R.A.S.); ricardo.lima.ufma@gmail.com (R.J.C.L.); freitasfacanha@gmail.com (P.d.F.F.F.)
[2] Departamento de Engenharias, Universidade Ceuma, Imperatriz-MA 65903-093, Brazil
[3] Departamento de Física, Universidade Federal do Ceará, Fortaleza-CE 60455-760, Brazil; tarso@fisica.ufc.br
* Correspondence: andreluiscavaignac@gmail.com; Tel.: +55-999-8108-4153

Received: 4 May 2018; Accepted: 5 June 2018; Published: 7 June 2018

Abstract: Single-crystal samples of the semi-organic compounds mono-L-alaninium nitrate and monoglycine nitrate have been studied by Raman spectroscopy in a diamond-anvil cell up to 5.5 GPa, in order to observe the behavior of the deformation mode of NH₃ units. It was observed for these semi-organic crystals that increasing pressure produces a decrease in the wavenumber of the band associated with the deformation vibration, differently from most of the modes. Comparatively, mono-L-alaninium has a higher dν/dP than monoglycine nitrate, for the band associated with the deformation vibration. The anomalous behavior is explained in terms of the effect of high pressure in the short and linear intermolecular hydrogen bonds.

Keywords: semi-organic crystals; high-pressures; raman spectroscopy

---

1. Introduction

Hydrogen bonds (H-bond) play a very special role in the interactions among molecules of biological interest, including those with large applications and importance for pharmaceutical and medical sciences [1,2]. One way to gain insights about the behavior of H-bonds is approximating molecules through the application of high pressure. Unfortunately, there is the problem of obtaining accurate information on the H-atom positions and a complete picture of the effect of high pressure on H-bonds, which is a difficult experimental task. In fact, an H atom has only one electron, which makes them very hard to detect with X-rays accurately because X-rays are scattered from the electron density. On the other hand, thermal neutrons diffract strongly from the nuclei of H atoms, which make them detectable by using nuclear reactors, but the accessibility for researchers is limited.

Regarding the studies of amino acids from a structural point of view, only a few works have focused on the question of H-bonds. These investigations attempted to furnish a description of this interaction and to understand the stability of the crystal structure. For example, in the orthorhombic phase of L-cysteine, the values of the S-H•••O and the S-H•••S angles remain unknown and a complete picture of the effect of high pressure on H-bonds was not obtained [3]. On the other hand, an X-ray crystallographic study in L-alanine furnished a good picture about the high pressure effect in H-bonds [4]. In that work high-pressure single-crystal X-ray measurements were shown, which were carried out using L-alanine-h₇ combined with high-pressure neutron powder diffraction measurements, carried out with L-alanine-d₇ and DFT (Density functional theory) calculations. One of the main effects of pressure in strong N-H●●●O bond is reducing significantly the bond angle, which varies from an average 155.75° at room pressure to 148.97° at 9.87 GPa [4].

Calculations on crystals using the DFT methodology have proven to be quite accurate [5]. A recent study on L-tyrosine under high pressure [6] showed that it is not possible to distinguish substantial differences between the theoretical spectrum and the experimental FT-IR spectrum. However, we are not using such calculations in the present work. Additionally, a complete picture of the effect of high pressure in H-bonds is not trivially obtained through high pressure crystallographic experiments. For example, deuteration, which requires a difficult procedure to obtain samples, is essential for neutron powder diffraction [4] due to the differences in scattering amplitude caused by the replaced H atom by the deuterium [7]. Thus, the main strength of Raman spectroscopy is the ability to provide a great wealth of easily analyzable qualitative information very rapidly, providing a powerful diagnostic of this process.

In other words, high-pressure Raman spectroscopy can provide interesting insights on the effect of high pressure in H-bonds. For example, we observed for the first time that the behavior of the NH_3^+ torsional mode in L-alanine is unusual because it presents a negative slope in wavenumber shifts with pressure ($dv/dp < 0$). We have interpreted this to be a consequence of the particular characteristics of H-bond geometry, where, on compression, the N-H●●●O bonds become less linear [8].

It is worth mentioning that a recent Raman scattering study, focusing on polymorphism [9], studied mono-L-alaninium nitrate under high pressure and observed a phase transition between 3.5 and 4.1 GPa. In another study, on monoglycine nitrate under high pressure [10], the occurrence of two phase transitions in the network modes between 1.1–1.6 GPa and 4.0–4.6 GPa was observed. In the present study, the focus is on the influence of the hydrogen bonds in a specific vibrational dynamic in the two semi-organic crystals. In addition, we try to show that the Raman spectroscopic technique is an easier way to obtain experimental results on these binding behaviors than crystallographic techniques.

This work is the continuation of a Raman spectroscopic investigation to monitor the effects of high pressure in H-bonds. Now we show a study about the behavior of the deformation mode of NH_3^+ unit, $\delta(NH_3^+)$, in two semi-organic compounds, mono-L-alaninium nitrate ($+NH_3$-CH_3-OH-COOH, NO_3^-), and monoglycine nitrate ($+NH_3$-CH_2-COOH, NO_3^-). The analysis allowed us to understand the change of geometry of H-bonds in the two compounds as a consequence of high pressure application.

2. Materials and Methods

Crystals of mono-L-alaninium nitrate were prepared by slow spontaneous evaporation of a solution of L-alanine 99%, Sigma-Aldrich (St. Louis, MO, USA) in nitric acid, Erba Lachema (Brun, Czech Republic), 1 moll 1-1 (in a molar ratio of 1:1) at laboratory temperature. Crystals of monoglycine nitrate were prepared by reaction equimolar of glycine 98% Sigma-Aldrich (St. Louis, MO, USA), and nitric acid 65% Dinâmica (Diadema, Brazil), at laboratory temperature. To confirm the orthorhombic $P2_12_12_1$ structures of both crystals, X-ray diffraction patterns of crushed powder were recorded using a Philips Analytical X-ray diffractometer (Model PW1710) with Cu Kα radiation. A Raman scattering experiment was performed from a small crystal in a membrane diamond anvil cell (MDAC), whose hole diameter in a stainless gasket was ~130 μm pre-indented to a thickness of ~54 μm. The pressure in the cell was monitored using the shifts of the Cr^{+3}:Al_2O_3 lines. The pressure calibration is expected to be accurate to ±0.08 GPa. The mineral oil Nujol was used as a hydrostatic pressure medium. The excited source was a semi-conductor Nd:YVO4 laser, model Verdi V-5 of Coherent, with a line beam of 532 nm operating at 250 mW. However, due to the experimental laser assembly scheme, only a small fraction of the output power is absorbed by the sample. Therefore, the effect of the temperature increase of the sample in our work is neglected, an approximation commonly used in this type of experiment. The backscattering light was analyzed using a Jobin Yvon Triplemate 64,000 micro-Raman system equipped with an N_2-cooled CCD system. The slits were set for a 2 cm^{-1} spectral resolution. The laser beam was focused on the sample surface using an Olympus BX40 microscope lens with f = 20.5 mm. Spectra with six 90-s acquisitions were made. From this point, increasing numbers or acquisition time did not result in better spectrum performance.

3. Results and Discussions

In mono-L-alaninium nitrate (MAN), the structure molecule is in the form of a protonated alanininum cation (+NH$_3$-CH$_3$-OH-COOH). Each NH$_3^+$ group forms three conventional H-bonds. The bonds with lengths 2.830 and 2.900(1) Å, connect NH$_3^+$ group and nitrate anions (NO$_3^-$). The intermolecular bond with length 2.877(1) Å, connects the NH$_3^+$ group and ⁻COOH group [11].

In monoglycine nitrate (MGN) the molecule exists as +NH$_3$-CH$_2$-COOH, NO$_3^-$. Because H-bonds are long-range interactions, a group N-H is bonded to more than one acceptor (oxygen atom) at the same time and two of the hydrogen atoms of the NH$_3$ group form bifurcated hydrogen bonds (BH-bond). The third hydrogen atom of the NH$_3^+$ group forms a strong intermolecular conventional hydrogen bond (H-bond) with lengths 2.792 Å and angle 169° [12]. Figure 1 presents the local H-bond environment, with length and angle, of the NH3$^+$ group in both (a) MGN and (b) MAN.

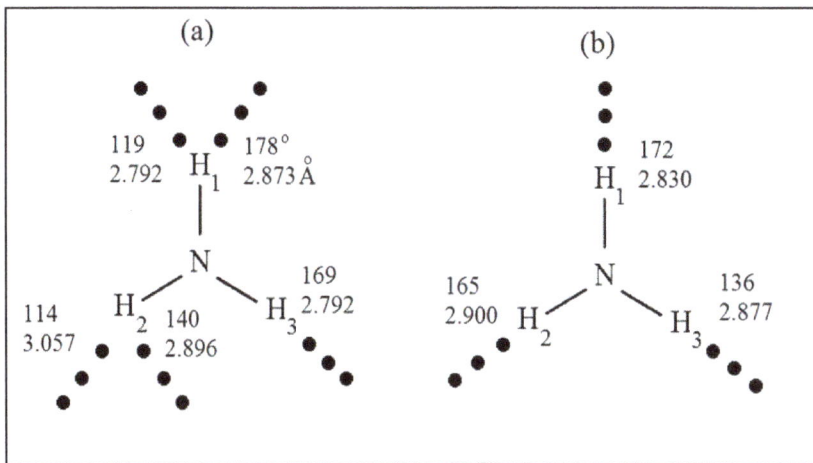

Figure 1. Representation of NH$_3^+$ group in (a) monoglycine nitrate and (b) mono-L-alaninium nitrate showing the hydrogen bonds with their respective lengths and angles.

An analysis and comparison of the H-bonds in both compounds reveals substantial differences. In fact, in MGN, the presence of two BH-bonds is observed, and these interactions are longer than most of the H-bonds in MAN, e.g., N-H•••O. As a consequence, usually, the BH-bond has the characteristics of weak H-bonds [13].

Figure 2 presents the Raman spectra of MGN and MAN in the wavenumber region where the observation of deformation bending vibration of NH$_3^+$, δ(NH$_3^+$) is expected, for two different pressure values. The pressure dependence of the wavenumber modes between 0.0 and 5.5 GPa is presented in Figure 3, where MGN is represented by the symbol (■) and MAN by the symbol (□). The wavenumber pressure dependence of the bands exhibits a linear behavior. In order to furnish a quantitative analysis of the behavior of the wavenumber pressure dependence, Table 1 shows the wavenumber values for δ(NH$_3^+$) (at atmospheric pressure) and the variation of the wavenumber as a function of pressure (dv/dP), for MGN and MAN.

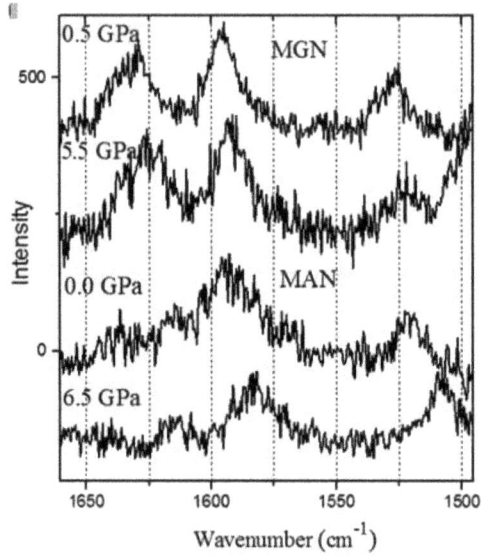

Figure 2. Raman spectra of monoglycine nitrate (MGN) and mono-L-alaninium nitrate (MAN) for two different pressures, showing the region where it is expected the observation of mode associated with the bending of NH_3^+.

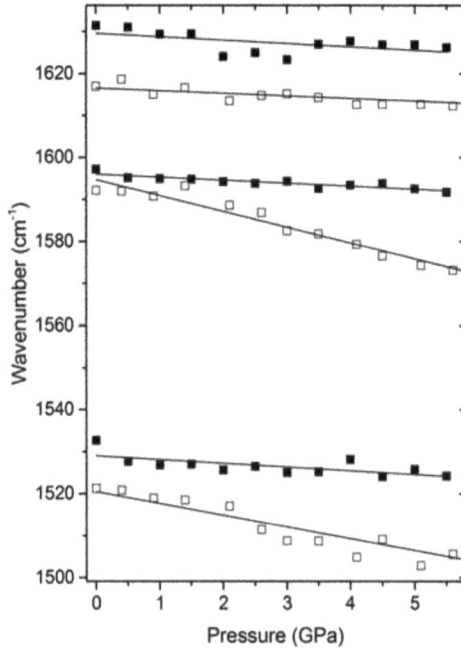

Figure 3. Wavenumber of the mode associated with bending of NH_3^+ as a function of pressure for monoglycine nitrate (full symbols) and mono-L-alaninium nitrate (empty symbols).

Table 1. The wavenumber at atmospheric pressure v (cm^{-1}) of the deformation mode of NH$_3^+$ unit and the variation of the wavenumber as a function of pressure (dv/dP).

MGN		MAN	
v (cm^{-1})	dv/dP (cm^{-1}/GPa)	v (cm^{-1})	dv/dP (cm^{-1}/GPa)
1533	−0.9	1521	−2.8
1597	−0.7	1592	−3.8
1631	−0.8	1617	−0.6

In general terms, the Raman bands are shifted to higher wavenumbers with increasing pressure because interatomic and intermolecular bonds stiffen as they shorten. The experimental result presented here indicated an anomalous behavior of the δ(NH$_3^+$) for both of the compounds with negative dv/dP values. It is observable that the δ(NH$_3^+$) of H bond is influenced by the pressure effects, making it less linear and weakening the bonds. Additionally, for MAN, dv/dP values vary more strongly. It is reasonable to affirm that the wavenumber of the NH$_3^+$ vibrations strongly depends on the local H-bond environment, as reported in [8] for the NH$_3^+$ torsional mode in L-alanine. In other words, the wavenumber behavior for the δ(NH$_3^+$) modes should be dominated by pressure effects on H-bonds set.

The NH$_3^+$ unit in MGN, is endowed with BH-bonds in two H-atoms (H1 and H2) and a strong H-bond (H3). Competitive interaction occurs between BH-bonds and strong H-bonds with compression having an effect on dv/dP ~−1.0 cm^{-1}/GPa for δ(NH$_3^+$) bands. On the other hand, the NH$_3^+$ unit in MAN, endowed with three relatively strong H-bonds, having as an effect a higher variation in dv/dP values than MGN, with dv/dP ~−4.0 cm^{-1}/GPa for δ(NH$_3^+$) bands.

These results suggest a model to explain the unusual behavior of the δ(NH$_3^+$) mode for both compounds, where strong H-bonds become less linear with increasing pressure pointing to a weakening of the force constant at higher pressures. A similar effect occurs with H-bonds in L-alanine as verified through the behavior of the NH$_3^+$ torsional mode. In MGN, pressure has the effect of strengthening BH-bonds and weakening strong H-bonds resulting in a small variation of dv/dP for δ(NH$_3^+$) bands. In MAN, which has three very strong H-bonds, the pressure increase causes the weakening not only of one h-bond, but of three, by decreasing linearity. Thus, this results in a large variation of dv/dP for δ (NH$_3^+$) bands.

In a previous Raman spectroscopic work on MGN [10], we have studied the phase transition phenomenon, but we were not able to detail the effect of high pressure in the H-bonds. In that study, we have presented the behavior of the NH$_3^+$ torsional mode (at 506 cm^{-1} at atmospheric pressure) in MGN, that shows a positive slope in wavenumber shifts with pressure (dv/dP = 1.1 cm^{-1}/GPa). According to our model, the NH$_3^+$ torsional mode in MGN is dominated by the effects of pressure on the weak BH-bonds, strengthening of the force constant at higher pressures. Unfortunately, the bands corresponding to the NH$_3^+$ torsional mode in MAN, which are very important for monitoring the hydrostatic pressure effects in the H-bonds and strengthen our model, are difficult to follow in the Raman spectra since they are too weak. In subsequent studies, the DFT investigation could contribute to elucidating what happens to the H bonds, completing the experimental observation.

The pressure or stress-induced wavenumber shifts $\Delta v/v$ are mainly determined by the associated deformation of volume, $\Delta V/V$. Pressure affects the equilibrium spacings between nuclei, distorts the electron clouds, and, through them, modifies the restoring forces. Interatomic potential containing power laws or exponential terms are strongly anharmonic, and are composed by attractive and repulsive terms [14]. Therefore, for short and linear N-H•••O bonds, the approximation of the N, H, and O atoms, due to compression, is impeded by the repulsive terms of the interatomic potential having the effect of reducing the angle of the bond.

On one hand, our study showed the behavior of the H-bond geometry due to pressure effects. On the other hand, very recently, a study has shown that certain guest molecular inclusions result in stronger H-bonds [15]. This means that H-bonds play a fundamental role in binding guest molecules to the respective host compounds, with a clear influence on the link of the host-guest complex [15]. So, we give as a suggestion for future works the investigation of host-guest compounds under high pressure conditions in order to draw a picture about the joint effects of strengthening H-bonds due to both the guest and the pressure effect itself.

4. Conclusions

We have presented a study on simple semi-organic compounds, mono-L-alaninium nitrate and monoglycine nitrate, in order to shed light on the problem of the effect of high pressure in H-bonds. Particularly, we have studied the pressure-dependence of the bending mode of the NH_3^+ unit under the scrutiny of a spectroscopic tool, an easier access tool than experimental methods used by [4]. The experimental results indicated that strong N-H•••O bonds, at room temperature, become less linear and weak with increasing pressure, and, comparatively, the wavenumber behavior for the $\delta(NH_3^+)$ modes should be dominated by pressure effects on the H-bonds set. Additionally, the present study has improved the discussion about the effect of high pressure on H-bonds, complementing a previous work where the torsional vibrations of the same group in certain amino acids crystals were investigated.

Author Contributions: In this work, A.L.d.O.C. and R.J.C.L conceived of the presented idea, R.A.S. and P.d.F.F.F. carried out the experiment, A.L.d.O.C. and R.J.C.L. wrote the manuscript with input from all authors. All authors discussed the results and contributed to the final manuscript.

Acknowledgments: Authors acknowledge funding support from CNPq, CAPES, FAPEMA and Uniceuma.

References

1. Boldyreva, E.V.; Sowa, H.; Ahsbahs, H.; Goryainov, S.V.; Chernyshev, V.V.; Dmitriev, V.P.; Seryotkin, Y.V.; Kolesnik, E.N.; Shakhtshneider, T.P.; Ivashevskaya, S.N.; et al. Pressure-induced phase transitions in organic molecular crystals: A combination of X-ray single-crystal and powder diffraction, Raman and IR-spectroscopy. *J. Phys. Conf. Ser.* **2008**, *121*, 022023. [CrossRef]
2. Masson, P.; Tonello, C.; Balny, C. High-pressure biotechnology in medicine and pharmaceutical science. *J. Biomed. Biotechnol.* **2001**, *1*, 85–88. [CrossRef] [PubMed]
3. Allan, D.R.; Clark, S.J.; Gutmann, M.J.; Sawyer, L.; Moggach, S.A.; Parsons, S.; Pulham, C.R. High-pressure polymorphism in L-cysteine: The crystal structures of L-cysteine-III and L-cysteine-IV. *Acta Crystallogr. B* **2006**, *62*, 296–309. [CrossRef]
4. Funnell, N.P.; Dawson, A.; Francis, D.; Lennie, A.R.; Marshall, W.G.; Moggach, S.A.; Warrenc, J.E.; Parsons, S. The effect of pressure on the crystal structure of L-alanine. *CrystEngComm* **2010**, *12*, 2573–2583. [CrossRef]
5. Szeleszczuk, Ł.; Pisklak, D.M.; Zielinska-Pisklak, M. Can we predict the structure and stability of molecular crystals under increased pressure? First-principles study of glycine phase transitions. *J. Comput. Chem.* **2018**. [CrossRef] [PubMed]
6. Dos Santos, C.A.A.S.; Carvalho, J.O.; da Silva Filho, J.G.; Rodrigues, J.L.; Lima, R.J.C.; Pinheiro, G.S.; Freire, P.T.C.; FaçanhaFilho, P.F. High-pressure Raman spectra and DFT calculations of L-tyrosinehydrochloride. *Phys. B Condens. Matter* **2017**. [CrossRef]
7. Jacrot, B. The study of biological structures by neutron scattering from solution. *Rep. Prog. Phys.* **1976**, *39*, 911. [CrossRef]
8. Freire, P.T.C.; Melo, F.E.A.; Filho, J.M.; Lima, R.J.C.; Teixeira, A.M.R. The behavior of NH3 torsional vibration of L-alanine, L-threonine and taurine crystals under high pressure: A Raman spectroscopic study. *Vib. Spectrosc.* **2007**, *45*, 99–102. [CrossRef]
9. Soares, R.A.; Lima, R.J.C.; Façanha Filho, P.F.; Freire, P.T.C.; Lima, J.A., Jr.; da Silva Filho, J.G. High-pressure Raman study of mono-L-alaninium nitrate crystals. *Phys. B Condens. Matter* **2017**, *521*, 317–322. [CrossRef]

10. Carvalho, J.O.; Moura, G.M.; Dos Santos, A.O.; Lima, R.J.C.; Freire, P.T.C.; Façanha Filho, P.F. High pressure Raman spectra of monoglycine nitrate single crystals. *Spectrochim. Acta A* **2016**, *161*, 109–114. [CrossRef] [PubMed]

11. Nemec, I.; Cısarova, I.; Micka, Z. The crystal structure, vibrational spectra and DSC measurement of mono-L-alaninium nitrate. *J. Mol. Struct.* **1999**, *476*, 243–253. [CrossRef]

12. Narayanan, P.; Venkataraman, S. Crystal structure analyses of some addition compounds of glycine. *J. Cryst. Mol. Struct.* **1975**, *5*, 15–26. [CrossRef]

13. Desiraju, G.R.; Steiner, T. The Weak Hydrogen Bond. In *Structural Chemistry and Biology*; Oxford University Press: New York, NY, USA, 1999.

14. Lucazeau, G. Effect of pressure and temperature on Raman spectra of solids: Anharmonicity. *J. Raman Spectrosc.* **2003**, *34*, 478–496. [CrossRef]

15. Usman, R.; Khan, A.; Wang, M.L. Study of H-bonded assemblies of the solvates of anthracene derivatives: Guest effect on the crystal symmetry and spectroscopic properties. *Supramol. Chem.* **2017**, *29*, 497–505. [CrossRef]

crystals

MDPI

Article

Structural Phase Transition and Compressibility of CaF$_2$ Nanocrystals under High Pressure

Jingshu Wang [1] [iD], Jinghan Yang [1], Tingjing Hu [1] [iD], Xiangshan Chen [1], Jihui Lang [1,*], Xiaoxin Wu [1], Junkai Zhang [1], Haiying Zhao [1], Jinghai Yang [1] and Qiliang Cui [2]

[1] Key Laboratory of Functional Materials Physics and Chemistry of the Ministry of Education, National Demonstration Center for Experimental Physics Education, Jilin Normal University, Siping 136000, China; wjs@jlnu.edu.cn (J.W.); jlsfdxchina@163.com (J.Y.); tjhumars@126.com (T.H.); 13134497345@163.com (X.C.); wuXiaoxin@126.com (X.W.); junkaizhang126@126.com (J.Z.); HaiyingZh@126.com (H.Z.); jhyang1@jlnu.edu.cn (J.Y.)
[2] State Key Laboratory of Superhard Materials, Jilin University, Changchun 130012, China; wjs_916_82@126.com
* Correspondence: langjihui80@126.com

Received: 19 March 2018; Accepted: 26 April 2018; Published: 3 May 2018

Abstract: The structural phase transition and compressibility of CaF$_2$ nanocrystals with size of 23 nm under high pressure were investigated by synchrotron X-ray diffraction measurement. A pressure-induced fluorite to α-PbCl$_2$-type phase transition starts at 9.5 GPa and completes at 20.2 GPa. The phase-transition pressure is lower than that of 8 nm CaF$_2$ nanocrystals and closer to bulk CaF$_2$. Upon decompression, the fluorite and α-PbCl$_2$-type structure co-exist at the ambient pressure. The bulk modulus B_0 of the 23 nm CaF$_2$ nanocrystals for the fluorite and α-PbCl$_2$-type phase are 103(2) and 78(2) GPa, which are both larger than those of the bulk CaF$_2$. The CaF$_2$ nanocrystals exhibit obviously higher incompressibility compare to bulk CaF$_2$. Further analysis demonstrates that the defect effect in our CaF$_2$ nanocrystals plays a dominant role in the structural stability.

Keywords: high pressure; CaF$_2$ nanocrystals; phase transitions; X-ray; bulk modulus

1. Introduction

Calcium fluoride (CaF$_2$)—with low absorption coefficient, high transmittance, anionic conductivity, and high resistivity—has become the focus of fundamental scientific research and industrial applications [1–3]. Especially because of its typical crystal structure, CaF$_2$ becomes the best material for high pressure research. At ambient conditions, bulk CaF$_2$ crystallizes in the cubic fluorite structure (*Fm3m*), in which Ca and F atoms occupy Wyckoff position 4a and 8c positions, respectively. Many theoretical and experimental studies on structural phase transition, optical, and electronic properties of CaF$_2$ under high pressure have been reported [4–13]. At high pressure, CaF$_2$ undergoes two structural phase transitions to highly coordinated structures. The first phase transition from the fluorite structure to an orthorhombic α-PbCl$_2$-type structure (*Fm3m* to *Pnma*) is reported to occur at 8–10 GPa [4–6]. Theoretical studies of CaF$_2$ predicted that the second phase transition from α-PbCl$_2$-type structure to Ni$_2$In-type structure (*Pnma* to *P6$_3$/mmc*) takes place at 68–278 GPa [6,7]. Experimental research by S M Dorfman et al. reported that bulk CaF$_2$ transformed from the α-PbCl$_2$-type structure to the Ni$_2$In-type structure at 79 GPa with heating to about 2000 K [8]. Although there are widely high-pressure studies on bulk CaF$_2$, yet very few experimental studies on nanosized CaF$_2$ exist for comparison.

Because of the surface effect and quantum confinement effect yielded by the significantly decrease in size, applications of high pressure have proven to be an important tool in tailoring the properties of nanomaterials [14–33]. Previous high-pressure studies on nanomaterials have revealed a set of novel high-pressure behaviors, which different from that of their corresponding bulk materials. The size effect has a great influence on the phase transition pressure [16–19], the course of amorphization [21–25], and phase transition routines [26]. Our previous high-pressure X-ray diffraction study revealed that CaF_2 nanocrystals with size of 8 nm transformed from fluorite structure into α-$PbCl_2$-type structure at 14 GPa, and the high-pressure structure was stable up to 46.5 GPa. The enhancement of structural stability is mainly due to the surface energy differences between the cubic and orthorhombic phases [34]. Beyond that, pressure-induced structural transition of nanosized CaF_2 has no other reports, and the bulk modulus of nanoscale CaF_2 is still not well known. In order to further research the high-pressure behaviors of nanosized CaF_2, and explore the important factors to affect the structural stability of nanomaterials, more experimental data for various sized CaF_2 is urgently needed.

Herein, we present our observations on structural phase transition and compressibility of 23 nm CaF_2 nanocrystals using synchrotron X-ray diffraction measurement. We found that the phase-transition pressure from fluorite to α-$PbCl_2$-type is lower than that previously reported of 8 nm CaF_2 nanocrystals and closer to bulk CaF_2. The bulk modulus of the 23 nm CaF_2 nanocrystals for the fluorite and α-$PbCl_2$-type structure are both larger than those of the bulk CaF_2. We believe that the enhancement of bulk modulus is due to higher surface energy. We observed that the as-synthesized CaF_2 nanocrystals contain visible dislocations and defects which could play a dominant role in structural stability.

2. Materials and Methods

The 23 nm-sized CaF_2 nanocrystals were synthesized using the solvothermal synthesis method. 0.5 mmol of $Ca(NO_3)_2$ and 1 mmol of NaF were dissolved in 10 mL distilled water to form clear solutions, respectively. The two solutions were mixed under full stirring to obtain opaque white suspension. The solution was transferred in to a 40ml autoclave which was filled to 85% of its total capacity with ethanol. Then, the system was kept in an oven at 140 °C for 15 h and cooled in air naturally. The final products were obtained after centrifugation and drying treatment.

The crystalline morphology and particle size of synthesized CaF_2 nanocrystals were examined by transmission electron microscopy TEM (200 KV, H-8100IV; HITACHI, Tokyo, Japan) and high-resolution transmission electron microscopy HRTEM (JEM-2100HR; JEOL, Tokyo, Japan). The sample was loaded into a symmetric diamond anvil cell (DAC) with a culet size of 300 μm for high-pressure characterization, and silicon oil was chosen as the pressure transmitting media. The synthesized CaF_2 nanocrystals were enclosed into a 100-μm-diameter hole of the T301 stainless steel gasket. The shift of the ruby R1 line was utilized to calibrate the pressure. A high-pressure X-ray diffraction experiment was performed at BL15U stations of Shanghai Synchrotron Radiation Facility26 (SSRF, λ = 0.6199 Å). A focused beam size of 3×3 μm² was used for data collection. MAR165 CCD detector was utilized to collect the X-ray diffraction data. The distance between the sample and CCD detectors was 173 mm which was calibrated using a CeO_2 standard. FIT2D software was used to process the two-dimensional X-ray diffraction images. The ORIGIN8 and MATERIAL STUDIO programs were adopted to refine high-pressure synchrotron XRD patterns.

3. Results and Discussion

Figure 1 shows that the shape of as-prepared CaF_2 nanocrystals is irregular particle with uniform morphology. Selected-area electron diffraction (SAED) pattern (inset in Figure 1a) shows that the major diffraction rings can be indexed into a cubic structure. Figure 1b displays the corresponding particle size distribution histograms. The distribution is narrow, and the average size is 23 ± 4 nm.

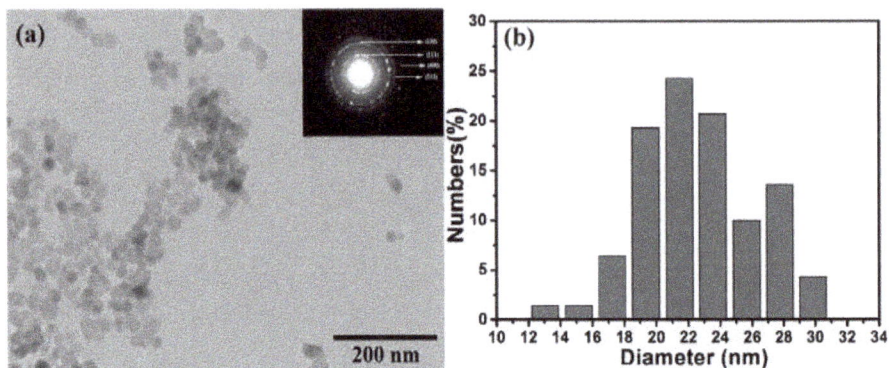

Figure 1. (a) TEM image of the as-synthesized CaF_2 nanocrystals. Inset in (a) represents the corresponding SAED pattern. (b) Particle size distribution of CaF_2 nanocrystals.

Figure 2b presents XRD spectra for the 23 nm CaF_2 nanocrystals collected by increasing the pressure gradually up to 23.5 GPa. Starting at ambient pressure, three diffraction peaks of CaF_2 nanocrystals together with one peak of the gasket material (marked by star) are observed. As shown in Figure 3a, the Rietveld refinement performed at ambient pressure suggests that the nanocrystals diffraction peaks is a good agreement with belong to the fluorite structure with space group of $Fm3m$ ($Rwp = 0.98\%$, $Rp = 0.62\%$). The fluorite structure was characterized by 'Ca' and 'F' atoms occupying the (0, 0, 0) and (0.25, 0.25, 0.25) position, respectively. We can clearly observe that a phase transition from fluorite structure to α-$PbCl_2$-type structure takes place at 9.5 GPa. This phase-transition pressure is lower than that the previously reported 8 nm CaF_2 nanocrystals and closes to bulk CaF_2. At pressures higher than 9.5 GPa, additional diffraction new peaks emerge and intensity increases with increasing pressure. At the same time, the diffraction peaks representing the cubic phase become weak. Figure 4 shows the unit-cell parameters as function of pressure. At 9.5 GPa, the cell parameter and unit cell volume of cubic structure were calculated to be a = 5.35(9) Å and V= 153.93 (3) Å^3. Meanwhile, we obtain the cell parameters and unit cell volume of the α-$PbCl_2$-type structure (a = 6.90(1), b = 5.83(7) Å, c = 3.45(6) Å, and V= 139.18 (5) Å^3). The volume change for the fluorite and α-$PbCl_2$-type structures is a decrease of 9.6% from the low-pressure structure. Complete phase transformation to the high-pressure phase occurs at 20.2 GPa. It is clearly observed that the cubic fluorite structure still exists in the 10–20 GPa range after the phase-transition pressure at 9.5 GPa. The phase transformation is quite sluggish. The Rietveld refinement performed at 20.2 GPa shows that the XRD experimental datum is in a great agreement with an α-$PbCl_2$-type cell with space group $Pnma$ ($Rwp = 1.09\%$, $Rp = 1.07\%$) with the unit cell volume of 129.95 (2) Å^3 (Figure 3b). The Ca and F ions of the α-$PbCl_2$-type structure occupy the positions Ca(0.253, 0.25, 0.109), F1(0.859, 0.25, 0.073), and F2 (0.438, 0.25, 0.834), respectively. The high-pressure α-$PbCl_2$-type structure can be maintained up to the highest studied pressure of 23.5 GPa in our XRD measurements.

Upon decompression, the original fluorite phase of 23 nm CaF_2 nanocrystals is not recovered when the pressure is reduced to 8.4 GPa. The experimental data point that the transition from orthorhombic to cubic phase exhibits strong hysteresis under decompression. After decompression to ambient condition, the sample retains the α-$PbCl_2$-type structure with for the peaks attributed to fluorite structure remain in the XRD datum. It indicates that the fluorite and α-$PbCl_2$-type structure co-exist at the ambient pressure after decompression. This result is in good accordance with the reports on the bulk CaF_2 [4], and the hysteresis might be explained as the inherent sluggish nature of CaF_2. For the reported 8 nm-sized CaF_2 nanocrystals, the high-energy hindrance prevents the transition to the fluorite structure which might be the reason for irreversibility at ambient pressure.

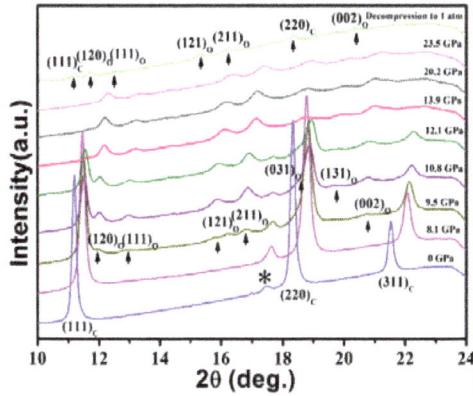

Figure 2. High-pressure X-ray diffraction patterns of CaF_2 nanocrystals. Upon being completely quenched to the ambient conditions. The peak (marked with asterisk) is derived from the T301 stainless steel gasket.

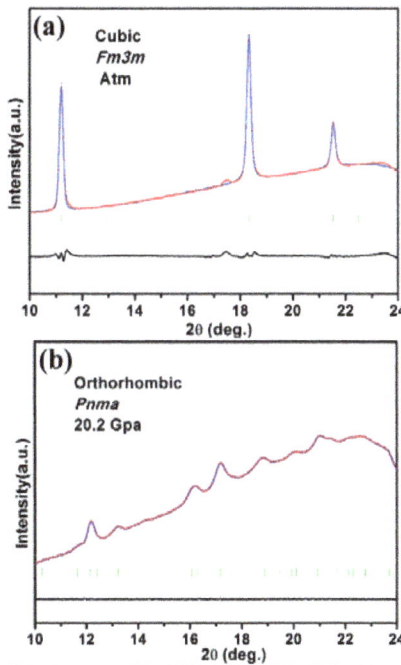

Figure 3. Refinements of the experiment (red dots), simulation (blue line), and difference (black line) ADXRD patterns of fluorite (*Fm3m*) and α-PbCl$_2$-type (*Pnma*) phase: (**a**) at ambient pressure; (**b**) at 20.2 GPa. Green bars mark the positions of corresponding Bragg reflections.

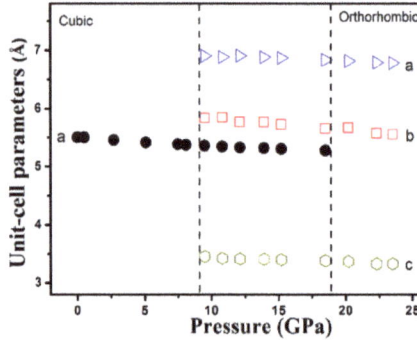

Figure 4. Pressure dependence of the lattice parameters in CaF_2 nanocrystals. Filled symbols denote cubic fluorite structure. Open symbols denote the orthorhombic $PbCl_2$-type structure.

Figure 5 demonstrates the EOS data of CaF_2 nanocrystals to pressures of 23.5 GPa. A third-order Birch-Murnaghan (BM) equation of state (EOS) was fitted to the experimental $P-V$ data [35].

$$P = (3/2)B_0[(V/V_0)^{-7/3} - (V/V_0)^{-5/3}]\{1 + (3/4)(B_0' - 4) \times [(V/V_0)^{-2/3} - 1]\} \tag{1}$$

where V and V_0 are the volumes at pressure P given in GPa and ambient pressure, respectively. For the CaF_2 nanocrystals, we yield the bulk modulus B_0 of 103(2) GPa with B_0' *fixed at* 5 for the fluorite structure and B_0 of 78(2) GPa with B_0'fixed at 4 for the α-$PbCl_2$-type structure. The isothermal bulk modulus of α-$PbCl_2$-type phase at ambient pressure is lower than the fluorite phase. The lower bulk modulus at high pressure indicates higher compressibility of the high-pressure phase of CaF_2 nanocrystals. This result is consistent with previous studies on bulk CaF_2, but it is different from bulk SrF_2 and BaF_2, which have higher incompressibility under high pressure [8,36]. The bulk modulus of CaF_2 nanocrystals for the fluorite and α-$PbCl_2$-type structure are both significantly larger than those of the bulk CaF_2 (B_0 = 87(5) and 74(5) GPa) [4,8], indicating the high incompressibility of nanosized CaF_2.

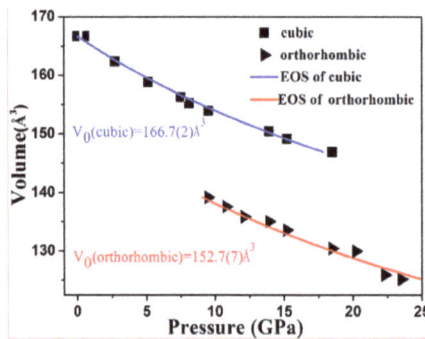

Figure 5. Unit-cell volume as a function of pressure determined for the CaF_2 nanocrystals. Solid curves are the Birch–Murnaghan EOS fits to the experimental data.

Table 1 exhibits the phase transition (P_T) starting pressure and the values of B_0 and B_0' of CaF_2 materials, which clearly revealed the differences between bulk and nanoscale CaF_2. It was found that the high-pressure behavior of the 23 nm CaF_2 nanocrystals is closer to that in bulk CaF_2. However, the bulk modulus of the CaF_2 nanocrystals for the fluorite and α-$PbCl_2$-type structure are both larger than those of the bulk CaF_2. This indicates that the CaF_2 nanocrystals exhibit obviously higher incompressibility compared to bulk CaF_2. This behavior is different form the results reported for nanocrystalline SnO_2 and TiO^2 [37,38]. Compared with bulk CaF_2, a relatively higher surface energy is expected in our CaF_2 nanocrystals. Based on the Hall–Petch effect [39,40], a continuous decrease of grain size could further elevate material hardness, and thus the enhancements in bulk modulus can be easily understood.

Table 1. Transition pressure (P_T), and equation of state parameters (B_0 and B_0') of fluorite-type and α-PbCl₂-type CaF_2

Morphology	Size	P_T (GPa)	B_0 (GPa)		B_0'	
			Fm3m	*Pnma*	*Fm3m*	*Pnma*
Bulk	Micro	9.5 [1]	87(5) [1]	74(5) [2]	5	4.7
		9 [3]	81(1) [3]		5.22	
		8.1 [4]	79.54 [4]	70.92 [4]	4.54	4.38
Nanocrystals	8 nm	14 [5]	–	–	–	–
This work	23 nm	9.5	103(2)	78(2)	5	4

[1] Ref. [4]; [2] Ref. [8]; [3] Ref. [41]; [4] Ref. [6]; [5] Ref. [34].

Extensive high pressure studies indicated that many nanomaterials—such as PbS, CdSe, and ZnS—exhibit obvious enhancement of structural stability compared with the corresponding bulk materials due to higher surface energy [16–19]. However, for our 23 nm CaF_2 nanocrystals, the higher surface energy does not contribute to the improvement of structural stability. To the best of our knowledge, defects are considered to be one of the most important factors to affect the transition pressure of nanomaterials. In order to prove our deduction, we present the HRTEM image of the 23 nm CaF_2 nanocrystals at ambient condition in Figure 6. It is clear that our sample contains visible structural impurities like dislocations and defects. These dislocations and defects could act as the weak points and induce stress concentration, so a new high-pressure phase prefers to nucleate at such defect sites. Therefore, 23 nm CaF_2 nanocrystals with structural defects has a reduced nucleation pressure. The distortions and defects of the crystal structure play a dominant role in the structural stability. Wang et al. [19] studied bulk and nano ZnS using synchrotron X-ray diffraction and found that when the grain size is larger than 15 nm, the hosted defect acts to behave similarly to that in bulk; below 15 nm, the defect activities turn silent, and surface energy begins directing the enhancement of structural stability. This high-pressure research on ZnS nanomaterials is similar to our studies on CaF_2 nanomaterials.

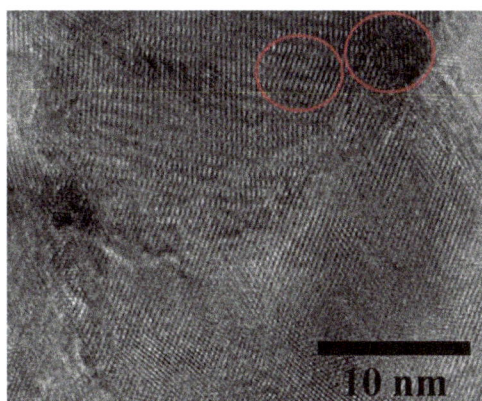

Figure 6. HRTEM image of the as-synthesized 23 nm CaF_2 nanocrystals.

4. Conclusions

The structural phase transition and compressibility of CaF$_2$ nanocrystals with size of 23 nm under high pressure were investigated by synchrotron X-ray diffraction measurement. A pressure-induced fluorite to α-PbCl$_2$-type phase transition starts at 9.5 GPa and completes at 20.2 GPa. The phase-transition pressure is lower than previously reported that of 8 nm CaF$_2$ nanocrystals and closes to bulk CaF$_2$. Upon decompression, the fluorite and α-PbCl$_2$-type structure co-exist at ambient pressure. The bulk modulus B_0 of the 23 nm CaF$_2$ nanocrystals for the fluorite and α-PbCl$_2$-type phase are 103(2) and 78(2) GPa, which are both larger than those of the bulk CaF$_2$. The enhancement of bulk modulus compared with the corresponding bulk materials is due to higher surface energy. The HRTEM image on the sample clearly shows that the as-synthesized CaF$_2$ nanocrystals contain visible dislocations and defects which act as the weak points and induce stress concentration. The distortions and defects of the crystal structure play a dominant role in structural stability.

Author Contributions: JingshuWang and Jihui Lang conceived and designed the experiments; Jinghan Yang, Haiying Zhao and Xiangshan Chen fabricated and characterized the sample; Tingjing Hu, Junkai Zhang and Xiaoxin Wu collaborated in XRD, TEM measurements; Jinghai Yang and Qiliang Cui analyzed the data. All authors discussed the experiment results and contributed to writing the paper.

Acknowledgments: Project supported by the National Natural Science Foundation of China (grant nos. 11404137, 61378085, 51608226, 21776110), 20th Five-Year Program for Science and Technology of Education Department of Jilin Province (item no. 20150221). Synchrotron XRD experiments were performed at beamline 15U1 at the Shanghai Synchrotron Radiation Facility (SSRF).

Conflicts of Interest: The authors declare no conflict of interest.

References

1. Wang, G.; Peng, Q.; Li, Y. Upconversion Luminescence of Monodisperse CaF$_2$: Yb^{3+}/Er^{3+} Nanocrystals. *J. Am. Chem. Soc.* **2009**, *131*, 14200–14201. [CrossRef] [PubMed]
2. Zhang, C.; Li, C.; Peng, C.; Chai, R.; Huang, S.; Yang, D.; Cheng, Z.; Lin, J. Facile and Controllable Synthesis of Monodisperse CaF$_2$ and CaF$_2$: Ce^{3+}/Tb^{3+} Hollow Spheres as Efficient Luminescent Materials and Smart Drug Carriers. *Chem. Eur. J.* **2010**, *16*, 5672–5680. [CrossRef] [PubMed]
3. Feldmann, C.; Roming, M.; Trampert, K. Polyol-Mediated Synthesis of Nanoscale CaF$_2$ and CaF$_2$: Ce, Tb. *Small* **2006**, *2*, 1248–1250. [CrossRef] [PubMed]
4. Gerward, L.; Olsen, J.S.; Steenstrup, S.; Sbrink, S.A.; Waskowska, A. X-ray diffraction investigations of CaF$_2$ at high pressure. *J. Appl. Crystallogr.* **1992**, *25*, 578–581. [CrossRef]
5. Kavner, A. Radial diffraction strength and elastic behavior of CaF$_2$ in low-and high-pressure phases. *Phys. Rev. B* **2008**, *77*, 224102. [CrossRef]
6. Cui, S.X.; Feng, W.X.; Hua, H.Q. Structural stabilities, electronic and optical properties of CaF$_2$ under high pressure: A first-principles study. *Comput. Mater. Sci.* **2009**, *47*, 41–45. [CrossRef]
7. Wu, X.; Qin, S.; Wu, Z.Y. First-principles study of structural stabilities, and electronic and optical properties of CaF$_2$ under high pressure. *Phys. Rev. B* **2006**, *73*, 134103. [CrossRef]
8. Dorfman, S.M.; Jiang, F.; Mao, Z.; Kubo, A.; Meng, Y.; Prakapenka, V.B.; Duffy, T.S. Phase transitions and equations of state of alkaline earth fluorides CaF$_2$, SrF$_2$, and BaF$_2$ to Mbar pressures. *Phys. Rev. B* **2010**, *81*, 174121. [CrossRef]
9. Boulfelfel, S.E.; Zahn, D.; Hochrein, O.; Grin, Y.; Leoni, S. Low-dimensional sublattice melting by pressure: Superionic conduction in the phase interfaces of the fluorite-to-cotunnite transition of CaF$_2$. *Phys. Rev. B* **2006**, *74*, 094106. [CrossRef]
10. Hu, T.; Cui, X.; Wang, J.; Zhong, X.; Chen, Y.; Zhang, J.; Li, X.; Yang, J.; Gao, C. The Electrical Properties of Tb-Doped CaF$_2$ Nanoparticles under High Pressure. *Crystals* **2018**, *8*, 98. [CrossRef]
11. Hu, T.; Cui, X.; Wang, J.; Zhang, J.; Li, X.; Yang, J.; Gao, C. Transport properties of mixing conduction in CaF$_2$ nanocrystals under high pressure. *Chin. Phys. B* **2018**, *27*, 016401. [CrossRef]
12. Cazorla, C.; Errandonea, D. Superionicity and polymorphism in calcium fluoride at high pressure. *Phys. Rev. Lett.* **2014**, *113*, 235902. [CrossRef] [PubMed]

13. Cazorla, C.; Errandonea, D. Giant Mechanocaloric Effects in Fluorite-Structured Superionic Materials. *Nano Lett.* **2016**, *16*, 3124–3129. [CrossRef] [PubMed]

14. Lin, Y.; Yang, Y.; Ma, H.; Cui, Y.; Mao, W.L. Compressional Behavior of Bulk and Nanorod $LiMn_2O_4$ under Nonhydrostatic Stress. *J. Phys. Chem. C* **2011**, *115*, 9844–9849. [CrossRef]

15. Wu, H.; Wang, Z.W.; Fan, H.Y. Stress-Induced Nanoparticle Crystallization. *J. Am. Chem. Soc.* **2014**, *136*, 7634–7636. [CrossRef] [PubMed]

16. Bian, K.; Wang, Z.; Hanrath, T. Comparing the Structural Stability of PbS Nanocrystals Assembled in fcc and bcc Superlattice Allotropes. *J. Am. Chem. Soc.* **2012**, *134*, 10787–10790. [CrossRef] [PubMed]

17. Tolbert, S.H.; Alivisatos, A.P. Size Dependence of a First Order Solid-Solid Phase Transition: The Wurtzite to Rock Salt Transformation in CdSe Nanocrystals. *Science* **1994**, *265*, 373–376. [CrossRef] [PubMed]

18. Tolbert, S.H.; Alivisatos, A.P. High-Pressure Structural Transformations in Semiconductor Nanocrystals. *Annu. Rev. Phys. Chem.* **1995**, *46*, 595–626. [CrossRef] [PubMed]

19. Wang, Z.; Guo, Q. Size-Dependent Structural Stability and Tuning Mechanism: A Case of Zinc Sulfide. *J. Phys. Chem. C* **2009**, *113*, 4286–4295. [CrossRef]

20. Jiang, J.Z.; Gerward, L.; Olsen, J.S. Pressure induced phase transformation in nanocrystal SnO_2. *Scr. Mater.* **2001**, *44*, 1983–1986. [CrossRef]

21. Wang, L.; Yang, W.; Ding, Y.; Ren, Y.; Xiao, S.; Liu, B.; Sinogeikin, S.V.; Meng, Y.; Gosztola, D.J.; Shen, G.; et al. Size-Dependent Amorphization of Nanoscale Y_2O_3 at High Pressure. *Phys. Rev. Lett.* **2010**, *105*, 095701. [CrossRef] [PubMed]

22. Varghese, S.; Alexei, K.; Dubrovinsky, L.S.; Mcmillan, P.F.; Prakapenka, V.B.; Shen, G.; Muddle, B.C. Size-Dependent Pressure-Induced Amorphization in Nanoscale TiO_2. *Phys. Rev. Lett.* **2006**, *96*, 135702. [CrossRef]

23. Li, Q.; Liu, B.; Wang, L.; Li, D.; Liu, R.; Zou, B.; Cui, T.; Zou, G. Pressure-Induced Amorphization and Polymorphism in One-Dimensional Single-Crystal TiO_2 Nanomaterials. *J. Phys. Chem. Lett.* **2010**, *1*, 309–314. [CrossRef]

24. Quan, Z.W.; Wang, Y.X.; Bae, I.T.; Loc, W.S.; Wang, C.Y.; Wang, Z.W.; Fang, J.Y. Reversal of Hall–Petch Effect in Structural Stability of PbTe Nanocrystals and Associated Variation of Phase Transformation. *Nano Lett.* **2011**, *11*, 5531–5536. [CrossRef] [PubMed]

25. Wang, J.; Cui, Q.; Hu, T.; Yang, J.; Li, X.; Liu, Y.; Liu, B.; Zhao, W.; Zhu, H.; Yang, L. Pressure-Induced Amorphization in BaF_2 Nanoparticles. *J. Phys. Chem. C* **2016**, *120*, 12249–12253. [CrossRef]

26. Lv, H.; Yao, M.; Li, Q.; Li, Z.; Liu, B.; Liu, R.; Lu, S.; Li, D.; Mao, J.; Ji, X.; et al. Effect of Grain Size on Pressure-Induced Structural Transition in Mn_3O_4. *J. Phys. Chem. C* **2012**, *116*, 2165–2171. [CrossRef]

27. Wieligor, M.; Wang, Y.; Zerda, T.W. Raman spectra of silicon carbide small particles and nanowires. *J. Phys. Condens. Matter* **2005**, *17*, 2387–2395. [CrossRef]

28. Wang, Y.; Zhang, J.; Wu, J.; Coffer, J.L.; Lin, Z.; Sinogeikin, S.V.; Yang, W.; Zhao, Y. Phase transition and compressibility in silicon nanowires. *Nano Lett.* **2008**, *8*, 2891–2895. [CrossRef] [PubMed]

29. Wang, Y.; Zhang, J.; Zhao, Y. Strength weakening by nanocrystals in ceramic materials. *Nano Lett.* **2007**, *7*, 3196–3199. [CrossRef] [PubMed]

30. Senter, R.A.; Pantea, C.; Wang, Y.; Liu, H.; Zerda, T.W.; Coffer, J.L. Structural Influence of Erbium Centers on Silicon Nanocrystal Phase Transitions. *Phys. Rev. Lett.* **2004**, *93*, 175502. [CrossRef] [PubMed]

31. Wu, J.; Coffer, J.L.; Wang, Y.; Schulze, R. Oxidized Germanium as a Broad-Band Sensitizer for Er-Doped SnO_2 Nanofibers. *J. Phys. Chem. C* **2009**, *113*, 12–16. [CrossRef]

32. Wang, Y.; Zhao, Y.; Zhang, J.; Xu, H.; Wang, L.; Luo, S.; Daemen, L. In situ phase transition study of nano- and coarse-grained TiO_2 under high pressure/temperature conditions. *J. Phys. Condens. Matter* **2008**, *20*, 125224. [CrossRef]

33. Wang, Y.; Yang, W.; Zou, G.; Wu, J.; Coffer, J.L.; Sinogeikin, S.V.; Zhang, J. Anomalous Surface Doping Effect in Semiconductor Nanowires. *J. Phys. Chem. C* **2017**, *121*, 11824–11830. [CrossRef]

34. Wang, J.; Hao, J.; Wang, Q.; Jin, Y.; Li, F.; Liu, B.; Li, Q.; Liu, B.; Cui, Q. Pressure-induced structural transition in CaF_2 nanocrystals. *Phys. Status Solidi B* **2011**, *248*, 1115–1118. [CrossRef]

35. Birch, F. Finite Strain Isotherm and Velocities for Single-Crystal and Polycrystalline NaCl at High Pressures and 300 K. *J. Geophys. Res.* **1978**, *83*, 1257–1268. [CrossRef]

36. Wang, J.; Ma, C.; Zhou, D.; Xu, Y.; Zhang, M.; Gao, W.; Zhu, H.; Cui, Q. Structural phase transitions of SrF_2 at high pressure. *J. Solid State Chem.* **2012**, *186*, 231–234. [CrossRef]

37. Popescu, C.; Sans, J.A.; Errandonea, D.; Segura, A.; Villanueva, R.; Sapiña, F. Compressibility and structural stability of nanocrystalline TiO$_2$ anatase synthesized from freeze-dried precursors. *Inorg. Chem.* **2014**, *53*, 11598–11603. [CrossRef] [PubMed]
38. Grinblat, F.; Ferrari, S.; Pampillo, L.G.; Saccone, F.; Errandonea, D.; Santamaria-Perez, D.; Segura, A.; Vilaplana, R.; Popescu, C. Compressibility and structural behavior of pure and Fe-doped SnO$_2$ nanocrystals. *Solid State Sci.* **2017**, *64*, 91–98. [CrossRef]
39. Hall, E.O. The Deformation and Ageing of Mild Steel: III Discussion of Results. Proceedings of the Physical Society. *Sect. B* **1951**, *64*, 747. [CrossRef]
40. Petch, N. The cleavage strength of polycrystals. *J. Iron Steel Inst.* **1953**, *174*, 25–28.
41. Angel, R.J. The high-pressure, high-temperature equation of state of calcium fluoride, CaF$_2$. *J. Phys. Condens. Matter* **1993**, *5*, L141. [CrossRef]

crystals

MDPI

Article

The Electrical Properties of Tb-Doped CaF$_2$ Nanoparticles under High Pressure

Tingjing Hu [1] , Xiaoyan Cui [1,*], Jingshu Wang [1], Xin Zhong [1], Yinzhu Chen [1], Junkai Zhang [1], Xuefei Li [1], Jinghai Yang [1] and Chunxiao Gao [2]

[1] Key Laboratory of Functional Materials Physics and Chemistry of the Ministry of Education, National Demonstration Center for Experimental Physics Education, Jilin Normal University, Siping 136000, China; tjhumars@126.com (T.H.); jingshuwang126@126.com (J.W.); zhongxin@calypso.cn (X.Z.); yinzhuchen126@126.com (Y.C.); junkaizhang126@126.com (J.Z.); xuefeili163@163.com (X.L.); jhyang1@jlnu.edu.cn (J.Y.)

[2] State Key Laboratory of Superhard Materials, Jilin University, Changchun 130012, China; chunxiaogao126@126.com

* Correspondence: xycuimail@163.com

Received: 9 January 2018; Accepted: 12 February 2018; Published: 14 February 2018

Abstract: The high-pressure transport behavior of CaF$_2$ nanoparticles with 3 mol% Tb concentrations was studied by alternate-current impedance measurement. All of the electrical parameters vary abnormally at approximately 10.76 GPa, corresponding to the fluorite-cotunnite structural transition. The substitution of Ca^{2+} by Tb^{3+} leads to deformation in the lattice, and finally lowers the transition pressure. The F$^-$ ions diffusion, electronic transport, and charge-discharge process become more difficult with the rising pressure. In the electronic transport process, defects at grains play a dominant role. The charge carriers include both F$^-$ ions and electrons, and electrons are dominant in the transport process. The Tb doping improves the pressure effect on the transport behavior of CaF$_2$ nanocrystals.

Keywords: high pressure; Tb doping; phase transitions; transport properties

---

1. Introduction

Recently, rare earth (RE)-doped nanomaterials have attracted much attention [1–5], due to their potential applications such as advanced phosphor [6], display monitors [7], light amplification [8], and biological labeling [9,10], etc. Among these host materials, calcium fluoride (CaF$_2$) is an attractive host for RE doping because of its high transparency in a wide wavelength region and low phonon energy [11–15].

As an important optical and optoelectronic functional material, a thorough study of the electrical transport properties is essential, and the underlying physical transport behaviors, such as charge carrier type and scattering processes, are worthy of exploration. The impedance spectrum measurement method has long been conventional in studies of electrical charge transportation and related physical properties [16–20]. Specially, using the impedance method, the presence of independent pathways for charge transportation in an inorganic material [21], and the mixed electronic and ionic conduction in various organic and inorganic materials have been satisfactorily addressed [22–26]. We have investigated the electrical properties of CaF$_2$ nanoparticles with Tb concentrations from 1 mol% to 5 mol% at atmospheric pressure, and it was found that the resistance of the sample with a concentration of 3 mol% Tb is the smallest. Therefore, in this work, the electrical properties of CaF$_2$ nanoparticles with 3 mol% Tb concentrations under high pressure were investigated by alternate-current (AC) impedance measurement up to 26 GPa. The underlying physical transport behaviors were discussed. Additionally, the pressure effect on the structural and electrical properties of Tb-doped CaF_2 nanocrystals was compared with that of un-doped nanocrystals.

2. Materials and Methods

A diamond anvil cell (DAC) was used to generate high pressure. The detailed configuration of the electrodes and sample has been illustrated in previous works [27–29]. The final microcircuit and the profile of our designed DAC are shown in Figure 1. Pressure was calibrated by using ruby fluorescence. The ruby measurement scale is 100 GPa [30] and the accuracy of our measurement is 0.1 GPa. To avoid additional error on the electrical transport measurements, no pressure-transmitting medium was used. This will cause non-hydrostatic conditions [31]; however, the effects on the transport measurements can be neglected in our experiment pressure range [32].

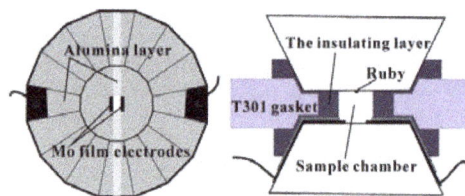

Figure 1. The completed microcircuit (**left**) on diamond anvil and the profile of our designed diamond anvil cell (DAC) (**right**).

Impedance spectroscopy was measured by a Solartron 1260 impedance analyzer (Solartron, Hampshire, UK) equipped with a Solartron 1296 dielectric interface. A voltage signal with an amplitude of 1 V was applied to the sample and its frequency ranged from 0.1 to 10^7 Hz.

The sample was prepared by the hydrothermal synthesis method as reported in our previous work [33]. The Tb doping concentrations were 3 mol%. The sample was characterized by transmission electron microscopy (TEM) (JEOL Ltd., Tokyo, Japan) and X-ray diffraction (XRD λ = 1.5406 Å) (Rigaku, Tokyo, Japan). Figure 2 exhibits the TEM image and the size distribution histogram. It can be seen that the shape of the sample is square with a mean dimension of 8 ± 2 nm.

Figure 2. *Cont.*

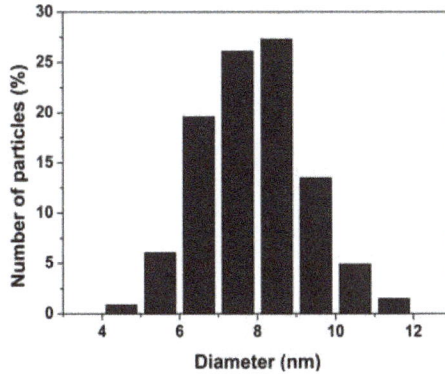

Figure 2. The TEM image and the size distribution histogram of 3 mol% Tb-doped CaF_2 nanoparticles.

3. Results and Discussion

Figure 3 shows the X-ray diffraction pattern of CaF_2 nanoparticles with 3 mol% Tb concentrations. The diffraction peaks of the sample match well with the pure cubic (space group: *Fm3m* (225) $\alpha = \beta = \gamma = 90°$) phase of CaF_2 (Joint Committee on Powder Diffraction Standards JCPDS Card No. 35-0816) and the lattice constant is 5.432 Å, which suggests that the original structure of CaF_2 was retained after doping. No impurity peaks are observed in the pattern, indicating that the Tb^{3+} ions were incorporated into the CaF_2 lattice and substitute Ca^{2+} ions. The average size estimated from the full width at half maximum (FWHM) using the Debye-Scherrer formula is 8.3 nm, which has good agreement with the TEM result.

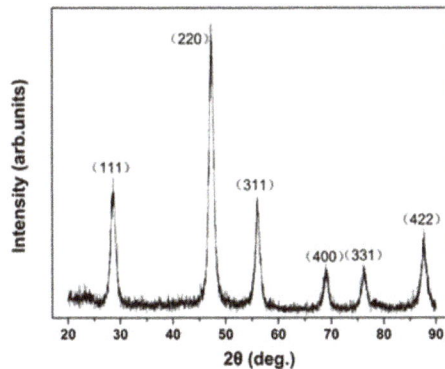

Figure 3. The X-ray diffraction pattern of CaF_2 nanoparticles with 3 mol% Tb concentrations at atmospheric pressure.

The Nyquist impedance spectra of CaF_2 nanoparticles with 3 mol% Tb concentrations under several pressures are presented in Figure 4.

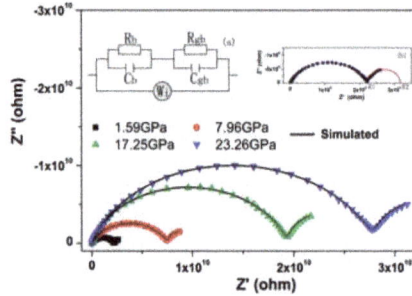

Figure 4. The Nyquist impedance spectra under several pressures. The inset (**a**) shows the equivalent circuit model, R_b and R_{gb} are grain and grain boundary resistance, C_b and C_{gb} are grain and grain boundary capacitance, and W_i is the Warburg impedance. The inset (**b**) is the spectroscopy at 1.59 GPa, R_1 and R_2 are two intercepts on the real impedance axis.

To analyze the ionic conduction, the impedance spectra were replotted into $Z'\sim\omega^{-1/2}$ plots, as shown in Figure 5.

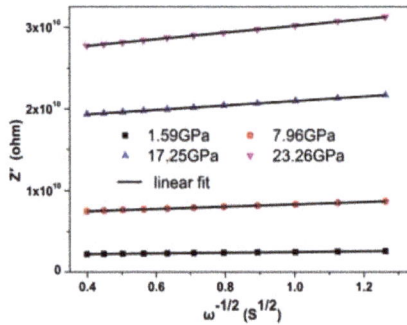

Figure 5. The $Z'\sim\omega^{-1/2}$ curves at low frequencies under several pressures.

In the low frequency region, the Z' can be expressed as:

$$Z' = Z'_0 + \sigma\omega^{-1/2}, \tag{1}$$

where Z'_0 is a parameter independent of frequency, σ is the Warburg coefficient, and ω is the frequency. By linear fitting the $Z'\sim\omega^{-1/2}$ plots, the Warburg coefficient of various pressure was obtained. The diffusion coefficient of the ions (D_i) can be obtained from:

$$D_i = 0.5(\frac{RT}{AF^2\sigma C})^2, \tag{2}$$

where R is the ideal gas constant, T is the temperature, A is the electrode area, F is the Faraday constant, and C is the F^- ions molar concentration. We set the F^- ion diffusion coefficient at 0 GPa as D_0, and the curve D_i/D_0 under different pressures was obtained and is shown in Figure 6a.

To quantify the pressure effect on the electrical transport properties, the impedance spectra were fitted with the equivalent circuit model (the inset (a) of Figure 4) on the Zview2 impedance analysis software. The obtained bulk and grain boundary resistances (R_b, R_{gb}) are plotted in Figure 6. The relaxation frequency of bulk (f_b) under different pressures was obtained from the $Z''\sim f$ curve and is presented in Figure 6d.

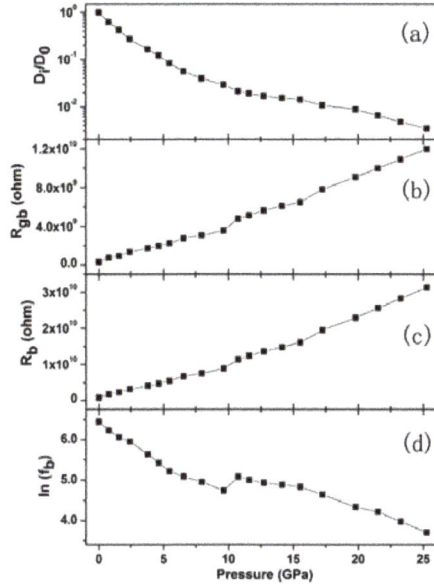

Figure 6. (a) the diffusion coefficient, (b) the bulk resistance, (c) the grain boundary resistance, (d) the bulk relaxation frequency under high pressure. D_0 represents the diffusion coefficient at 0 GPa.

From Figure 6, it can be seen that all of the parameters vary discontinuously at approximately 10.76 GPa, corresponding to the fluorite-cotunnite (*Fm3m–Pnma*) structural transition of the sample. According to our previous works [29,33], this phase transition of un-doped CaF_2 nanocrystals occurs at about 14 GPa. The variation in the phase transition pressure with the substitution of Ca^{2+} by Tb^{3+} can be discussed as follows: the ionic radius of Tb^{3+} (0.092 nm) is smaller than that of Ca^{2+} (0.099 nm), and the valence of Tb^{3+} is different with that of Ca^{2+}; these result in deformation in the lattice and the increasing of the deformation potential, and finally make the transition pressure lower.

In the whole pressure range, the diffusion coefficient decreases with pressure; however, the grain and grain boundary resistance increase, indicating that the F^- ions diffusion and electronic transport become more difficult with the rising pressure. The grain resistance is larger than the grain boundary resistance, which indicates that defects at grains play a dominant role in the electronic transport process.

The pressure dependence of grain activation energy (dH/dP) can be obtained from:

$$d(\ln f_b)/dP = -(1/k_B T)(dH/dP), \qquad (3)$$

where k_B is the Boltzmann constant and T is the temperature. By linear fitting to the curve $lnf_b\sim P$, the dH/dP of the *Fm3m* and *Pnma* phases were obtained and are listed in Table 1. The dH/dP of un-doped CaF_2 nanocrystals were obtained by the data of Reference [27] and are also shown in Table 1.

Table 1. Pressure dependence of the grain activation energy of Tb-doped and un-doped CaF_2 nanocrystals.

Phase	dH/dp (meV/GPa) (Tb-Doped)	dH/dp (meV/GPa) (Un-Doped)
Fm3m	4.70	3.12
Pnma	2.97	1.44

The positive values of dH/dP in *Fm3m* and *Pnma* phases indicate that the charge-discharge process becomes more difficult under compression. In the *Fm3m* and *Pnma* phases, the dH/dP values of the

Tb-doped CaF_2 nanocrystals are larger than those of un-doped CaF_2 nanocrystals. This indicates that pressure has a larger effect on the charge-discharge process of the Tb-doped sample.

To distinguish the contributions of F^- ions and electrons to the transport process, the transference number were calculated by the following equations [34]:

$$t_i = (R_2 - R_1)/R_2, \tag{4}$$

$$t_e = R_1/R_2, \tag{5}$$

where t_i is the transference number of F^- ions, t_e is the transference number of electrons, and R_1 and R_2 are the intercepts on the real impedance axis as shown in the inset (b) of Figure 4. t_i and t_e under various pressures are shown in Figure 7. It can be seen that electrons play a dominant role in the transport process and the electron transference number slightly increases as the pressure rises.

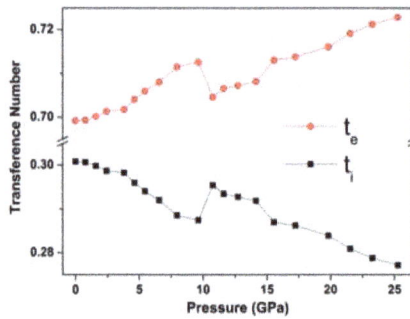

Figure 7. t_i and t_e under various pressures.

To further revealing the effect of Tb doping on the high-pressure transport behavior, the resistance variation of Tb-doped CaF_2 nanocrystals is compared with that of un-doped CaF_2 nanocrystals. The bulk and grain boundary resistances at 0 GPa were set as R_{b0} and R_{gb0}, then the R_b/R_{b0} and R_{gb}/R_{gb0} of Tb-doped and un-doped CaF_2 nanocrystals were obtained and are shown in Figure 8. It can be observed that both in the bulk and grain boundary, the resistance variation of the Tb-doped sample is larger than that of the un-doped sample. This indicates that the Tb doping improves the pressure effect on the transport behavior of CaF_2 nanocrystals.

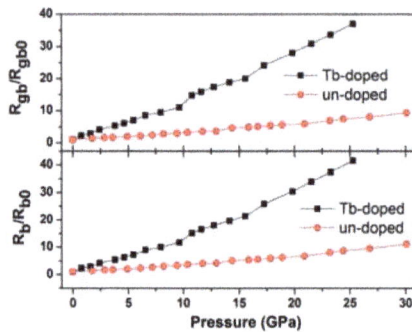

Figure 8. The R_b/R_{b0} and R_{gb}/R_{gb0} of Tb-doped and un-doped CaF_2 nanocrystals. R_{b0} and R_{gb0} represent bulk and grain boundary resistances at 0 GPa.

4. Conclusions

The electrical properties of CaF_2 nanoparticles with 3 mol% Tb concentrations under high pressure were investigated by impedance measurement. All of the electrical parameters vary abnormally at approximately 10.76 GPa, corresponding to the *Fm3m–Pnma* structural transition. The substitution of Ca^{2+} by Tb^{3+} leads to deformation in the lattice, and finally lowers the transition pressure. The F^- ions diffusion, electronic transport, and charge-discharge process become more difficult with the rising pressure. In the electronic transport process, defects at grains play a dominant role. The charge carriers include both F^- ions and electrons, and electrons are dominant in the transport process. The Tb doping improves the pressure effect on the transport behavior of CaF_2 nanocrystals. Other lanthanides such as Yb, Er, Ce, etc. would cause similar effects and should be explored in the future.

Acknowledgments: This work was financially supported by the National Natural Science Foundation of China (Grant Nos. 11674404, 11704151 and 11404137), the Twentieth Five-Year Program for Science and Technology of Education Department of Jilin Province, China (Grant No. 20150221), and the Open Project of State Key Laboratory of Superhard Materials (Jilin University) (Grant No. 201710)

Author Contributions: Tingjing Hu conceived and designed the experiments; Jingshu Wang and Yinzhu Chen fabricated and characterized the sample; Xin Zhong and Junkai Zhang collaborated in XRD, TEM measurements; Xiaoyan Cui, Jinghai Yang and Chunxiao Gao analyzed the data. All authors discussed the experiment results and contributed to writing the paper.

Conflicts of Interest: The authors declare no conflict of interest.

References

1. Bazzi, R.; Flores, M.A.; Louis, C.; Lebbou, K.; Zhang, W.; Dujardin, C.; Roux, S.; Mercier, B.; Ledoux, G.; Bernstein, E.; et al. Synthesis and properties of europium-based phosphors on the nanometer scale: Eu_2O_3, Gd_2O_3: Eu, and Y_2O_3: Eu. *J. Colloid Interface Sci.* **2004**, *273*, 191–197. [CrossRef] [PubMed]

2. Nishi, M.; Tanabe, S.; Inoue, M.; Takahashi, M.; Fujita, K.; Hirao, K. Optical-telecommunication-band fluorescence properties of Er^{3+}-doped YAG nanocrystals synthesized by glycothermal method. *Opt. Mater.* **2005**, *27*, 655–662. [CrossRef]

3. Rulison, A.J.; Flagan, R.C. Synthesis of yttria powders by electrospray pyrolysis. *J. Am. Ceram. Soc.* **1994**, *77*, 3244–3250. [CrossRef]

4. Patra, A.; Friend, C.S.; Kapoor, R.; Prasad, P.N. Fluorescence upconversion properties of Er^{3+}-doped TiO_2 and $BaTiO_3$ nanocrystallites. *Chem. Mater.* **2003**, *15*, 3650–3655. [CrossRef]

5. Masenelli, B.; Melinon, P.; Nicolas, D.; Bernstein, E.; Prevel, B.; Kapsa, J.; Boisron, O.; Perezl, A.; Ledoux, G.; Mercier, B.; et al. Rare earth based clusters for nanoscale light source. *Eur. Phys. J. D* **2005**, *34*, 139–143. [CrossRef]

6. Wei, Z.G.; Sun, L.D.; Liao, C.S.; Yan, C.H. Fluorescence intensity and color purity improvement in nanosized YBO_3: Eu. *Appl. Phys. Lett.* **2002**, *80*, 1447–1449. [CrossRef]

7. Matsuura, D. Red, green, and blue upconversion luminescence of trivalent-rare-earth ion-doped Y_2O_3 nanocrystals. *Appl. Phys. Lett.* **2002**, *81*, 4526–4528. [CrossRef]

8. Barber, D.B.; Pollock, C.R.; Beecroft, L.L.; Ober, C.K. Amplification by optical composites. *Opt. Lett.* **1997**, *22*, 1247–1249. [CrossRef] [PubMed]

9. Yi, G.; Lu, H.; Zhao, S.; Ge, Y.; Yang, W.; Chen, D.; Guo, L. Synthesis, characterization, and biological application of size-controlled nanocrystalline $NaYF_4$:Yb, Er infrared-to-visible up-conversion phosphors. *Nano Lett.* **2004**, *4*, 2191–2196. [CrossRef]

10. Chen, Z.; Chen, H.; Hu, H.; Yu, M.; Li, F.; Zhang, Q.; Zhou, Z.; Yi, T.; Huang, C. Versatile synthesis strategy for carboxylic acid−functionalized upconverting nanophosphors as biological labels. *J. Am. Chem. Soc.* **2008**, *130*, 3023–3029. [CrossRef] [PubMed]

11. Fujihara, S.; Kadota, Y.; Kimura, T. Role of organic additives in the sol-gel synthesis of porous CaF_2 anti-reflective coatings. *J. Sol-Gel Sci. Technol.* **2002**, *24*, 147–154. [CrossRef]

12. McKeever, S.W.S.; Brown, M.D.; Abbundi, R.J.; Chan, H.; Mathur, V.K. Characterization of optically active sites in CaF_2: Ce, Mn from optical spectra. *J. Appl. Phys.* **1986**, *60*, 2505–2510. [CrossRef]

13. Fukuda, Y. Thermoluminescence in sintered CaF_2:Tb. *J. Radiat. Res.* **2002**, *43*, S67–S69. [CrossRef] [PubMed]

14. Pote, S.S.; Joshi, C.P.; Moharil, S.V.; Muthal, P.L.; Dhopte, S.M. Luminescence of Ce^{3+} in $Ca_{0.65}$ $La_{0.35}F_{2.35}$ host. *J. Lumin.* **2010**, *130*, 666–668. [CrossRef]

15. Cazorla1, C.; Errandonea, D. Superionicity and polymorphism in calcium fluoride at high pressure. *Phys. Rev. Lett.* **2014**, *113*, 235902. [CrossRef] [PubMed]

16. Sinclair, D.C.; West, A.R. Impedance and modulus spectroscopy of semiconducting $BaTiO_3$ showing positive temperature coefficient of resistance. *J. Appl. Phys.* **1989**, *66*, 3850–3856. [CrossRef]

17. Sinclair, D.C.; Adams, T.B.; Morrison, F.D.; West, A.R. $CaCu_3Ti_4O_{12}$: One-step internal barrier layer capacitor. *Appl. Phys. Lett.* **2002**, *80*, 2153–2155. [CrossRef]

18. Dutta, S.; Choudhary, R.N.P.; Sinha, P.K.; Thakur, A.K. Microstructural studies of $(PbLa)(ZrTi)O_3$ ceramics using complex impedance spectroscopy. *J. Appl. Phys.* **2004**, *96*, 1607–1613. [CrossRef]

19. Hsu, H.S.; Huang, J.C.A.; Chen, S.F.; Liu, C.P. Role of grain boundary and grain defects on ferromagnetism in Co : ZnO films. *Appl. Phys. Lett.* **2007**, *90*, 102506. [CrossRef]

20. Dualeh, A.; Moehl, T.; Tetreault, N.; Teuscher, J.; Gao, P.; Nazeeruddin, M.K.; Gratzel, M. Impedance spectroscopic analysis of lead iodide perovskite-sensitized solid-state solar cells. *ACS Nano* **2013**, *8*, 362–373. [CrossRef] [PubMed]

21. Huggins, R.A. Simple method to determine electronic and ionic components of the conductivity in mixed conductors: A review. *Ionics* **2002**, *8*, 300–313. [CrossRef]

22. Teraoka, Y.; Zhang, H.; Okamoto, K.; Yamazoe, N. Mixed ionic-electronic conductivity of $La_{1-x}Sr_xCo_{1-y}Fe_yO_{3-\delta}$ perovskite-type oxides. *Mater. Res. Bull.* **1988**, *23*, 51–58. [CrossRef]

23. Riess, I. Measurements of electronic and ionic partial conductivities in mixed conductors, without the use of blocking electrodes. *Solid State Ionics* **1991**, *44*, 207–214. [CrossRef]

24. Riess, I. Review of the limitation of the Hebb-Wagner polarization method for measuring partial conductivities in mixed ionic electronic conductors. *Solid State Ion.* **1996**, *91*, 221–232. [CrossRef]

25. Adler, S.B.; Lane, J.; Steele, B. Electrode kinetics of porous mixed-conducting oxygen electrodes. *J. Electrochem. Soc.* **1996**, *143*, 3554–3564. [CrossRef]

26. Riess, I. Mixed ionic–electronic conductors—Material properties and applications. *Solid State Ion.* **2003**, *157*, 1–17. [CrossRef]

27. Cui, X.Y.; Hu, T.J.; Wang, J.S.; Zhang, J.K.; Zhao, R.; Li, X.F.; Yang, J.H.; Gao, C.X. Mixed conduction in BaF_2 nanocrystals under high pressure. *RSC Adv.* **2017**, *7*, 12098–12102. [CrossRef]

28. Cui, X.Y.; Hu, T.J.; Wang, J.S.; Zhang, J.K.; Li, X.F.; Yang, J.H.; Gao, C.X. High pressure impedance spectroscopy of SrF_2 nanocrystals. *High Press. Res.* **2017**, *37*, 312–318. [CrossRef]

29. Hu, T.J.; Cui, X.Y.; Wang, J.S.; Zhang, J.K.; Li, X.F.; Yang, J.H.; Gao, C.X. Transport properties of mixing conduction in CaF_2 nanocrystals under high pressure. *Chin. Phys. B* **2018**, *27*, 016401. [CrossRef]

30. Mao, H.K.; Bell, P.M. High-pressure physics: The 1-megabar mark on the ruby R1 static pressure scale. *Science* **1976**, *191*, 851–852. [CrossRef] [PubMed]

31. Errandonea, D.; Muñoz, A.; Gonzalez-Platas, J. Comment on "High-pressure X-ray diffraction study of YBO_3/Eu^{3+}, $GdBO_3$, and $EuBO_3$: Pressure-induced amorphization in $GdBO_3$" [J. Appl. Phys. 115, 043507 (2014)]. *J. Appl. Phys.* **2014**, *115*, 216101. [CrossRef]

32. Errandonea, D.; Segura, A.; Martínez-García, D.; Muñoz-San Jose, V. Hall-effect and resistivity measurements in CdTe and ZnTe at high pressure: Electronic structure of impurities in the zinc-blende phase and the semimetallic or metallic character of the high-pressure phases. *Phys. Rev. B* **2009**, *79*, 125203. [CrossRef]

33. Wang, J.S.; Hao, J.; Wang, Q.S.; Jin, Y.X.; Li, F.F.; Liu, B.; Li, Q.J.; Liu, B.B.; Cui, Q.L. Pressure-induced structural transition in CaF_2 nanocrystals. *Phys. Status Solidi B* **2011**, *248*, 1115–1118. [CrossRef]

34. Wang, Q.L.; Liu, C.L.; Gao, Y.; Ma, Y.Z.; Han, Y.H.; Gao, C.X. Mixed conduction and grain boundary effect in lithium niobate under high pressure. *Appl. Phys. Lett.* **2015**, *106*, 132902. [CrossRef]

crystals

MDPI

Article

The Jahn-Teller Distortion at High Pressure: The Case of Copper Difluoride

Dominik Kurzydłowski [1,2]

[1] Centre of New Technologies, University of Warsaw, 02-097 Warsaw, Poland; d.kurzydlowski@cent.uw.edu.pl
[2] Faculty of Mathematics and Natural Sciences, Cardinal Stefan Wyszyński University, 01-038 Warsaw, Poland

Received: 14 February 2018; Accepted: 16 March 2018; Published: 19 March 2018

Abstract: The opposing effects of high pressure (in the GPa range) and the Jahn-Teller distortion led to many intriguing phenomena which are still not well understood. Here we report a combined experimental-theoretical study on the high-pressure behavior of an archetypical Jahn-Teller system, copper difluoride (CuF_2). At ambient conditions this compound adopts a distorted rutile structure of $P2_1/c$ symmetry. Raman scattering measurements performed up to 29 GPa indicate that CuF_2 undergoes a phase transition at 9 GPa. We assign the novel high-pressure phase to a distorted fluorite structure of $Pbca$ symmetry, iso-structural with the ambient-pressure structure of AgF_2. Density functional theory calculations indicate that the $Pbca$ structure should transform to a non-centrosymmetric $Pca2_1$ polymorph above 30 GPa, which, in turn, should be replaced by a cotunnite phase ($Pnma$ symmetry) at 72 GPa. The elongated octahedral coordination of the Cu^{2+} cation persists up to the $Pca2_1$–$Pnma$ transition upon which it is replaced by a capped trigonal prism geometry, still bearing signs of a Jahn-Teller distortion. The high-pressure phase transitions of CuF_2 resembles those found for difluorides of transition metals of similar radius (MgF_2, ZnF_2, CoF_2), although with a much wider stability range of the fluorite-type structures, and lower dimensionality of the high-pressure polymorphs. Our calculations indicate no region of stability of a nanotubular polymorph observed for the related AgF_2 system.

Keywords: copper; fluorides; Jahn-Teller effect; high pressure; polymorphism

---

1. Introduction

In 1937, Jahn and Teller showed that non-linear molecules exhibiting orbital degeneracy will undergo a distortion leading to a lower-energy, and orbitally non-degenerate, structure [1]. The so-called Jahn-Teller (JT) effect is particularly strong in systems containing divalent copper ($3d^9$ electron count). Due to operation of the JT effect the first coordination sphere of the Cu^{2+} cation is distorted and most often forms an elongated instead of a regular octahedron with four shorter equatorial bonds and two longer axial ones [2–4].

The Jahn-Teller effect, also present in compounds containing the iso-electronic Ag^{2+} cation ($4d^9$) [5], has a large impact on material properties such as magnetism [6], and electronic structure [7–9]. Even subtle distortions in the first coordination sphere of the JT-active cation can lead to large changes in material properties, as exemplified by the case of Ag^{2+}-bearing fluorides [10,11]. Due to the fluxional nature of the Jahn-Teller effect many studies have been devoted to tuning this distortion either by chemical substitution [12–16], or by high external pressure [17–22].

In the latter case pressures above 1 GPa (=10 kbar) are used to induce substantial volume reduction which in turn leads to changes in the electronic and structural properties of the studied system. Large compression generally leads to the reduction of the JT distortion; it was found, however, that in compounds containing both Cu^{2+} and Mn^{3+} cations (the latter has a $3d^4$ configuration and is JT-active in the high-spin state) this distortion is surprisingly robust. In $LaMnO_3$ JT-distorted domains persist up to the insulator-to-metal transition at 34 GPa [18], while for $CsMnF_4$ it was found that the effect is quenched only above 37 GPa when Mn^{3+} cations enter the low-spin state [17]. The JT distortion seems to be even more stable in the case of divalent copper [19,20]. For $CuWO_4$ it initially decreases upon compression, but then increases abruptly during a phase transition at 9.9 GPa, and remains in place up to at least 20 GPa [19]. For Rb_2CuCl_4 it was found that the JT-distorted first coordination sphere of Cu^{2+} is stiffer than the rest of the crystal structure, which leads to tilting distortions at high pressure [20].

In order to elucidate the complex interplay between the effect of large compression and the Jahn-Teller distortions we studied the high pressure phase transitions of copper difluoride (CuF_2). This compound is one of the simplest binary connections containing the Cu^{2+} cation. It belongs to the family of metal difluorides, which have been extensively studied at high pressure [23–37]. Due to the operation of the JT effect CuF_2 adopts at ambient conditions a rarely-encountered crystal structure found only in one other compound (CrF_2) [38].

Here we present experimental and computational evidence that up to 100 GPa CuF_2 undergoes three phase transitions. The four lowest-enthalpy structures can be assigned to the rutile, fluorite, and cotunnite structure families, and the general phase transition sequence found for CuF_2 (rutile → fluorite → cotunnite) resembles that observed in other difluorides. However due to the operation of the Jahn-Teller effect the coordination of Cu^{2+}, as well as the dimensionality of the structures, differs from that found for other MF_2 systems. The high-pressure phase transitions of CuF_2 are reminiscent of those recently reported for its heavier analogue, AgF_2 [37], with the exception that a nanotubular phase found for AgF_2 is not observed [36].

2. Materials and Methods

Copper difluoride supplied by Sigma-Aldrich (Saint Louis, Missouri, United States) in a form of a powder (98% purity) was used in the study. Due to its hygroscopic nature all loadings were performed in an argon-filled glovebox with both water and oxygen content below 0.5 ppm. Measurements at ambient condition were performed on samples flame-sealed in quartz capillaries (OD 0.3 mm). The purity of the sample was confirmed by powder X-ray diffraction measurements (see Figure S1 in Supplementary Materials).

Raman spectra were acquired with the use of the Alpha300M+ system (Witec Gmbh, Ulm, Germany). We used a 532 nm laser line (35 mW power at sample) delivered to a confocal microscope by a single-mode optical fiber. The signal was collected through a 20× long working distance objective, and passed through a multi-mode optical fiber to a lens-based spectrometer (Witec UHTS 300, f/4 aperture and a focal length of 300 mm) coupled with an Andor iDUS 401 detector (Oxford Instruments, Abingdon-on-Thames, UK). The spectra were collected with the use of a 1800 mm grating resulting in a 1.5 cm^{-1} spectral resolution.

A total of three high-pressure runs were conducted with the use of a diamond anvil cell (DAC) supplied by D'Anvils (Hod-Hasharon, Israel). The DAC was equipped with low-fluorescence Ia diamonds (single-beveled with culet sizes of 400 µm and 500 µm) and a pre-indented stainless-steel gasket (35 µm thick). The gasket hole of 250 µm was drilled by spark-erosion. The pressure was determined from the shift of the R1 ruby fluorescence line [39]. The position of Raman bands was established with Fityk 0.9.8 software (Marcin Wojdyr, Poland) by background subtraction and fitting of the observed spectra with Lorentzian profiles [40].

Periodic DFT calculations utilized the rotationally-invariant DFT+U method [41], with the PBE exchange-correlation functional [42]. We set the U and J values of the DFT+U method to 7 eV and 0.9 eV, respectively, as suggested in a recent study on $KCuF_3$ [43]. These value are similar to those used in other studies [44,45]. The employed method yielded lattice constants and Cu-F bond lengths overestimated by less than 2% compared to the experimental structure of CuF_2 determined at low temperature [46].

The projector–augmented-wave (PAW) method was used [47], as implemented in the VASP 5.4 code. The cutoff energy of the plane waves was set to 920 eV with a self-consistent-field convergence criterion of 10^{-6} eV. Valence electrons (Zn, Cu: 3d, 4s; F: 2s, 2p) were treated explicitly, while standard VASP pseudopotentials (accounting for scalar relativistic effects) were used for the description of core electrons. The k-point mesh was set at $2\pi \times 0.03$ Å^{-1}. All structures were optimized using a conjugate-gradient algorithm until the forces acting on the atoms were smaller than 5 meV/Å. For each structure the optimization was performed for the lowest-energy spin state, that is: (i) AFM ordering within [CuF$_{4/2}$] sheets for $P2_1/c$, $Pbca$, and $Pca2_1$; (ii) FM ordering within chains of the cotunnite $Pnma$ structure; and (iii) AFM ordering within nanotubes present in the $Pbcn$ polymorph.

Evolutionary algorithm searches were performed at 20, 60, and 100 GPa for $Z = 8$ with XtalOpt software (version r9 [48]) which was coupled with the DFT+U method described above. The searches yielded the $Pbca/Pca2_1/Pnma$ structures as the lowest-enthalpy polymorphs of CuF$_2$ at 20/60/100 GPa, in accordance with results presented in this work.

Calculations of Γ-point vibration frequencies were conducted in VASP within the DFT finite-displacement method (0.007 Å displacement was used) and a tighter SCF convergence (10^{-8} eV). Visualization of all structures was performed with the VESTA software package [49]. For symmetry recognition we used the FINDSYM program [50]. Group theory analysis of the vibrational modes was performed with the use of the Bilbao Crystallographic Server [51].

3. Results

3.1. Ambient Pressure

At ambient conditions CuF$_2$ crystallizes in a structure belonging to the rutile-type family. The rutile (TiO$_2$) aristotype, adopted by most of the first row transition metal difluorides, has tetragonal ($P4_2/mnm$) symmetry, and features a six-fold coordination of the metal center (Figure 1a). Due to operation of a collective JT distortion copper difluoride adopts a structure with lower symmetry (monoclinic, $P2_1/n$), exhibiting a 4 + 2 coordination of Cu^{2+} [46,52–55]. This structure, shown in Figure 1b, can be also transformed to a $P2_1/c$ setting (Figure 1c) which more clearly illustrates the presence of puckered sheets of [CuF$_{4/2}$] stoichiometry. These sheets host a relatively strong antiferromagnetic (AFM) interaction between the neighboring Cu^{2+} sites [56], which together with a weak ferromagnetic (FM) inter-layer coupling leads to a spin-canted 2D AFM state below 70 K [46,57,58]. Hereinafter when referring to the ambient pressure rutile-type structure of CuF$_2$ we will use the $P2_1/c$ setting.

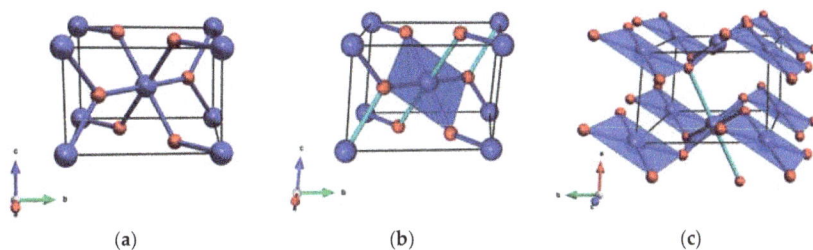

(a)   (b)   (c)

Figure 1. (a) The rutile aristotype ($P4_2/mnm$); (b) The CuF$_2$ structure in the $P2_1/n$ setting; and (c) the same structure in the $P2_1/c$ setting. Blue/red balls mark metal/ligand atoms (Cu/F in case of CuF$_2$); for CuF$_2$ dark blue cylinders depict short Cu-F bonds (1.9 Å); light blue cylinders depict long bonds (2.3 Å).

Up to date various techniques have been employed in the characterization of CuF$_2$, but to our knowledge there are no reports concerning the Raman spectrum of this material. Group theory analysis of the $P2_1/c$ structure of copper difluoride ($Z = 2$), performed with the use of the Bilbao

Crystallographic Server [51], indicates that among the 18 Γ-point vibrational modes ($3A_g + 6A_u + 3B_g + 6B_u$) six are Raman-active ($3A_g + 3B_g$).

We performed calculations of the Γ-point frequencies for the $P2_1/c$ structure of CuF_2 with the use of the density functional theory with the inclusion of the on-site Coulomb repulsion (DFT+U method, for calculation details see Materials and Methods). The resulting values are compared in Table 1 with those obtained from ambient-pressure Raman measurements (Figure 2). The lowest-energy A_g mode is not observed experimentally as its predicted frequency of 70 cm^{-1} lies below the detection limit of our Raman setup. Two B_g and two A_g modes with calculated frequencies in the 200–350 cm^{-1} range can be assigned to the four strongest Raman bands found in experiment (Figure 2). The frequency of these four bands is on average only 4.7% higher than those predicted theoretically. Finally, the highest-frequency B_g mode is found experimentally at a Raman shift 8.0% higher than predicted from DFT+U. One additional band at 496 cm^{-1} is observed in the Raman spectrum of powder CuF_2. This transition, marked by a star in Figure 2, can be tentatively assigned as an overtone of the strongest A_g mode at 254 cm^{-1}, or as a combination mode of two B_g vibrations at 221 and 293 cm^{-1}.

Table 1. Comparison of calculated (ω_{th}) and experimental (ω_{exp}) Γ-point Raman frequencies (in cm^{-1}) of the rutile-type $P2_1/c$ structure of CuF_2 at ambient pressure. No scaling was applied to the calculated frequencies.

Symmetry	ω_{th}	ω_{exp}
B_g	524	566
A_g	338	355
B_g	280	293
A_g	245	254
B_g	210	221
A_g	70	n.d.

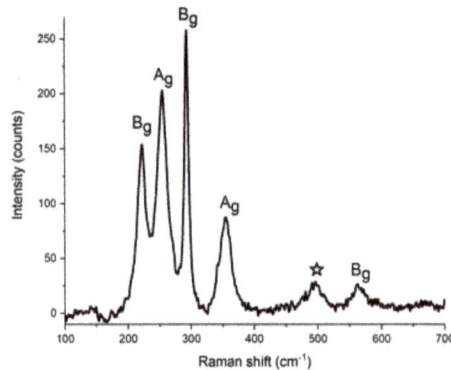

Figure 2. Raman spectrum of powder CuF_2 at ambient condition.

The good accordance between the experimental and theoretical Raman frequencies gives confidence that DFT+U method employed here can accurately simulate the pressure dependence of the frequencies of Raman active modes of CuF_2. In particular the theoretical values should fall close to the experimental ones after scaling by 1.047. This is indeed the case, as will be shown in the subsequent section

3.2. Raman Scattering up to 29 GPa

Powder samples of CuF_2 were loaded into the DAC and compressed to 29.4 GPa with Raman spectra taken upon compression in ca. 2 GPa intervals (for more details see Materials and Methods section). At high pressure all of the observed Raman modes shift to higher frequencies and broaden

(Figure 3a). Around 9 GPa a splitting of the highest-frequency A_g band is observed, as well as a new band appears at 185 cm^{-1} signaling changes in the structure of CuF$_2$ (Figure 3b, see also Figure S2 in Supplementary Materials, for a deconvolution of the Raman spectra at 19.6 GPa). As we will argue below the changes in the Raman pattern at 9 GPa are a result of a phase transition from the ambient pressure rutile-type $P2_1/c$ structure to a fluorite-like polymorph of $Pbca$ symmetry.

Before we discuss this transition we note that the pressure dependence of the Raman frequencies below 9 GPa is in very good agreement with that predicted theoretically for the $P2_1/c$ structure. Interestingly, the lowest-frequency A_g mode (not observed experimentally) is predicted to soften upon compression. This behavior resembles the one found in compounds adopting at ambient conditions the rutile aristotype, for example ZnF$_2$ [24], CoF$_2$ [28,31], FeF$_2$ [30], and MnF$_2$ [33]. In these systems the pressure-induced softening of a low-frequency B_{1g} mode leads to a second order phase transition from the $P4_2/mnm$ structure to a CaCl$_2$-type polymorph ($Pnnm$ symmetry, $Z = 2$). The latter structure can be obtained from the rutile aristotype by introducing tilts of the MF$_6$ octahedra about the c axis (compare Figure 1a).

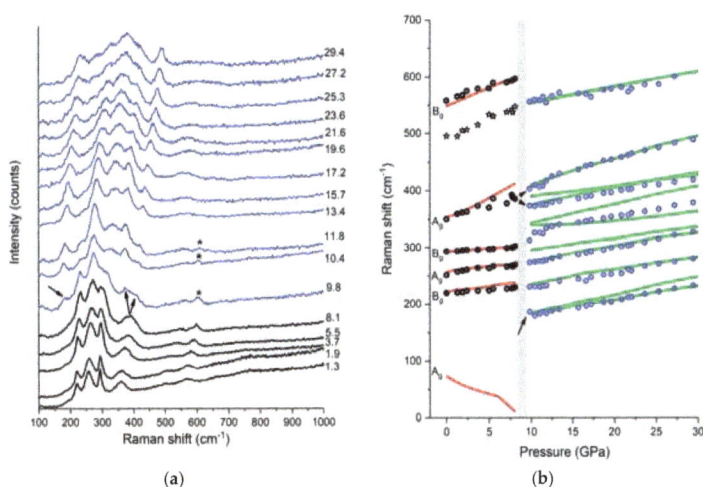

Figure 3. (**a**) Raman spectrum of powder CuF$_2$ at selected pressures (their values are given in GPa), the spectra corresponding to the rutile $P2_1/c$ phase are shown in black while those assigned to the fluorite-type $Pbca$ phase are shown in blue; and (**b**) pressure dependence of the frequency of the Raman bands (circles for experiment: black—$P2_1/c$; blue—$Pbca$; lines for DFT+U calculations: red—$P2_1/c$; green—$Pbca$). Arrows mark the appearance of a new low-frequency band and splitting of the A_g band. Asterisks in (**a**) mark the B_g band originating from traces of the $P2_1/c$ structure still present above the phase transition, while stars in (**b**) indicate the pressure dependence of the A_g overtone or B_g combination mode of rutile CuF$_2$ (see text). The calculated frequencies were scaled by 1.047.

One might, therefore, expect that the ambient-pressure $P2_1/c$ structure of CuF$_2$ will undergo a similar transition. Indeed in our calculations we find another structure of $P2_1/c$ symmetry and $Z = 2$ which is related to the ambient-pressure structure by rotation of the CuF$_6$ units about the a axis (compare Figure 1c). At 9 GPa this polymorph, which we will refer to as $P2_1/c$ (I), has a marginally lower enthalpy than $P2_1/c$ ($\Delta H = -1.4$ meV per f.u.). We note, however, that the frequency of its Raman modes is very similar to that of the original $P2_1/c$ structure (differences not exceeding 3%), with the exception of the lowest-frequency A_g mode which is shifted from 11 cm^{-1} for $P2_1/c$ to 74 cm^{-1} for $P2_1/c$ (I). Hence the Raman bands predicted for $P2_1/c$ (I) cannot account for the changes observed in the spectral region above 100 cm^{-1}.

A possible candidate for the high-pressure polymorph of CuF$_2$ is a fluorite type structure of *Pbca* symmetry (Z = 4, Figure 4a), which is adopted at ambient conditions by AgF$_2$ [59,60]. Indeed, as can be seen in Figure 3b, in the whole pressure range studied there is a good match between the frequencies of the Raman-active modes predicted for this structure and those observed in experiment. Therefore the phase transition at 9 GPa can be assigned to the transformation from *P2$_1$/c* to *Pbca*. This notion is further corroborated by DFT+U calculation which predict a phase transition between these two CuF$_2$ polymorphs at the same pressure (vide infra). Interestingly the 2D puckered sheets present in *P2$_1$/c* are retained in the *Pbca* polymorph (Figure 4a).

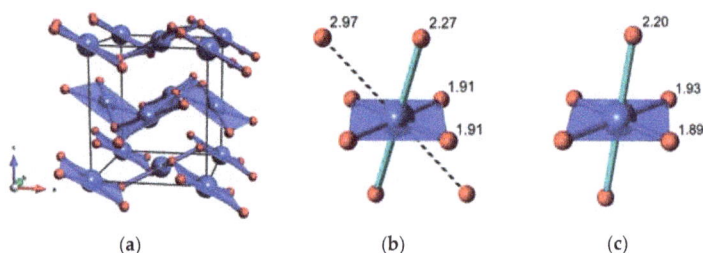

(a) (b) (c)

Figure 4. (**a**) The fluorite-type *Pbca* structure CuF$_2$ (for clarity only the four shortest Cu-F bonds are shown); (**b**) the coordination of Cu^{2+} in *Pbca* calculated at 9 GPa; and (**c**) the coordination of Cu^{2+} in *P2$_1$/c* calculated at 9 GPa; Cu-F distances are given in Å.

The *Pbca* structure can be related to the fluorite aristotype (CaF$_2$, *Fm-3m* symmetry, Z = 4) [37]. Therefore the high-pressure transition from rutile-type *P2$_1$/c* to *Pbca* is analogous to the rutile-fluorite transition found in difluorides containing non-JT ions (e.g., MgF$_2$ [23], ZnF$_2$ [26], CoF$_2$ [31]). The *Pbca* structure exhibits a 4 + 2 + 2 coordination of Cu^{2+} with two Cu-F contacts considerably longer ($\approx$30%) than the remaining six. Therefore the number of neighbors in the first coordination sphere of Cu^{2+} (6) remains unchanged upon transition from *P1/c* to *Pbca*.

3.3. Calculations up to 100 GPa

In order to further validate the interpretation of experiment, and to extend our study to higher pressures we performed DFT+U calculations for various CuF$_2$ phases up to a pressure of 100 GPa. Apart from the *P2$_1$/c* and *Pbca* polymorphs mentioned earlier we took into account three other possible structures: *Pca2$_1$* (Z = 4, Figure 5a), *Pbcn* (Z = 8, Figure 5b), and *Pnma* (Z = 4, Figure 5c). These structure were proposed as high-pressure polymorphs of AgF$_2$ with *Pca2$_1$* and *Pbcn* indeed observed experimentally [36,37]. We also searched for other candidate structures with the use of the XtalOpt evolutionary algorithm [48], but did not find any structure competitive in terms of enthalpy with the five mentioned above.

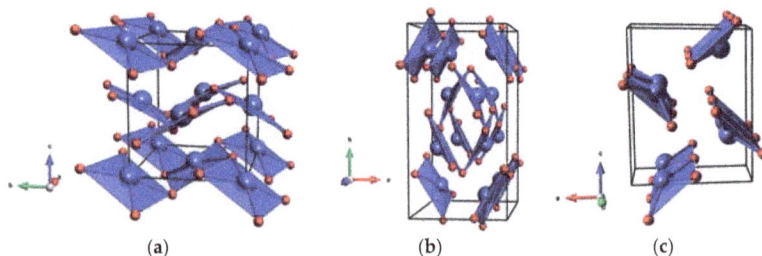

(a) (b) (c)

Figure 5. Possible high-pressure structures of CuF$_2$ (**a**) *Pca2$_1$* (Z = 4); (**b**) *Pbcn* (Z = 8); and (**c**) *Pnma* (Z = 4). For clarity only the four shortest Cu-F bonds are shown.

For AgF$_2$, the *Pca2$_1$* polymorph (HP1-AgF$_2$) is stable between 9 and 14 GPa (Figure 5a). This structure arises from a phonon instability of the ambient-pressure *Pbca* polymorph stable up to 9 GPa [37]. These two fluorite-type structures are closely related and both feature 2D sheets. The main difference between *Pca2$_1$* and *Pbca* is that, in the former structure, the metal cations are displaced out of the plane formed by the four nearest F atoms which results in a non-centrosymmetric coordination of the metal cation

The *Pbcn* polymorph (HP2-AgF$_2$), observed for AgF$_2$ from 15 GPa up to at least 36 GPa [36,37], features nanotubes built from AgF$_4$ plaquettes distorted in the same way as in *Pca2$_1$* (Figure 5b). Finally, the *Pnma* structure consists of chains built from analogous AgF$_4$ units (Figure 5c). The *Pnma* phase is isostructural with the cotunnite (α-PbCl$_2$) aristotype, a structure featuring nine-fold coordination of the metal center. The α-PbCl$_2$ polytype is adopted by many metal difluorides at large compression [27]. The *Pbcn* phase also belongs to the cotunnite structure family [37].

Optimization of the *Pca2$_1$*, *Pbcn*, and *Pnma* structures assuming a CuF$_2$ stoichiometry does not lead to changes in the bonding topology between fluorine atoms and metal centers with respect to that found in the respective AgF$_2$ polymorphs. By performing calculations at various pressures we were able to extract and compare the enthalpy of each of the five studied phases up to 100 GPa. In accordance with experiment we find that at ambient conditions (p $\approx$ 0 GPa) the *P2$_1$/c* rutile-type structure is the lowest energy polymorph of CuF$_2$ (Figure 6a). Calculations indicate that at 9 GPa CuF$_2$ should undergo a phase transition from *P2$_1$/c* to *Pbca*, in accordance with the high-pressure experimental results presented in the previous section. We predict a substantial volume decrease (14%) at this transition (Figure 6b).

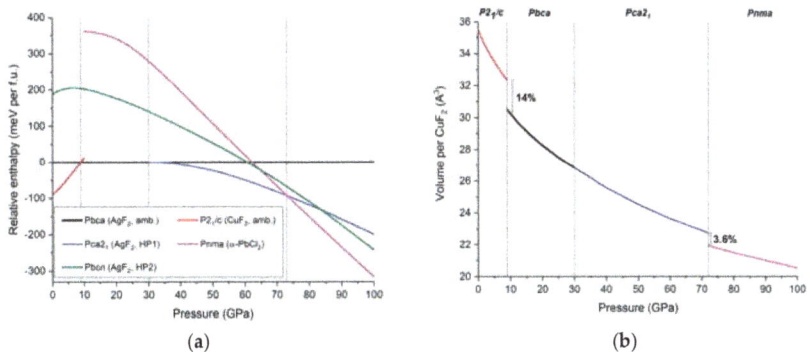

Figure 6. (**a**) The pressure dependence of the relative enthalpies (referenced to that of *Pbca*) of various CuF$_2$ high-pressure polymorphs; and (**b**) the pressure dependence of the volume per one CuF$_2$ unit. Dotted lines mark *P2$_1$/c* $\rightarrow$ *Pbca*, *Pbca* $\rightarrow$ *Pca2$_1$*, and *Pca2$_1$* $\rightarrow$ *Pnma* phase transition predicted at 9, 30, and 72 GPa, respectively.

Upon further compression *Pbca* is predicted to transform into the *Pca2$_1$* polymorph at 30 GPa. The smooth enthalpy change upon the transition, as well as the lack of a volume discontinuity suggests that this is a second order transition, in analogy with what was previously reported for an analogous transition in AgF$_2$ [37]. The last structural transition, between *Pca2$_1$* and *Pnma* is predicted to occur at 72 GPa with a 3.6% volume reduction. We note that in contrast to the *P2$_1$/c*, *Pbca*, and *Pca2$_1$* polymorphs *Pnma* features 1D chains. The calculations indicate no region of stability for the nanotubular *Pbcn* phase which is observed for AgF$_2$.

For the rutile (*P2$_1$/c*) and fluorite (*Pbca*) phases of CuF$_2$ we fitted the calculated volumes with the Birch-Murnaghan equation of state [61]. The obtained values of the bulk modulus (B$_0$), given in Table 2, indicate that, surprisingly, the low-pressure *P2$_1$/c* structure is less compressible than the rutile-like polymorph (at the same time *P2$_1$/c* has a larger volume than *Pbca*). The B$_0$ values calculated for the CuF_2 phases are about 30% lower than those calculated for the rutile and fluorite phases of ZnF_2 (Table 2). Given the fact that Zn^{2+} has nearly identical radius to Cu^{2+} ($R_{oct}(Zn^{2+}) = 0.88$ Å; $R_{oct}(Cu^{2+}) = 0.87$ Å [62]), one would expect a similar value of B_0 for both CuF_2 and ZnF_2. The lower bulk moduli found for copper difluoride phases most likely stems from the 2D character of its structures which results in facile compression in the direction perpendicular to the sheets. This notion is corroborated by the fact that both $P2_1/c$ and *Pbca* exhibit anisotropic compression with the inter-sheet distances more compressible than the intra-sheet ones (see Figure S3 in the Supplementary Materials).

Table 2. The bulk modulus in GPa (B_0), and its derivative (B_0') calculated for CuF_2 phases. Results obtained for the rutile ($P4_2/mnm$) and fluorite ($Fm-3m$) structures of ZnF_2 are shown for comparison.

Phase	B_0	B_0'
$P2_1/c$	75	6.1
Pbca	71	5.4
ZnF_2 ($P4_2/mnm$)	101 (105) [1]	4.3
ZnF_2 ($Fm-3m$)	116 (120) [1]	4.7

[1] DFT calculations with the PBE functional from ref. [32].

We now move to the analysis of the bonding pattern in the high-pressure polymorphs of CuF_2. As can be seen in Figure 7a the Jahn-Teller distortion in $P2_1/c$ is reduced upon compression. This observation is further corroborated by comparing the compressibility of M-F distances in CuF_2 and ZnF_2 (see Figures S4 and S5 in Supplementary Materials). As mentioned earlier the number of neighbors in the first coordination sphere of Cu^{2+} remains at six upon the $P2_1/c$ to *Pbca* transition. This can be well seen in the pressure dependence of Cu-F contacts shown in Figure 7a. It is noteworthy to point out that the distortion of the CuF_6 octahedron becomes larger at the transition. This signals an increase of the JT effect upon the $P2_1/c$–*Pbca* phase transition in analogy to what was found for $CuWO_4$ [19].

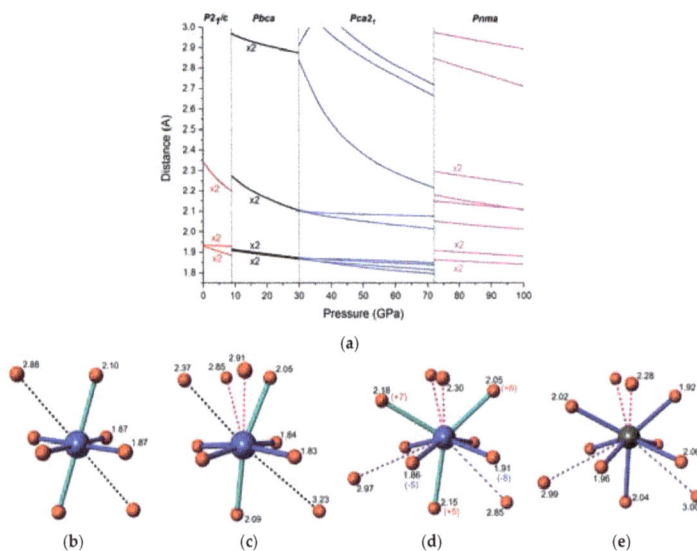

Figure 7. (a) Calculated pressure dependence of the Cu-F distances in the high-pressure polymorphs of CuF_2. The coordination of the Cu^{2+} cation in (b) *Pbca* at 30 GPa; (c) $Pca2_1$ at 50 GPa; and (d) *Pnma* at 72 GPa; together with (e) the Zn^{2+} coordination in the *Pnma* phase of ZnF_2 optimized at 72 GPa. Distances are given in Å; numbers in parentheses indicate the percentage difference between the Cu-F and Zn-F distances in the *Pnma* polymorphs.

The elongated octahedral coordination is also retained during the *Pbca-Pca2₁* transition, although due to additional secondary contacts the CuF$_6$ units become more distorted in *Pca2₁* (compare Figure 7b,c). Upon compression of *Pca2₁* one of the Cu-F contacts in the second coordination sphere of Cu^{2+} shortens considerably (by 22% from 30 to 72 GPa), and at 72 GPa is only 6.7% longer than the longer of the two Cu-F axial bonds.

The most dramatic changes in the coordination of Cu^{2+} are seen upon the *Pca2₁-Pnma* transition. The four short equatorial bonds, and the two axial ones elongate upon the transition. Additionally, the longer axial bond becomes nearly equal in length with one of the secondary Cu-F contacts (compare Figure 7c,d). As a result the first coordination sphere of Cu^{2+} can no longer be described as a distorted octahedron, but rather as a capped trigonal prism (coordination number equal to 7). In fact, it closely resembles that of the Zn^{2+} cation in the same *Pnma* phase of ZnF$_2$ (Figure 7d,e). This might suggest that the Jahn-Teller effect, present in the *P2₁/c*, *Pbca*, and *Pca2₁* phases, is quenched in the *Pnma* phase.

However, the four shortest Cu-F bonds in *Pnma* (dark blue cylinders in Figure 7d) are 5% to 8% shorter than the corresponding distances in ZnF$_2$, while the three longer bonds (light blue cylinders) are longer by approximately the same amount. Those differences in the coordination spheres of Cu^{2+} and Zn^{2+} resemble the Jahn-Teller effect found for the octahedral environment. Therefore, it is highly probable that the JT effect is still operational in the *Pnma* phase of CuF$_2$, although in a different coordination environment. We note that in our calculations that magnetic moments on Cu^{2+} atoms (m_{Cu}), as well as a substantial the band gap (E_g) are retained in the *Pnma* polymorph even at 100 GPa (m_{Cu} = 0.83 μ$_B$, E_g = 2.4 eV). Moreover the shape of the spin-density of *Pnma* at this pressure (Figure S6 in Supplementary Materials) suggests occupation of a local $d(x^2 - y^2)$ orbital on each Cu^{2+} site, in analogy with the situation found for an elongated octahedral coordination of a d^9 cation.

4. Discussion

The high-pressure transformations of CuF$_2$ can be compared to that of other metal difluorides, in particular those containing cations of similar size [62]: Mg^{2+} (0.86 Å), Zn^{2+} (0.88 Å), and Co^{2+} (0.89 Å in the high-spin state). The MF$_2$ systems (M = Mg, Zn, Co), all adopting the undistorted rutile structure at ambient conditions, exhibit a similar phase transition sequence upon compression [23,26,31]: rutile ($P4_2/mnm$) → CaCl$_2$-type (distorted rutile, *Pnnm*) → HP-PdF$_2$ (distorted fluorite, *Pa-3*) → α-PbCl$_2$ (cotunnite, *Pnma*). Only in the case of CoF$_2$ an additional undistorted fluorite phase (*Fm-3m*) exhibits a region of stability between the HP-PdF$_2$ and α-PbCl$_2$ phases [31].

The corresponding transition pressures are summarized and compared with that of CuF$_2$ in Figure 8. The subsequent high-pressure transitions of CuF$_2$ from rutile *P2₁/c* to fluorite *Pbca* and *Pca2₁* up to cotunnite *Pnma* matches that found for MF$_2$ (M = Mg, Zn, Co). The differences between copper difluoride and other systems lies in the lower symmetry of CuF$_2$ phases, which is a result of the JT effect. Moreover, for CuF$_2$, the stabilization pressure of the cotunnite structure is shifted to much higher pressures compared to the MF$_2$ systems.

The *Pbca* → *Pca2₁* phase transition predicted to occur at 30 GPa for CuF$_2$ is analogous to that found at 9 GPa for its heavier analogue, AgF$_2$ [37]. The difference between the two compounds lies in the fact that for CuF$_2$ the *Pca2₁* polymorph is predicted to transform to the cotunnite *Pnma* phase at 72 GPa, while for AgF$_2$ *Pca2₁* transforms to a nanotubular cotunnite-like *Pbcn* structure at 14 GPa. Calculations on the AgF$_2$ system indicate that *Pnma* and *Pbcn* polymorphs become nearly degenerate in terms of enthalpy above 50 GPa [37]. We do find for CuF$_2$ that *Pbcn* is more stable than *Pnma* below 64 GPa (see Figure 6a), but at this pressure both are less stable than the *Pca2₁* polymorph, and above this pressure *Pnma* is more stable.

In conclusion, Raman measurements indicate that CuF$_2$ undergoes a phase transition at 9 GPa between the rutile-type *P2₁/c* structure and the fluorite-type *Pbca* structure. This result is corroborated by DFT+U calculations, which further indicate that, at 30 GPa, it should transform to a structurally-related *Pca2₁* polymorph. Upon further compression copper difluoride should adopt a cotunnite *Pnma* structure at 72 GPa. Due to the low dimensionality of its high-pressure phases CuF$_2$

should be more compressible than ZnF_2. Surprisingly for CuF_2 high pressure induces a transition from 2D structure ($P2_1/c$, $Pbca$, $Pca2_1$) t0 a 1D polymorph ($Pnma$).

The classical Jahn-Teller effect leading to an elongated octahedral coordination of Cu^{2+} can be observed in the $P2_1/c$, $Pbca$, and $Pca2_1$ phases up to 72 GPa. Upon entering the $Pnma$ phase at that pressure the first coordination sphere of Cu^{2+} changes substantially, but the Jahn-Teller effect seems to be still operational. We hope that our results will motivate further studies into CuF_2 subject to high pressure, in particular measurements which will enable direct probing of the local electronic structure of the Cu^{2+} cations.

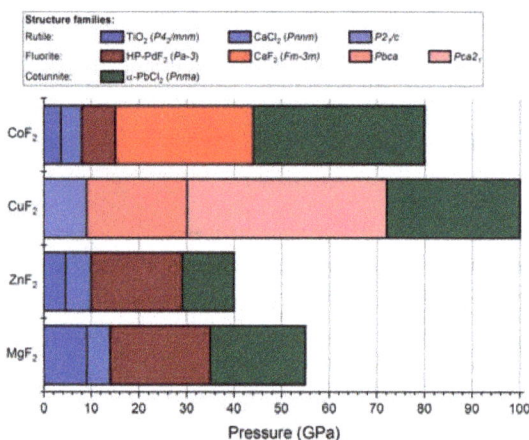

Figure 8. Bar diagram showing the pressure stability intervals of the different structural modifications of MF_2 fluorides. Experimental results for CoF_2, ZnF_2, and MgF_2 are taken from [23,26,31], respectively. The HP-PdF$_2$ to cotunnite phase transition for ZnF_2 at 29 GPa is taken from our calculations (see Figure S7 in Supplementary Materials).

Supplementary Materials: The following are available online at nline at http://www.mdpi.com/2073-4352/8/3/140/s1, Table S1: Comparison of Γ-point Raman-active modes of ZnF_2 and CuF_2, Figure S1: The experimental powder X-ray diffraction pattern of a sample of CuF_2 together with patterns simulated for the CuF_2 crystal and $CuF_2 \cdot 2H_2O$, Figure S2: The experimental Raman spectrum of CuF_2 together with the deconvolution into Lorentzian profiles, Figure S3: A comparison of the compressibility of inter-sheet and intra-sheet Ag-Ag distances in $P2_1/c$ and $Pbca$, Figure S4: A comparison of the eigenvectors of the B_{2g} mode of ZnF_2, and the symmetry-related B_g mode of CuF_2, Figure S5: Calculated pressure evolution of the difference between the Zn-F/Cu-F bonds together with the predicted differences in the frequencies of the highest B_g mode of CuF_2 and the B_{2g} mode of ZnF_2, Figure S6: A comparison of the spin-density calculated for $Pbca$ at 30 GPa and $Pnma$ at 100 GPa, Figure S7: The pressure dependence of the relative enthalpy of the cotunnite phase of ZnF_2 referenced to that of the HP-PdF$_2$ phase.

Acknowledgments: The author acknowledges the support from the Polish National Science Centre (NCN) within grant no. UMO-2014/13/D/ST5/02764. This research was carried out with the support of the Interdisciplinary Centre for Mathematical and Computational Modelling (ICM) University of Warsaw under grant no. GA67-13. Comments from Jakub Gawraczyński and Adam Grzelak are greatly appreciated.

Conflicts of Interest: The author declares no conflict of interest. The founding sponsors had no role in the design of the study; in the collection, analyses, or interpretation of data; in the writing of the manuscript; or in the decision to publish the results.

References

1. Jahn, H.A.; Teller, E. Stability of Polyatomic Molecules in Degenerate Electronic States. I. Orbital Degeneracy. *Proc. R. Soc. A Math. Phys. Eng. Sci.* **1937**, *161*, 220–235. [CrossRef]
2. Reinen, D.; Friebel, C. Local and Cooperative Jahn-Teller Interactions in Model Structures Spectroscopic and Structural Evidence. In *Structural Problems*; Springer: Berlin/Heidelberg, Germany, 1979; Volume 37, pp. 1–60.

3. Falvello, L.R. Jahn–Teller effects in solid-state co-ordination chemistry. *Dalton Trans.* **1997**, 4463–4476. [CrossRef]

4. Halcrow, M.A. Jahn-Teller distortions in transition metal compounds, and their importance in functional molecular and inorganic materials. *Chem. Soc. Rev.* **2013**, *42*, 1784–1795. [CrossRef] [PubMed]

5. Mazej, Z.; Kurzydłowski, D.; Grochala, W. Unique Silver(II) Fluorides: The Emerging Electronic and Magnetic Materials. In *Photonic and Electronic Properties of Fluoride Materials*; Tressaud, A., Poeppelmeier, K., Eds.; Elsevier: Amsterdam, The Netherlands, 2016; pp. 231–260. ISBN 9780128016398.

6. Kugel', K.I.; Khomskii, D.I. The Jahn-Teller effect and magnetism: Transition metal compounds. *Sov. Phys. Uspekhi* **1982**, *25*, 231–256. [CrossRef]

7. Guennou, M.; Bouvier, P.; Toulemonde, P.; Darie, C.; Goujon, C.; Bordet, P.; Hanfland, M.; Kreisel, J. Jahn-Teller, Polarity, and Insulator-to-Metal Transition in $BiMnO_3$ at High Pressure. *Phys. Rev. Lett.* **2014**, *112*, 75501. [CrossRef] [PubMed]

8. Reinen, D. The Modulation of Jahn–Teller Coupling by Elastic and Binding Strain Perturbations—A Novel View on an Old Phenomenon and Examples from Solid-State Chemistry†. *Inorg. Chem.* **2012**, *51*, 4458–4472. [CrossRef] [PubMed]

9. Bersuker, I.B. The Jahn-Teller and pseudo Jahn-Teller effect in materials science. *J. Phys. Conf. Ser.* **2017**, *833*, 12001. [CrossRef]

10. McLain, S.E.; Dolgos, M.R.; Tennant, D.A.; Turner, J.F.C.; Barnes, T.; Proffen, T.; Sales, B.C.; Bewley, R.I. Magnetic behaviour of layered Ag(II) fluorides. *Nat. Mater.* **2006**, *5*, 561–566. [CrossRef] [PubMed]

11. Kurzydłowski, D.; Derzsi, M.; Mazej, Z.; Grochala, W. Crystal, electronic, and magnetic structures of M_2AgF_4 (M = Na–Cs) phases as viewed from the DFT+U method. *Dalton Trans.* **2016**, *45*, 16255–16261. [CrossRef] [PubMed]

12. Alonso, J.A.; Martínez-Lope, M.J.; Casais, M.T.; Fernández-Dáz, M.T. Evolution of the Jahn-Teller distortion of MnO_6 octahedra in $RMnO_3$ perovskites (R = Pr, Nd, Dy, Tb, Ho, Er, Y): A neutron diffraction study. *Inorg. Chem.* **2000**, *39*, 917–923. [CrossRef] [PubMed]

13. Lufaso, M.W.; Woodward, P.M. Jahn-Teller distortions, cation ordering and octahedral tilting in perovskites. *Acta Crystallogr. B.* **2004**, *60*, 10–20. [CrossRef] [PubMed]

14. Goodenough, J.B. Electronic and ionic transport properties and other physical aspects of perovskites. *Rep. Prog. Phys.* **2004**, *67*, 1915–1993. [CrossRef]

15. Kurzydłowski, D.; Mazej, Z.; Grochala, W. Na_2AgF_4: 1D antiferromagnet with unusually short $Ag^{2+}\cdots Ag^{2+}$ separation. *Dalton Trans.* **2013**, *42*, 2167–2173. [CrossRef] [PubMed]

16. Kurzydłowski, D.; Jaroń, T.; Ozarowski, A.; Hill, S.; Jagličić, Z.; Filinchuk, Y.; Mazej, Z.; Grochala, W. Local and Cooperative Jahn–Teller Effect and Resultant Magnetic Properties of M_2AgF_4 (M = Na–Cs) Phases. *Inorg. Chem.* **2016**, *55*, 11479–11489. [CrossRef] [PubMed]

17. Aguado, F.; Rodríguez, F.; Núñez, P. Pressure-induced Jahn-Teller suppression and simultaneous high-spin to low-spin transition in the layered perovskite $CsMnF_4$. *Phys. Rev. B* **2007**, *76*, 94417. [CrossRef]

18. Baldini, M.; Struzhkin, V.V.; Goncharov, A.F.; Postorino, P.; Mao, W.L. Persistence of Jahn-Teller Distortion up to the Insulator to Metal Transition in $LaMnO_3$. *Phys. Rev. Lett.* **2011**, *106*, 66402. [CrossRef] [PubMed]

19. Ruiz-Fuertes, J.; Segura, A.; Rodríguez, F.; Errandonea, D.; Sanz-Ortiz, M.N. Anomalous High-Pressure Jahn-Teller Behavior in $CuWO_4$. *Phys. Rev. Lett.* **2012**, *108*, 166402. [CrossRef] [PubMed]

20. Aguado, F.; Rodríguez, F.; Valiente, R.; Itiè, J.-P.; Hanfland, M. Pressure effects on Jahn-Teller distortion in perovskites: The roles of local and bulk compressibilities. *Phys. Rev. B* **2012**, *85*, 100101. [CrossRef]

21. Calestani, G.; Orlandi, F.; Mezzadri, F.; Righi, L.; Merlini, M.; Gilioli, E. Structural evolution under pressure of $BiMnO_3$. *Inorg. Chem.* **2014**, *53*, 8749–8754. [CrossRef] [PubMed]

22. Friedrich, A.; Winkler, B.; Morgenroth, W.; Perlov, A.; Milman, V. Pressure-induced spin collapse of octahedrally coordinated Mn^{3+} in the tetragonal hydrogarnet henritermierite $Ca_3Mn_2[SiO_4]_2 [O_4 H_4]$. *Phys. Rev. B* **2015**, *92*, 014117. [CrossRef]

23. Haines, J.; Léger, J.M.; Gorelli, F.; Klug, D.D.; Tse, J.S.; Li, Z.Q. X-ray diffraction and theoretical studies of the high-pressure structures and phase transitions in magnesium fluoride. *Phys. Rev. B* **2001**, *64*, 134110. [CrossRef]

24. Perakis, A.; Lampakis, D.; Boulmetis, Y.C.; Raptis, C. High-pressure Raman study of the ferroelastic rutile-to-$CaCl_2$ phase transition in ZnF_2. *Phys. Rev. B* **2005**, *72*, 144108. [CrossRef]

25. Wu, X.; Wu, Z. Theoretical calculations of the high-pressure phases of ZnF_2 and CdF_2. *Eur. Phys. J. B* **2006**, *50*, 521–526. [CrossRef]

26. Kusaba, K.; Kikegawa, T. In situ X-ray observation of phase transitions in under high pressure and high temperature. *Solid State Commun.* **2008**, *145*, 279–282. [CrossRef]

27. Dorfman, S.M.; Jiang, F.; Mao, Z.; Kubo, A.; Meng, Y.; Prakapenka, V.B.; Duffy, T.S. Phase transitions and equations of state of alkaline earth fluorides CaF_2, SrF_2, and BaF_2 to Mbar pressures. *Phys. Rev. B* **2010**, *81*, 174121. [CrossRef]

28. Wang, H.; Liu, X.; Li, Y.; Liu, Y.; Ma, Y. First-principles study of phase transitions in antiferromagnetic XF_2 (X = Fe, Co and Ni). *Solid State Commun.* **2011**, *151*, 1475–1478. [CrossRef]

29. Liu, G.; Wang, H.; Ma, Y.; Ma, Y. Phase transition of cadmium fluoride under high pressure. *Solid State Commun.* **2011**, *151*, 1899–1902. [CrossRef]

30. López-Moreno, S.; Romero, A.H.; Mejía-López, J.; Muñoz, A.; Roshchin, I.V. First-principles study of electronic, vibrational, elastic, and magnetic properties of FeF_2 as a function of pressure. *Phys. Rev. B* **2012**, *85*, 134110. [CrossRef]

31. Barreda-Argüeso, J.A.; López-Moreno, S.; Sanz-Ortiz, M.N.; Aguado, F.; Valiente, R.; González, J.; Rodríguez, F.; Romero, A.H.; Muñoz, A.; Nataf, L.; et al. Pressure-induced phase-transition sequence in CoF_2: An experimental and first-principles study on the crystal, vibrational, and electronic properties. *Phys. Rev. B* **2013**, *88*, 214108. [CrossRef]

32. Torabi, S.; Hammerschmidt, L.; Voloshina, E.; Paulus, B. Ab initio investigation of ground-state properties of group-12 fluorides. *Int. J. Quantum Chem.* **2014**, *114*, 943–951. [CrossRef]

33. López-Moreno, S.; Romero, A.H.; Mejía-López, J.; Muñoz, A. First-principles study of pressure-induced structural phase transitions in MnF_2. *Phys. Chem. Chem. Phys.* **2016**, *18*, 33250–33263. [CrossRef] [PubMed]

34. Stavrou, E.; Yao, Y.; Goncharov, A.F.; Konôpková, Z.; Raptis, C. High-pressure structural study of MnF_2. *Phys. Rev. B* **2016**, *93*, 54101. [CrossRef]

35. Barreda-Argüeso, J.A.; Aguado, F.; González, J.; Valiente, R.; Nataf, L.; Sanz-Ortiz, M.N.; Rodríguez, F. Crystal-Field Theory Validity Through Local (and Bulk) Compressibilities in CoF_2 and $KCoF_3$. *J. Phys. Chem. C* **2016**, *120*, 18788–18793. [CrossRef]

36. Grzelak, A.; Gawraczyński, J.; Jaroń, T.; Kurzydłowski, D.; Mazej, Z.; Leszczyński, P.J.; Prakapenka, V.B.; Derzsi, M.; Struzhkin, V.V.; Grochala, W. Metal fluoride nanotubes featuring square-planar building blocks in a high-pressure polymorph of AgF_2. *Dalton Trans.* **2017**, *46*, 14742–14745. [CrossRef] [PubMed]

37. Grzelak, A.; Gawraczyński, J.; Jaroń, T.; Kurzydłowski, D.; Budzianowski, A.; Mazej, Z.; Leszczyński, P.J.; Prakapenka, V.B.; Derzsi, M.; Struzhkin, V.V.; et al. High-Pressure Behavior of Silver Fluorides up to 40 GPa. *Inorg. Chem.* **2017**, *56*, 14651–14661. [CrossRef] [PubMed]

38. Maitland, R.; Jack, K.H. The Crystal Structure and Interatomic Bonding of Chromous and Chromic Fluorides. *Proc. Chem. Soc.* **1957**, *1957*, 232–233.

39. Dewaele, A.; Torrent, M.; Loubeyre, P.; Mezouar, M. Compression curves of transition metals in the Mbar range: Experiments and projector augmented-wave calculations. *Phys. Rev. B* **2008**, *78*, 104102. [CrossRef]

40. Wojdyr, M. Fityk: A general-purpose peak fitting program. *J. Appl. Crystallogr.* **2010**, *43*, 1126–1128. [CrossRef]

41. Liechtenstein, A.I.; Zaanen, J. Density-functional theory and strong interactions: Orbital ordering in Mott-Hubbard insulators. *Phys. Rev. B* **1995**, *52*, R5467. [CrossRef]

42. Perdew, J.P.; Burke, K.; Ernzerhof, M. Generalized Gradient Approximation Made Simple. *Phys. Rev. Lett.* **1996**, *77*, 3865–3868. [CrossRef] [PubMed]

43. Legut, D.; Wdowik, U.D. Vibrational properties and the stability of the $KCuF_3$ phases. *J. Phys. Condens. Matter* **2013**, *25*, 115404. [CrossRef] [PubMed]

44. Pavarini, E.; Koch, E.; Lichtenstein, A.I. Mechanism for Orbital Ordering in $KCuF_3$. *Phys. Rev. Lett.* **2008**, *101*, 266405. [CrossRef] [PubMed]

45. Binggeli, N.; Altarelli, M. Orbital ordering, Jahn-Teller distortion, and resonant x-ray scattering in $KCuF_3$. *Phys. Rev. B* **2004**, *70*, 85117. [CrossRef]

46. Fischer, P.; Hälg, W.; Schwarzenbach, D.; Gamsjäger, H. Magnetic and crystal structure of copper(II) fluoride. *J. Phys. Chem. Solids* **1974**, *35*, 1683–1689. [CrossRef]

47. Blöchl, P.E. Projector augmented-wave method. *Phys. Rev. B* **1994**, *50*, 17953–17979. [CrossRef]

48. Falls, Z.; Lonie, D.C.; Avery, P.; Shamp, A.; Zurek, E. XtalOpt version r9: An open-source evolutionary algorithm for crystal structure prediction. *Comput. Phys. Commun.* **2016**, *199*, 178–179. [CrossRef]
49. Momma, K.; Izumi, F. VESTA: A three-dimensional visualization system for electronic and structural analysis. *J. Appl. Crystallogr.* **2008**, *41*, 653–658. [CrossRef]
50. Stokes, H.T.; Hatch, D.M. FINDSYM: Program for identifying the space-group symmetry of a crystal. *J. Appl. Crystallogr.* **2005**, *38*, 237–238. [CrossRef]
51. Kroumova, E.; Aroyo, M.I.; Perez-Mato, J.M.; Kirov, A.; Capillas, C.; Ivantchev, S.; Wondratschek, H. Bilbao Crystallographic Server: Useful Databases and Tools for Phase-Transition Studies. *Phase Transit.* **2003**, *76*, 155–170. [CrossRef]
52. Billy, C.; Haendler, H.M. The Crystal Structure of Copper(II) Fluoride. *J. Am. Chem. Soc.* **1957**, *79*, 1049–1051. [CrossRef]
53. Taylor, J.C.; Wilson, P.W. The structures of fluorides VI. Precise structural parameters in copper difluoride by neutron diffraction. *J. Less Common Met.* **1974**, *34*, 257–259. [CrossRef]
54. Chatterji, T.; Hansen, T.C. Magnetoelastic effects in Jahn–Teller distorted CrF_2 and CuF_2 studied by neutron powder diffraction. *J. Phys. Condens. Matter* **2011**, *23*, 276007. [CrossRef] [PubMed]
55. Burns, P.C.; Hawthorne, F.C. Rietveld Refinement of the Crystal Structure of CuF_2. *Powder Diffr.* **2013**, *6*, 156–158. [CrossRef]
56. Reinhardt, P.; Moreira, I.D.P.R.; de Graaf, C.; Dovesi, R.; Illas, F. Detailed ab-initio analysis of the magnetic coupling in CuF_2. *Chem. Phys. Lett.* **2000**, *319*, 625–630. [CrossRef]
57. Joenk, R.J.; Bozorth, R.M. Magnetic Properties of CuF_2. *J. Appl. Phys.* **1965**, *36*, 1167–1168. [CrossRef]
58. Boo, W.O.J.; Stout, J.W. Heat capacity and entropy of CuF_2 and CrF_2 from 10 to 300 °K. Anomalies associated with magnetic ordering and evaluation of magnetic contributions to the heat capacity. *J. Chem. Phys.* **1979**, *71*, 9–16. [CrossRef]
59. Fischer, P.; Roult, G.; Schwarzenbach, D. Crystal and magnetic structure of silver difluoride-II. Weak 4d-ferromagnetism of AgF2. *J. Phys. Chem. Solids* **1971**, *32*, 1641–1647. [CrossRef]
60. Jesih, A.; Lutar, K.; Žemva, B.; Bachmann, B.; Becker, S.; Müller, B.G.; Hoppe, R. Einkristalluntersuchungen an AgF2. *Z. Anorg. Allg. Chem.* **1990**, *588*, 77–83. [CrossRef]
61. Birch, F. Finite Elastic Strain of Cubic Crystals. *Phys. Rev.* **1947**, *71*, 809–824. [CrossRef]
62. Shannon, R.D. Revised effective ionic radii and systematic studies of interatomic distances in halides and chalcogenides. *Acta Crystallogr. Sect. A* **1976**, *32*, 751–767. [CrossRef]

crystals

MDPI

Review

Layered Indium Selenide under High Pressure: A Review

Alfredo Segura [ORCID]

Departamento de Física Aplicada-ICMUV, Malta-Consolider Team, Universitat de València, 46100 Burjassot, Spain; alfredo.segura@uv.es; Tel.: +349-63544792

Received: 11 April 2018; Accepted: 7 May 2018; Published: 9 May 2018

Abstract: This paper intends a short review of the research work done on the structural and electronic properties of layered Indium Selenide (InSe) and related III–VI semiconductors under high pressure conditions. The paper will mainly focus on the crucial role played by high pressure experimental and theoretical tools to investigate the electronic structure of InSe. This objective involves a previous revision of results on the pressure dependence of the InSe crystal structure and related topics such as the equation of state and the pressure-temperature crystal phase diagram. The main part of the paper will be devoted to reviewing the literature on the optical properties of InSe under high pressure, especially the absorption experiments that led to the identification of the main optical transitions, and their assignment to specific features of the electronic structure, with the help of modern first-principles band structure calculations. In connection with these achievements we will also review relevant results on the lattice dynamical, dielectric, and transport properties of InSe, as they provided very useful supplementary information on the electronic structure of the material.

Keywords: InSe; layered semiconductors; III–VI semiconductors; high pressure; optical properties; magnetoabsorption; electronic structure; lattice dynamics; dielectric properties; transport properties

1. Introduction

Layered Indium Selenide (InSe) has been the object of scientific interest for nearly 50 years. Early studies focused on the investigation of the effect of crystal anisotropy on its transport and optical properties [1,2]. The availability of high-quality large single crystals [3], and the possibility of n and p doping [4], encouraged studies on its applications in photovoltaic solar energy conversion resulting in solar cells with conversion efficiencies of up to 10% [5,6]. Later on, several groups showed possible applications to nonlinear optical devices in the mid and far infrared spectrum [7]. The development of the Van der Waals epitaxy in the early 1990s [8], with the increase in the thin film crystal quality, brought some renewed interest in photovoltaic devices [9]. More recently, as the interest in 2D materials was triggered by graphene's remarkable properties [10], InSe has become the object of intensive research as an ideal semiconductor for a large variety of single layer nano-devices [11–14].

The historical role of high pressure techniques in the investigation of the electronic structure of semiconductors can be hardly overstated. As early as 1961, Paul's empirical rule [15] on the pressure coefficients of electronic transitions was a crucial tool to unravel the order of the conduction bands in zinc-blende semiconductors. The development of the diamond anvil cell (DAC) [16], and the ruby pressure scale [17,18], opened the way to the use of a large variety of optical spectroscopy and X-ray structural techniques under high pressure conditions. Accurate optical absorption spectra could be so obtained and quantitatively interpreted using sophisticated physical models, as reviewed by Goñi and Syassen [19]. The use of DAC in third generation synchrotron radiation facilities also produced very accurate X-ray diffraction and absorption spectra under high pressure, giving access to precise Equations of State (EOS) and crystal phase diagrams, as reviewed by Nelmes and McMahon [20].

Those experiment results have therefore become a rigorous experimental test for modern ab-initio electronic structure calculations [21].

In this context, high pressure studies have been a tool of choice for investigating the electronic structure of InSe and related III–VI layered semiconductors (GaS, GaSe, and GaTe). This paper reviews the literature on this subject in the last 40 years. This review mainly focuses on InSe, but, when relevant for the discussion, results on other III–VI semiconductors will be also presented. Section 2 is devoted to results on the pressure evolution of InSe crystal structure, its EOS, and its pressure-temperature crystal phase diagram. Section 3 will be devoted to studies on InSe electronic structure under pressure, showing how optical absorption and reflection experiments have contributed to clarifying the nature of electronic states and transitions, with the invaluable aid of ab-initio band structure calculations. Section 4 will be devoted to the literature on the lattice dynamics of low and high pressure InSe phases. Section 5 will discuss some results on the dielectric properties on layered InSe under pressure, and its correlation to electronic transitions and ionicity of the material. Finally, Section 6 will review some papers on the transport properties of InSe under high pressure, which have given supplementary information on relevant aspects of its electronic structure.

2. Crystal Structure, EOS, and Pressure-Temperature Phase Diagram of InSe

At ambient conditions, InSe crystallizes in the layered rhombohedral phase [22], built as a stack of 2D layers formed by two honeycomb In-Se sheets bound by strong In-In covalent bonds, as shown in Figure 1a. As in other semiconductors of the III–VI family (GaSe, GaS), layers are bound by weak Van der Waals forces. Several polytypes have been described, with different layer stacking sequence, as shown in Figure 1b. Four stacking sequences have been detected for the related compound GaSe [23]. InSe single crystals grown by the Bridgmann method [3] crystallize at ambient conditions in the so-called γ-polytype (InSe-I), with two chemical formulas per primitive unit cell, which belongs to the space group C_{3v} (R3m) [22]. Most of the high-pressure experiments here discussed have been obtained with γ-InSe samples. A few results on lattice dynamics in ε-InSe under high pressure will be discussed in Section 4.

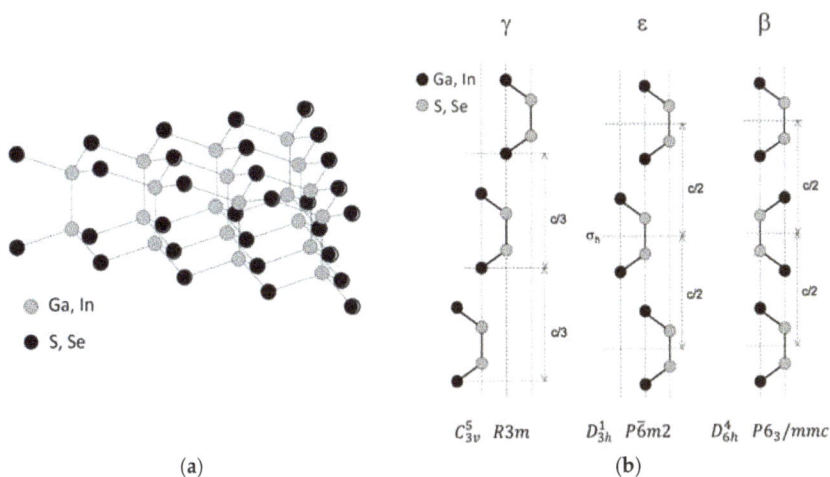

Figure 1. Crystal structure of III–VI layered semiconductors: (**a**) structure of a single layer and (**b**) stacking sequence of the single layers for three different polytypes.

A first approach to the EOS of γ-InSe at low pressure, through ultrasonic measurement, was done by Gatulle et al. [24], who reported the pressure dependence of all elastic moduli, as well as the compressibility tensor, whose components are the linear compressibilities parallel ($\chi_\parallel$) and perpendicular ($\chi_\perp$) to the *c*-axis. Values are given in Table 1, exhibiting a large anisotropy ratio ($\chi_\parallel/\chi_\perp \approx 7$). It must be stressed that the compressibility values obtained from ultrasonic measurements are affected by very large relative errors (Table 1).

In a paper on the optical properties of InSe and GaSe under pressure, Kuroda et al. [25] report the pressure dependence of the *c* parameter of γ-InSe, as obtained from a non-published X-ray diffraction (XRD) experiment in DAC. They assumed a Murnaghan-type [26] pressure dependence for the *c* hexagonal unit cell parameter

$$c(P) = c_0 \left(1 + \frac{B_0'}{B_0} P \right)^{-\frac{1}{3B_0'}}, \tag{1}$$

In which B_0 is the bulk modulus and B_0' is its pressure derivative. By assuming $B_0 = 1/3\chi_\parallel$ and taking the compressibility value from Ref. [22], they estimate the pressure derivative of the bulk modulus $B_0' = 10.8$, stressing that such high value illustrates the extremely nonlinear pressure behavior of the material compression along the *c*-axis.

Schwartz et al. [27,28] reported for the first time the EOS and high-pressure crystal phase diagram of γ-InSe up to 30 GPa. From a powder XRD experiment in DAC, using an X-ray tube as source, they obtained the EOS of the layered phase and showed that it is stable up to 10 GPa (Figure 1a). Experimental data were interpreted through the Murnaghan EOS [26]

$$V(P) = V_0 \left(1 + \frac{B_0'}{B_0} P \right)^{-\frac{1}{B_0'}}, \tag{2}$$

yielding the bulk modulus and its derivative given in Table 1. The anisotropy ratio of the low-pressure compressibility tensor, $\chi_\parallel/\chi_\perp$, as obtained from the pressure dependence of the *a* and *c* parameters, is much smaller than the one resulting from ultrasound experiments [24] (Table 1). Single crystal X-ray diffraction [29] and X-ray in DAC, using a synchrotron source, provided a more detailed picture of the low-pressure range of the EOS, as shown in Figure 1b. The values of the EOS parameters and compressibility tensor so obtained are affected by smaller relative errors, as shown in Table 1. In the low pressure range a larger compressibility is observed, as compared to the previous powder XRD results [25,27,28], as shown in Figure 2b. It must be emphasized that, in spite of the dispersion of the values given in Table 1, they are all compatible within the experimental error. When the low-pressure compressibility is determined using XRD data for $P < 1$ GPa, the fitting procedure yields larger compressibility values, closer to those of Reference [24] but with a very large relative error, larger than 50%, associated with the large relative errors of pressure, as determined through the ruby scale in the very low-pressure range ($P < 1$ GPa) [17,18].

Table 1. Equation of state parameters and compressibility tensor for InSe crystal phases.

Crystal	V_0 (Å³)	B_0 (GPa)	B_0'	$\chi_\perp$ (GPa⁻¹)	$\chi_\parallel$ (GPa⁻¹)	$\chi_\parallel/\chi_\perp$
γ-InSe [1]	-	35(10)	-	0.0033(19)	0.022(7)	6.7
γ-InSe [2]	-	-	10.8(8)	0.0033(19) [3]	-	-
γ-InSe [4]	350.8	36(10)	4.05(30)	0.005(1) [5]	0.014(2) [5]	2.8
γ-InSe [6]	350.4	24(3)	8.6(8)	0.0063(6)	0.016(2)	2.5
RS-InSe [4]	190.5	51.2	4	-	-	-
MC(T)-InSe [7]	207	44	5.4	-	-	-

[1] Ultrasounds [24]; [2] XRD [25]; [3] taken from [24]; [4] powder XRD [27,28]; [5] estimated from Figure 1a in [28]; [6] single crystal XRD [29]; [7] powder XRD [30].

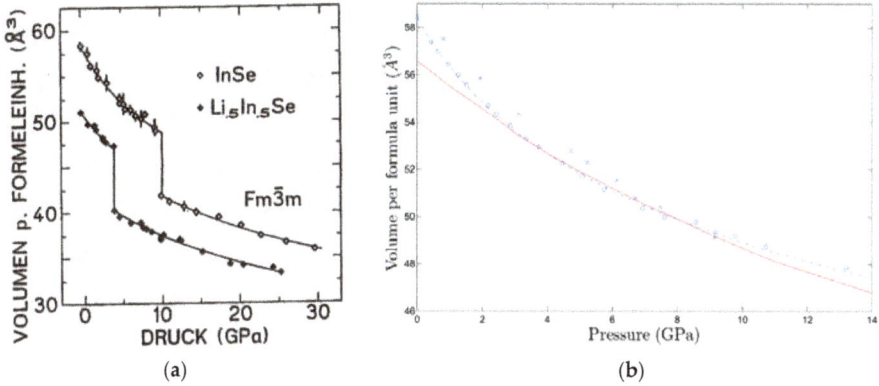

Figure 2. Results of XRD experiments in InSe-I DAC: (a) EOS of InSe-I and InSe-II from powder diffraction. The *x*-axis corresponds to the pressure and the *y*-axis to the volume per formula [27]. (b) EOS of of InSe-I from single crystal diffraction (circles). Crosses represent the data of Figure 1a [29].

Results of X-ray absorption (XAS) [31] in DAC, using γ-InSe single crystals, added relevant information concerning the evolution of the whole crystal structure. Figure 3a shows the pressure dependence of the Se-In bond-length ($d_{In\text{-}Se}$), as obtained from XAS measurements. Compared with the pressure dependence of the *a*-parameter [27–29], it turns out that the linear compressibility of *a* is much larger than that of $d_{In\text{-}Se}$. This implies that the angle of the In-Se covalent bond with the layer plane increases with pressure. The combined analysis of XRD and XAS results in DAC for other III–VI layered compounds (GaS, GaSe. GaTe) [32–36] showed that this behavior can result in an unexpected effect: the thickness of the layer can actually increase under pressure, as shown in Figure 3b [37]. This behavior is very relevant to the discussion of the reliability of deformation potential models that were proposed to give quantitative account of the extremely non-linear pressure dependence of the bandgap in III–VI semiconductors, as we will discuss in Section 3.

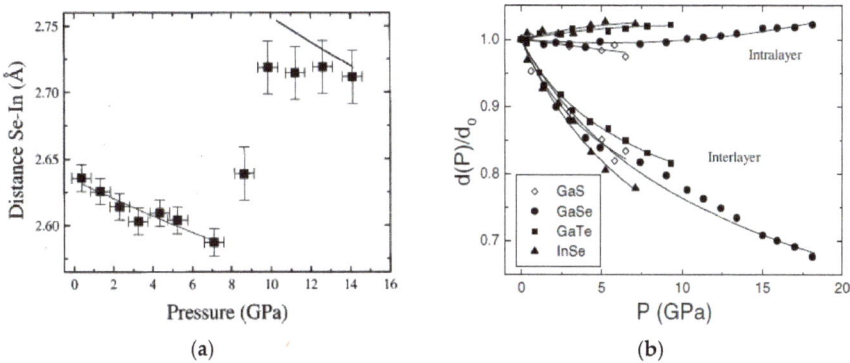

Figure 3. (a) Pressure dependence of the InSe bond-length as obtained from EXAFS measurements [31]. (b) Relative variation of the layer thickness (intralayer) and the Van der Waals gap (interlayer) as a function of pressure [37].

As concerns InSe crystal phase diagram, Schwarz et al. [27,28] showed that, around 10 GPa at room temperature, InSe-I transits to a rock-salt (RS) cubic phase (InSe-III) that is stable up to 30 GPa (Figure 2a). The transition to the rock-salt phase was also detected by EXAFS measurements [31].

The comparison of Figures 2a and 3a illustrates the fact that, in spite of the volume collapse from the layered to the rock-salt phase, the first neighbor distance increases as a result of the increase of the first-neighbor coordination number.

The existence of a metastable phase introduces more complexity in the pressure-temperature phase diagram of InSe. This phase was first synthesized by Vezzoli [38], who did not report its crystal structure. A more systematic study by Iwasaki and coworkers [39,40] reported that InSe-I transforms into a monoclinic (MC) structure material (InSe-II) with space group C_{2h} (P2/m), at relatively low pressure (1–3 GPa) and high temperature (500–700 K). This is an InS-like layered phase in which all In-In covalent bonds are virtually parallel to the layer plane, as shown in Figure 4. InSe-II is a semiconductor, which is metastable in ambient conditions [39–41].

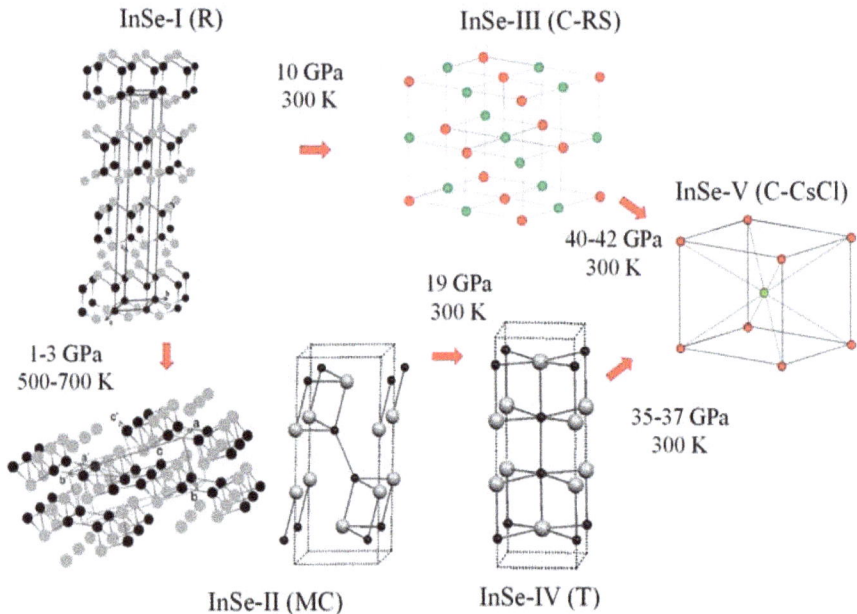

Figure 4. Basic scheme of crystal phase transitions between the different InSe crystal phases.

The P-T phase boundary between the rhombohedral and monoclinic phases of InSe was explored by means of transport measurements under high pressure and temperatures [41], and by in-situ XRD measurements at high pressure and temperatures [42] in a Paris-Edinburgh press. [43] An inverse correlation was found between the pressure and temperature at which MC InSe-II grows from R InSe-I [41,42]. While at pressures below 0.6 GPa the temperature must be 750 K, above 7 GPa, only 400 K are needed for completing the phase transition.

Under high pressure, MC InSe-II progressively increases its symmetry and gradually approaches a Hg_2Cl_2-like tetragonal phase (InSe-IV), with symmetry D_{4h} (I4/mmm) [30]. This seems to be a second order fully reversible phase transition, occurring at 19 GPa. The second sketch of the MC unit cell shown in Figure 4 illustrates the similarity between the MC and T structures, and the fact that they can transform into each other in a continuous way. InSe-IV was shown to be stable up to 30 GPa [41]. More recently it has been shown that, under further pressure increase, both RS InSe-III and T InSe-IV transform into a cubic CsCl phase, InSe-V (Figure 4) [44].

3. Electronic Structure under High Pressure

3.1. Optical Measurements and ab-Initio Band Structure Calculations

At ambient conditions InSe-I is a semiconductor with a bandgap of 1.27 eV, exhibiting intense excitonic effects in its fundamental absorption edge [2,45], as also observed in ε-GaSe, with a direct badgap of 2 eV at RT [46]. In β-GaS, with an indirect bandgap of 2.5 eV, the resonant direct exciton (at 3 eV) is observed only at low temperature [47].

The pressure dependence of the absorption edge of these III–VI semiconductors was first investigated by Besson and coworkers [47–49] by means of optical absorption measurements in large volume piston cells, with the aim of extending Paul's empirical rule [15] to non-tetrahedrally coordinated semiconductors. These authors reported complex behavior including (i) a nonlinear pressure dependence of the direct bangap, exhibiting a low-pressure interval with negative pressure coefficient and then increasing with pressure after a minimum, as shown in Figure 5b; (ii) a large negative pressure coefficient for the indirect gap of GaS; and (iii) a progressive widening and disappearing of the exciton peak, as shown in Figure 5a. Concerning ε-GaSe, Panfilov et al. [50] suggested that nonlinear behavior was the result of a phase transition to a different polytype occurring at 0.6 GPa, a hypothesis that was not supported by later XRD experiments [32,34].

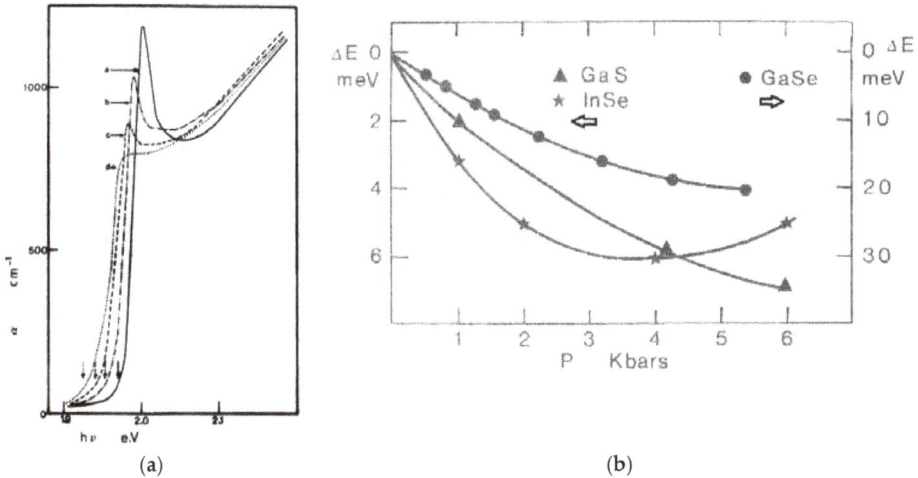

Figure 5. Pressure dependence of the direct absorption edge in III–VI semiconductors. (**a**) Absorption edge of GaSe at different pressures [48]. (**b**) Pressure dependence of the direct gap shift for GaSe [48], InSe [49], and GaS [47].

The nonlinearity in the pressure dependence of the bandgap was explained by the interplay between intra- and inter-layer interactions, the latter dominating in the low-pressure range due to the large compressibility of the interlayer distances, determined by Van der Waals interactions. The widening and disappearance of the exciton absorption peak, i.e., the quick decrease of the exciton lifetime, was explained by the strong carrier inter-valley scattering resulting from the direct-to-indirect crossover [47,48].

Attempts to give a theoretical account for this behavior were first done by means of empirical pseudopotential band structure calculations [51] that correctly predicted the bandgap coefficients but could hardly render a realistic description of the complex electronic structure of III–VI semiconductors. The first investigation of γ-InSe band structure using modern ab-initio methods, based in density functional theory (DFT) in the local density approximation (LDA), was done by Gomez da Costa et al. [52]. Results are shown in Figure 6 (notice that, as explained in their paper, the authors shifted the conduction bands upwards to match the experimental value of the bandgap). Even if this band structure was reported after some of the experimental papers that we will discuss later in this section, we will describe its main features so as to facilitate the identification of the main electronic transitions involved in the optical absorption experiments.

The valence band maximum (VBM) and conduction band minimum (CBM) are at the Z point of the first Brillouin zone (BZ) (also shown in Figure 6). This calculation gives account of the most relevant features of InSe electronic structure and especially the unexpected anisotropy of the electron and hole effective masses, which turns out to be smaller in the direction of the c-axis (Table 2).

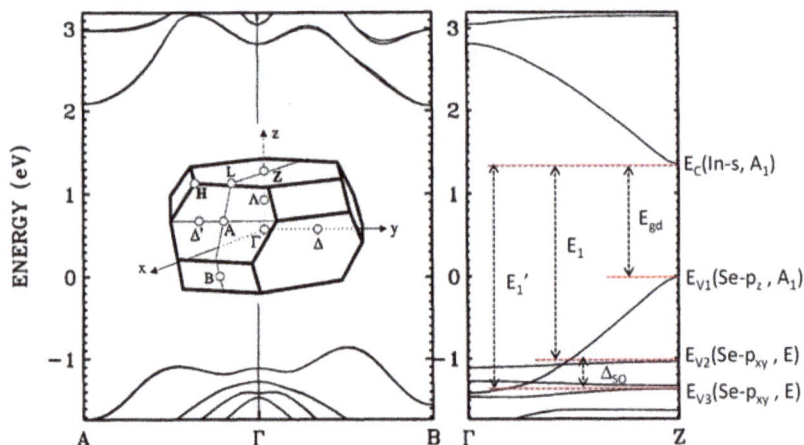

Figure 6. Band structure of γ-InSe as obtained from ab-initio calculations, with assignment of the three lowest energy direct transitions [52].

Table 2. Electron and hole effective mass in γ-InSe.

Title	$m_{e\perp}/m_0$	$m_{e\parallel}/m_0$	$m_{h\perp}/m_0$	$m_{h\parallel}/m_0$
Theory [1]	0.18	0.11	5	0.11
Theory [2]	0.12	0.03	3.1	0.03
Experiment [3,4]	0.141(2)	0.081(9)	0.73(9)	0.17(3)

[1] Empirical pseudopotential [51]; [2] ab-initio FTD-LDA [52]; [3] electrons, from cyclotron resonance experiments [53]; [4] holes, from photoluminescence experiments [54].

Figure 6 also shows the assignment of the main direct transitions observed in the absorption spectrum of γ-InSe, as well as the main orbital character and symmetry of the initial and final states. It is relevant to notice that the fundamental transition is fully allowed only for polarization parallel to the c-axis. In the framework of the $k \cdot p$ model [55], effective masses in a band extremum are inversely proportional to the squared dipole matrix element with other extrema and proportional to the energy difference between them. Effective mass values in Table 2 are then consistent with the observed features of the main direct transition [2,45,56]. The low values of the electron and hole effective masses along the c-axis are correlated to the strong allowed character of the fundamental transition for polarization parallel to the c-axis. For polarization perpendicular to the c-axis transitions, E_1 and E_1' are allowed, while the fundamental transition at E_{gd} becomes partially allowed by spin-orbit interaction mixing Se-p_z states at the VBM with two deeper valence bands with Se-p_{xy} character [52].

Further studies on γ-InSe and ε-GaSe absorption edge under pressure in DAC by Kuroda et al. [25] confirmed the nonlinear behavior of the fundamental edge (B-edge in Figure 7, corresponding to transition E_{gd} in Figure 6), as well as the pressure-induced quenching of the exciton peak in both compounds. The use of very thin samples also led these authors to investigate the behavior of a more intense direct transition at larger photon energies (A-edge in Figure 7, corresponding to transition E_1 in Figure 6), assigned to a transition from a deeper valence band to the CBM, which exhibits a quasi-lineal pressure dependence. The main features of ε-GaSe band structure can be imagined by folding γ-InSe bands along the ΓZ direction and shifting the CBM by about 0.7 eV, which makes the direct transition in ε-GaSe very close in energy to the indirect transition from the VBM to the conduction band minima at the Brillouin zone edge (points A and B in Figure 6).

This paper was the first one to report that, while in ε-GaSe the energy difference E_{V1}-E_{V2} increases under pressure in the whole explored pressure range, in γ-InSe it slightly increases up to about 1 GPa and then quickly decreases as pressure increases. This different behavior of E_{V1}-E_{V2} is most probably a consequence of the different symmetry of the ε and γ polytypes, leading to marked differences between both compounds regarding the pressure effects on the shape of the VBM, as we will discuss in Section 6. These authors also proposed a detailed empirical model including four deformation potentials to give quantitative account of the nonlinear dependence of the fundamental gap.

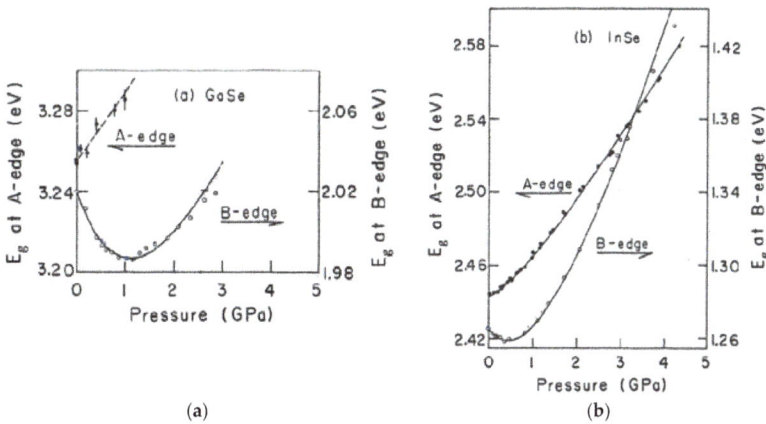

Figure 7. Pressure dependence of the fundamental absorption edge (B-edge) and the second direct transition (A-edge) in GaSe (**a**) and InSe (**b**) [25].

In a paper on the pressure effects on the lattice dynamics and optical properties of ε-GaSe, Gauthier et al. [57] reported a very detailed analysis of absorption edge that included a sound decomposition of it in contributions of direct and indirect transitions, and allowed them to obtain the pressure dependence of both the direct and indirect gaps (Figure 8a). The non-linearity of the pressure dependence of the direct gap was also explained (like in [25,48]) by the interplay between intra- and interlayer interactions. The direct gap contribution was analyzed by applying the Elliott-Toyozawa models [58,59], that gives account of the effect of the electron-hole electrostatic interaction on the absorption edge [58] and explains the widening of the exciton peaks from exciton-phonon scattering processes [59]. From that complete analysis, the authors obtained the pressure dependence of the direct exciton parameters (binding energy and absorption peak width) and the direct transition dipole-matrix-element. Under pressure, the matrix element linearly decreases (Figure 8b) as a consequence of the decrease of the Se-p_{xy} contribution to the VBM, consistently with the increase of E_{V1}-E_{V2} reported by Kuroda et al. [25], as discussed in the previous paragraph. The observed quick decrease of the exciton binding energy (Figure 8b) was proposed to be due to the increase of the static dielectric constant, an issue that will be discussed in Sections 4 and 5. The large increase of the exciton peak width was explained through the pressure induced direct-to-indirect crossover, with a detailed discussion on the pressure dependence of the different mechanisms intervening in the exciton scattering [46,60].

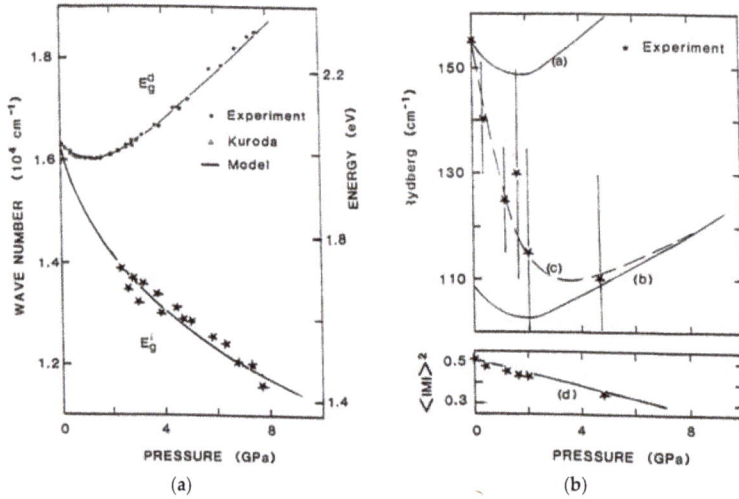

Figure 8. (a) Pressure dependence of the direct and indirect absorption edges in GaSe. **(b)** Pressure dependence of effective exciton Rydberg (exciton binding energy) and the dipole matrix element (exciton absorption intensity) in GaSe fundamental edge [57].

The first full interpretation of the pressure evolution of γ-InSe absorption edge in terms of the Elliot-Toyozawa model [58,59] was done by Goñi et al. [61] by means of low-temperature and high pressure optical measurements in DAC. Figure 9a shows the pressure evolution of the absorption edge of InSe at 10 K. Figure 9b shows the pressure dependence of the exciton binding energy.

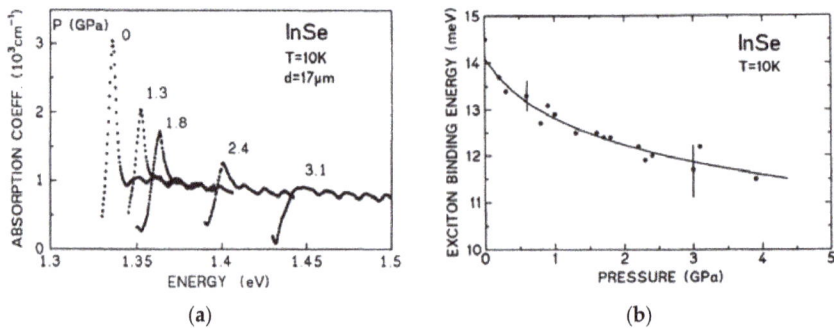

Figure 9. InSe optical properties under pressure. **(a)** InSe fundamental absorption edge at 10 K and different pressures. **(b)** Pressure dependence of the exciton binding energy in InSe [58].

As previously found in GaSe, [57] and in spite of the expected increase of the effective mass under pressure, the exciton binding energy decreases as pressure increases. This effect was attributed to a large increase of the static dielectric constant under pressure. As we will discuss in Section 5, this was later confirmed by capacitance measurements under high pressure. The authors also reported a detailed analysis of the pressure dependence of the exciton peak width, which was explained through a direct-to-indirect crossover attributed to the shift to lower energy of conduction band minima in points A or B of the Brillouin zone in the band structure shown in Figure 6, analogously to the large negative pressure coefficient of the indirect gap found in ε-GaSe [57] and β-GaS. [47] Below in this section we will discuss how this direct-to-indirect crossover in the conduction band is not enough to give account of the exciton peak widening in the low-pressure range. Changes in the valence band maximum at the Z point must also be taken into account.

A series of later experiments on the optical properties of γ-InSe under high pressure contributed to improve the picture of its electronic structure with the crucial help of modern DFT-LDA ab-initio electronic structure calculations. Ulrich et al. [62] reported the pressure dependence of the three direct transitions described in Figure 6, as obtained from photo-modulated reflectance (PMR) measurements in DAC. Figure 10a shows the PMR spectrum of InSe in the range of the high photon energy fully allowed direct transitions E_1 and E_1', whose pressure dependence, along with that of the fundamental gap is shown in Figure 10b. These results confirm the previously discussed ones by Kuroda et al. [23], showing that E_1 and E_1' direct transitions do not exhibit the extreme nonlinear behavior of the fundamental gap. From the pressure dependence of the three transitions given in Figure 10b, one can determine the pressure dependence of the energy differences E_{V1}-E_{V2} and E_{V1}-E_{V3} between the upper VBM and the second and third valence bands at the Z point. These energies correspond to E_1-E_{gd} and E_1'-E_{gd}. They slightly increase in the low-pressure range, but quickly decrease above 1 GPa. At 7 GPa they are nearly 20% below their value at ambient pressure. As previously discussed, this behavior increases the contribution of Se p_{xy} states to the VBM.

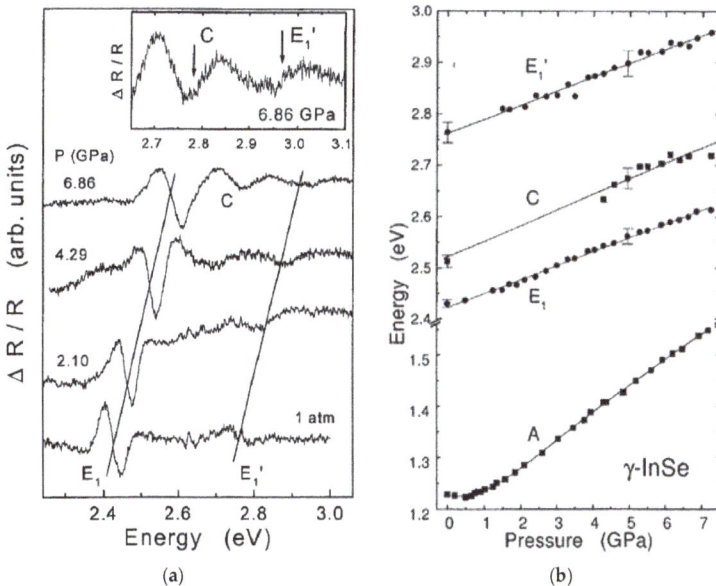

Figure 10. Optical properties of γ-InSe under pressure. (a) PMR spectra of γ-InSe at different pressures in the spectral range of direct allowed transitions. (b) Pressure dependence of the direct transitions in InSe [62].

The role of the direct-to-indirect crossovers in InSe was further investigated in a series of systematic experiments combined with DFT-LDA ab-initio calculations by Manjón et al. [63,64]. Through a detailed analysis of the absorption edge, illustrated in Figure 11a, it was shown that two different direct-to-indirect cross-over occur in InSe in the pressure range up to 4 GPa. These crossovers are consistently reflected in the shape of the absorption edge and in the pressure dependence of the exciton absorption width, which increases under pressure with two clearly defined onsets [64].

In this way, the pressure dependence of the fundamental direct gap (Z in Figure 11a) and the two indirect transitions (I_1 and I_2 in Figure 11a) could be determined, as shown in Figure 11b.

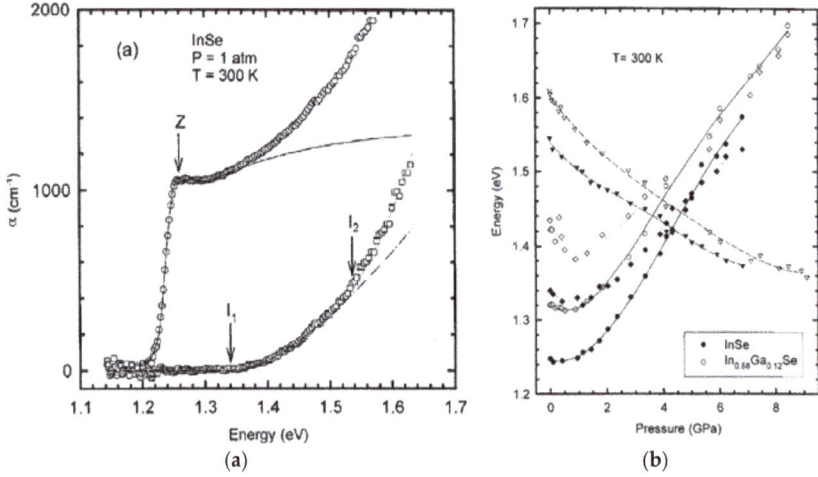

(a)

(b)

Figure 11. InSe absorption edge under pressure. (**a**) Decomposition of the absorption spectrum in direct and two indirect absorption contributions. (**b**) Pressure dependence of the direct and indirect transitions in InSe and $In_{0.86}Ga_{0.12}Se$ [64].

The band structure of γ-InSe was calculated using ab-initio DFT-LDA methods, including paths in the BZ that had not been explored in previous calculations [62] but were included in later ones [65,66]. As pressure increases, the VBM undergoes a dramatic change: a new maximum develops, close to the Z point, in the direction ZH. This maximum becomes the absolute VBM at about 2.5 GPa.

The transition assignment shown in Figure 12a was proposed by comparing the calculated pressure dependence of the VBM and CBM at Z, the toroidal maximum, the CBM at H, and the experimental pressure dependence of the direct and indirect transitions shown in Figure 11b.

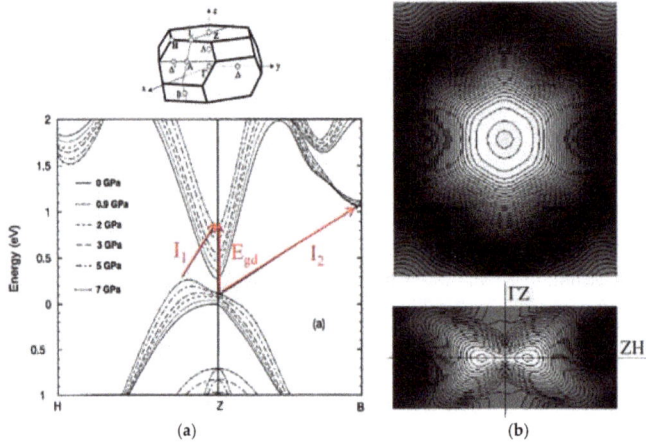

(a)

(b)

Figure 12. Band structure of γ-InSe under presssure. (**a**) Assignment of the indirect transitions [64]. (**b**) Constant energy surface plots around the toroidal valence band maximum, for a plane perpendicular to the *c*-axis at the Γ point (upper figure) and a plane parallel to the *c*-axis at the Γ point (lower figure) [37].

A detailed analysis of the new VBM at 4 GPa showed that it has quasi-cylindrical symmetry around the *c*-axis and mirror symmetry with respect to the ZHL plane, giving rise to toroidal constant energy surfaces [37].

This assignment is fully consistent with later experiments on intrinsic photoluminescence (PL) under high pressure [67]. On the one side, the pressure dependence of the PL peak width clearly exhibits two onsets, as shown in Figure 13a and previously observed for the exciton peak absorption edge [64], corresponding to the direct to indirect crossovers, at the predicted pressures, as shown in Figure 12a. On the other side, the exponential quenching of the intrinsic PL intensity for pressures beyond 4 GPa, shown in Figure 13b, is consistent with the direct to indirect crossover occurring in the conduction band at that pressure. As we will see in Section 6, this crossover is also responsible for the behavior of the transport properties of n-type InSe under high pressure.

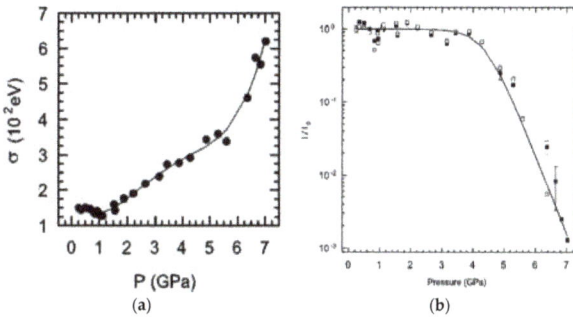

Figure 13. Intrinsic photoluminescence in γ-InSe under high pressure. (**a**) Pressure dependence of the PL peak width. (**b**) Pressure dependence of the photoluminescence peak intensity [67].

Further details on the electronic structure of InSe were obtained through magneto-optic experiments at low temperature and high pressure [68,69].

Magneto-absorption oscillations in the absorption spectrum in pulsed magnetic fields up to 56 T [69] allowed for a detailed measurement of the of the Landau levels structure as a function of pressure (Figure 14a). This led to the determination of the pressure dependence of the reduced effective mass in the layer plane ($m_{\perp}$) that was shown to increase linearly with pressure. In the framework of a simple $k \cdot p$ model [52], this is an unexpected behavior, as the effective mass should be proportional to the bandgap and then exhibit its nonlinear behavior.

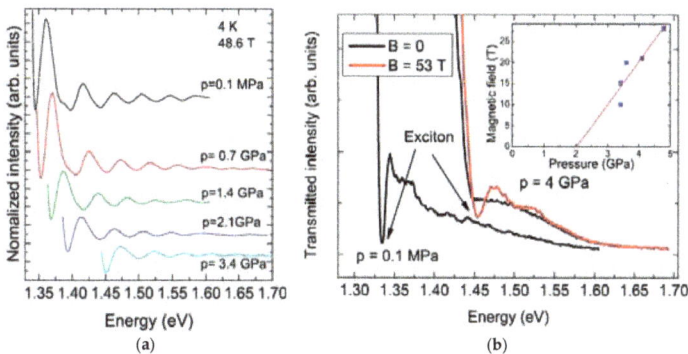

Figure 14. Magnetoabsorption experiments in InSe. (**a**) Magnetoabsorption spectra at different pressures. (**b**) Reemergence of the exciton peak at 4 GPa under high magnetic field. Inset: magnetic field at which the exciton peak reappears as a function of pressure [69].

This apparent inconsistency was explained through a $k \cdot p$ model adapted to the specific features of InSe band structure [34,61,66]. Given that that most intense transitions for polarization perpendicular to the c-axis are those named as E_1 and E_1' in Figure 6, the electron effective mass in the layer plane must follow their pressure behavior and increase linearly as they actually do (Figures 7b and 10b) [25,62].

On the other side, a remarkable behavior of the exciton peak was observed at high pressure and high magnetic field. In the absence of a magnetic field, the exciton peak is no longer observed at 4 GPa. Under high magnetic field, the exciton peak reappears, as shown in Figure 14b, indicating that Landau levels associated with the toroidal valence band maximum shift to lower energies quicker than those associated to the maximum at Z, that so becomes the absolute maximum. This behavior indicates that the hole effective mass at the toroidal maximum is much smaller than the effective mass at Z. The magnetic field at which the exciton peak reappears increases with pressure, as the inset in Figure 14b shows. This effect allows for an estimation of the hole effective mass in the toroidal maximum $m_{hT} < 0.03m_0$ [69].

It is important to notice that none of the ab-initio band structure calculations discussed in this section gives quantitative account of the extreme non-linear behavior of the pressure dependence of the bandgap. This seems to be related to the inability of DFT-LDA calculations to deal with Van der Waals interactions. Several attempts to give quantitative account of the nonlinear behavior of the bandgap, based in empirical deformation potential models, were proposed in References [25,57] or [61]. All these empirical models share simple assumptions about intra and interlayer compressibilities that are hardly compatible with the complex pressure behavior of the intralayer bond-lengths and bond-angles revealed by XRD diffraction and absorption experiments under pressure, as discussed in Section 1.

3.2. Electronic Structure of High Pressure Phases

As regards the electronic structure of high pressure phases, RS-InSe was shown to be a metal, as expected from the odd number of electrons per primitive unit cell and clearly confirmed by its Drude-like plasma reflection in the near infrared [27,28], as shown in Figure 15a, with a plasma frequency of about 2 eV, corresponding to a carrier concentration below 10^{22} cm^{-3}.

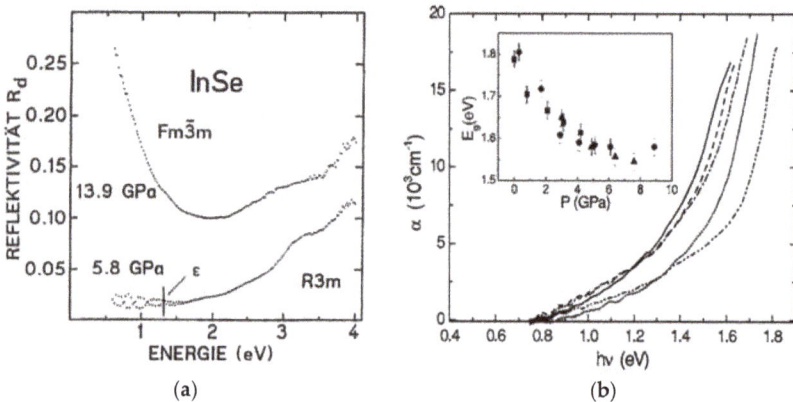

Figure 15. (a) Reflectivity versus photon energy in the plasma spectral range of RS-InSe [27]. (b) Absorption edge of monoclinic InSe under high pressure. Inset: pressure dependence of the direct bandgap in MC InSe [41].

Both band structure calculations and optical measurements indicate that monoclinic InSe is a semiconductor with a bandgap of the order of 1.6–1.8 eV at ambient pressure, as shown in Figure 15(b) [41,70]. Band structure calculations predict a small band overlapping in tetragonal InSe [41], but no bandgap closure was observed in the transition from the MC to the T phase [30].

Also, as we will discuss in Section 4, no indication of metallization was observed in the Raman effect spectrum at the transition pressure [30]. Thus, tetragonal InSe is most probably a low gap semiconductor. As regards CsCl cubic InSe [44], it must necessarily be a metal, given its odd number of electrons per primitive unit cell.

4. Lattice Dynamics under High Pressure

Raman effect experiments in III–VI layered materials under high pressure have been used as a tool to investigate the structure stability and the evolution of the chemical bond anisotropy, especially the relative intensity of intra- and interlayer bonds. Lattice dynamics of β-GaS and ε-GaSe under high pressure was investigated by Polian et al. [71], Kuroda et al. [72], and Gauthier et al. [57]. The pressure coefficients of phonon modes were found to be very dependent on their intra- or interlayer character. The largest pressure coefficients were found for low-frequency, rigid-layer interlayer modes in which restoring forces are mainly determined by weak Van der Waals bonds, whose strength quickly increases under high pressure. The frequency of these modes, which only occurs in polytypes with two or more layers per primitive unit cell, nearly doubles in the pressure range up to 6 GPa. [57,71]. This behavior was also reported for ε-InSe in Raman effect measurements up to 1 GPa [73,74].

The first systematic study on γ-InSe lattice dynamics under pressure was carried out by Ulrich et al. [75], as shown in Figure 16b. Later on, Choi and Yu [76] reported a Raman experiment under pressure in ε-InSe, but they do not report the behavior of the low-frequency rigid-layer mode that is the main signature of the ε-polytype. The primitive unit cell of γ-InSe contains only one layer per unit cell, and it does not present the low-frequency rigid-layer mode. The lowest frequency mode is the $E^{(1)}$ mode (Figure 16), a rigid-half-layer mode in which the restoring forces receive contributions from interlayer forces and from In-In intra-layer bond bending.

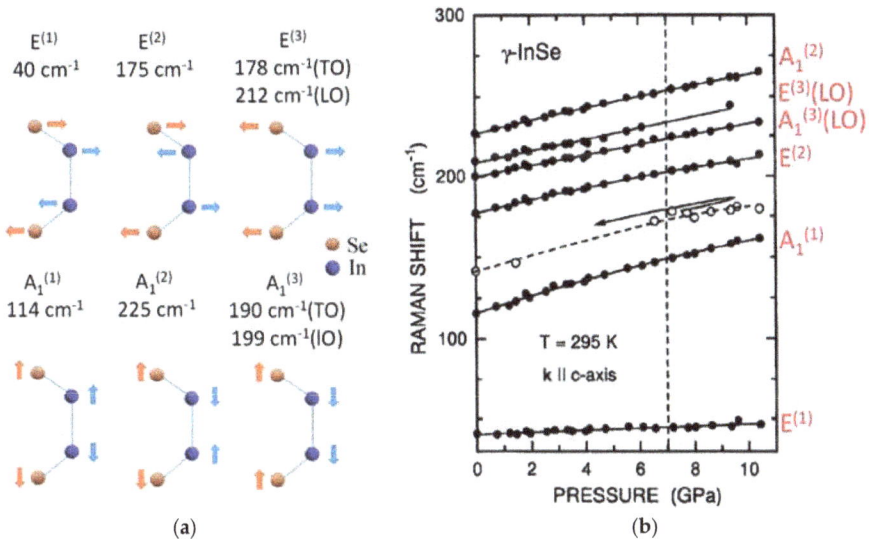

Figure 16. Phonons in γ-InSe. (a) Vibration schemes of the normal modes at the Γ point. (b) Pressure dependence of the normal modes frequencies [75].

The pressure coefficient of this mode is remarkably low, as also reported for the similar modes in β-GaS [71] and ε-GaSe [57]. This behavior indicates that the increase of the interlayer forces under pressure is accompanied by a weakening of the In-In bond, suggesting the existence of some kind of charge transfer from intra-layer covalent bonds to the interlayer space under high pressure. In the case of GaSe [57], this effect was proposed to be correlated to the observed marked decrease of the transverse dynamic charge associated with the LO-TO splitting of polar phonons vibrating perpendicular to the *c*-axis (E phonons). Based on this consideration, it was proposed that the transverse dynamic charge of polar phonons vibrating parallel to the *c*-axis (A_1 phonons) should dramatically increase under pressure, resulting in a strong increase of the static dielectric constant for polarization parallel to the *c*-axis [57,77]. This issue will be discussed in Section 5.

An ab-initio investigation of the lattice dynamics of γ-InSe under pressure was published by Rushchanskii [78]. This calculation accurately predicts the experimental pressure dependence of the Raman modes [75]. It is relevant to notice that the calculation underestimates the LO-TO splitting of A_1 polar mode at ambient pressure. Also, even if it predicts its increase under high pressure, this pressure-enhanced LO-TO splitting is not be enough to reproduce the pressure increase of the static dielectric constant under pressure predicted by the charge transfer model proposed by Gauthier et al. [57,77].

The lattice dynamics of metastable and high-pressure phases of InSe have been investigated in the context of the crystal phase transition from MC-InSe to T-InSe [30]. Figure 17a shows the vibration scheme of Raman active modes in both crystal phases. Figure 17b shows the pressure dependence of its frequencies and how four non-degenerate Raman active modes of MC-InSe (A_g-B_g modes) converge into two doubly degenerate E_g modes in T-InSe.

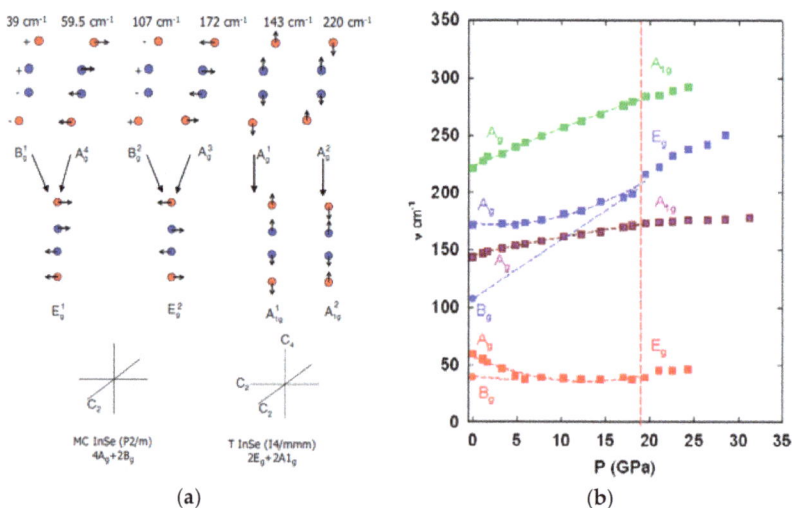

Figure 17. Phonon modes in MC- and T-InSe. (**a**) Vibration schemes of Raman active modes at the Γ point. (**b**) Pressure dependence of the normal modes frequencies [30].

The fact that no discontinuity in the frequency or intensity of the Raman peaks is observed through the monoclinic-to-tetragonal phase transition is consistent with the semiconductor character of T-InSe, as previously discussed in Section 3.1.

5. Dielectric Properties under High Pressure

The pressure dependence of the electronic and lattice contributions to the static dielectric constant in InSe and other layered III–VI materials has been investigated through different experimental techniques. The pressure effect on the electronic contribution to the dielectric constant has been determined through refractive index measurements under pressure.

Polian et al. [78] used Brillouin effect experiments in DAC to measure the pressure dependence of refractive index in β-GaS, for light polarization perpendicular ($n_\perp$) and parallel ($n_\parallel$) to the *c*-axis (named n_o and n_e respectively in Figure 18a), reporting a large increase of both indexes under pressure. The pressure increase was shown to be much larger for $n_\parallel$ and, as a result, at 15 GPa the difference $n_\perp - n_\parallel$ vanishes, as shown in Figure 18a.

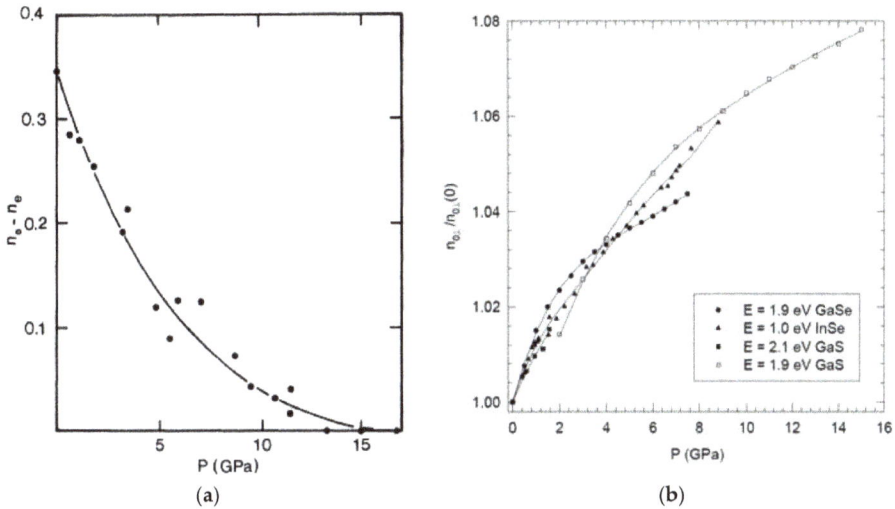

Figure 18. (**a**) Pressure dependence of the birefringence in β-GaS [79]. (**b**) Relative change of the refractive index $n_\perp$ as a function of pressure for β-GaS, ε-GaSe, and γ-InSe [80].

For ε-GaSe [57,72] and γ-InSe [80,81], the interference fringe pattern of the transmitted light was used to determine the pressure dependence of the refractive index for polarization perpendicular to the *c*-axis ($n_\perp$). A large increase of $n_\perp$ was found for both materials. Figure 18b illustrates those findings. Results were interpreted in the framework of the Phillips-Van Vechten model [82,83] for semiconductor dielectric response.

Once the large compressibility of these materials is taken into account, it turns out that the electronic polarizability for light polarization perpendicular to the *c*-axis decreases under pressure due to the positive pressure coefficient of the material Penn gap, i.e., the average most intense dipole allowed transition between the valence and conduction bands [84].

Errandonea et al. [85–87] investigated the pressure dependence of the static dielectric constant for polarization parallel to the *c*-axis ($\varepsilon_{0\parallel}$) for the three compounds by means of capacitance measurements on insulating samples, in Bridgman anvil cells, as shown in Figure 19a. The steep increase at 1.5 GPa for β-GaS is related to a reversible crystal phase transition to a denser layered phase [23,88]. For this polarization, the material compressibility accounts only for a small part of the large increase of the static dielectric constant [87]. A large increase of the total polarizability must be assumed. Given that Brillouin effect experiments in β-GaS confirm a large increase of $n_\parallel$, it seems clear that the electronic contribution is responsible for most of the increase of the static dielectric constant for polarization parallel to the *c*-axis. The pressure behavior of the Penn gap for this light polarization was found to be correlated to the pressure behavior of the indirect gap in the three compounds, as shown in Figure 19b. It was proposed that the strong decrease under pressure of both electronic transitions has the same origin: the quick shift to lower energies of a CBM that is the final state in both the indirect gap and the Penn gap transitions [87].

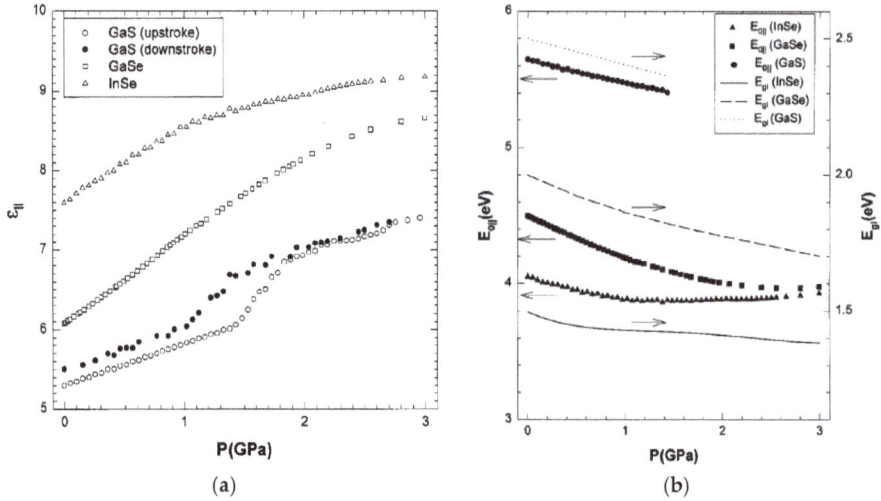

Figure 19. (a) Pressure dependence of the static dielectric constant $\varepsilon_{0\parallel}$ for β-GaS, ε-GaSe, and γ-InSe [87]. (b) Correlation between the pressure behavior of the Penn gap $E_{0\parallel}$ and the indirect gap in the three compounds [87].

6. Electronic Transport Properties under High Pressure

The pressure behavior of electronic transport parameters of layered III–VI semiconductors has been investigated through resistivity, Hall effect, and thermos-power measurements under high pressure. Given the extrinsic character of most samples in these materials, obtaining reliable information on intrinsic parameters (like effective masses or impurity ionization energies) or specific features of the electronic structure involves very systematic experiments, using well characterized samples with carrier concentration extending over several order of magnitude.

In the case of n-type InSe doped with Sn, a clear correlation was found between the ambient pressure carrier concentration and its pressure behavior, with a more accused, pressure-induced exponential quenching of the free carrier concentration for larger carrier concentrations at ambient pressure, as shown in Figure 20a [89–91]. This behavior could be consistently explained by assuming the existence of an electron trap associated with an excited minimum of the conduction band, moving down with a pressure coefficient of -100 meV/GPa, and trapping the free electrons as they enters the forbidden band and cross the Fermi level, as shown in Figure 20b.

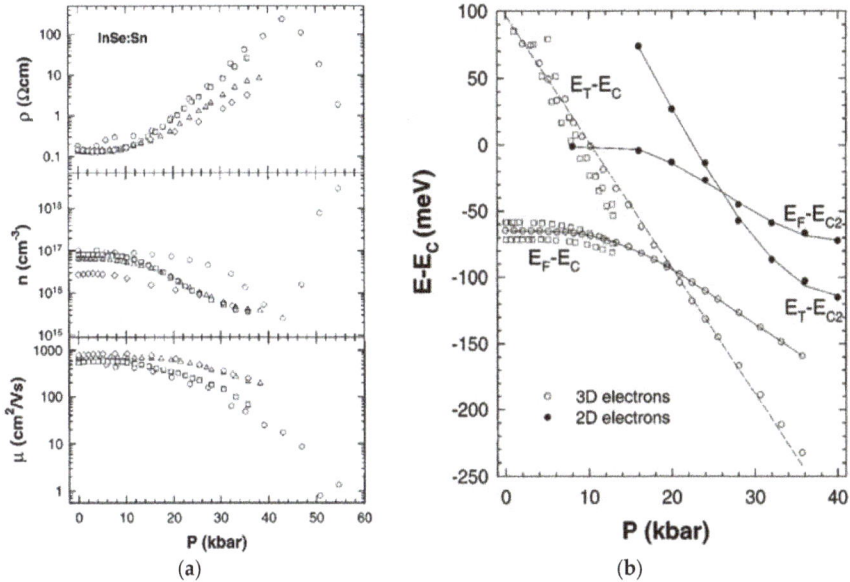

Figure 20. (a) Pressure dependence of the transport parameters (resistivity, electron concentration, and electron mobility) in Sn-doped n-type γ-InSe. (b) Pressure dependence of the Fermi level and the deep trap (both with respect to the conduction band minimum) as determined from the data in Figure 20a [90].

This picture is consistent with the findings of the pressure-induced changes in the band structure of γ-InSe discussed in Section 3.1. Optical measurements and ab initio band structure calculations showed a direct to indirect crossover in the conduction band [64]. Impurity levels associated with the different minima in the conduction band have different ionization energies. Given that all band structure calculations [64–66] predict a large electron effective mass for the zone-edge excited minimum of the conduction band, moving downwards under pressure (minimum at B in Figure 12a), its related donor can reasonably be assumed to be a deep level. Then, it can be assimilated into an electron trap moving down in energy, trapping free electrons as it approaches the Fermi level, as depicted in Figure 20b. This also explains why the electron trapping onset occurs around 1.2 GPa, a pressure lower than the direct to indirect crossover pressure (4 GPa).

It is also relevant to mention that transport measurements have also been used to investigate precursor effects of the structural phase transitions. It was noticed [90] that at about 4.5 GPa, some irreversible changes occur in the material, which were detected first in the transport properties as an irreversible increase of the carrier concentration, as shown in Figure 20a. Then, at about 7 GPa, dark lines start appearing in the monocrystalline samples [92] and a new Raman active mode is observed, which remains at ambient pressure after the pressure down-stroke [75,92]. All these effects can be considered as precursor effects of the crystal phase transition to the RS phase occurring at 10 GPa [27–29]. The appearance of these dark lines was attributed to a local increase of the pressure, associated with the stress field of edge dislocations. Along the edge dislocation lines, the rhombohedral phase would become locally unstable at a lower macroscopic pressure [92].

Transport measurements in p-type InSe were also shown to be consistent with the findings of the pressure-induced changes occurring in the band structure around the VBM at the Z point, unraveled by optical measurements and band structure calculations.

The hole concentration in p-type γ-InSe samples doped with different acceptors has been reported to increase by a factor 40 between 1 and 3 GPa, while the hole mobility increases by a factor 2, as shown in Figure 21 [93–96]. The onset pressure at which the hole concentration starts rising (1 GPa) is practically the same as the onset pressure at which the width of the exciton absorption peak [61,64] and the width of the PL peak [67] start increasing. The behavior of hole transport parameters (concentration and mobility) is consistent with the emergence of a new VBM with a *larger* effective density of states and *lower* effective mass. These are distinctive features of the toroidal VBM are shown in Figure 12.

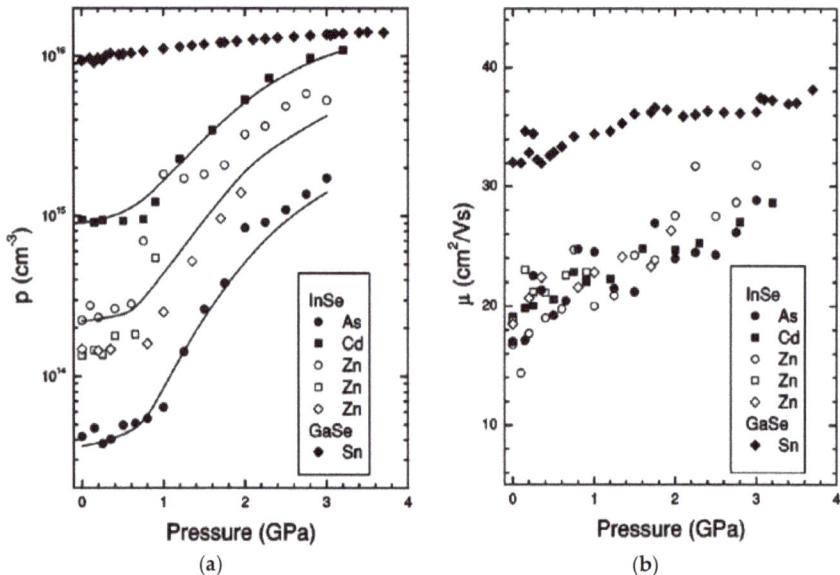

Figure 21. (a) Pressure dependence of the hole concentration in p-type γ-InSe and ε-GaSe. (b) Pressure dependence of the hole mobility in p-type γ-InSe and ε-GaSe.

In contrast, both the hole concentration and mobility increase monotonously under pressure in p-type ε-GaSe and do not exhibit any dramatic changes (Figure 21). Ab-initio band structure calculations [32,94] do not show any trace of a toroidal maximum or any other dramatic modification of the VBM (occurring at the Γ point) in ε-GaSe, at 7 GPa [94], or at 16 GPa [32]. The different pressure behavior of the VBM in ε-GaSe with respect to γ-InSe has been attributed to differences in the band mixing and dipole matrix elements between the highest energy valence bands imposed by symmetry elements of the P6m2 group (to which ε-GaSe belongs) [94].

7. Conclusions and Perspectives

In spite of the remarkable advances here reviewed, some relevant features of the electronic structure of InSe and related III–VI semiconductors are not yet well understood. Concerning ab-initio band structure calculations, as we stressed in Section 3, they give a quantitative account of the linear pressure dependence of the main electronic transitions in the high pressure range but fail to predict the extremely non-linear behavior of some physical parameters in the low pressure range. The problem stems from the well-known inability of DFT-LDA ab-initio calculations to deal with van der Waals interactions. New techniques, like the so called van der Waals corrected-DFT [97–99], which were successfully applied to layered materials like graphene or MoS_2, do not seem to have yet been used to investigate III–VI layered materials.

From an experimental point of view, the main structural feature of InSe and related compounds (the weakness of the inter-layer forces) makes it especially difficult to prepare thin samples with faces containing the *c*-axis. This has led to a lack of results for the pressure dependence of some important physical parameters like the absorption coefficient for polarization parallel to the *c*-axis, the static dielectric constant for polarization perpendicular to the *c*-axis, and the electron and hole mobility along the *c*-axis.

We will finally mention some perspectives on the investigation of impurity levels, a crucial issue for the design of electronic devices. Shallow donors [100] and acceptors [54] in InSe are quite well characterized at ambient pressure. High pressure studies on impurity levels here reviewed [89–96] are based on transport measurements and do not give direct information on the internal electronic structure of impurity levels, as the one provided by Fourier transform infrared spectroscopy (FTIR) in DAC. FTIR spectroscopy investigations would also provide information on the pressure behavior of relevant band structure parameters, like carrier effective mass tensors. In an extended spectral range, FTIR experiments in DAC would also serve to complete the understanding of high pressure lattice dynamics of III–VI materials by exploring the pressure behavior of polar phonons, for which Raman Effect measurements give relatively limited information, as we have seen in Section 4.

In summary, this review has shown that high pressure experimental techniques and ab-initio band structure calculations have been an exceptional tool, leading to a deep understanding of the electronic structure of InSe and related III–VI layered semiconductors.

Acknowledgments: This work has been supported by the Spanish MINECO/FEDER under Contract No. MAT2016-75586-C4-1-P.

Conflicts of Interest: The authors declare no conflict of interest.

References

1. Damon, R.W.; Redington, R.W. Electrical and optical properties of indium selenide. *Phys. Rev.* **1954**, *96*, 1498–1500. [CrossRef]
2. Andriyashik, M.V.; Sakhnovskit, M.Y.; Timofeev, V.B.; Yakimova, A.S. Optical transitions in the spectra of the fundamental absorption and reflection of InSe single crystals. *Phys. Status Solidi B* **1968**, *28*, 277–285. [CrossRef]
3. Chevy, A.; Kuhn, A.; Martin, M.S. Large InSe monocrystals grown from a nonstoichiometric melt. *J. Cryst. Growth* **1977**, *38*, 118–122. [CrossRef]
4. Chevy, A. Segregation of dopants in melt-grown indium selenide crystals. *J. Appl. Phys.* **1984**, *56*, 978–982. [CrossRef]
5. Segura, A.; Guesdon, J.P.; Besson, J.M.; Chevy, A. Photoconductivity and photo-voltaic effect in indium selenide. *J. Appl. Phys.* **1983**, *54*, 876–888. [CrossRef]
6. Martinez-Pastor, J.; Segura, A.; Valdes, J.L.; Chevy, A. Electrical and photovoltaic properties of indium-tin-oxide/p-InSe/Au solar cells. *J. Appl. Phys.* **1987**, *62*, 1477–1483. [CrossRef]
7. Bringuier, E.; Bourdon, A.; Piccioli, N.; Chevy, A. Optical 2nd-harmonic generation in lossy media-application to GaSe and InSe. *Phys. Rev. B* **1994**, *49*, 16971–16982. [CrossRef]
8. Koma, A. Van der Waals epitaxy: A new epitaxial-growth method for a highly lattice-mismatched system. *Thin Solid Films* **1992**, *216*, 72–76. [CrossRef]
9. Sanchez-Royo, J.F.; Segura, A.; Lang, O.; Schaar, E.; Pettenkofer, C.; Jaegermann, W.; Roa, L.; Chevy, A. Optical and photovoltaic properties of indium selenide thin films prepared by van der Waals epitaxy. *J. Appl. Phys.* **2001**, *90*, 2818–2823. [CrossRef]
10. Novoselov, K.S.; Geim, A.K.; Morozov, S.V.; Jiang, D.; Zhang, Y.; Dubonos, S.V.; Grigorieva, I.V.; Firsov, A.A. Electric field effect in atomically thin carbon films. *Science* **2004**, *306*, 666–669. [CrossRef] [PubMed]
11. Mudd, G.W.; Svatek, S.A.; Ren, T.; Patanè, A.; Makarovsky, O.; Eaves, L.; Beton, P.H.; Kovalyuk, Z.D.; Lashkarev, G.V.; Kudrynskyi, Z.R.; et al. Tuning the bandgap of exfoliated InSe nanosheets by quantum confinement. *Adv. Mater.* **2013**, *25*, 5714–5718. [CrossRef] [PubMed]

12. Sánchez-Royo, J.F.; Muñoz-Matutano, G.; Brotons-Gisbert, M.; Martínez-Pastor, J.P.; Segura, A.; Cantarero, A.; Mata, R.; Canet-Ferrer, J.; Tobias, G.; Canadell, E.; et al. Gerardot, Electronic structure, optical properties, and lattice dynamics in atomically thin indium selenide flakes. *Nano Res.* **2014**, *7*, 1556–1568. [CrossRef]

13. Sucharitakul, S.; Goble, N.J.; Kumar, U.R.; Sankar, R.; Bogorad, Z.A.; Chou, F.C.; Chen, Y.T.; Gao, X.P.A. Intrinsic Electron Mobility Exceeding 10 cm^2/(V s) in Multilayer InSe FETs. *Nano Lett.* **2015**, *15*, 3815–3819. [CrossRef] [PubMed]

14. Bandurin, D.A.; Tyurnina, A.V.; Yu, G.L.; Mishchenko, A.; Zolyomi, V.; Morozov, S.V.; Kumar, R.K.; Gorbachev, R.V.; Kudrynskyi, Z.R.; Pezzini, S.; et al. High electron mobility, quantum Hall effect and anomalous optical response in atomically thin InSe. *Nat. Nanotechnol.* **2017**, *12*, 223–227. [CrossRef] [PubMed]

15. Paul, W. Band structure of intermetallic semiconductors from pressure experiments. *J. Appl. Phys. Suppl.* **1961**, *32*, 2092–2095. [CrossRef]

16. Jayaraman, A. Diamond anvil cell and high-pressure physical investigations. *Rev. Mod. Phys.* **1983**, *55*, 65–108. [CrossRef]

17. Forman, R.A.; Piermarini, G.J.; Barnett, J.D.; Block, S. Pressure Measurement Made by the Utilization of Ruby Sharp-Line Luminescence. *Science* **1972**, *176*, 284–285. [CrossRef] [PubMed]

18. Barnett, J.D.; Block, S.; Piermarini, G.J. An Optical Fluorescence System for Quantitative Pressure Measurement in the Diamond-Anvil Cell. *Rev. Sci. Instrum.* **1973**, *44*, 1–9. [CrossRef]

19. Goñi, A.R.; Syassen, K. Optical Properties of Semiconductors under pressure. In *Semiconductors and Semimetals*; Suskiy, T., Paul, W., Eds.; Academic Press: New York, NY, USA, 1998; Volume 54, pp. 247–425.

20. Nelmes, R.J.; McMahon, M.I. Structural transitions in the group IV, III_V and II-VI semiconsuctors under pressure. In *Semiconductors and Semimetals*; Suskiy, T., Paul, W., Eds.; Academic Press: New York, NY, USA, 1998; Volume 54, pp. 145–246.

21. Mujica, A.; Rubio, A.; Munoz, A.; Needs, R.J. High-pressure phases of group-IV, III–V, and II–VI compounds. *Rev. Mod. Phys.* **2003**, *75*, 863–912. [CrossRef]

22. Likforman, A.; Carre, D.; Etienne, J.; Bachet, B. Crystal-structure of indium monoselenide (InSe). *Acta Crystallogr. Sect. B Struct. Crystallogr. Cryst. Chem.* **1975**, *31*, 1252–1254. [CrossRef]

23. Polian, A.; Kunc, K.; Kuhn, A. Low-frequency lattice-vibrations of δ-GaSe compared to epsilon-polytypes and gamma-polytypes. *Solid State Commun.* **1976**, *19*, 1079–1082. [CrossRef]

24. Gatulle, M.; Fischer, M.; Chevy, A. Elastic-constants of the layered compounds GaS, GaSe, InSe, and their pressure-dependence. 1. Experimental part. *Phys. Status Solidi B* **1983**, *119*, 327–336. [CrossRef]

25. Kuroda, N.; Ueno, O.; Nishina, Y. Supernonlinear shifts of optical-energy gaps in InSe and GaSe under hydrostatic-pressure. *J. Phys. Soc. Jpn.* **1986**, *55*, 581–589. [CrossRef]

26. Murnaghan, F.D. The compressibility of media under extreme pressures. *Proc. Natl. Acad. Sci. USA* **1944**, *30*, 244–247. [CrossRef] [PubMed]

27. Schwarz, U. Zustandgleichungen unf Phasenumwaldungen von Halbleitenden Verbindungen unter Hohem Druck: Rötgenpulveruntersuchungen und Optische Spektroskopie, Ph.D. Thesis, Technishen Hochschule Darmstadt, Darmstadt, Germany, 1992.

28. Schwarz, U.; Goñi, A.R.; Syassen, K.; Cantarero, A.; Chevy, A. Structural and optical properties of InSe under pressure. *High Press. Res.* **1991**, *8*, 396–398. [CrossRef]

29. Pellicer-Porres, J.; Machado-Charry, E.; Segura, A.; Gilliland, S.; Canadell, E.; Ordejon, P.; Polian, A.; Munsch, P.; Chevy, A.; Guignot, N. GaS and InSe equations of state from single crystal diffraction. *Phys. Status Solidi B* **2007**, *244*, 169–173. [CrossRef]

30. Errandonea, D.; Martínez-García, D.; Segura, A.; Haines, J.; Machado-Charry, E.; Canadell, E.; Chervin, J.C.; Chevy, A. High-pressure electronic structure and phase transitions in monoclinic InSe: X-ray diffraction, Raman spectroscopy, and density functional theory. *Phys. Rev. B* **2008**, *77*, 045208. [CrossRef]

31. Pellicer-Porres, J.; Segura, A.; Muñoz, V.; San Miguel, A. High-pressure x-ray absorption study of InSe. *Phys. Rev. B.* **1999**, *50*, 3757–5763. [CrossRef]

32. Schwarz, U.; Olguin, D.; Cantarero, A.; Hanfland, M.; Syassen, K. Effect of pressure on the structural properties and electronic band structure of GaSe. *Phys. Status Solidi B* **2007**, *244*, 244–255. [CrossRef]

33. Schwarz, U.; Syassen, K.; Kniep, R. Structural phase-transition of GaTe at high-pressure. *J. Alloy. Compd.* **1995**, *224*, 212–216. [CrossRef]

34. Takumi, M.; Hirata, A.; Ueda, T.; Koshio, Y.; Nishimura, H.; Nagata, K. Structural Phase Transitions of Ga$_2$Se$_3$ and GaSe under High Pressure. *Phys. Status Solidi B* **2001**, *223*, 423–426. [CrossRef]

35. Pellicer-Porres, J.; Segura, A.; Munoz, V.; San Miguel, A. High-pressure X-ray absorption study of GaTe including polarization. *Phys. Rev. B* **2000**, *61*, 125–131. [CrossRef]

36. Pellicer-Porres, J.; Segura, A.; Ferrer, C.; Munoz, V.; San Miguel, A.; Polian, A.; Itie, J.P.; Gauthier, M.; Pascarelli, S. High-pressure X-ray-absorption study of GaSe. *Phys. Rev. B* **2002**, *65*, 174103. [CrossRef]

37. Segura, A.; Manjón, F.J.; Errandonea, D.; Pellicer-Porres, J.; Muñoz, V.; Tobias, G.; Ordejón, P.; Canadell, E.; San Miguel, A.; Sánchez-Portal, D. Specific features of the electronic structure of III−VI layered semiconductors: Recent results on structural and optical measurements under pressure and electronic structure calculations. *Phys. Status Solidi B* **2003**, *235*, 267–276. [CrossRef]

38. Vezzoli, G.C. Synthesis and properties of a pressure-induced and temperature-induced phase of indium selenide. *Mater. Res. Bull.* **1971**, *6*, 1201–1204. [CrossRef]

39. Iwasaki, H.; Watanabe, Y.; Kuroda, N.; Nishina, Y. Pressure-induced layer-nonlayer transformation in InSe. *Physica B C* **1981**, *105B*, 314–318. [CrossRef]

40. Kuroda, N.; Nishina, Y.; Iwasaki, H.; Watanabe, Y. Raman scatterings of layered and non-layered phases of InSe. *Solid State Commun.* **1981**, *38*, 139–142. [CrossRef]

41. Errandonea, D.; Martínez-García, D.; Segura, A.; Chevy, A.; Tobias, G.; Canadell, E.; Ordejon, P. High-pressure, high-temperature phase diagram of InSe: A comprehensive study of the electronic and structural properties of the monoclinic phase of InSe under high pressure. *Phys. Rev. B.* **2006**, *73*, 235202. [CrossRef]

42. Ferlat, G.; Martinez-Garcia, D.; San Miguel, A.; Aouizerat, A.; Muñoz-Sanjosé, V. High pressure-high temperature phase diagram of InSe. *High Press. Res.* **2004**, *4*, 111–116. [CrossRef]

43. Besson, J.M.; Nelmes, R.J.; Hamel, G.; Loveday, J.S.; Weill, G.; Hull, S. Neutron Powder Diffraction above 10-GPa. *Physica B* **1992**, *180*, 907–910. [CrossRef]

44. Segura, A.; Pellicer-Porres, J.; Martinez-García, D.; Rodríguez-Hernández, P.; Muñoz, A.; Itié, J.P. CsCl-InSe as End-Phase of Metastable Rock-Salt and Tetragonal InSe, Book of Abstracts of the Joint AIRAPT-25 and EHPRG-53 (Abstract P2-67). In Proceedings of the Joint AIRAPT-25 and EHPRG-53, Madrid, Spain, 30 August–4 September 2015.

45. Camassel, J.; Merle, P.; Mathieu, H.; Chevy, A. Excitonic absorption-edge of indium selenide. *Phys. Rev. B* **1978**, *17*, 4718–4725. [CrossRef]

46. Le Toullec, R.; Piccioli, N.; Chervin, J.C. Optical properties of the band-edge exciton in GaSe crystals at 10 K. *Phys. Rev. B* **1980**, *22*, 6162–6170. [CrossRef]

47. Mejatty, M.; Segura, A.; Letoullec, R.; Besson, J.M.; Chevy, A.; Fair, H. Optical absorption edge of GaS under hydrostatic pressure. *J. Phys. Chem. Solids* **1978**, *39*, 25–28. [CrossRef]

48. Besson, J.M.; Jain, K.P.; Kuhn, A. Optical-absorption edge in GaSe under hydrostatic-pressure. *Phys. Rev. Lett.* **1974**, *32*, 936–939. [CrossRef]

49. Segura, A. Phototransport Dans InSe: Application à La Conversion Photovoltaique de L'énergie Solaire. Thèse de Trosième Cycle, Univerité Pierre et Marie Curie, Paris, France, 1977.

50. Panfilov, V.V.; Subbotin, S.I.; Vereshchagin, L.F.; Ivanov, I.I.; Molchanova, R.T. Exciton absorption, band-structure, and phase-transformation of GaSe under pressure. *Phys. Status Solidi B* **1975**, *72*, 823–831. [CrossRef]

51. Bourdon, A.; Chevy, A.; Besson, L. Band Structure of Indium Selenide: Physics of Semiconductors 1978. In Proceedings of the 14th International Conference on the Physics of Semiconductors, Edinburgh, Scotland, 4–8 September 1978; Wilson, H., Ed.; Institute of Physics and Physical Society: London, UK, 1979; pp. 1371–1374.

52. Gomes da Costa, P.; Dandrea, R.G.; Wallis, R.F.; Balkanski, M. First principles of the electronic structure study of γ-InSe and β-InSe. *Phys. Rev. B* **1993**, *48*, 14135–14141. [CrossRef]

53. Kress-Rogers, E.; Nicholas, R.J.; Portal, J.C.; Chevy, A. Cyclotron-resonance studies on bulk and two-dimensional conduction electrons in InSe. *Solid State Commun.* **1982**, *44*, 379–383. [CrossRef]

54. Ferrer-Roca, C.; Segura, A.; Andres, M.V.; Pellicer, J.; Munoz, V. Investigation of nitrogen-related acceptor centers in indium selenide by means of photoluminescence: Determination of the hole effective mass. *Phys. Rev. B* **1997**, *55*, 6981–6987. [CrossRef]

55. Kane, E.O. Band structure of indium antimonide. *J. Phys. Chem. Solids* **1957**, *1*, 249–261. [CrossRef]

56. Piccioli, N.; Le Toullec, R.; Piccioli, N.; Chervin, J.C. Constantes optiques de InSe entre 10,500 cm^{-1} (1.30 eV) et 22,500 cm^{-1} (2.78 eV). *J. Physique* **1981**, *42*, 1129–1135. [CrossRef]

57. Gauthier, M.; Polian, A.; Besson, J.M.; Chevy, A. Optical-properties of gallium selenide under high-pressure. *Phys. Rev. B* **1989**, *40*, 3837–3854. [CrossRef]

58. Elliott, R.J. Intensity of optical absorption by excitons. *Phys. Rev.* **1957**, 108. [CrossRef]

59. Toyozawa, Y. Theory of line-shapes of the exciton absorption bands. *Prog. Theor. Phys.* **1958**, *20*, 53–81. [CrossRef]

60. Piccioli, N.; Letoullec, R. Exciton-phonon interaction in GaDe. *J. Phys.* **1989**, *50*, 3395–3406. [CrossRef]

61. Goi, A.R.; Cantarero, A.; Schwarz, U.; Syassen, K.; Chevy, A. Low temperature exciton absorption in InSe under pressure. *Phys. Rev. B* **1992**, *45*, 4221–4226. [CrossRef]

62. Ulrich, C.; Olguin, D.; Cantarero, A.; Goñi, A.R.; Syassen, K.; Chevy, A. Effect of pressure on direct optical transitions of gamma-InSe. *Phys. Status Solidi B* **2000**, *221*, 777–787. [CrossRef]

63. Errandonea, D.; Manjon, F.J.; Pellicer, J.; Segura, A.; Munoz, V. Direct to indirect crossover in III–VI layered compounds and alloys under pressure. *Phys. Status Solidi B* **1999**, *211*, 33–38. [CrossRef]

64. Manjón, F.J.; Errandonea, D.; Segura, A.; Muñoz, V.; Tobías, G.; Ordejón, P.; Canadell, E. Experimental and theoretical study of band structure of InSe and In$_{1-x}$Ga$_x$Se (x < 0.2) under high pressure: Direct to indirect crossovers. *Phys. Rev. B* **2001**, *63*, 125330. [CrossRef]

65. Ferlat, G.; Xu, H.; Timoshevskii, V.; Blasé, X. Ab initio studies of structural and electronic properties of solid indium selenide under pressure. *Phys. Rev. B* **2002**, *66*, 085210. [CrossRef]

66. Olguín, D.; Cantarero, A.; Ulrich, C.; Syassen, K. Effect of pressure on structural properties and energy band gaps of γ-InSe. *Phys. Status Solidi B* **2003**, *235*, 456–463. [CrossRef]

67. Manjón, F.J.; Segura, A.; Muñoz-Sanjosé, V.; Tobías, G.; Ordejón, P.; Canadell, E. Band structure of indium selenide investigated by intrinsic photoluminescence under high pressure. *Phys. Rev. B* **2004**, *70*, 125201. [CrossRef]

68. Millot, M.; Broto, J.M.; George, S.; Gonzalez, J.; Segura, A. High pressure and high magnetic field behavior of free and donor-bound-exciton photoluminescence in InSe. *Phys. Status Solidi B* **2009**, *246*, 532–535. [CrossRef]

69. Millot, M.; Broto, J.M.; George, S.; Gonzalez, J.; Segura, A. Electronic structure of indium selenide probed by magneto-absorption spectroscopy under high pressure. *Phys. Rev. B* **2010**, *81*, 205211. [CrossRef]

70. Ghalouci, L.; Taibi, F.; Ghalouci, F.; Bensaid, M.O. Ab initio investigation into structural, mechanical and electronic properties of low pressure, high pressure and high pressure-high temperature phases of Indium Selenide. *Comput. Mater. Sci.* **2016**, *124*, 62–77. [CrossRef]

71. Polian, A.; Chervin, J.C.; Besson, J.M. Phonon modes and stability of GaS up to 200 kilobars. *Phys. Rev. B* **1980**, *22*, 3049–3058. [CrossRef]

72. Kuroda, N.; Ueno, O.; Nishina, Y. Lattice-dynamic and photoelastic properties of GaSe under high-pressures studied by raman-scattering and electronic susceptibility. *Phys. Rev. B* **1987**, *35*, 3860–3870. [CrossRef]

73. Allahverdi, K.; Babaev, S.; Ellialtioğlu, Ş.; Ismailov, A. Raman scattering in layer indium selenide under pressure. *Solid State Commun.* **1993**, *87*, 675–678. [CrossRef]

74. Allakhverdiev, K.; Ellialtioglu, S.; Ismailov, Z. Raman scattering and Hall effect in layer InSe under pressure. *High Press. Res.* **1994**, *13*, 121–125. [CrossRef]

75. Ulrich, C.; Mroginski, M.A.; Goñi, A.; Cantarero, A.; Schwarz, U.; Muñoz, V.; Syassen, K. Vibrational Properties of InSe under Pressure: Experiment and Theory. *Phys. Status Solidi B* **1996**, *198*, 121–127. [CrossRef]

76. Choi, I.H.; Yu, P.Y. Pressure dependence of phonons and excitons in InSe films prepared by metal-organic chemical vapor deposition. *Phys. Rev. B.* **2003**, *68*, 165339. [CrossRef]

77. Gauthier, M. Pressure-induced charge transfer. *High Press. Res.* **1992**, 9330–9342. [CrossRef]

78. Rushchanskii, K.Z. The influence of hydrostatic pressure on the static and dynamic properties of an InSe crystal: A first-principles study. *Phys. Solid State* **2006**, *46*, 179–187. [CrossRef]

79. Polian, A.; Besson, J.M.; Grimsditch, M.; Vogt, H. Elastic properties of GaS under high-pressure by brillouin-scattering. *Phys. Rev. B* **1982**, *25*, 2767–2775. [CrossRef]

80. Manjón, F.J. Estudio de la estructura de bandas del seleniuro de indio mediante medidas ópticas bajo presión hidrostática. Tesis Doctoral, Universitat de València, Valencia, Spain, 1999.

81. Manjon, F.J.; Van der Vijver, Y.; Segura, A.; Muñoz, V. Pressure dependence of the refractive index in InSe. *Semicond. Sci. Technol.* **2000**, *15*, 806–812. [CrossRef]

82. Van Vechten, J.A. Quantum dielectric theory of electronegativity in covalent systems. I. electronic dielectric constant *Phys. Rev.* **1969**, *182*, 891–905. [CrossRef]

83. Phillips, J.C. Ionicity of chemical bond in crystals. *Rev. Mod. Phys.* **1970**, *42*, 317–356. [CrossRef]

84. Penn, D.R. Wave-number-dependent dielectric function of semiconductors. *Phys. Rev.* **1962**, *128*, 2093–2097. [CrossRef]

85. Segura, A.; Chevy, A. Large increase of the low-frequency dielectric-constant of gallium sulfide under hydrostatic-pressure. *Phys. Rev.* **1994**, *49*, 4601–4604. [CrossRef]

86. Errandonea, D.; Segura, A.; Munoz, V.; Chevy, A. Pressure dependence of the low-frequency dielectric constant in III-VI semiconductors. *Phys. Status Solidi B* **1999**, *211*, 201–206. [CrossRef]

87. Errandonea, D.; Segura, A.; Muñoz, V.; Chevy, A. Effects of pressure and temperature on the dielectric constant of GaS, GaSe, and InSe: Role of the electronic contribution. *Phys. Rev. B* **1999**, *60*, 15866–15874. [CrossRef]

88. D'Amour, H.; Holzapfel, W.B.; Polian, A.; Chevy, A. Crystal-structure of a new high-pressure polymorph of gas. *Solid State Commun.* **1982**, *44*, 853–855. [CrossRef]

89. Errandonea, D.; Segura, A.; Sanchez-Royo, J.F.; Muñoz, V.; Ulrich, C.; Grima, P.; Chevy, A. Effects of Conduction Band Structure and Dimensionality of the Electron Gas on Transport Properties of InSe under Pressure. *Phys. Status Solidi B* **1996**, *198*, 129–134. [CrossRef]

90. Errandonea, D.; Segura, A.; Sanchez-Royo, J.F.; Muñoz, V.; Grima, P.; Chevy, A.; Ulrich, C. Investigation of conduction-band structure, electron-scattering mechanisms, and phase transitions in indium selenide by means of transport measurements under pressure. *Phys. Rev. B* **1997**, *55*, 16217–16225. [CrossRef]

91. Errandonea, D.; Segura, A.; Manjón, F.J.; Chevy, A. Transport measurements in InSe under high pressure and high temperature: Shallow-to-deep donor transformation of Sn related donor impurities. *Semicond. Sci. Technol.* **2003**, *18*, 241–246. [CrossRef]

92. Manjon, F.J.; Errandonea, D.; Segura, A.; Chervin, J.C.; Muñoz, V. Precursor effects of the rhombohedral-to-cubic phase transition in indium selenide. *High Press. Res.* **2002**, *22*, 261. [CrossRef]

93. Errandonea, D.; Sanchez-Royo, J.F.; Segura, A.; Chevy, A.; Roa, L. Investigation of acceptor levels and hole scattering mechanisms in *p*-gallium selenide by means of transport measurement under pressure. *High Press. Res.* **1998**, *16*, 13–26. [CrossRef]

94. Errandonea, D.; Segura, A.; Manjón, F.J.; Chevy, A.; Machado, E.; Tobias, G.; Ordejón, P.; Canadell, E. Crystal symmetry and pressure effects on the valence band structure of γ-InSe and γ-GaSe: Transport measurements and electronic structure calculations. *Phys. Rev. B* **2005**, *71*, 125206. [CrossRef]

95. Errandonea, D.; Martínez-García, D.; Segura, A.; Ruiz-Fuertes, J.; Lacomba-Perales, R.; Fages, V.; Chevy, A.; Roa, L.; Muñoz-San José, V. High-pressure electrical transport measurements on p-type GaSe and InSe. *High Press. Res.* **2006**, *26*, 513–518. [CrossRef]

96. Segura, A.; Errandonea, D.; Martínez-García, D.; Manjón, F.J.; Chevy, A.; Tobias, G.; Ordejón, P.; Canadell, E. Transport measurements under pressure in III–IV layered semiconductors. *Phys. Status Solidi B* **2007**, *244*, 162–168. [CrossRef]

97. Rydberg, H.; Dion, M.; Jacobson, N.; Schroder, E.; Hyldgaard, P.; Simak, S.I.; Langreth, D.C.; Lundqvist, B.I. Van der Waals density functional for layered structures. *Phys. Rev. Lett.* **2003**, *91*, 126402. [CrossRef] [PubMed]

98. Cazorla, C.; Boronat, J. Simulation and understanding of atomic and molecular quantum crystals. *Rev. Mod. Phys.* **2017**, *89*, 035003. [CrossRef]

99. Tawfik, S.A.; Gould, T.; Stampfl, C.; Ford, M.J. Evaluation of van der Waals density functionals for layered materials. *Phys. Rev. Mater.* **2018**, *2*, 034005. [CrossRef]

100. Martinez-Pastor, J.; Segura, A.; Julien, C.; Chevy, A. Shallow-donor impurities in indium selenide investigated by means of far-infrared spectroscopy. *Phys. Rev. B* **1992**, *46*, 4607–4616. [CrossRef]

crystals

MDPI

Article

The Structure of Ferroselite, $FeSe_2$, at Pressures up to 46 GPa and Temperatures down to 50 K: A Single-Crystal Micro-Diffraction Analysis

Barbara Lavina [1,*], **Robert T. Downs** [2] and **Stanislav Sinogeikin** [3]

[1]   Center for High Pressure Science & Technology Advanced Research (HPSTAR), Beijing 100094, China
[2]   Department of Geosciences, University of Arizona Tucson, Tucson, AZ 85721-0077, USA;
      rdowns@u.arizona.edu
[3]   High Pressure Collaborative Access Team, Geophysical Laboratory, Carnegie Institute of Washington,
      Argonne, IL 60439, USA; ssinogeikin@carnegiescience.edu
*   Correspondence: barbara.lavina@icloud.com

Received: 11 June 2018; Accepted: 6 July 2018; Published: 13 July 2018

Abstract: We conducted an in situ crystal structure analysis of ferroselite at non-ambient conditions. The aim is to provide a solid ground to further the understanding of the properties of this material in a broad range of conditions. Ferroselite, marcasite-type $FeSe_2$, was studied under high pressures up to 46 GPa and low temperatures, down to 50 K using single-crystal microdiffraction techniques. High pressures and low temperatures were generated using a diamond anvil cell and a cryostat respectively. We found no evidences of structural instability in the explored P-T space. The deformation of the orthorhombic lattice is slightly anisotropic. As expected, the compressibility of the Se-Se dumbbell, the longer bond in the structure, is larger than that of the Fe-Se bonds. There are two octahedral Fe-Se bonds, the short bond, with multiplicity two, is slightly more compressible than the long bond, with multiplicity four; as a consequence the octahedral tetragonal compression slightly increases under pressure. We also achieved a robust structural analysis of ferroselite at low temperature in the diamond anvil cell. Structural changes upon temperature decrease are small but qualitatively similar to those produced by pressure.

Keywords: $FeSe_2$; high pressure; low temperature; single-crystal diffraction

1. Introduction

Iron selenides form economically important ore deposits and are relevant to the geochemical cycle of chalcogenides. In material science, dichalcogenides are extensively explored for solar energy applications because of their suitable thermoelectric and optical properties along with their availability and low toxicity [1–3]. Compounds with the marcasite crystal structure display a variety of intriguing physical properties intimately related to their structural arrangements [4]. Furthermore, the marcasite structure type is adopted by several interesting high-pressure phases such as Fe, Rh and Os pernitrides [5–7].

Ferroselite is a mineral of the chalcogenide series with end-member composition $FeSe_2$ and with the marcasite-type crystal structure. Ferroselite is the stable phase of iron diselenide at ambient conditions. Upon heating at ambient pressure, $FeSe_2$ does not show phase transitions until its decomposition at 850 K [8]; upon heating at 1200 K under moderately high pressure (65 Kbar) iron diselenide adopts the pyrite structure type [9,10]. Iron is bonded to six selenium atoms in ferroselite, whereas selenium forms a monatomic bond and 3 bonds with iron (Figure 1). The $FeSe_6$ edge-sharing octahedra form chains along the c-direction, while the Se-Se dumbbell bond lies in the ab plane connecting octahedral chains. In the marcasite-type structure, symmetry constrains impose some degree of distortion to the coordination geometries. The octahedron shows angular distortion and features two different bond lengths, in ferroselite those with multiplicity two are slightly shorter than the four equatorial bonds. Iron is located in $2a$ with all symmetry-constrained coordinated whereas selenium, located in $4g$, shows variable coordinates x and y.

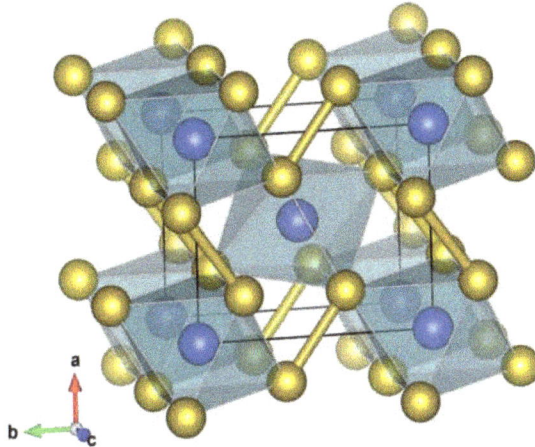

Figure 1. Representation of the crystal structure of marcasite-type $FeSe_2$ generated with VESTA [11]. Iron atoms (blue) are coordinated to 6 selenium atoms (yellow) defining a tetragonally compressed octahedron. Selenium dumbbells lie in the ab plane and connect the chains of octahedra running parallel to the c-axis.

The thermal expansion of the ferroselite lattice has been explored from ambient conditions up to decomposition temperatures [8], while magnetism and electrical properties were explored in a broad range of temperatures [2]. Mechanically alloyed ferroselite nanocrystalline materials were studied under pressure via absorption spectroscopy up to 19 GPa, showing no evidence of phase transitions [12]. The elastic properties of $FeSe_2$ were recently determined by means of first-principles calculations [13].

To study the bulk and atomic response of $FeSe_2$ to external high pressure and to low temperature we performed synchrotron single-crystal microdiffraction experiments using a diamond anvil cell and a cryostat to generate target conditions. We performed three different experiments: (i) compression at ambient temperature up to ~46 GPa; (ii) cooling down to 50 K at ambient pressure; (iii) a combined high-pressure low-temperature experiment down to 110 K at 3.8 GPa. Both low-temperature experiments were conducted with single crystals loaded in a diamond anvil cell (DAC) contained in a cryostat (hereafter DAC and cryostat).

2. Materials and Methods

The specimen investigated in this study is a mineral from Paradox Valley, Uravan District, Montrose County, Colorado, USA obtained from the RRUFF collection (RRUFF.info/R070461). The composition reported in the RRUFF database [14] was determined via electron microprobe analysis. Measured elements were Se, Fe, Pb, S, Zn, Cu, Ag; within experimental resolution, the sample is pure and stoichiometric.

Conditions of high pressure and of low temperature were generated with a 4-post diamond anvil cell (DAC) and with a liquid-flow helium cryostat (Figure 2). The DAC was equipped with conical diamonds anvils [15] of 85° aperture and 0.3 and 0.6 mm culet diameter for high pressure and low temperature experiments respectively. Gaskets were fabricated from pure Re or W foils; 160 and

360 μm diameter holes in the center of 35 μm thick indentations provided the sample chambers for the high-pressure and the low-temperature experiments respectively. The sample chambers were filled with pre-pressurize neon in order to maintain quasi-hydrostatic stress on the crystals in the whole range of experimental conditions.

X-ray microdiffraction data (Figure S1) were collected at the insertion device station 16ID-B of HPCAT, Sector 16, Advanced Photon Source, Argonne National Laboratory. Experiments were performed using hard X-rays (λ = 0.40662, 0.36793 Å) focused to about 5 × 5 μm FWHM at the sample position. The experimental station was equipped with a heavy-duty motorized sample stage suitable for measurements with a cryostat for high pressure studies. The cryostat provided wide angular X-ray access and was equipped with a gas membrane for pressure control. Diffracted X-rays were collected with the rotation method (vertical ω axis) using a MAR165-CCD area detector (marXperts, Norderstedt, Schleswig-Holstein, Germany) and a MAR345IP detector (marXperts, Norderstedt, Schleswig-Holstein, Germany), both were calibrated using powder patterns of CeO_2 standard and the GSASII software [16]. The rotation range was 72° for high-pressure data and 60° for low-temperature data, the X-ray access was reduced by the gas membrane and cryostat body in the latter case. For the samples in the cryostat diffraction images were collected with the detector (CCD) at three different locations in the horizontal direction perpendicular to the beam. The maximum resolution achieved was 0.6 Å.

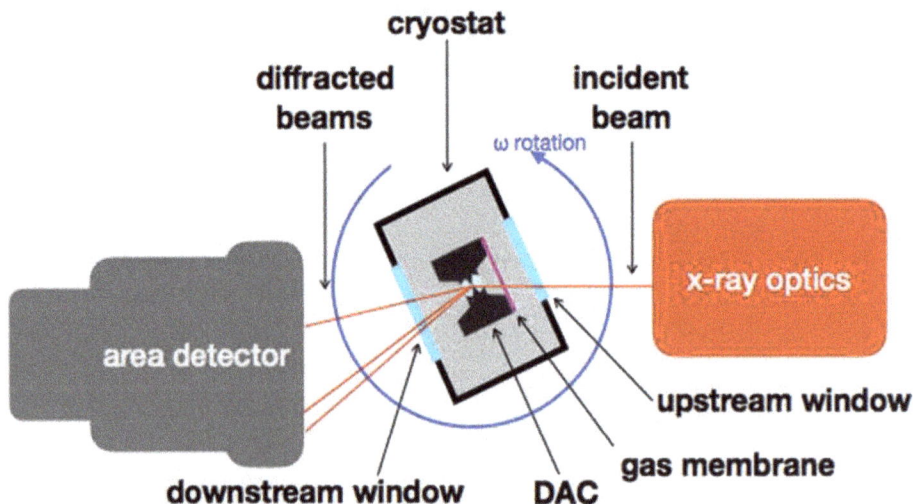

Figure 2. Schematic representation of the high-pressure low-temperature setup at beamline 16ID-B of the APS, ANL.

Pressure was calibrated using the equation of state of platinum [17] for ambient temperature data and ruby [18] for low-temperature data. Temperature was measured using silicon diode temperature sensors positioned on the DAC body and on the copper block, the two differ by less than two degrees during data collection. Data reduction was performed GSE_ADA & RSV [19], WinGX [20], and DIOPTAS [21]. Structural refinements were carried out using Shelxl [22]. Standard powder patterns were analyzed with GSASII [16].

3. Results and Discussion

3.1. Ambient Conditions Structural Refinements

Measuring structure factors in the diamond anvil cell with a micron-sized X-ray beam is an established technique providing extremely valuable and robust results in spite of its challenges.

In addition to high-pressure measurements, here we collected data from small single crystals loaded in the DAC which was then loaded in a cryostat (Figure 2). The cryostat we used for high pressure measurements is a relatively bulky device with several connections, the most cumbersome being the liquid helium supply line and the vacuum line. These connections caused an increase in the sphere of confusion of the rotation axis. Because crystals are roughly 20 μm in diameter and the beam is around 5 μm FWHM, we anticipated that the X-ray flux on the crystals would vary more dramatically during data collection compared with ambient temperature measurements. As a consequence, the set of observed structure factors would not be in scale. Furthermore, for relatively low-symmetry crystals, a reliable empirical correction calculated from comparing sets of equivalent reflections might be difficult to define due to the low data redundancy. To compensate for such effects we loaded three crystals with different crystallographic orientations in the diamond anvil cell for low temperature work (Table S1). Furthermore, we collected several rotation images in a small grid pattern around the crystal center for a few datapoints. In Table 1 literature data are compared with some of our results. All data collections were performed with samples in environmental cells, DAC and DAC and cryostat, before conditions were changed. In addition to the scale factor, we refined the two symmetry-unconstrained fractional coordinates of Se and isotropic displacement parameters for both atoms for a total of 5 variables. Overall results are in good agreement. We note that: (i) because we always measured a good number of reflections and we only had 2 positional parameters to refine for the heavier element, these were always reasonable, including high R factors refinements; (ii) merging grid diffraction patterns provided in most cases excellent results and low disagreements between equivalent reflections and better results than empirical corrections; (iii) datasets are not uniform because, we infer, the increase in the sphere of confusion caused by the cryostat is not fully reproducible. In conclusion, it appears beneficial to acquire redundant datasets, this allows for adopting different data reduction strategies and provides the best likelihood that robust refinements can be obtained.

Table 1. Ambient-conditions unit-cell parameters and atomic fractional coordinates from the literature and from this work (see text).

Sample	N_{all}, N_{ind}	a (Å)	b (Å)	c (Å)	R_{eq}, R_1 (%)	x (Se)	y (Se)	Ueq-Fe (Å²)	Ueq-Se (Å²)
Ref. [23]		4.8002(4)	5.7823(5)	3.5834(4)		0.2127(6)	0.3701(5)		
Ref. [24]		4.804(2)	5.784(2)	3.586(2)		0.2134(2)	0.3690(1)		
RRUFF [a]		4.795(3)	5.777(4)	3.584(1)					
Ref. [2]		4.8031(6)	5.7849(2)	3.5840(4)		0.2127(2)	0.3691(7)		
Measurement in the DAC									
DAC	204, 98	4.801(3)	5.787(2)	3.5859(7)	12, 4.9	0.213593)	0.3692(2)	0.0095(5)	0.0099(5)
Measurement in the DAC and cryostat									
C1 [b]	205, 82	4.804(4)	5.781(3)	3.5814(7)	8.4, 4.4	0.2133(3)	0.3692(1)	0.0056(5)	0.0059(4)
C2 [c]	253,103	4.8016(9)	5.777(1)	3.5850(8)	9.6, 5.4	0.2137(2)	0.36921(13)	0.0090(5)	0.0090(4)
C3 [b]	214,94	4.799(1)	5.785(2)	3.5820(4)	33, 12	0.2130(7)	0.3699(3)	0.0067(12)	0.0086(10)

[a] same specimen of the present work RRUFF.info/R070461; [b] no grid scan, no empirical corrections; [c] three detector positions, *hkl* from merged grid scans, no empirical corrections.

3.2. High Pressure

Diffraction data of ferroselite were collected up to 46 GPa (Table 2). There are no indications of phase transitions but the highest pressure pattern shows moderate peak broadening that might be the result of non-hydrostatic stress as well as the manifestation of an incipient phase transition. The datapoint at 3.63 GPa was collected with the DAC and cryostat before decreasing temperature, structural parameters at this pressure are in the same trend of data collected in the DAC, confirming our ability to collect full datasets in these conditions without introducing systematic errors in our analysis. The anisotropy of the deformation of the orthorhombic cell of ferroselite under quasi-hydrostatic compression is clear after 17 GPa. As shown in Figure 3 the lattice is more compressible in the direction of the *a*-axis and is stiffer along the *b*-axis. This observation is consistent with the recently predicted behavior [13]. The bulk compression of marcasite-$FeSe_2$ can be modeled by a second-order Birch–Murnaghan EoS with an ambient bulk modulus, K_0, of 121.6 GPa (blue curve in Figure 4). Although we collected too few datasets to reliably fit a third order EoS, it can be inferred from the plot in Figure 4 that a third order EoS might be more appropriate to describe the compressibility of this material, data in the low-pressure range fall below the blue curve whereas the highest pressure datapoint falls above the curve. Also, values significantly larger than 4 for the first derivative of the bulk modulus, K_0', have been suggested for FeS_2 marcasite [25] and several other marcasite-type chalcogenides. Fixing K_0' to 4.6, a value that both leads to best fitting and is close to the value in marcasite, and using the third-order Birch–Murnaghan EoS results in a K_0 of 114.1 GPa. More data ought to be collected in order to better constrain the bulk modulus and its first derivative. We could not however satisfactorily fit our data with the same EoS equation and using the recently proposed bulk modulus of 74.7 GPa [13].

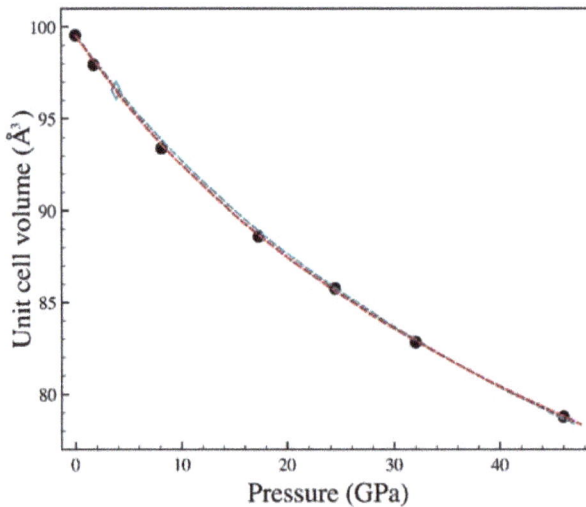

Figure 3. Relative compression of the orthorhombic unit-cell edges.

Table 2. Unit-cell parameters and atomic fractional coordinates of ferroselite at high pressure.

P (GPa)	a (Å)	b (Å)	c (Å)	x (Se)	y (Se)	Ueq-Fe (Å^2)	Ueq-Se (Å^2)
1.7	4.771(2)	5.756(1)	3.5663(6)	0.2129(4)	0.36884(13)	0.0079(6)	0.0083(5)
3.83 [a]	4.751(2)	5.729(2)	3.5467(11)	0.2125(60	0.3687(7)	0.014(2)	0.013(2)
8.1	4.698(3)	5.667(2)	3.5088(7)	0.2123(6)	0.3688(2)	0.0083(8)	0.0090(7)
17.3	4.603(3)	5.581(2)	3.4492(9)	0.2123(3)	0.36854(9)	0.0085(5)	0.0088(5)
24.4	4.555(2)	5.524(12)	3.4090(5)	0.2114(4)	0.3683(2)	0.0066(9)	0.0057(7)
32	4.488(4)	5.471(3)	3.3734(11)	0.2111(6)	0.3676(2)	0.0063(13)	0.0064(12)
46	4.404(3)	5.391(2)	3.3185(9)	0.2103(4)	0.36689(13)	0.0066(9)	0.0070(8)

[a] DAC and cryostat measurement, refinement from merged grid scan images.

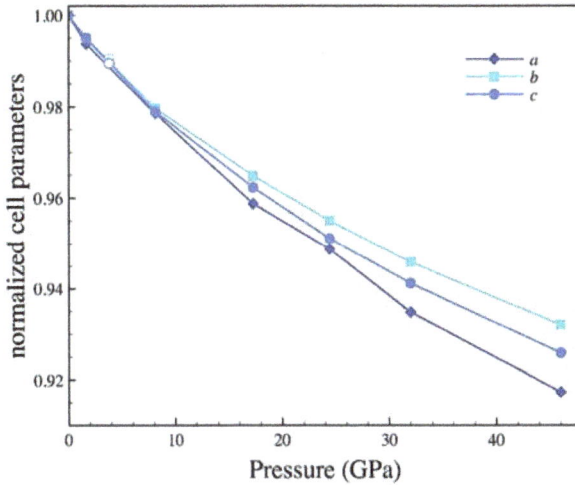

Figure 4. Bulk compressibility of FeSe$_2$. The blue symbol shows data collected at ambient pressure in DAC and cryostat. The blue line shows a second order Birch–Murnaghan EoS fit, the red line a third order Birch–Murnaghan EoS fit (see text).

Robust crystal structure analysis allows for exploring the changes in atomic arrangement with pressure. Figure 5 shows the pressure dependence of interatomic distances and their relative variations. As could be expected, the longest bond, the Se-Se dumbbell, shows the greatest compressibility. The octahedral bonds however show a less obvious behavior, with the shorter bond being more compressible than the longer bond, hence the octahedral distortion increases with pressure.

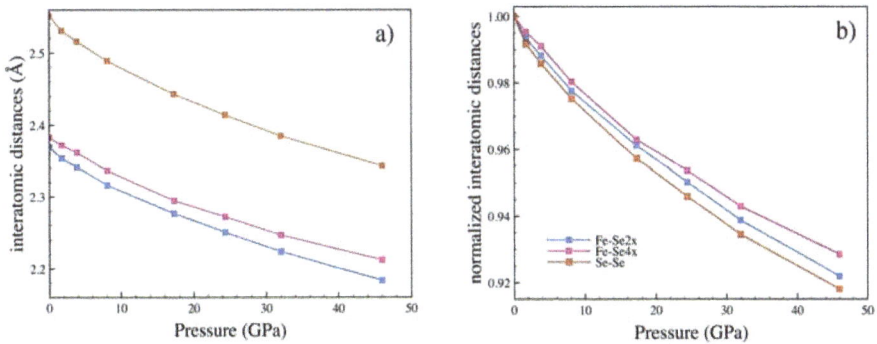

Figure 5. Absolute (**a**) and relative (**b**) interatomic distances as a function of pressure. Uncertainties are smaller than the symbols' sizes.

3.3. Low Temperature

Structural data of FeSe$_2$ were collected ambient pressure at three temperatures 198.2, 148.4, and 50.4 K and at about 3.7 GPa at 197 and 116 K (Table 3). Upon attempting to maintain a constant pressure in the sample chamber during further cooling by increasing the gas membrane pressure, the experiment failed abruptly when the load on the DAC applied with the gas membrane was rapidly transferred to the diamond anvils upon overcoming the DAC friction.

Table 3. Unit-cell parameters and atomic fractional coordinates at low temperature and both ambient pressure and high pressure. For ambient pressure, weighed averages of three crystals are reported.

T (K)	P (GPa)	a (Å)	b (Å)	c (Å)	x (Se)	y (Se)
198.2	10^{-4}	4.7885	5.7782	3.5821	0.2148(4)	0.3692(2)
148.4	10^{-4}	4.7893	5.7775	3.5795	0.2135(2)	0.3691(2)
50.4	10^{-4}	4.7834	5.7767	3.5789	0.2138(3)	0.3694(2)
197	3.68	4.751(2)	5.723(2)	3.5501(11)	0.2131(7)	0.3699(6)
116	3.64	4.7446(11)	5.772(2)	3.5512(8)	0.2130(6)	0.3695(7)

We expected the effect on structural parameters of lowering temperature from 300 to 50 K to be small, close to the resolution of our experiment. Hence we loaded three crystals with different orientations for this experiment in order to obtain a more complete sampling of the reciprocal space and increase data redundancy. Because of the limited access to the reciprocal space different crystallographic directions can be probed with different precision in differently oriented crystals (Table S1) as can be inferred inspecting error bars in Figure 6A–C. As a consequence, unit cell variations of individual crystals hardly show discernible patterns; however trends are appreciable when weighted average are considered (Figure 6, black symbols). As for the high pressure behavior and for the high temperature behavior [8], the greatest lattice parameters variations are observed in the direction of the *a*-axis (Figure 6D). Unit-cell edges variations along the other principal axes are close to uncertainties.

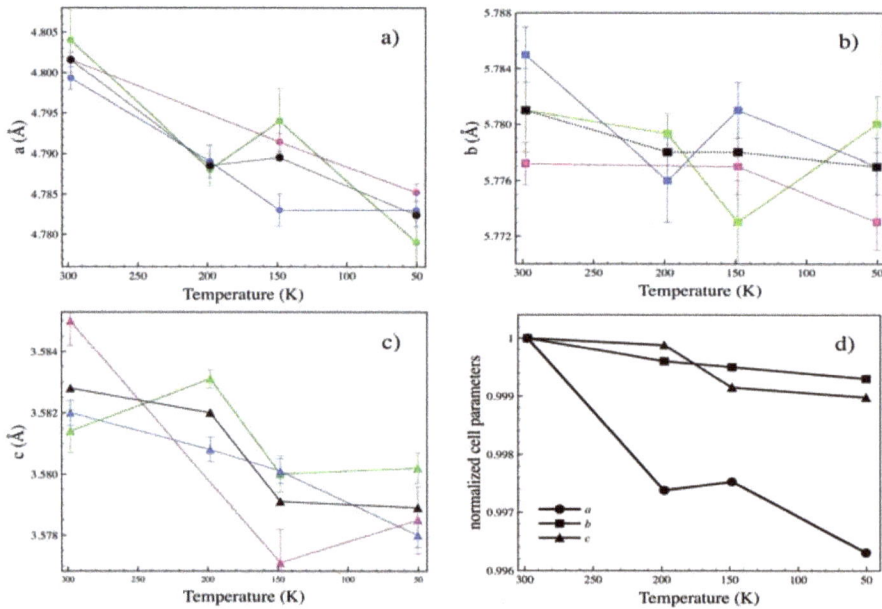

Figure 6. Unit-cell parameters as a function of temperature at ambient pressure. Absolute variations of lattice vectors *a*, *b* and *c* are shown in (**a**), (**b**) and (**c**) respectively; relative variations of the three vectors are shown in (**d**). Red, Blue and green symbols: different crystals in the same sample chamber; black: weighted average of the three crystals.

The anisotropy of the lattice variations observed with temperature decrease is qualitatively similar to that observed with pressure. The unit-cell volume decreases by ~0.6% upon cooling from ambient temperatures down to 50 K (Figure 7, Table 3), a change that corresponds to a pressure of 0.8–0.9 GPa depending on the EoS adopted (see above). The volume-pressure trend is appreciably non linear, and it might in part reflect changes in the physical properties of the material upon cooling [2].

Variations of the Fe-Se bond lengths are also within uncertainties as shown in Figure 8. The Se-Se bond shrinkage is clearer, so it could be argued that variations in the unit-cell length along the *a*-axis and volume are mostly attributed to the monoatomic bond.

We were able to collected just two pressure points at combined low temperature and high pressure (Table 3). Variations on the structure of ferroselite induced by temperature at this pressure are within uncertainties.

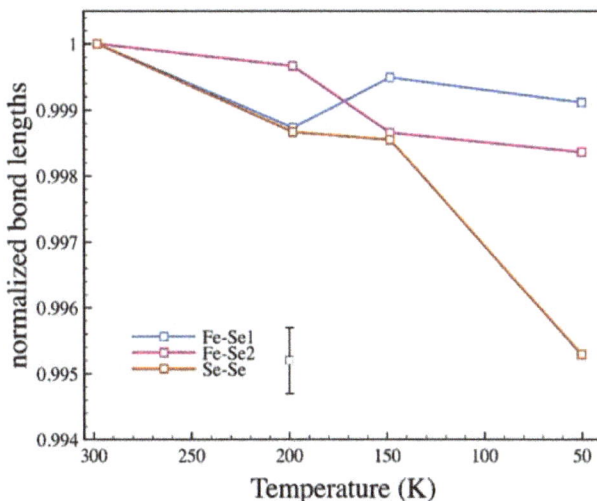

Figure 7. Variation of the unit-cell volume as a function of temperature at ambient pressure. Red, green and blue symbols show values of three different crystals whereas black symbols refer to weighted average values.

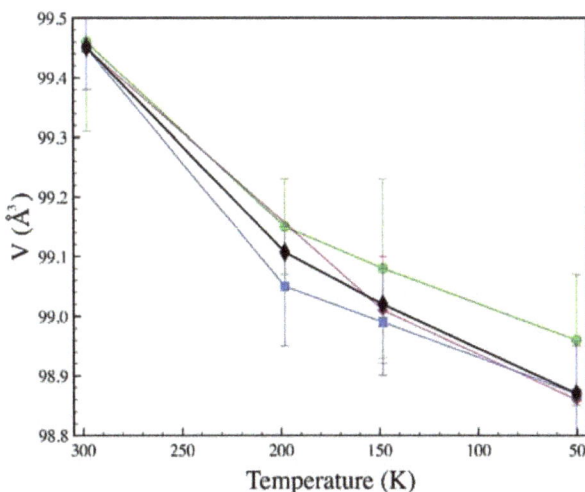

Figure 8. Normalized bond lengths vs temperature at ambient pressure. A representative error bar is shown in black.

4. Conclusions

The physical properties of materials important for critical technologies such as solar energy ought to be defined in great detail. Studying a material's behavior in a broad range of conditions allows more stringent constraints to modeling hence a better general understanding of the material. We conducted a detailed examination of the crystal structure of marcasite-type iron diselenide at pressures up to 46 GPa, temperatures down to 50.4 K and combined high pressure and low temperature conditions of ~3.8 GPa and 197 and 116 K. The phase shows no clear signs of phase transitions in this range, even though it is probably metastable at the highest pressures, considering that it transforms to the pyrite structure above ~6.5 GPa upon heating [9]. We described in details the anisotropy of the lattice response to external conditions and changes in the atomic arrangement.

Maintaining sufficient centering of microcrystals while performing rotation data collections of samples in bulky environmental cells such as combined DAC and cryostat is challenging and not necessarily reproducible. Hence we find that collecting redundant datasets, in this case grid scans, is the safest way to gather datasets from which reliable structure factors can be extracted. Multiple crystal orientations allow obtaining a uniform precision in lattice parameters determinations in addition to better coverage of the reciprocal space.

Supplementary Materials: The following are available online at http://www.mdpi.com/2073-4352/8/7/289/s1, Figure S1: Representative diffraction pattern of ferroselite collected in the DAC, Table S1: Orientation matrices of the three crystals (C1, C2, C3) studied at low temperature.

Author Contributions: Conceptualization, B.L.; Methodology, B.L.; Formal Analysis, B.L.; Investigation, B.L.; Resources, B.L., R.T.D. and S.S.; Writing-Original Draft Preparation, B.L.; Writing-Review & Editing, B.L.; Visualization, B.L.; Project Administration, B.L.; Funding Acquisition, B.L.

Funding: This research was sponsored by the DOE-NNSA under Cooperative Agreement No. DE-FC52-06NA262740. This work was conducted at HPCAT (Sector 16), Advanced Photon Source (APS), Argonne National Laboratory. HPCAT operations are supported by DOE-NNSA under Award No. DE-NA0001974, with partial instrumentation funding by NSF. The Advanced Photon Source is a U.S. Department of Energy (DOE) Office of Science User Facility operated for the DOE Office of Science by Argonne National Laboratory under Contract No. DE-AC02-06CH11357. Use of the COMPRES-GSECARS gas loading system was supported by COMPRES under NSF Cooperative Agreement EAR11-57758 and by GSECARS through NSF Grant EAR-1128799 and DOE Grant DE-FG02- 94ER14466.

Conflicts of Interest: The authors declare no conflict of interest.

References

1. Gudelli, V.K.; Kanchana, V.; Vaitheeswaran, G.; Valsakumar, M.C.; Mahanti, S.D. Thermoelectric properties of marcasite and pyrite FeX_2 (X = Se, Te): A first principle study. *RSC Adv.* **2014**, *4*, 9424–9431. [CrossRef]
2. Li, G.; Zhang, B.; Rao, J.; Gonzalez, D.H.; Blake, G.R.; de Groot, R.A.; Palstra, T.T.M. Effect of vacancies on magnetism, electrical transport, and thermoelectric performance of marcasite $FeSe_{2-\delta}$ (δ = 0.05). *Chem. Mater.* **2015**, *27*, 8220–8229. [CrossRef]
3. Ghosh, A.; Thangavel, R. Electronic structure and optical properties of iron-based chalcogenide FeX_2 (X = S, Se, Te) for photovoltaic applications: A first principle study. *Indian J. Phys.* **2017**, *91*, 1339–1344. [CrossRef]
4. Goodenough, J.B. Energy bands in TX_2 compounds with pyrite, marcasite, and arsenopyrite structures. *J. Solid State Chem.* **1972**, *5*, 144–152. [CrossRef]
5. Wang, Y.X.; Arai, M.; Sasaki, T. Marcasite osmium nitride with high bulk modulus: First-principles calculations. *Appl. Phys. Lett.* **2007**, *90*, 061922. [CrossRef]
6. Niwa, K.; Dzivenko, D.; Suzuki, K.; Riedel, R.; Troyan, I.; Eremets, M.; Hasegawa, M. High pressure synthesis of marcasite-type rhodium pernitride. *Inorg. Chem.* **2014**, *53*, 697–699. [CrossRef] [PubMed]
7. Wang, Z.; Li, Y.; Li, H.; Harran, I.; Jia, M.; Wang, H.; Chen, Y.; Wang, H.; Wu, N. Prediction and characterization of the marcasite phase of iron pernitride under high pressure. *J. Alloys Compd.* **2017**, *702*, 132–137. [CrossRef]
8. Kjekshus, A.; Rakke, T. Compounds with the marcasite type crystal structure. XI. High temperature studies of chalcogenides. *Acta Chem. Scand.* **1975**, *A29*, 443–452. [CrossRef]

9. Bither, T.A.; Prewitt, C.T.; Gillson, J.L.; Bierstedt, P.E.; Flippen, R.B.; Young, H.S. New transition metal dichalcogenides formed at high pressure. *Solid State Commun.* **1966**, *4*, 533–535. [CrossRef]

10. Bither, T.A.; Bouchard, R.J.; Cloud, W.H.; Donohue, P.C.; Siemons, W.J. Transition metal pyrite dichalcogenides. High-pressure synthesis and correlation of properties. *Inorg. Chem.* **1968**, *7*, 2208–2220. [CrossRef]

11. Momma, K.; Izumi, F. *VESTA3* for three-dimensional visualization of crystal, volumetric and morphology data. *J. Appl. Crystallogr.* **2011**, *44*, 1272–1276. [CrossRef]

12. Campos, C.E.M.; de Lima, J.C.; Grandi, T.A.; Machado, K.D.; Itié, J.P.; Polian, A. Pressure-induced effects on the structural properties of iron selenides produced by mechano-synthesis. *J. Phys. Condens. Matter* **2004**, *16*, 8485–8490. [CrossRef]

13. Tian, X.-H.; Zhang, J.-M. The structural, elastic, electronic and optical properties of orthorhombic FeX_2 (X = S, Se, Te). *Superlattices Microstruct.* **2018**, *119*, 201–211. [CrossRef]

14. Lafuente, B.; Downs, R.T.; Yang, H.; Stone, N. The power of databases: The RRUFF project. In *Highlights in Mineralogical Crystallography*; Armbruster, T., Danisi, R.M., Eds.; W. De Gruyter: Berlin, Germany, 2015; pp. 1–30.

15. Boehler, R.; De Hantsetters, K. New anvil designs in diamond-cells. *High Press. Res.* **2004**, *24*, 391–396. [CrossRef]

16. Toby, B.H.; Von Dreele, R.B. GSAS-II: The genesis of a modern open-source all purpose crystallography software package. *J. Appl. Crystallogr.* **2013**, *46*, 544–549. [CrossRef]

17. Matsui, M.; Ito, E.; Katsura, T.; Yamazaki, D.; Yoshino, T.; Yokoyama, A.; Funakoshi, K.I. The temperature-pressure-volume equation of state of platinum. *J. Appl. Phys.* **2009**, *105*, 013505. [CrossRef]

18. Mao, H.; Xu, J.; Bell, P. Calibration of the ruby pressure gauge to 800-Kbar under quasi-hydrostatic conditions. *J. Geophys. Res. Solid Earth Planets* **1986**, *91*, 4673–4676. [CrossRef]

19. Dera, P.; Zhuravlev, K.; Prakapenka, V.; Rivers, M.L.; Finkelstein, G.J.; Grubor-Urosevic, O.; Tschauner, O.; Clark, S.M.; Downs, R.T. High pressure single-crystal micro X-ray diffraction analysis with GSE_ADA and RSV software. *High Press. Res.* **2013**, *33*, 466–484. [CrossRef]

20. Farrugia, L.J. WinGX and ORTEP for Windows: An update. *J. Appl. Crystallogr.* **2012**, *45*, 849–854. [CrossRef]

21. Prescher, C.; Prakapenka, V.B. DIOPTAS: A program for reduction of two-dimensional X-ray diffraction data and data exploration. *High Press. Res.* **2015**, *35*, 223–230. [CrossRef]

22. Sheldrick, G.M. A short history of SHELX. *Acta Crystallogr. Sect. A* **2008**, *64*, 112–122. [CrossRef] [PubMed]

23. Kjekshus, A.; Rakke, T.; Andresen, A.F. Compounds with the marcasite type crystal structure. IX. Structural data for $FeAs_2$, $FeSe_2$, $NiAs_2$, $NiSb_2$, and $CuSe_2$. *Acta Chem. Scand.* **1974**, *A28*, 996–1000. [CrossRef]

24. Pickardt, J.; Reuter, B.; Riedel, E.; Söchtig, J. On the formation of $FeSe_2$ single crystals by chemical transport reactions. *J. Solid State Chem.* **1975**, *15*, 366–368. [CrossRef]

25. Tian, X.-H.; Zhang, J.-M. The structural, elastic and electronic properties of marcasite FeS_2 under pressure. *J. Phys. Chem. Solids* **2018**, *118*, 88–94. [CrossRef]

crystals

MDPI

Article
P-T Phase Diagram of LuFe$_2$O$_4$

Maria Poienar [1], Julie Bourgeois [2], Christine Martin [2], Maryvonne Hervieu [2], Françoise Damay [3], Gaston Garbarino [4], Michael Hanfland [4], Thomas Hansen [5], François Baudelet [6], Jean Louis Bantignies [7], Patrick Hermet [1], Julien Haines [1] and Jérôme Rouquette [1],*

[1] Institut Charles Gerhardt UMR CNRS 5253, Université de Montpellier, Place E Bataillon, cc1504, 34095 Montpellier CEDEX, France; maria_poienar@yahoo.com (M.P.); patrick.hermet@umontpellier.fr (P.H.); julien.haines@umontpellier.fr (J.H.)

[2] Laboratoire CRISMAT, UMR 6508 CNRS, ENSICAEN, 6 Boulevard du Maréchal Juin, 14050 Caen CEDEX, France; julie.bourgeois@ensicaen.fr (J.B.); christine.martin@ensicaen.fr (C.M.); maryvonne.hervieu@ensicaen.fr (M.H.)

[3] Laboratoire Léon Brillouin, CEA-CNRS UMR 12, 91191 Gif-sur-Yvette CEDEX, France; francoise.damay@cea.fr

[4] European Synchrotron Radiation Facility, BP 220, 38043 Grenoble CEDEX, France; gaston.garbarino@esrf.fr (G.G.); hanfland@esrf.fr (M.H.)

[5] Institut Laue-Langevin, 6 rue Jules Horowitz, Boîte Postale 156, 38042 Grenoble CEDEX 9, France; hansen@ill.fr

[6] Synchrotron Soleil, L'Orme des Merisiers, Saint-Aubin BP 48 91192 Gif-sur-Yvette CEDEX, France; francois.baudelet@synchrotron-soleil.fr

[7] Laboratoire Charles Coulomb, UMR 5221 CNRS-Université de Montpellier, 34095 Montpellier, France; Jean-Louis.Bantignies@umontpellier.fr

* Correspondence: jerome.rouquette@umontpellier.fr

Received: 7 March 2018; Accepted: 20 April 2018; Published: 24 April 2018

Abstract: The high-pressure behavior of LuFe$_2$O$_4$ is characterized based on synchrotron X-ray diffraction and neutron diffraction, resistivity measurements, X-ray absorption spectroscopy and infrared spectroscopy studies. The results obtained enabled us to propose a *P-T* phase diagram. In this study, the low pressure charge-ordering melting could be detected by synchrotron XRD in the *P-T* space. In addition to the ambient pressure monoclinic $C2/m$ and rhombohedral $R\bar{3}m$ phases, the possible $P\bar{1}$ triclinic phase, the monoclinic high pressure form Pm and metastable modulated monoclinic phases were observed; the latter modulated monoclinic phases were not observed in the present neutron diffraction data. Furthermore, the transition to the Pm phase which was already characterized by strong kinetics is found to be favored at high temperature (373 K). Based on X-ray absorption spectroscopy data the Pm phase, which could be recovered at atmospheric pressure, can be explained by a change in the Fe-local environment from a five-fold coordination to a distorted 5 + 1 one.

Keywords: LuFe$_2$O$_4$; pressure-temperature phase diagram

---

1. Introduction

Magnetoelectric multiferroics combine ferroelectric and ferromagnetic properties in a single material, providing a possible route for controlling electric polarization with a magnetic field and magnetic order with an electric field [1]. Among the multiferroics, LuFe$_2$O$_4$ is considered as a prototype material in which ferroelectricity is driven by the electronic process of frustrated (Fe^{2+}/Fe^{3+}) charge ordering (CO) which is also coupled to magnetic order and magnetic fields [2]. The existence and/or origin of ferroelectricity in LuFe$_2$O$_4$ are still the subject of many debates and will not be considered in this paper [3–5]. Note however the strong importance of the CO in the physical behavior of LuFe$_2$O$_4$.

The mixed valence compound $LuFe_2O_4$ has been reported in the literature to crystallize at room temperature in the (R) $R\bar{3}m$ space group [6]. The layered crystal structure (2D) is described as an alternative stacking along the c axis of triangular lattice of Lu, the Fe and O ions forming double triangular $Fe_2O_{2.5}$ layers separated by single $LuO_{1.5}$ layers. A monoclinic distortion (M1) was evidenced for the CO state, which disappears above T_{CO}, and the crystal structure was refined in the monoclinic C2/m space group [7]. Along with the previously reported q_2 and q_3 modulations vectors distinctive of the charge-ordering (CO) of the iron species, an incommensurate order was observed characterized by a third vector q_1 associated with a tiny oxygen deviation from the O_4 stoichiometry [7]. The oxygen storage ability of $LuFe_2O_{4+x}$ has been demonstrated over a large x-range [0–0.5] associated with a complex oxygen intercalation/de-intercalation process with several intermediate metastable states which was found to be perfectly reversible [8]. In addition to the structural complexity of $LuFe_2O_4$, the coexistence of different magnetic ground states have been evidenced [9–12] using X-ray and neutron scattering techniques.

Pressure was shown to induce a monoclinic crystal structure (M2) [13], i.e., space group Pm, with a stacking of rectangular $[Fe]_\infty$ and buckled triangular $[Lu]_\infty$ layers in a misfits structure. This high-pressure form exhibits antiferromagnetic ordering similar to that of wustite FeO up to 380 K [14].

Dependence of $LuFe_2O_4$ on P and T is investigated based on synchrotron X-ray diffraction (XRD) and neutron diffraction (ND) studies, resistivity measurements, infrared spectroscopy and X-ray absorption spectroscopy studies.

2. Materials and Methods

2.1. Sample Preparation

$LuFe_2O_4$ was prepared by solid state reaction, starting from a 10 g mixture of 0.485 Lu_2O_3:0.815 Fe_2O_3:0.37 Fe, pressed in the shape of a rod (6 mm diameter and several centimetre length) and heated at 1180 °C for 12 h in an evacuated silica ampoule [7]. Laboratory XRD indicates that the sample presents the nominal composition and it is a single phase and well crystallized. The refined cell parameters are a = 3.44051(2) Å and c = 25.2389(2) Å in the $R\bar{3}m$ space group (hexagonal setting) [6].

The magnetic phase transition temperature (T_N = 240 K) and charge-ordering transition temperature T_{CO} = 330 K have been determined previously [7,9] and are in agreement with the literature [10].

The samples were carefully crushed, and sieved (at 60 μm and then at 20 μm) in order to have a homogeneous powder.

2.2. Synchrotron XRD

High-pressure synchrotron XRD experiments were performed at the European Synchrotron Radiation Facility (ESRF) on the ID09A beam line, with the X-ray beam collimated to about 10 μm × 10 μm. Angle-dispersive powder XRD of the sample was measured using monochromatic synchrotron radiation with a wavelength λ = 0.415811 Å and the patterns were collected with an online MAR345 image plate detector. The pressure was measured using the ruby fluorescence method [15].

In order to analyze the evolution of structure in function of T and P, three set-ups were used: first, at room temperature where the powder sample was loaded in a gas-membrane diamond anvil cell (DAC) and with helium as a pressure-transmitting medium to assure good hydrostatic conditions up to the highest investigated pressure of ~19 GPa.

A gas-membrane DAC was used also for the high temperature cell with neon as transmitting medium. The highest temperature of 473 K was reached in a resistive furnace and the pressure was up to ~13.5 GPa. For the low temperature measurements a screw-type DAC with helium as transmitting medium was placed inside a He flow cryostat (T = 50 K, P up to ~22 GPa).

The two-dimensional diffraction images were analyzed using the Fit2D software (ESRF, Grenoble, France) [16], yielding intensity vs. 2θ diffraction patterns (the peaks coming from the diamond cell were removed) and full profile matching was performed using the FullProf software (ILL, Grenoble, France) [17]. Unfortunately, statistics could not enable us to perform Rietveld refinements.

The pressure was increased/decreased in steps of approximately 1–2 GPa and the system was allowed to equilibrate for 5 to 10 min at each pressure point and the acquisition time was several seconds. For low temperature measurements, the pressure was always increased at room temperature in order to avoid stress in the sample.

2.3. Resistivity Measurements

The electrical resistivity measurements on a pressed sample with a size of 50 µm were performed in a sintered diamond Bridgman anvil apparatus using a pyrophyllite gasket and two steatite disks as the pressure medium and by using a Keithley (Solon, OH, USA) 2400 source meter and a Keithley 2182 nanovoltmeter [18,19]. The isobar data up to ~20 GPa were recorded between 150 K and 290 K.

2.4. Neutron Powder Diffraction

For the high-pressure ND experiment on the D20 diffractometer [20], at ILL Grenoble with λ = 1.36 Å, the sample was loaded in a Paris-Edinbourg press equipped with anvils of cubic boron nitride (c-BN) and a Ti-Zr gasket. The data were recorded at RT up to 12 GPa using 4:1 deutereted methanol: ethanol as pressure transmitting medium. Pressure is determined based on the equation of state obtained using Synchrotron XRD data.

2.5. Infrared Spectroscopy

Far infrared experiments under hydrostatic pressure conditions were performed in the $35-650$ cm^{-1} range, at room temperature with a resolution of 2 cm^{-1}. The powder sample of $LuFe_2O_4$ was placed in a 250 µm diameter hole drilled in a strainless steel gasket preindented to 50 µm, along with a small ruby as a pressure gauge. We used a diamond-anvil cell (equipped with a gas membrane for pressurization) and a Bruker (Karlsruhe, Germany) IFS66S/V infrared spectrometer. The latter is equipped for these measurements with a liquid He bolometer detector, a Ge-coated Mylar (6 µm beam splitter, a Bruker beam condenser system with two ×15 NA 0.4 cassegrain objectives, and a Mercury arc discharge source. Pressure was determined with the ruby luminescence method with a BETSA photoluminescence system (laser wavelength, λ = 532 nm) coupled to an Ocean Optics BV/HR2000+ spectrometer (bandwidth: 675−575 nm).

2.6. Ab-Initio Calculations

Zone-center phonon frequencies are calculated within the DFT+U formalism as implemented inside VASP [21–23] at the GGA/PBE level [24]. We used the direct method in the harmonic approximation associated to an atomic displacement of 0.03 A. Positive and negative displacements are considered to minimize the anharmonic effects. The computational details (DFT parameters, supercell, magnetism, ...) are given in [25].

2.7. X-Ray Absorption Spectroscopy

High-pressure XANES measurements at the Fe *K*-edge (7112 eV) at 300 K, were performed at the ODE beamline at synchrotron SOLEIL, Saint-Aubin, France. The $LuFe_2O_4$ sample, together with the pressure transmitting medium, silicone oil, were subject to high pressure up to 25 GPa using a diamond-anvil cell (DAC). Ab-initio simulations were performed using the FDMNES code [26] using the finite difference method with the full multiple scattering theory. We chose in this study this theory which works within the muffin-tin approximation on the potential shape, but is more tractable with large systems.

3. Results

3.1. Synchrotron XRD

3.1.1. 373 K

Figure 1 shows the synchrotron XRD data obtained at 373 K upon increasing pressure up to 12 GPa. The high pressure behavior of $LuFe_2O_4$ appears to be quite complex with four distinct pressure ranges. No superlattice reflection can be observed in the low angle part of the low pressure diffraction patterns, i.e., 0 GPa $\leq P \leq$ 5.1 GPa in Figure 1b, in agreement with the melting of the CO above 330 K under atmospheric pressure. Additionally, no monoclinic splitting could be detected in this pressure range. As loss of the CO state was found to be accompanied by a loss of the monoclinic distortion, these low pressure data were analyzed using the $R\bar{3}m$ space group. Figure 2a shows the pressure dependence of the $LuFe_2O_4$ volume. The B_0 value calculated using the Birch-Murnaghan state equation ($B'_0 = 4$) is 140 (2) GPa which is in agreement with the value previously reported at 298 K (138 (2) GPa). Interestingly some strong diffuse scattering (DS) appears between 10° and 12° (2θ) and between 6° and 8° (2θ) for 3 GPa $\leq P \leq$ 6.4 GPa, Figure 1a,b. This diffuse scattering is generally associated with some strong dynamic disorder within the structure, which disappears with additional increases in pressure, above 6.4 GPa, to be replaced by many superlattice reflections (SR), Figure 1a,b, and the onset of a monoclinic distortion, Figure 2b.

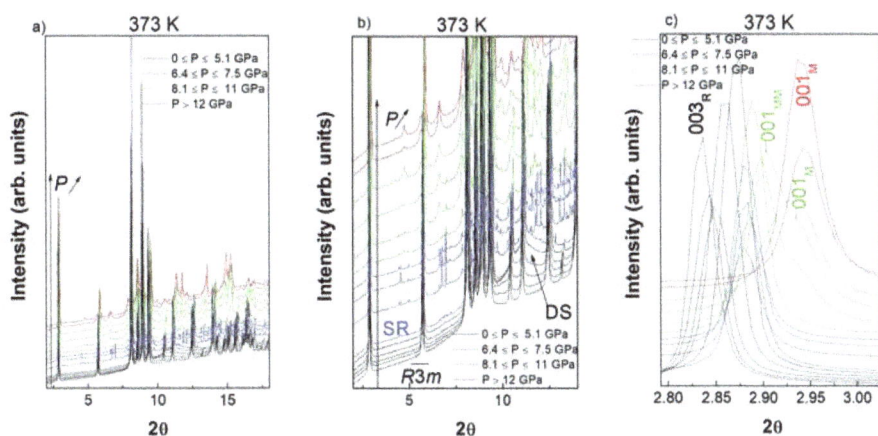

Figure 1. Pressure dependence of the $LuFe_2O_4$ X-ray data at 373 K, (**a**) full-2θ-range, (**b**) 2°–14°-2θ-range, (**c**) 2.8°–3.02°-2θ-range. Appearance of superlattice reflections (SR) and diffuse scattering (DS) are indicated.

For these reasons, we describe this new phase as a modulated monoclinic structure (MM). Note that the superlattice reflections positions and intensities are incoherently changing with increasing pressure probably due to the bad powder statistics and/or due to the fact that MM corresponds to metastable states; this is why, we could not describe with more accuracy this modulated phase. This "physical pressure" behavior is similar to that described in the oxygen storage ability of $LuFe_2O_4$ for which the intercalation process, i.e., "chemical pressure", occurred through several intermediate metastable phases. Additional increases in pressure, 8.1 GPa $\leq P \leq$ 11 GPa, will lead to a mixing of this MM phase with the previously reported high pressure monoclinic form M2 (space group *Pm*), the MM phase completely transformed above 11 GPa. Phase mixing between MM and M2 can unambiguously be observed based on the coexistence of the 001_{MM} and 001_{M2} reflections, Figure 1c; at 298 K *c* parameter of the M2 *Pm* high pressure phase is known to be 2.7% more compressible than that of the $C2/m$ M1 phase [25,27]. For $P \geq 12$ GPa, the pure M2 phase can be obtained, Figure 1. As already mentioned, M2 can be recovered at atmospheric pressure. M2 phase exhibits a B_0 value ($B'_0 = 4$) of 153 (4) GPa at 473 K ($V_0 = 0.914(1)$) similar to the values reported at 300 K. In this experiment the temperature was increased from 373 K to 473 K in order to (unsuccessfully) attempt to back-transform the HP M2 phase to the initial LP phase.

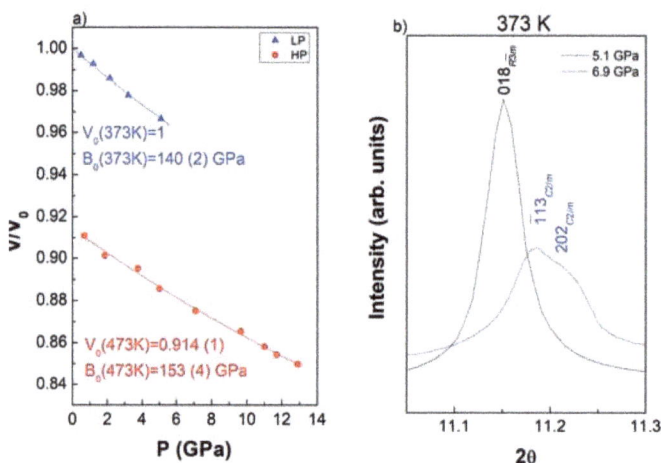

Figure 2. (**a**) Pressure dependence of the LuFe$_2$O$_4$ rhombohedral low pressure (at 373 K) and monoclinic M2 high pressure forms (at 473 K); (**b**) evidence of the apparition of a monoclinic distortion with increasing pressure. The obtained monoclinic reflections are indexed in the ambient temperature monoclinic $C2/m$ space group.

3.1.2. 298 K

Patm $\leq P \leq 6.6$ GPa

Figure 3 shows the synchrotron XRD data obtained at 298 K upon increasing pressure.

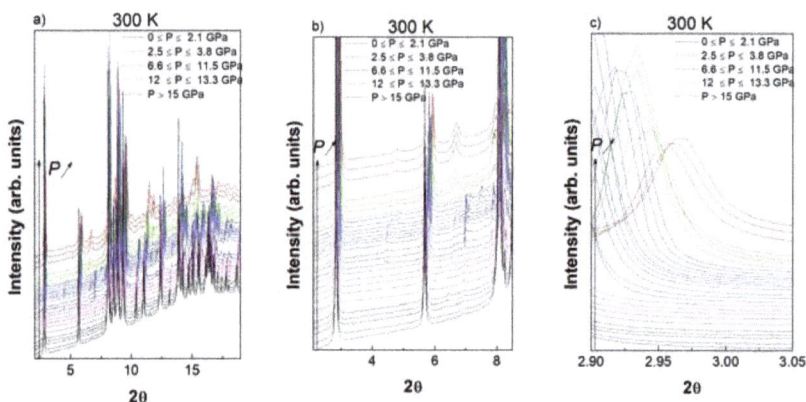

Figure 3. Pressure dependence of the LuFe$_2$O$_4$ X-ray data at 298 K, (**a**) full-2θ-range, (**b**) 2°–8.5°-2θ-range, (**c**) 2.9°–3.05°-2θ-range.

In the literature as already mentioned the CO state was evidenced by electron diffraction and Mössbauer spectroscopy. Additionally, the monoclinic distortion linked to the CO form was observed below T_{CO} = 330 K based on XRD and electron diffraction investigations [7]. Further, Blasco et al. proposed a monoclinic supercell ($C2/m$) which permitted to take into account the CO satellites below T_{CO} and showed that a transition to a triclinic phase ($P\bar{1}$) occurred below 170 K [28].

The data obtained at 300 K, Figure 3b, exhibit also some additional low intensity reflections in the low 2θ region for 0 GPa $\leq P \leq$ 2.1 GPa. A further inspection of the 0.5 (3) GPa isobar X-ray data confirmed that these reflections were intrinsic from the material as they exhibited a corresponding temperature dependence, Figure 4a: they are present at 50 K and 298 K, while they disappear above T_{CO} at 373 K. As additional satellites reflections can be observed at 50 K on Figure 4a, the $P\bar{1}$ low-temperature transition previously proposed is also confirmed in this study [28]; note that the proposed $C2/m$ and $P\bar{1}$ large supercell respectively fit the data obtained at 298 K and 50 K, Figure 4a.

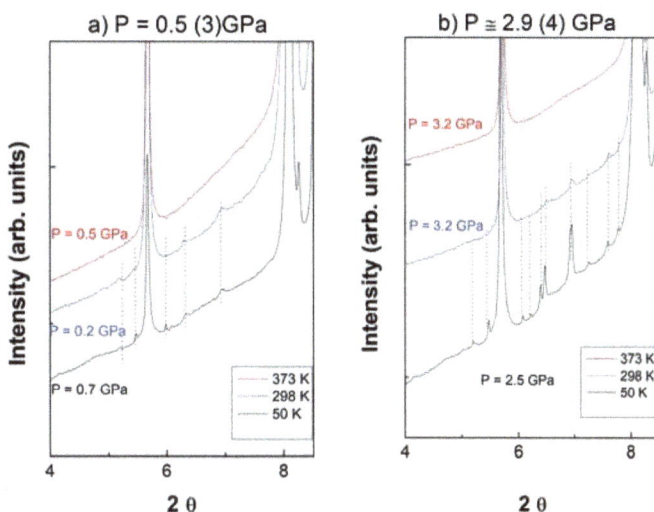

Figure 4. Temperature dependence of the superlattice reflections at (**a**) 0.5 (3) GPa and (**b**) 2.9 (4) GPa.

Upon pressure increase, 2.5 GPa $\leq P \leq$ 3.8 GPa, some additional superlattice reflections appear at 298 K, Figures 3b and 4b, which can be interpreted as a transition to the triclinic form; at 50 K, compression resulted as an intensity increase of the triclinic superlattices as a result of an increase in the distortion. Again, it is possible to check that these reflections are intrinsic to the material as they will disappear at 373 K, Figure 4b. The appearance of these additional superlattice reflections coincide with some changes observed in the LuFe$_2$O$_4$ structure (ND data).

At 298 K, the B_0 value calculated, 0.2 GPa $\leq P \leq$ 3.8 GPa using the Birch-Murnaghan state equation (B'_0 = 4) is 132 (3) GPa, Figure 5a; although a transition to the $P\bar{1}$ is suggested in this study, fitting this pressure dataset using two equations of state would be doutful. The lower B_0 value than that obtained at 373 K can be explained by the existence of the lower symmetry ($C2/m$) at 298 K which leaves more flexibility for the compression behavior; note however that the B_0 value has still the same order of magnitude as that obtained using an in-house high pressure X-ray diffractometer. The pressure dependence of the unit cell parameters using the $C2/m$ space group, Figure 5b–e, tends to confirm an anomaly in the compression behavior of LuFe$_2$O$_4$ at about 2.5 GPa; above 2.5 GPa the β-angle is found to saturate to a value of 103.47°. Note that Figure 5b–e exhibit also unit cell parameters obtained based on ND (see below).

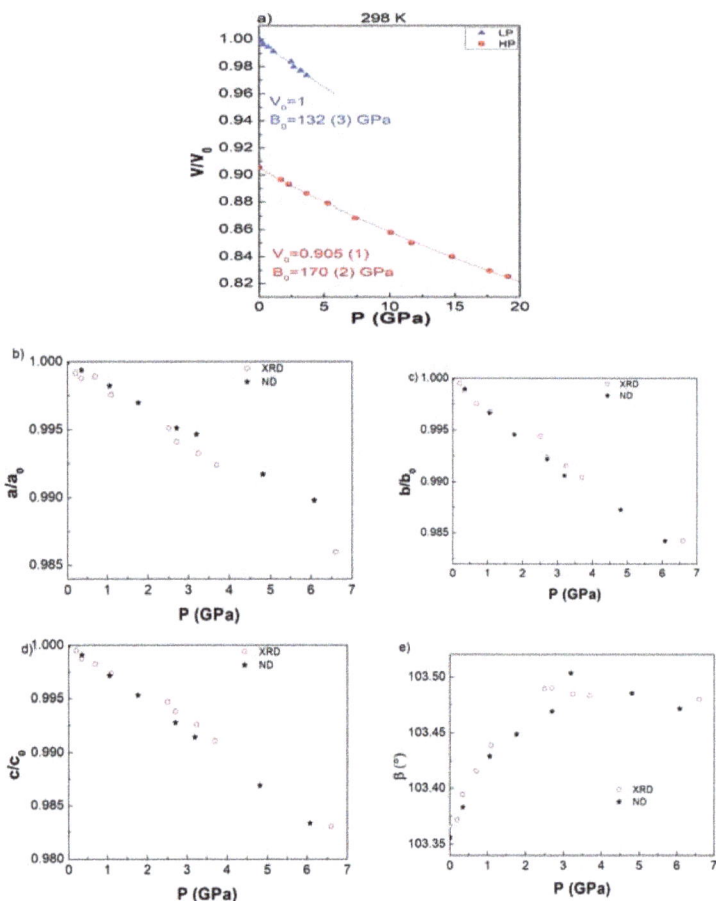

Figure 5. Pressure dependence at 298 K of (a) the LuFe$_2$O$_4$ low pressure monoclinic M1 and M2 high pressure forms, the relative monoclinic cell parameters (b) a/a_0; (c) b/b_0; (d) c/c_0 and (e) the monoclinic β angle.

3.1.2.2. GPa $\leq P \leq$ 19.1 GPa

As already shown in the 373 K isotherm X-ray data, the appearance of many superlattice reflections can be observed for 6.6 GPa $\leq P \leq$ 11.5 GPa, Figure 3a,b. As the monoclinic distortion is conserved, a transition to the modulated MM form is proposed. Compared with the 373 K data and the direct R-MM transformation, the absence of diffuse scattering at 298 K could be explained by the existence of the intermediate M1 monoclinic phase (and possibly T1) which would preclude the dynamic disorder. Note that the MM form could be associated with different metastable states as the superlattice reflections do not necessarily occur in the same 2θ-range as already observed at 373 K. Upon increasing pressure, 12 GPa $\leq P \leq$ 13.3 GPa, phase coexistence between MM and the high-pressure M2 phase is observed, Figure 3a-b. However, mechanisms implying the transition to the M2 phase probably exhibit particular pressure induced strain along the *c* direction as coexistence of $001_{C2/m}$ and 001_{Pm} reflections already existing at 373 K, Figure 3c, cannot be observed at 298 K, Figure 3c. For $P \geq$ 15 GPa, the pure M2 phase can be obtained, Figure 3, which can be recovered at atmospheric pressure as already mentioned. The B_0 value calculated of the M2 phase at 298 K is 170 (2) GPa with $V_0 = 0.905$ (1), Figure 5a.

3.1.3. 50 K

Figure 6a shows the synchrotron XRD data obtained at 50 K upon increasing pressure. As already mentioned on Figure 4, at 50 K, compression resulted as an intensity increase of the triclinic superlattices as a result of an increase in the distorsion. For this isotherm, bulk modulus of the low pressure T1 phase is 187 GPa (38). At 7 GPa, the modulated MM phase(s) is characterized by the appearance of many additional superlattice reflections, which are not the same at 7 GPa and 10.7 GPa respectively as an additional evidence of the metastability of the MM states.

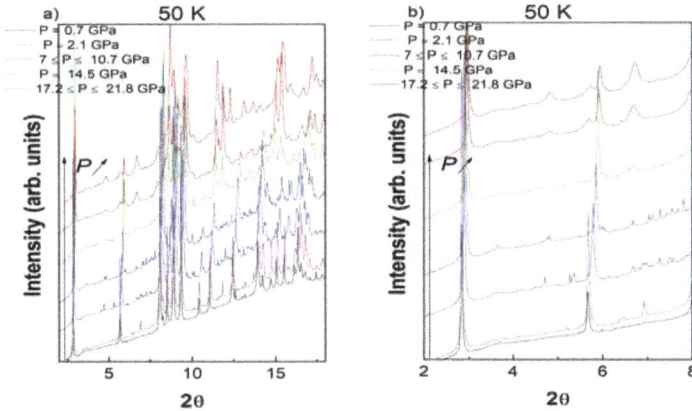

Figure 6. Pressure dependence of the LuFe$_2$O$_4$ X-ray data at 50 K, (a) full-2θ-range, (b) 2°–8°-2θ-range.

As already observed at 298 K, transformation to MM is not accompanied by the existence of diffuse scattering. Increasing the pressure to 14.5 GPa lead to a phase coexistence between the MM and the M2 phase. The later M2 phase is found to be pure above 17.2 GPa. Mechanisms implying the transition to the M2 phase, which exhibit pressure induced strain along the c direction as coexistence of $001_{C2/m}$ and 001_{Pm} reflections already shown at 373 K, Figure 1c, cannot be observed at 50 K, Figure 6b. The M2 phase can be recovered at atmospheric pressure and exhibits the following properties: B_0 = 157 (1) GPa and V_0 = 0.911 (1), Figure 7.

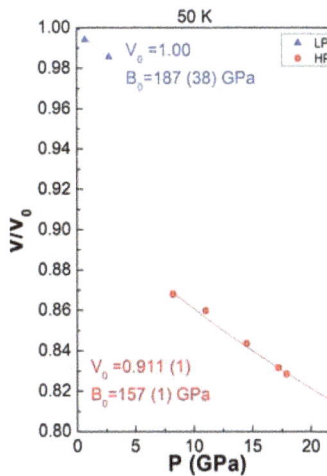

Figure 7. Pressure dependence of the LuFe$_2$O$_4$ low pressure T1 and monoclinic M2 high pressure forms at 50 K.

3.2. Neutron Powder Diffraction

As powder statistics did not permit us to follow the quantitative structure dependence of $LuFe_2O_4$, high pressure ND was performed. Figure 8 shows the ND data obtained at 2.7 GPa and the results of a refinement using the Rietveld method with the $C2/m$ space group, Table 1. This structural model was used for $0\ GPa \leq P \leq 6.1\ GPa$ although a smooth structural change is suggested above 2.5 GPa. Compared to the above high resolution X-ray data, the resolution neutron data at $\lambda = 1.36$ Å did not reasonably justify the use of lower symmetry.

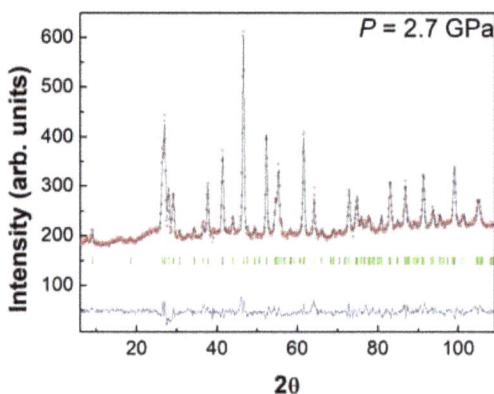

Figure 8. Rietveld refinement of $LuFe_2O_4$ neutron powder diffraction data at $P = 2.7$ GPa. Experimental data are represented by open circles using the calculated profile by a continuous line, and the allowed structural Bragg reflections by vertical marks. The difference between the experimental and calculated profiles is displayed at the bottom of the graph.

Table 1. Structural data of $LuFe_2O_4$ at $P = 2.7$ GPa obtained by neutron powder diffraction data.

	$C2/m$ (n° 12) $a_m = 5.9415$ (4) Å $b_m = 3.4181$ (2) Å $c_m = 8.6087$ (5) Å $\beta_m = 103.505$ (5)° $V = 170.002$ (2) Å^3				$R_{Bragg} = 13.9\ \%$ $\chi^2 = 4.75$
Atom	**Wyckoff Site**	x	y	z	**Isotropic B Factors**
Lu	(2a)	0	0	0	2.266
Fe	(4i)	0.218 (2)	0	0.6443(5)	1.100
O_1	(4i)	0.300 (2)	0	0.8760(7)	1.658
O_2	(4i)	0.129 (3)	0	0.3815 (6)	2.750

Pressure in the ND experiment was calibrated based on the equation of state obtained in the synchrotron XRD experiment, Figure 5a. Bond lengths calculation obtained using the Rietveld refined structure from NPD data tends to confirm the smooth structural change above 2.5 GPa to the triclinic T1 phase for the Fe-Fe, Lu-Lu, Lu-Fe, Lu-O and Fe-O distances, Figure 9.

Looking at the high pressure ND data, the transition to the M2 high pressure form can be evidenced in the 6–8 GPa range by (i) the decrease of the most intense reflection of the LP phase (M1) at about 46–48° (2θ) and by (ii) the appearance of a new extra low-2θ HP peak (M2), Figure 10, as a result of antiferromagnetic ordering as already previously reported [25].

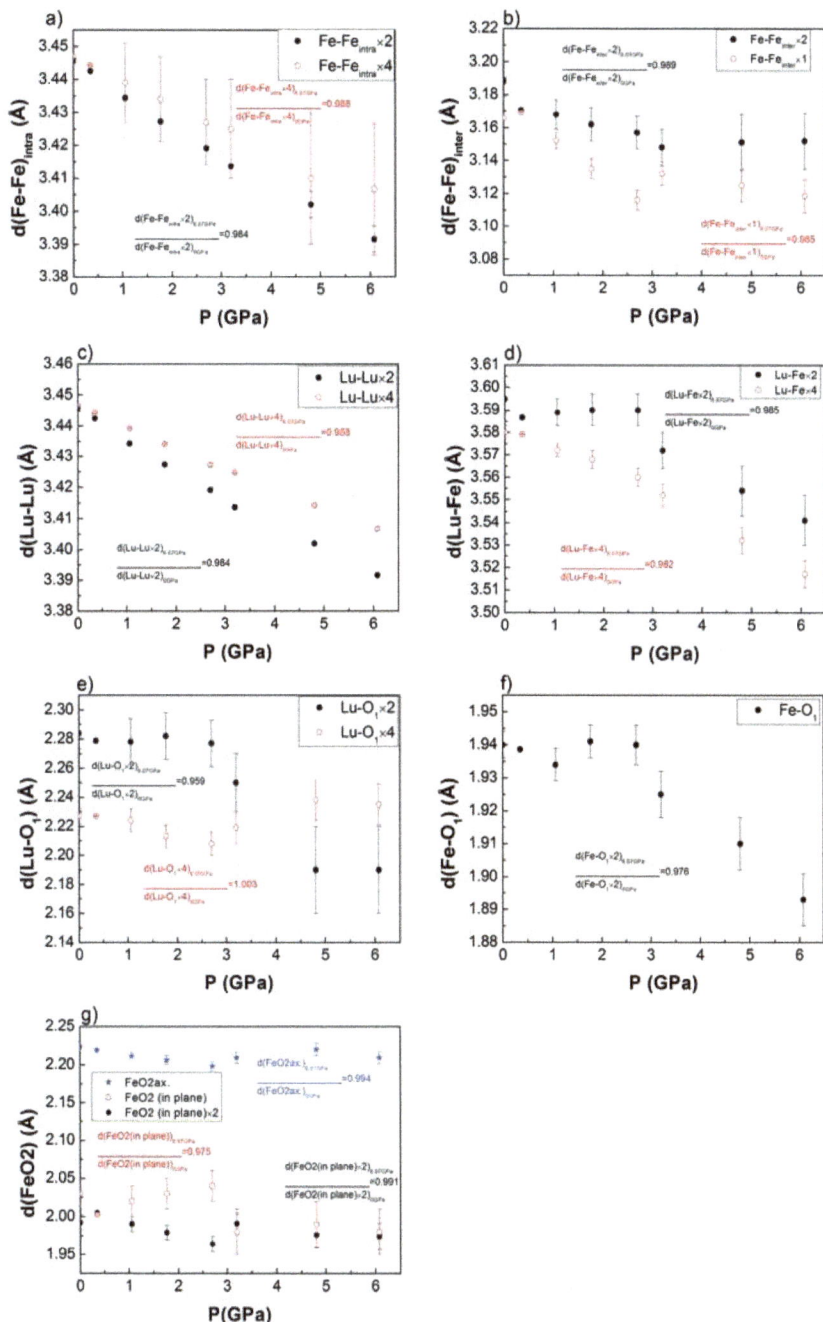

Figure 9. Pressure dependence of the (**a**) Fe-Fe$_{intra}$; (**b**) Fe-Fe$_{inter}$; (**c**) Lu-Lu; (**d**) Lu-Fe; (**e**) Lu-O$_1$; (**f**) Fe-O$_1$; (**g**) Fe-O$_2$ distances based on Rietveld refinements using ND data of the M1 phase. For each distance, the different relative compressibilities are calculated.

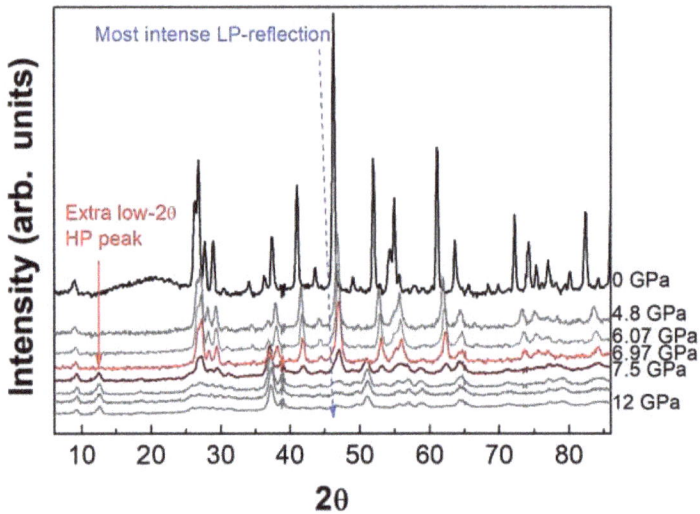

Figure 10. Pressure dependence of the ND data. Note the appropriate hydrostatic conditions which are confirmed by the presence of the diffuse scattering from the ethanol-methanol pressure transmitting medium close to 20° at 0 GPa. The M2 high-pressure phase is also characterized by the extra low 2θ peak arising from antiferromagnetic ordering.

Note that no structural information can be obtained above 6 GPa due to the phase coexistence between the M1 and the M2 phase. Additionally, it is of great importance to note that no supplementary MM phase is present in the ND data which clearly evidence that this latter form is metastable and is only observed in the synchrotron XRD, due to the different acquisition time used in these studies; 1 s and 2–3 h for the synchrotron and the ND studies respectively. At the highest pressure reached, i.e., 12 GPa, the transformation to the M2 phase is not totally complete. However, as the HP phase can be recovered at ambient pressure, one could estimate the exact completeness of the transition to the HP M2 phase (90%) based on an XRD investigation and the 2.7% compression difference between the *c* parameter of the M1 phase and the M2 one.

3.3. Resistivity Measurements

Figure 11 shows results obtained based on resistivity measurements. Figure 9a describes the temperature dependence of the resistivity, i.e., isobar measurements, as a function of pressure; it is consistent with semiconductor behavior over the entire pressure range. However, it is obvious that above 6 GPa the resistivity data tend to exhibit a similar temperature behavior. This is confirmed with Figure 11b which shows the pressure dependence of the resistance at 298 K and 150 K respectively, with a clear change in slope at 6 GPa, which could be linked with the transition to the M2 phase as described above. It is interesting to compare the present resistivity measurements with those previously reported which were obtained at ambient temperature with increasing pressure, i.e., isotherm measurements; in these measurements, the change in behavior occurred at about 11 GPa [27]. This apparent distinct behavior can be understood based on the acquisition time of the respective experiments similarly to what we observed in the synchrotron XRD and ND studies; the present isobar measurements required one day per pressure point, i.e., the entire dataset was performed over a month, whereas the isotherm measurements from atmospheric pressure to 15 GPa were performed within several hours. As we previously observed that the transition to the M2 phase could imply (several) intermediate metastable state(s) (MM), the kinetics are of the greatest importance in this transformation; these two sets of resistivity measurements represent furter evidence of the slow kinetics of this phase transition.

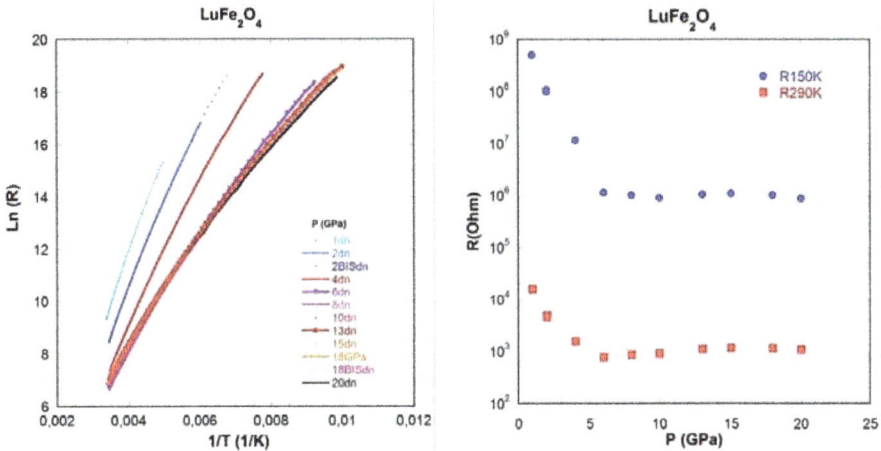

Figure 11. (**a**) Pressure dependence of resistivity measurements as a function temperature; (**b**) resistivity measurements as a function of pressure at 290 K and 150 K respectively.

3.4. Infrared Spectroscopy

Figure 12a shows the far infrared spectra of $LuFe_2O_4$ in the low-pressure M1 stability range, i.e., below 7 GPa. Infrared spectrum obtained at about 1 GPa is similar to that previously reported in the literature [29–32]. However, when pressure is increased above 2.4 GPa, some subtle changes noted A,B,C can be observed (mode assignment was indicated in [33]), Figure 12a, which could be interpreted as an indication of a phase transition. It is interesting to note that these changes occur in the same pressure range as modifications previously reported based on ND data, Figure 9, and the appearance of satellites reflections on XRD patterns, Figure 3. Such a signature could reinforce the probable existence of the T1 phase. When pressure increases above 6.3 GPa, Figure 12b, a clear change occurs and the infrared spectra saturate in intensity between 370 cm^{-1} and 600 cm^{-1}. This saturation of the absorption spectra has to be correlated with the change in resistivity observed at the transition to the M2 phase at close to 6 GPa, Figure 11. The saturation of the $LuFe_2O_4$ absorption spectrum therefore probably indicates that the bandgap of the M2 phase is in the infrared energy range. However, when pressure is further increased up to the highest pressure, the saturated absorption intensity between 370 cm^{-1} and 600 cm^{-1} progressively decreases and can even be measured during decompression, Figure 12b. Figure 12c shows the absorption spectrum of the HP M2 phase recovered at atmospheric pressure. The frequency positions and mode assignment of the *Pm* M2 phase were determined based on DFT calculations. Group theory shows us that the phonons of the M2 phase can be classified into $\Gamma = 28\,A'\oplus14A''$ where $2A'\oplus A''$ are accoustic modes. The A' and A'' irreducible representations are both Raman and infrared active.

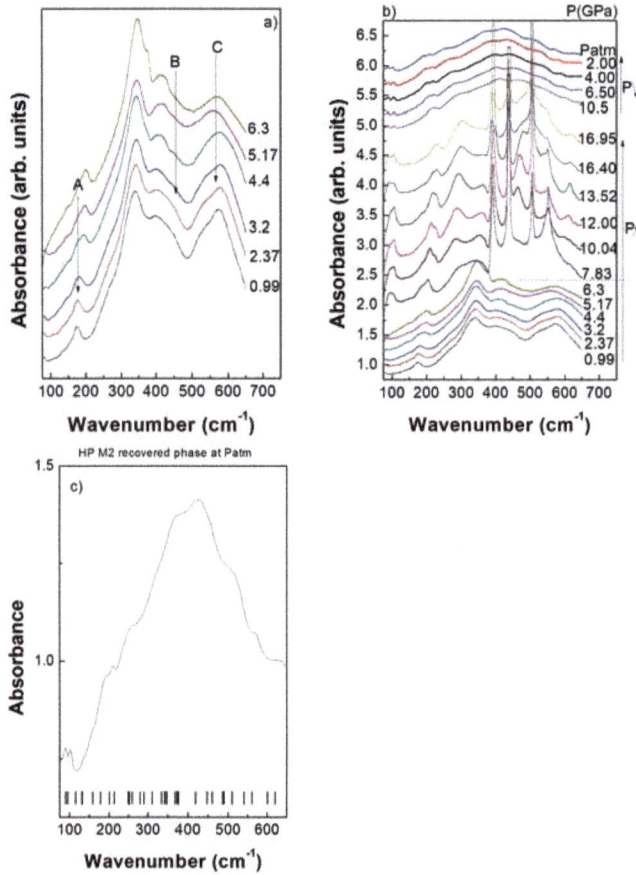

Figure 12. (a) Pressure dependence of the infrared spectra of LuFe$_2$O$_4$ (vertically shifted) in the low pressure regime, i.e., within the M1 phase stability range; A,B,C correspond to the labelled infrared modes which could be associated to a phase transition as previously observed based on diffraction techniques **(b)** vertically shifted infrared spectra during compression-decompression; the dotted line marks the transition to the M2 phase; **(c)** high-pressure M2 phase recovered at ambient pressure; vertical ticks represent calculated optical infrared modes. Note on Figure 10c that the frequency positions of the calculated infrared modes are upshifted as they were determined at 0 K whereas the experimental spectra are obtained at room temperature.

3.5. X-Ray Absorption Spectroscopy

Low pressure spectrum is consistent with those previously reported in the literature [13], Figure 13. High quality, high-pressure X-Ray absorption spectra data of LuFe$_2$O$_4$ at the Fe *K* edge were normalized to the jump at the absorption edge and a clear change is observed at about 16 GPa, Figure 13. In agreement with above results, such a high-pressure spectrum can be recovered to atmospheric pressure implying that the change in the Fe-local environment obtained under pressure is preserved during decompression. In order to verify the structural change to the M2 phase previously proposed, ab-initio simulations were performed using the FDMNES code [26], Figure 14, resulting in a good agreement between experiments and simulated X-Ray absorption spectra. 10 Å and 6 Å cluster radii were respectively used for the M1 [7] and M2 [25] phases. The 5-fold to distorted 5 + 1-fold change in the Fe-local environment obtained under pressure is therefore proposed to be at the origin of both the slow kinetics of the M1–M2 phase transition and the recovering of the high-pressure form at ambient pressure.

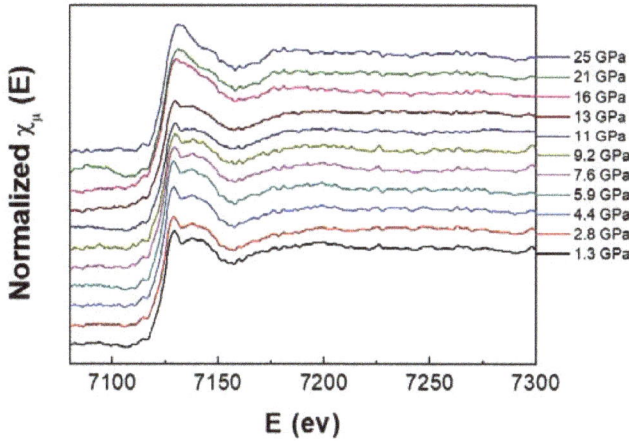

Figure 13. Pressure dependence of the x-ray absorption spectra of $LuFe_2O_4$ obtained at the *K*-edge.

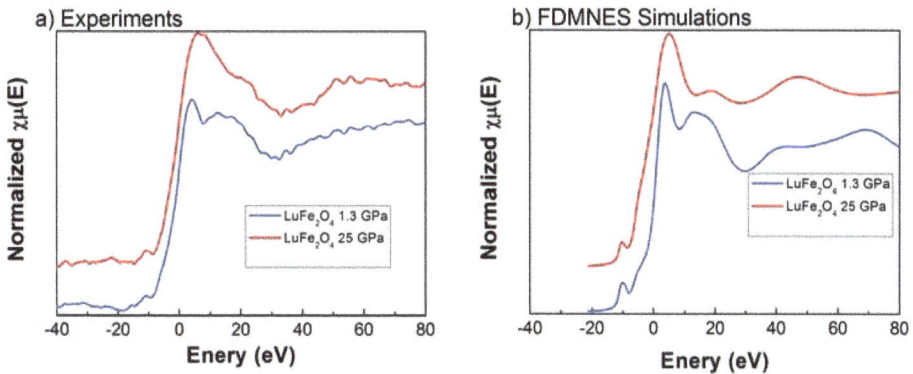

Figure 14. Comparison between low-pressure (1.3 GPa) and high-pressure (25 GPa) (**a**) experimental and (**b**) simulated X-ray absorption spectra of $LuFe_2O_4$ at the *K*-edge.

4. Conclusions

The present study allows us to construct a *P-T* phase diagram for $LuFe_2O_4$, Figure 15.

A clear phase boundary to the high pressure M2 phase was established. An additional transition to the rhombohedral high temperature phase are proposed along with a potential M1–T1 phase transition based on [28]. As the transition to the M2 high pressure phase is reconstructive with a change of five-fold to 5 + 1-fold Fe-local environment, the MM metastable phases characterized by synchrotron XRD are also included. It is interesting to note the parallel between physical pressure behavior in the present study and chemical pressure one in the oxygen (dis)-insertion previously reported [8] which was also characterized by several intermediate metastable states and a change in the Fe-local environment.

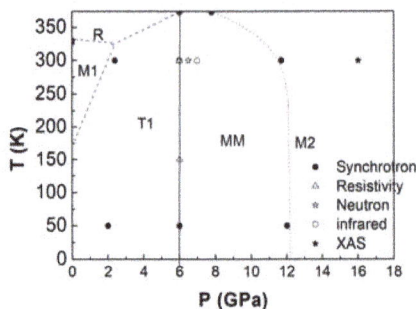

Figure 15. Pressure dependence of the $LuFe_2O_4$. The R ($R\bar{3}m$, M1 ($C2/m$) and T1 (proposed smooth charge/structural change of M1), MM (metastable modulated monoclinic phases) and M2 (*Pm* high-pressure phase) phase stability regions are shown. Solid line corresponds to the determined phase boundary to the high pressure M2 phase. Additional potential M1–R M1–T1 [28] and T1-R frontiers are proposed using dashed lines. Finally, the metastable MM phase(s) obtained from synchrotron X-ray study are also shown.

Author Contributions: The samples were prepared by J.B., synchrotron XRD data by M.P., J.B., C.M., F.D., M.H., J.H., J.R, resistivity measurements by G.G., NPD by M.P., J.B., C.M., F.D., T.H., J.H., J.R, infrared spectroscopy by M.P., J.L.B., J.R., ab-initio calculations by P.H., XAS by M.P., C.M., F.B., J.R. The manuscript was written by J.R. and revised by J.H., F.D., C.M. The projects direction were developed by J.R. and C.M.

Acknowledgments: This work was supported by financial support from the French Agence Nationale de la Recherche (JC08-331297 and ANR-08-BLAN-0005-01).

Conflicts of Interest: The authors declare no conflict of interest.

References

1.	Cheong, S.W.; Mostovoy, M. Multiferroics: A magnetic twist for ferroelectricity. *Nat. Mat.* **2007**, *6*, 13–20. [CrossRef] [PubMed]

2.	Ikeda, N.; Ohsumi, H.; Ohwada, K.; Ishii, K.; Inami, T.; Kakurai, K.; Murakami, Y.; Yoshii, K.; Mori, S.; Horibe, Y.; et al. Ferroelectricity from iron valence ordering in the charge-frustrated system Lufe2o4. *Nature* **2005**, *436*, 1136–1138. [CrossRef] [PubMed]

3.	Mundy, J.A.; Brooks, C.M.; Holtz, M.E.; Moyer, J.A.; Das, H.; Rebola, A.F.; Heron, J.T.; Clarkson, J.D.; Disseler, S.M.; Liu, Z.Q.; et al. Atomically engineered ferroic layers yield a room—Temperature magnetoelectric multiferroic. *Nature* **2016**, *537*, 523. [CrossRef] [PubMed]

4.	Niermann, D.; Waschkowski, F.; de Groot, J.; Angst, M.; Hemberger, J. Dielectric properties of charge-ordered lufe2o4 revisited: The Apparent influence of contacts. *Phys. Rev. Lett.* **2012**, *109*, 016405. [CrossRef] [PubMed]

5.	Lafuerza, S.; García, J.; Subías, G.; Blasco, J.; Conder, K.; Pomjakushina, E. Intrinsic electrical properties of Lufe2o4. *Phys.Rev. B* **2013**, *88*, 085130. [CrossRef]

6.	Isobe, M.; Kimizuka, N.; Iida, J.; Takekawa, S. Structures of Lufecoo4 and Lufe2o4. *Acta Crystallogr. Sect. C Cryst. Struct. Commun.* **1990**, *46*, 1917–1918. [CrossRef]

7.	Bourgeois, J.; Hervieu, M.; Poienar, M.; Abakumov, A.M.; Elkaim, E.; Sougrati, M.T.; Porcher, F.; Damay, F.; Rouquette, J.; Van Tendeloo, G.; et al. Evidence of oxygen-dependent modulation in Lufe2o4. *Phys. Rev. B* **2012**, *85*, 064102. [CrossRef]

8.	Hervieu, M.; Guesdon, A.; Bourgeois, J.; Elkaim, E.; Poienar, M.; Damay, F.; Rouquette, J.; Maignan, A.; Martin, C. Oxygen storage capacity and structural flexibility of $LuFe_2O_{4+x}$ ($0 \leq x \leq 0.5$). *Nat. Mater.* **2014**, *13*, 74–80. [PubMed]

9.	Bourgeois, J.; André, G.; Petit, S.; Robert, J.; Poienar, M.; Rouquette, J.; Elkaïm, E.; Hervieu, M.; Maignan, A.; Martin, C.; Damay, F. Evidence of magnetic phase separation in Lufe2o4. *Phys. Rev. B* **2012**, *86*, 024413. [CrossRef]

10. Iida, J.; Tanaka, M.; Nakagawa, Y.; Funahashi, S.; Kimizuka, N.; Takekawa, S. Magnetization and spin correlation of 2-dimensional triangular antiferromagnet Lufe2o4. *J. Phys. Soc. Jpn.* **1993**, *62*, 1723–1735. [CrossRef]

11. Christianson, A.D.; Lumsden, M.D.; Angst, M.; Yamani, Z.; Tian, W.; Jin, R.; Payzant, E.A.; Nagler, S.E.; Sales, B.C.; Mandrus, D. Three-dimensional Magnetic correlations in multiferroic Lufe(2)O(4). *Phys. Rev. Lett.* **2008**, *100*, 107601. [CrossRef] [PubMed]

12. Angst, M.; Hermann, R.P.; Christianson, A.D.; Lumsden, M.D.; Lee, C.; Whangbo, M.H.; Kim, J.W.; Ryan, P.J.; Nagler, S.E.; Tian, W.; Jin, R.; Sales, B.C.; Mandrus, D. Charge order in Lufe2o4: Antiferroelectric ground state and coupling to magnetism. *Phys. Rev. Lett.* **2008**, *101*, 227601. [CrossRef] [PubMed]

13. Lafuerza, S.; Garcia, J.; Subias, G.; Blasco, J.; Cuartero, V. Strong local lattice instability in hexagonal ferrites Rfe2o4 (R = Lu, Y, Yb) Revealed by X-ray absorption spectroscopy. *Phys. Rev. B* **2014**, *89*, 045129. [CrossRef]

14. Roth, W.L. Magnetic structures of Mno, Feo, Coo, and Nio. *Phys. Rev.* **1958**, *110*, 1333–1341. [CrossRef]

15. Piermarini, G.J.; Block, S.; Barnett, J.D.; Forman, R.A. Calibration of pressure-dependence of R1 ruby fluorescence line to 195 Kbar. *J. Appl. Phys.* **1975**, *46*, 2774–2780. [CrossRef]

16. Hammersley, A.P.; Svensson, S.O.; Hanfland, M.; Fitch, A.N.; Hausermann, D. Two-dimensional detector software: From Real detector to idealised image or two-theta scan. *High Press. Res.* **1996**, *14*, 235–248. [CrossRef]

17. Rodriguezcarvajal, J. Recent advances in magnetic-structure determination by neutron powder diffraction. *Physica B* **1993**, *192*, 55–69. [CrossRef]

18. Garbarino, G.; Weht, R.; Sow, A.; Sulpice, A.; Toulemonde, P.; Alvarez-Murga, M.; Strobel, P.; Bouvier, P.; Mezouar, M.; Nunez-Regueiro, M. Direct observation of the influence of the as-Fe-as angle on the T-C of superconducting Smfeaso1-Xfx. *Phys. Rev. B* **2011**, *84*, 024510. [CrossRef]

19. Sanfilippo, S.; Elsinger, H.; Nunez-Regueiro, M.; Laborde, O.; Le Floch, S.; Affronte, M.; Olcese, G.L.; Palenzona, A. Superconducting high pressure Casi2 phase with T-C up to 14 K. *Phys. Rev. B* **2000**, *61*, R3800–R3803. [CrossRef]

20. Hansen, T.C.; Henry, P.F.; Fischer, H.E.; Torregrossa, J.; Convert, P. The D20 instrument at the ILL: A versatile high-intensity two-axis neutron diffractometer. *Meas. Sci. Technol.* **2008**, *19*, 034001. [CrossRef]

21. Kresse, G.; Furthmuller, J. Efficient iterative schemes for ab-initio total-energy calculations using a plane-wave basis set. *Phys. Rev. B* **1996**, *54*, 11169–11186. [CrossRef]

22. Kresse, G.; Furthmüller, J. Efficiency of ab-initio total energy calculations for metals and semiconductors using a plane-wave basis set. *Comput. Mater. Sci.* **1996**, *6*, 15–50. [CrossRef]

23. Kresse, G.; Hafner, J. Abinitio molecular-dynamics for liquid-metals. *Phys. Rev. B* **1993**, *47*, 558–561. [CrossRef]

24. Perdew, J.P.; Burke, K.; Ernzerhof, M. Generalized gradient approximation made simple. *Phys. Rev. Lett.* **1996**, *77*, 3865–3868. [CrossRef] [PubMed]

25. Damay, F.; Poienar, M.; Hervieu, M.; Guesdon, A.; Bourgeois, J.; Hansen, T.; Elkaim, E.; Haines, J.; Hermet, P.; Konczewicz, L.; et al. High-pressure polymorph of Lufe2o4 with room-temperature antiferromagnetic order. *Phys. Rev. B* **2015**, *91*, 214111. [CrossRef]

26. Joly, Y. X-ray absorption near-edge structure calculations beyond the muffin-tin approximation. *Phys. Rev. B* **2001**, *63*, 125120. [CrossRef]

27. Rouquette, J.; Haines, J.; Al-Zein, A.; Papet, P.; Damay, F.; Bourgeois, J.; Hammouda, T.; Doré, F.; Maignan, A.; Hervieu, M.; Martin, C. Pressure-induced structural transition in Lufe2o4: Towards a new charge ordered state. *Phys. Rev. Lett.* **2010**, *105*, 237203. [CrossRef] [PubMed]

28. Blasco, J.; Lafuerza, S.; Garcia, J.; Subias, G. Structural properties in Rfe2o4 compounds (R = Tm, Yb, and Lu). *Phys. Rev. B* **2014**, *90*, 094119. [CrossRef]

29. Lee, C.; Kim, J.; Cheong, S.W.; Choi, E.J. Infrared optical response of Lufe2o4 under dc electric field. *Phys. Rev. B* **2012**, *85*, 014303. [CrossRef]

30. Vitucci, F.M.; Nucara, A.; Mirri, C.; Nicoletti, D.; Ortolani, M.; Schade, U.; Calvani, P. Infrared and transport properties of Lufe2o4 under electric fields. *Phys. Rev. B* **2011**, *84*, 153105. [CrossRef]

31. Vitucci, F.M.; Nucara, A.; Nicoletti, D.; Sun, Y.; Li, C.H.; Soret, J.C.; Schade, U.; Calvani, P. Infrared study of the charge-ordered multiferroic Lufe2o4. *Phys. Rev. B* **2010**, *81*, 195121. [CrossRef]

32. Xu, X.S.; de Groot, J.; Sun, Q.C.; Sales, B.C.; Mandrus, D.; Angst, M.; Litvinchuk, A.P.; Musfeldt, J.L. Lattice dynamical probe of charge order and antipolar bilayer stacking in Lufe2o4. *Phys.Rev. B* **2010**, *82*, 014304. [CrossRef]

33. Harris, A.B.; Yildirim, T. Charge and spin ordering in the mixed-valence compound Lufe2o4. *Phys. Rev. B* **2010**, *81*, 134417. [CrossRef]

crystals

MDPI

Article

High-Pressure Elastic, Vibrational and Structural Study of Monazite-Type GdPO$_4$ from Ab Initio Simulations

Alfonso Muñoz *[iD] and Placida Rodríguez-Hernández [iD]

Departamento de Física, Instituto de Materiales y Nanotecnología, MALTA Consolider Team,
Universidad de La Laguna, La Laguna, 38200 Tenerife, Spain; plrguez@ull.edu.es
*   Correspondence: amunoz@ull.edu.es; Tel.: +34-922-318-275

Received: 19 April 2018; Accepted: 8 May 2018; Published: 10 May 2018

Abstract: The GdPO$_4$ monazite-type has been studied under high pressure by first principles calculations in the framework of density functional theory. This study focuses on the structural, dynamical, and elastic properties of this material. Information about the structure and its evolution under pressure, the equation of state, and its compressibility are reported. The evolution of the Raman and Infrared frequencies, as well as their pressure coefficients are also presented. Finally, the study of the elastic constants provides information related with the elastic and mechanical properties of this compound. From our results, we conclude that monazite-type GdPO$_4$ becomes mechanically unstable at 54 GPa; no evidence of soft phonons has been found up to this pressure at the zone center.

Keywords: ab initio; monazite; orthophosphate; equation of state; elasticity; phonons; Raman

---

1. Introduction

The APO_4 orthophosphates compounds (A = trivalent cation) are materials analogous to orthoarsenates, orthovanadates, and orthosilicates. The size of the ionic radio of the A cation defines the two different structures where this family of compounds crystallizes. APO_4 compounds with an ionic radio lower than Gd crystallize in the tetragonal zircon structure ($I4_1/amd$, space group 141 with Z = 4). All other orthophosphates (A = La to Gd) crystallize into a monoclinic lower symmetry phase, that is, a monazite structure ($P2_1/n$, space group 14 with Z = 4) [1]. The monazite structure is isostructural to cerium phosphate mineral (monazite). This structure can be viewed as being composed by alternating edge-sharing AO_9 polyhedra, with a structural distortion derived from a rotation of the PO_4 tetrahedra and a lateral shift of the (100) planes that reduce the symmetry from the tetragonal of the zircon structure to the monoclinic symmetry.

Due to their numerous potential applications, the orthophosphate family has attracted a lot of interest in the last decades. The study of the mechanical properties of these materials can be relevant, since some orthophosphates have shown to be promising materials for oxidation-resistant ceramic toughening [2] and to have interesting luminescent and optical properties with a wide range of potential applications [3–6].

This family of minerals appears in nature as accessory mineral in rhyolites and granitoid rocks [1]. Therefore, the study of orthophosphates may be of importance in petrology [7], chemistry, and mineral physics. These materials can be employed as geochronometers for geological dating [8], because of their doping with radioactive elements that possess a very large half-life. Additionally, since orthophosphates incorporate rare earth elements, they can also be used to determine the distribution of rare earth elements in our Planet [9]. Another interesting application of these compounds is related to their use as potential matrix for nuclear waste disposal [5,10–12]. GdPO$_4$ and other orthophosphates have been suggested as inclusion compounds to made nuclear fuel canisters due to their high neutron absorption [13]. Therefore, the study of these orthophosphates under high pressure is very promising for future practical applications. Although there are several studies focused on pressure-induced transition and on the structural and vibrational properties of monazite phosphates under pressure [7,14–18], these studies were carried out only up to 30 GPa. A systematic understanding of their properties has not yet been achieved; the study of their elastic and mechanical properties is scarce [5,19,20]. Clearly, additional researches are needed to deepen our knowledge of this group of materials. A systematic investigation of the properties of monazite phosphates (particularly at high pressure) is necessary, especially if we want to use them in efficient and safe applications.

The ab initio methods have been successfully applied to study the thermodynamic properties of monazite rare earth orthophosphates [21–24]. Moreover, the use of ab initio simulations has become a strong and very well supported method to investigate the properties of materials under high pressures [25–28]. In particular, in the case of monazite-type orthophosphates under pressure, the ab initio simulations have been employed with good results [29,30].

The aim of the present paper is to contribute to the knowledge of the properties of orthophosphates at high pressure. In this work, we perform a detailed study of the structural, dynamic, and elastic properties of monazite-type $GdPO_4$ under pressures up to 60 GPa, going from first principles simulations. This method has proven to be quite efficient, in order to study a variety of properties on related compounds.

2. Simulations Details

The influence of pressure on the crystal structure, mechanical and vibrational properties of monazite-type $GdPO_4$ has been analyzed using ab initio simulations. The study was based on the Density Functional Theory (DFT) [31] employing the Vienna Ab initio Simulation Package (VASP) [32–34] and pseudopotentials generated with the Projector-Augmented Wave scheme (PAW) [35,36], with the f electrons included into the core. This has proven to give good results in the study of monazite-type structures [20,30,37]. To reach accurate results, the set of plane waves was extended up to a 520-eV cutoff energy and the exchange-correlation energy was expressed by means of the Generalized-Gradient Approximation (GGA) with the Perdew-Burke-Ernzerhof for solids (PBEsol) functional [38]. A dense grid of Monkhorst-Pack [39] k-special points was utilized to perform the integrations on the Brillouin Zone (BZ). The achieved convergence was 1 meV per formula unit in the total energy. All the structural parameters were optimized by minimizing the forces on the atoms and the stress tensor, at selected volumes. This method has been successfully applied to study the phase stability, and structural properties of semiconductors under high pressure [40]. To carry out the study of the mechanical properties of $GdPO_4$, we have evaluated the elastic constants, which describe the mechanical properties of a material in the region of small deformations. The elastic constants can be obtained by computing the macroscopic stress of a small strain with the use of the stress theorem [41,42]. In the present study, we employ the method implemented in the VASP code: the ground state and fully optimized structures were strained in different directions taking into account their symmetry [43].

The phonons study was performed using the supercell method [44]. The diagonalization of the dynamical matrix provides the frequencies of the normal modes. Furthermore, these calculations also provide the symmetry and eigenvectors of the phonon modes at the Γ point. The calculation of the phonon dispersion curves along the main directions within the BZ was carried out with a supercell of size $2 \times 2 \times 2$. Following the conclusions of our previous studies and those of Blanca-Romero et al. [45], in all the calculations, we neglect the spin–orbit interaction, because the structural properties are barely affected [15,17,26,29,30,37].

3. Results and Discussion

3.1. Structural Properties

The monoclinic structure of monazite, space group $P2_1/n$ [1] has been described in the introduction. A schematic view of the crystal structure is presented in Figure 1. It can be seen as an alternating chain of phosphorus-oxygen PO_4 tetrahedra and trivalent cation-oxygen GdO_9 polyhedra.

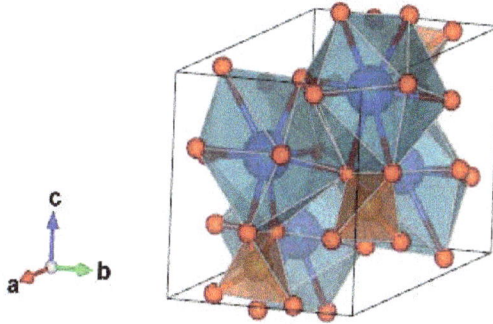

Figure 1. (**Color online**) Crystal structure of monazite-type $GdPO_4$. Oxygen, phosphorus, and gadolinium atoms are shown in red, orange, and blue, respectively. The PO_4 tetrahedral units and the GdO_9 polyhedral units are also shown.

The calculated structural parameters and atomic coordinates at ambient pressure are given in Tables 1 and 2, and are in a very close agreement with the available theoretical and experimental results [11,14,18,20,21]. Previous theoretical results [20] differ slightly due to the exchange-correlation functional used in the calculations.

Table 1. Structural parameters of monazite-type $GdPO_4$ at 0 GPa.

	This Study	Experiments	Theory
a (Å)	6.6276	6.623 [a] 6.6516(3) [b] 6.62(2) [c]	6.4152 [d] 6.713 [e]
b (Å)	6.8145	6.829 [a] 6.84840(7) [b] 6.823(2) [c]	6.6103 [d] 6.887 [e]
c (Å)	6.2930	6.335 [a] 6.33571(12) [b] 6.319(2) [c]	6.0953 [d] 6.358 [e]
β (°)	104.18°	103.80 [a] 104.023(2) [b] 104.16(2) [c]	104.6 [d] 104.2 [e]

[a] Reference [14]. [b] Reference [18]. [c] Reference [11]. [d] Reference [20]. [e] Reference [21]. Data from References [18,20] were obtained in a different setting than the one employed in the present work.

Table 2. Atomic positions of monazite-type $GdPO_4$ at 0 GPa.

Atom	Site	x	y	z
Gd	4e	0.28394	0.14513	0.08817
P	4e	0.29933	0.16215	0.61291
O_1	4e	0.2510	0.0000	0.4330
O_2	4e	0.3850	0.3381	0.5017
O_3	4e	0.4735	0.1010	0.8184
O_4	4e	0.1182	0.2100	0.7148

The evolution of the structural parameters with pressure is displayed in Figure 2a. The calculations have been already compared with experiments [14,18], showing an excellent agreement. The obtained linear axial compressibilities of each axis are: $K_a = \frac{-1}{a}\frac{da}{dp} = 2.76 \times 10^{-3}$ GPa^{-1}; $K_b = \frac{-1}{b}\frac{db}{dp} = 2.25 \times 10^{-3}$ GPa^{-1}; $K_c = \frac{-1}{c}\frac{dc}{dp} = 2.03 \times 10^{-3}$ GPa^{-1}. Therefore, from the present results and previous studies [14,18], it can be concluded that the compression of GdPO$_4$, as in other monazite-type phosphates, is not isotropic. The *a*-axis is the most compressible one and the *c*-axis the least compressible one. As shown in Figure 2a, there is a tendency for the unit-cell parameter *a* to approach the value of *c*. Moreover, in Figure 2b, it can be seen that the monoclinic β angle decreases under compression. As a result of these two facts, there is a gradual symmetrization of the monazite structure under compression.

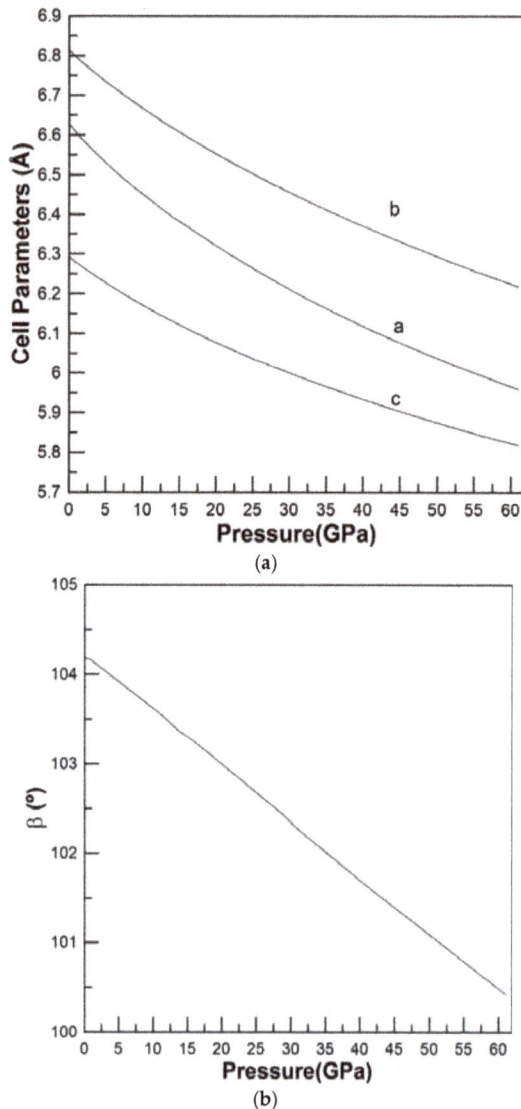

Figure 2. (a) Pressure dependence of the lattice parameters. (b) Pressure dependence of the β angle.

The bulk modulus, B_0, and its first pressure derivative were obtained by fitting the set of theoretical energy-volume data with a third-order Birch-Murnaghan (BM) Equation of State (EOS) [46]. The results ($B_0 = 138.3$ GPa, and $B'_0 = 4.02$) are summarized in Table 3 and are compared with previous results of B_0 reported for GdPO$_4$: the agreement between experiments is good.

Table 3. Volume at 0 GPa, bulk modulus, B_0, and its pressure derivative, B'_0, from a fit with a 3th order Birch Murnaghan EOS.

	This Work	Experiments	Theory
V (Å^3)	275.55	279.1 [a] 280.008(4) [b] 276.4(4) [c]	250.20 [e] 284.96 [f]
B$_0$ (GPa)	138.3	160 [a] 128.1(8) [b] 137 [d]	149 [g] 121.0 [h]
B'$_0$	4.07	5.8(2) [b]	

[a] Reference [14]. [b] Reference [18]. [c] Reference [11]. [d] Reference [47]. [e] Reference [20]. [f] Refence [21]. [g] Reference [48], [h] Reference [24].

From our calculations, we have determined the pressure dependence of the polyhedral volumes and distortions for monazite GdPO$_4$. Figure 3a presents the relative volume of the PO$_4$ and GdO$_9$ polyhedra and that of the unit–cell. As it can be observed, the PO$_4$ tetrahedron is highly incompressible, while the GdO$_9$ polyhedron is much more compressible. Actually, in the monazite-type GdPO$_4$, the volume change of the GO$_9$ polyhedra seems to be responsible for most of the volume reduction induced by pressure within the structure. In fact, the phosphate tetrahedra can be treated as a rigid unit in the monazite-type phosphate [17,18,48]. The P–O distances do not vary greatly within the tetrahedra. At zero pressure, two distances are 1.55 Å, and the two other 1.56 Å, and 1.54 Å, while at 53 GPa these distances decrease to 1.51 Å × 2, 1.52 Å, and 1.49 Å, respectively. Indeed, the distortion index, calculated with VESTA [49], for the PO$_4$ polyhedron (see Figure 3b) is very small, and it increases slowly with compression. Regarding the GdO$_9$ polyhedra at zero pressure, the central gadolinium atom is connected to nine oxygen atoms with eight bonds whose lengths vary from 2.34 to 2.55 Å, with one length at 2.78 Å. At 53 GPa the lengths range from 2.16 to 2.46 Å and the largest one changes to 2.53 Å. The GdO$_9$ distortion index do not vary linearly, as seen in Figure 3b. There is a decrease around 4 GPa, followed by a considerable increase in pressure, in a similar manner than in the isostructural LaPO$_4$ and CePO$_4$. The GdO$_9$ polyhedron is more distorted than the PO$_4$ tetrahedron: its distortion index changes from approximately 0.0381 at ambient pressure to 0.0440 at 54 GPa.

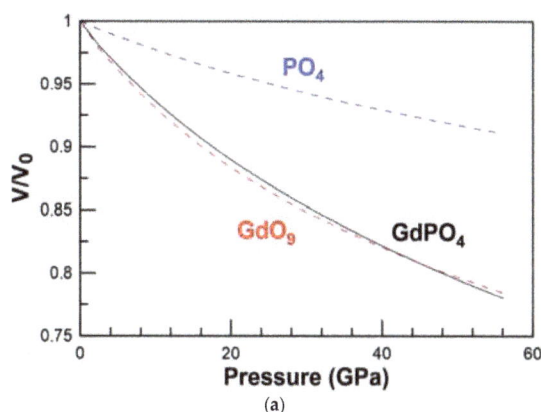

(a)

Figure 3. *Cont.*

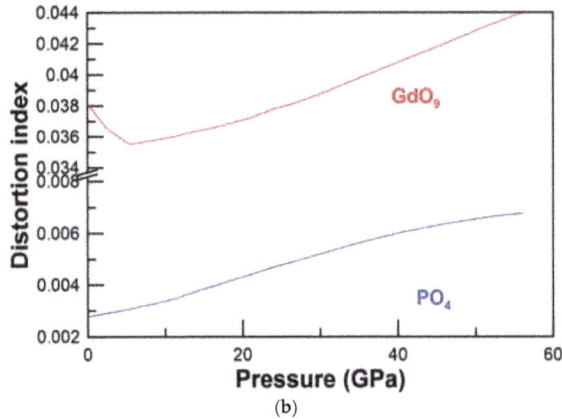

Figure 3. (**Color online**) (**a**) Relative variation with pressure of the polyhedral and unit cell volumes of monazite-type GdPO$_4$. (**b**) Distortion index of the PO$_4$ tetrahedron and the GdO$_9$ polyhedron.

3.2. Elastic Properties

GdPO$_4$ belongs to the monoclinic space group n° 14; it therefore has 13 independent elastic constants [50]: C_{11}, C_{12}, C_{13}, C_{15}, C_{22}, C_{23}, C_{25}, C_{33}, C_{35}, C_{44}, C_{46}, C_{55}, and C_{66} in the Voigt notation. Table 4 summarizes our results for the set of C_{ij} elastic constants at 0 GPa. The values here reported are in overall good agreement with those of Feng et al. [20]. The diagonal elastic constants C_{11}, C_{22}, C_{33}, are the highest ones, while C_{25}, C_{35}, C_{46} are the lowest ones. It must be noted that in [20], the cell parameters are underestimated, as commented above. When a non-zero uniform stress is applied to a crystal, the relevant magnitudes that describe their elastic properties are the elastic stiffness coefficients B_{ij} [51]. In the special case of a hydrostatic pressure, P, applied to a monoclinic crystal, the elastic stiffness coefficients are related to the elastic constants through the following relationships [51,52]:

$$B_{11} = C_{11} - P,\ B_{12} = C_{12} + P,\ B_{13} = C_{13} + P,\ B_{15} = C_{15},\ B_{22} = C_{22} - P,\ B_{23} = C_{23} + P,\ B_{25} = C_{25},$$
$$B_{33} = C_{33} - P,\ B_{35} = C_{35},\ B_{44} = C_{44} - P,\ B_{46} = C_{46},\ B_{55} = C_{55} - P\ \text{and}\ B_{66} = C_{66} - P.$$

Table 4. Elastic constants (in GPa) of monazite-type GdPO$_4$ at 0 GPa.

C_{11}	C_{22}	C_{33}	C_{44}	C_{55}	C_{66}	C_{12}	C_{13}	C_{15}	C_{23}	C_{25}	C_{25}	C_{46}
262.19	251.61	205.01	65.19	60.32	58.16	71.41	90.99	5.28	88.50	−18.17	−18.17	−13.70

Figure 4 shows the pressure dependence of the elastic stiffness coefficients, B_{ij}. It can be seen that, B_{11}, B_{22}, B_{33}, B_{12}, B_{13}, and B_{23} increase with pressure in the whole pressure range studied, especially in the diagonal components, while B_{44}, B_{55} and B_{66} decrease under compression. On the other hand, B_{15} is almost unaffected by pressure. It should be noted that B_{25}, B_{35}, B_{46} have similar negatives values and that they decrease under compression.

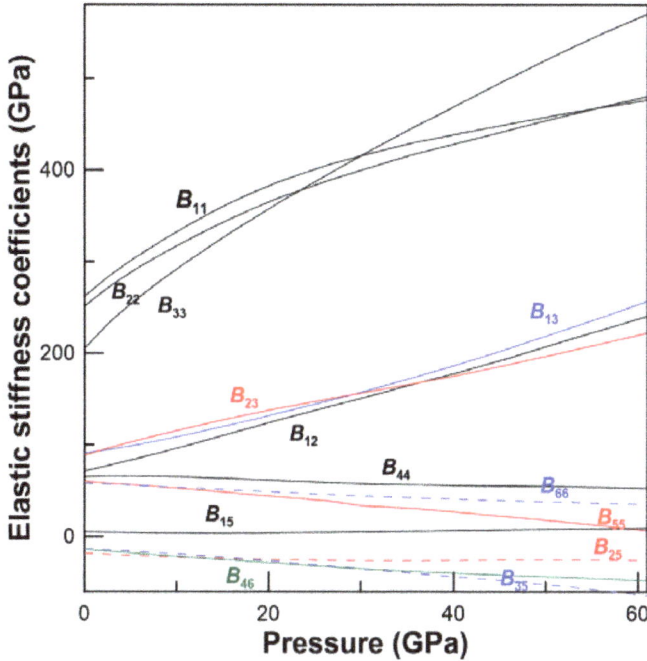

Figure 4. (Color online) Pressure evolution of the elastic stiffness coefficients, B_{ij}.

At zero pressure, a crystal is mechanically stable when the Born stability criteria are fulfilled [53]. However, when a hydrostatic pressure is applied to the crystal, the generalized Born stability criteria [52,54] must be employed to study the mechanical stability. This means that the matrix B_{ij} must be positive definite. Consequently, the stability conditions for a monoclinic crystal are given by the following conditions:

$$M_1 = B_{11} > 0$$

$$M_2 = B_{11}B_{22} - B_{12}^2 > 0$$

$$M_3 = (B_{22}B_{33} - B_{23}^2)B_{11} - B_{33}B_{12}^2 + 2B_{23}B_{12}B_{13} - B_{22}B_{13}^2 > 0$$

$$M_4 = B_{44} > 0$$

$$\begin{aligned}M_5 = {} & B_{12}^2 B_{35}^2 - B_{33}B_{55}B_{12}^2 + 2B_{55}B_{12}B_{13}B_{23} - 2B_{12}B_{13}B_{25}B_{35} \\ & -2B_{12}B_{15}B_{23}B_{35} + 2B_{33}B_{12}B_{15}B_{25} + B_{13}^2 B_{25}^2 - B_{22}B_{55}B_{13}^2 \\ & -2B_{13}B_{15}B_{23}B_{25} + 2B_{22}B_{13}B_{15}B_{35} + B_{15}^2 B_{23}^2 - B_{22}B_{33}B_{15}^2 \\ & -B_{11}B_{55}B_{23}^2 + 2B_{11}B_{23}B_{25}B_{35} - B_{11}B_{33}B_{25}^2 - B_{11}B_{22}B_{35}^2 \\ & +B_{11}B_{22}B_{33}B_{55} > 0\end{aligned}$$

$$M_6 = B_{44}B_{66} - B_{46}^2 > 0$$

The simulations of the elastic constants have been extended up to 60 GPa, well above the maximum experimental pressure reported [14,15]. The generalized stability criteria for GdPO$_4$ as function of the pressure are presented in Figure 5. At 54 GPa, one of the M_6 criterion is violated, making the system mechanically unstable at this pressure. No evidence of soft phonons was found at the Γ point in this pressure range. These results are consistent with the experimental and theoretical reports that confirm that there is no phase transition induced by pressure in the monazite structure up to the highest pressure reached in those studies (30 GPa) [14,17].

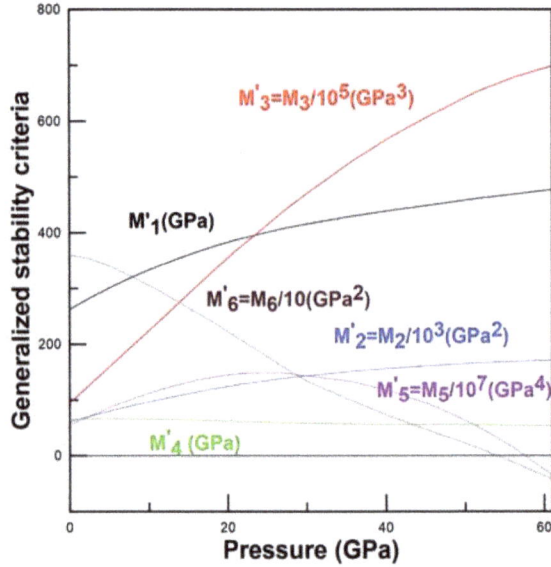

Figure 5. (**Color online**) Generalized stability criteria for monazite-type GdPO$_4$.

The bulk modulus (B), the shear modulus (G), the Young modulus (E), and the Poisson's ratio (ν) describe the major elastic properties of a material. Analytical expressions can be obtained for these moduli in the Voigt, Reuss and Hill approximations [55–57] from the elastic stiffness coefficients, B_{ij}. Hill has proved that the Voigt and Reuss approximations are limits and pointed out that the arithmetic mean of the two bounds can be considered as the actual B and G elastic moduli. In the case of a monoclinic crystal, the bulk and shear moduli can be expressed in the three approximations as [58]:

$$B_V = \frac{B_{11} + B_{22} + B_{33} + 2\,(B_{12} + B_{23} + B_{13})}{9}$$

$$\frac{1}{B_R} = S_{11} + S_{22} + S_{33} + 2\,(S_{12} + S_{23} + S_{13})$$

$$B_H = \frac{B_V + B_R}{2}$$

$$G_V = \frac{B_{11} + B_{22} + B_{33} - (B_{12} + B_{23} + B_{13}) + 3\,(B_{44} + B_{55} + B_{66})}{15}$$

$$\frac{1}{G_R} = \frac{4(S_{11} + S_{22} + S_{33}) - 4\,(S_{12} + S_{23} + S_{13}) + 3\,(S_{44} + S_{55} + S_{66})}{15}$$

$$G_H = \frac{G_V + G_R}{2}$$

where S_{ij} refers to components of the elastic compliances tensor, and subscripts V, R, and H stand for Voigt, Reuss and Hill respectively. The Young modulus, E, and the Poisson's ratio, ν, are obtained with the expressions [59]:

$$E_X = \frac{9B_X\,G_X}{G_X + 3B_X}$$

$$\nu_X = \frac{1}{2}\left(\frac{3B_X - 2G_X}{3B_X + G_X}\right)$$

where the subscript X refers to the symbols V, R, and H.

In order to discuss the elastic properties of GdPO$_4$ at ambient pressure, the elastic moduli at 0 GPa are summarized in Table 5. The bulk modulus, B_H = 134.47 GPa is in rather good agreement with our theoretical value of B_0 = 138.3 GPa, obtained from a fit with a third order Birch-Murnaghan equation of state [46], and with the experimental values of B_0 = 137 GPa previously reported [47]. As for *E*, *G*, and *B/G* elastic moduli, as well as the Poisson's ratio, *v*, at zero applied pressure, it should be noted once more that there is good agreement between our calculations and the experimental results. These facts demonstrate the quality of our simulations.

Table 5. Elastic moduli *B*, *E* and *G*, the *B/G* ratio, and the Poisson's ratio, *v*, in the Hill approximation, at 0 GPa.

B_H (GPa)	E_H (GPa)	G_H (GPa)	B_H/G_H	v
134.47 [a]	169.62 [a]	65.575 [a]	2.05 [a]	0.289 [a]
137 [b]	172 [b]	67 [b]	2.045 [b]	0.290 [b]
121.0 [c]	165.2 [c]	64.9 [c]	1.86 [c]	

[a] This work, [b] Reference [47], [c] Reference [24].

The pressure dependence of the elastic moduli up to 56.1 GPa is presented in Figure 6a–d. It can be observed that B_H, and v_H increase with pressure. However, E_H and G_H increase under compression reaching maximum values at 11 GPa and 9 GPa, respectively; above that pressure they decrease. All the elastic moduli change dramatically around 54 GPa, since the monazite structure becomes mechanically unstable. The Poisson's ratio gives information about the characteristics of the bonding forces and the chemical bonding. The value of the Poisson's ratio in the Hill approximation is v_H = 0.29 at 0 GPa. A value of v > 0.25 indicates that the interatomic bonding forces are predominantly central and that ionic bonding is predominant against covalent bonding [60,61]. The increase of v upon pressure can be interpreted as an increase of the metallization.

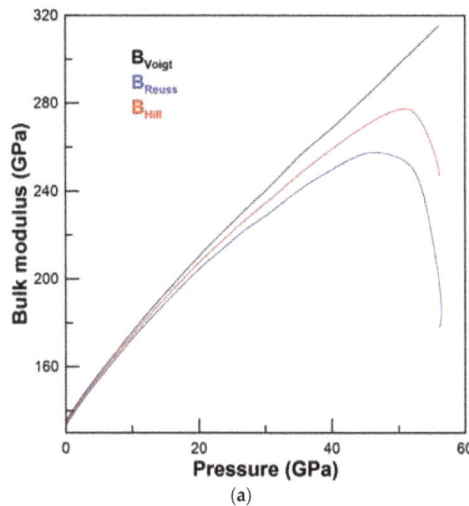

Figure 6. *Cont.*

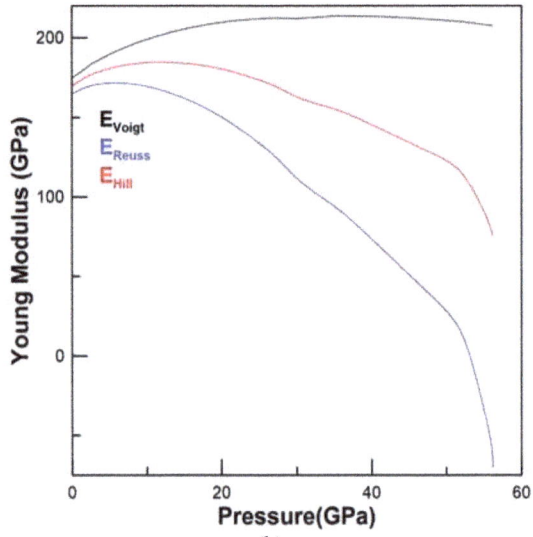

(b)

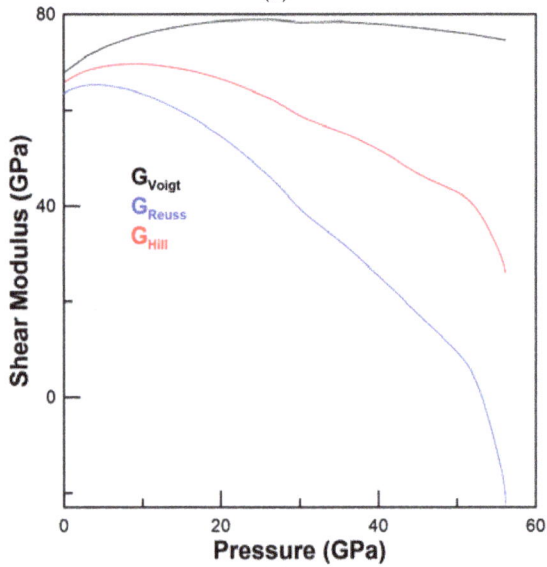

(c)

Figure 6. *Cont.*

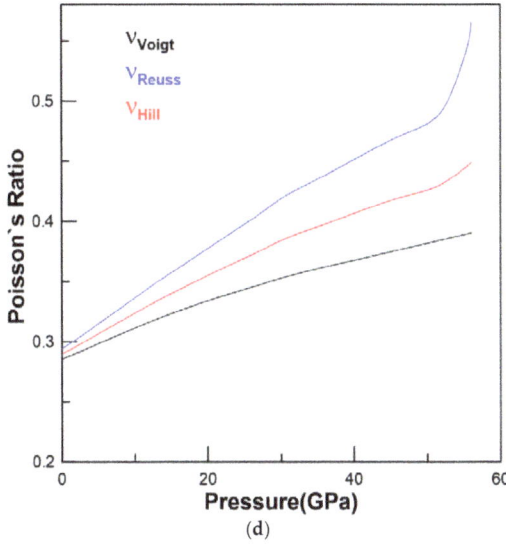

Figure 6. (Color online) (a) Pressure dependence of the bulk modulus, B. (b) Pressure dependence of the Young modulus, E. (c) Pressure dependence of the shear modulus, G. (d) Pressure dependence of the Poisson's, ν, in the approximation of Voigt, Reuss and Hill.

The B/G ratio is 2.05 and it increases with pressure (see Figure 7). Therefore, GdPO$_4$ monazite is more resistant to volume compression than to shear deformation ($B > G$). Moreover, according to Pugh criterion, monazite-type GdPO$_4$ behaves like a ductile material, since B/G is smaller than 1.75 [62], in the whole pressure range studied (materials with $B/G < 1.75$ behave as brittle materials). Our B/G value is consistent with the experimental result of Du et al. [47], although it is somewhat larger than the theoretical one of reference [24]. This is probably due to the small value of B_0 reported in that reference. However, it is clear that monazite GdPO$_4$ is a ductile material.

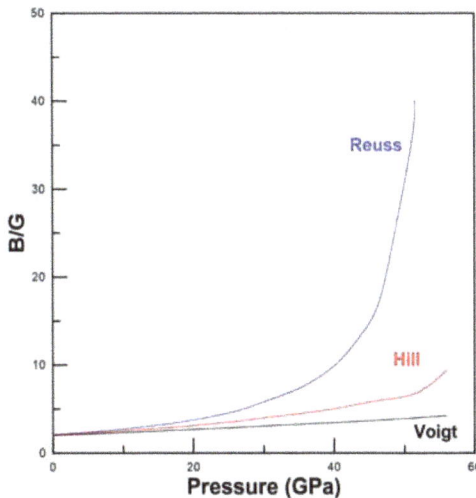

Figure 7. (Color online) Pressure dependence of the B/G ratio.

3.3. Vibrational Properties

The phonon dispersion of GdPO$_4$ along the main paths on the Brillouin zone is presented in Figure 8. Three frequency regions are separated by two gaps: a little one between the low- and the medium-frequency regions, and a bigger one between the medium- and the high-frequencies regions. The Partial Density Of States (PDOS) for phonons (see Figure 9) allows us to analyze these vibrations. The PDOS presents three zones separated by two gaps. The first zone, corresponding to the low-frequencies, down to 311 cm^{-1}, is composed of vibrations of Gd atoms and GdO$_9$ polyhedra, as well as a small contribution of PbO$_4$ tetrahedra. The intermediate-frequency zone, from 381 cm^{-1} to 600 cm^{-1}, is mainly due to vibrations of the PbO$_4$ tetrahedra with a very small contribution from the GdO$_9$ polyhedra. A large gap of $\approx$340 cm^{-1} separates this zone with the high-frequency region, which corresponds to frequencies between 942 cm^{-1} and 1073 cm^{-1} that are only due to vibrations from PbO$_4$ units. Therefore, the vibrational spectra of the GdPO$_4$ monazite could be interpreted in term of the modes of the PbO$_4$ tetrahedra, which can be considered as independent units in the monazite structure, and GdO$_9$ polyhedra. Indeed, the vibrational spectrum of monazites ABO_4 has been interpreted in terms of internal modes (bending and stretching modes) associated with the tetrahedron BO_4 and external modes (translational and rotational) which involves movement of both A and BO_4 ions [17,63].

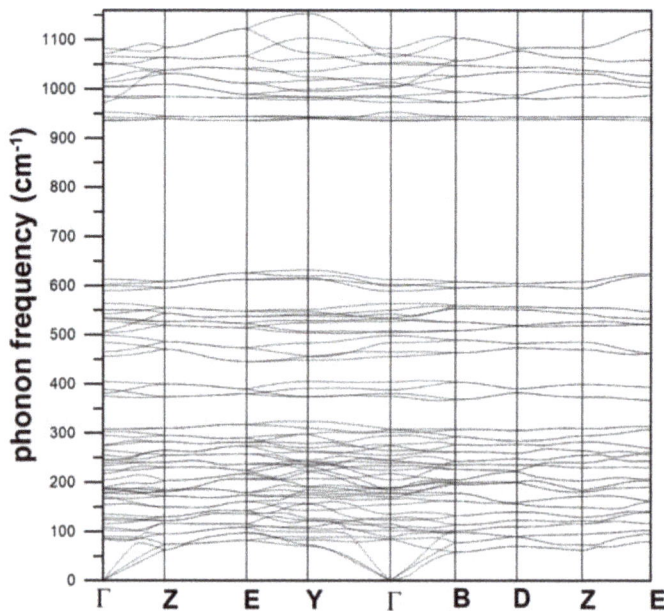

Figure 8. Phonon dispersion along the Brillouin zone of the of monazite-type GdPO$_4$.

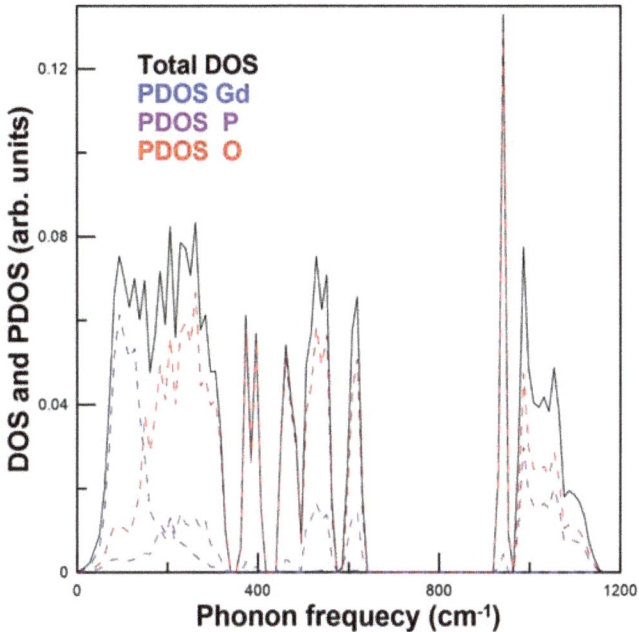

Figure 9. (Color online) Partial and total phonon density of monazite-type GdPO$_4$ states.

Through group theory analysis, it can be established that the monazite structure has 72 vibrational modes at the zone center: 36 optical Raman-active modes (18A$_g$ + 18B$_g$), 33 optical IR-active modes (17A$_u$ + 16B$_u$), and 3 acoustic modes (1A$_u$ + 2B$_u$). These vibrational modes can be interpreted as 36 internal (ν_1, ν_2, ν_3 and ν_4) and 36 external (translational (T) and rotational (R)) modes [17,63].

The Raman spectrum of many monazite-type phosphates (e.g., BiPO$_4$ LaPO$_4$, CePO$_4$, PrPO$_4$ and GdPO$_4$) at ambient conditions has been previously studied by different authors [18,29,64,65]. BiPO$_4$ LaPO$_4$, CePO$_4$, and PrPO$_4$ have also been studied under compression [17,66–68]. The frequency distribution of Raman modes is very similar for all of them. As in the case of GdPO$_4$, phonon gaps are observed between the three frequency regions in monazite phosphates; they are also observed in the monazite structure of chromates and selenates [30].

GdPO$_4$ has a distribution of Raman modes similar to other monazite phosphates [17]. The Raman mode frequencies and their pressure coefficients at room pressure are summarized in Table 6. The mode Grüneisen parameters [69], that provide a dimensionless representation of the response to compression are also reported.

It can be observed that the Raman spectrum of GdPO$_4$ monazite could be divided into three regions: the low-frequency region, up to 310.91 cm^{-1}, corresponds to the eighteen external or lattice T and R modes (9A$_g$ + 9B$_g$); the medium-frequency region, between 381.30 and 603.46 cm^{-1}, corresponds to the ten internal bending modes (5A$_g$ + 5B$_g$); and the high-frequency region, above 942.72 cm^{-1}, corresponds to the eight internal stretching modes (4A$_g$ + 4B$_g$).

As seen in Figure 10, which shows the theoretical pressure dependence of the Raman-active mode frequencies, most modes harden under compression, as is usual in most materials. However, there are a few modes between 84.78 and 153.94 cm^{-1} whose frequency changes very slightly under compression.

Table 6. Raman frequencies at ambient pressure, their pressure coefficients, and Grüneisen parameters (γ). The pressure dependence of frequency has been fitted with a second order polynomial: $\omega = \omega_0 + \left(\frac{\partial \omega}{\partial P}\right)P + \left(\frac{\partial^2 \omega}{\partial P^2}\right)P^2$.

Mode	ω_0 (cm^{-1})	$\frac{\partial \omega}{\partial P}$ (cm^{-1}/GPa)	$\frac{\partial^2 \omega}{\partial P^2}$ (cm^{-1}/GPa^2)	γ
Bg	84.78	0.762	0.0180	1.24
Ag	86.33	0.807	0.0128	1.29
Ag	108.61	0.846	0.0216	1.07
Ag	124.99	0.765	0.008	0.84
Bg	129.32	0.830	0.016	0.88
Bg	137.84	0.852	0.002	0.85
Bg	153.93	0.059	0.027	0.05
Ag	178.30	1.354	0.029	1.05
Bg	180.54	2.610	0.007	1.99
Ag	190.93	1.648	0.016	1.19
Ag	189.67	3.477	0.0343	2.53
Bg	235.29	3.887	0.057	2.28
Ag	240.14	3.014	0.0233	1.736
Bg	246.86	3.411	0.012	1.911
Ag	264.93	5.089	0.069	2.656
Bg	276.56	3.176	0.011	1.588
Bg	295.93	3.982	0.044	1.861
Ag	310.90	2.209	0.000	0.982
Bg	381.29	2.722	0.001	0.987
Ag	404.71	2.524	0.007	0.862
Ag	455.84	2.592	0.018	0.786
Bg	499.31	2.653	0.035	0.734
Ag	506.02	0.589	0.010	0.161
Bg	533.07	1.190	0.017	0.308
Ag	542.17	1.677	0.003	0.427
Bg	563.55	1.750	0.004	0.429
Ag	600.11	1.482	0.007	0.341
Bg	603.45	1.624	0.007	0.372
Bg	942.72	4.215	0.028	0.618
Ag	954.05	4.277	0.036	0.620
A	984.01	4.300	0.013	0.604
Ag	1007.23	5.005	0.039	0.687
Bg	1020.45	3.972	0.016	0.538
A	1051.30	4.269	0.014	0.561
Bg	1055.21	4.837	0.050	0.633
Bg	1072.33	4.480	0.008	0.577

External modes (low frequency region) involve movements of the trivalent Gd atoms, and in phosphate monazites ABO_4, their frequencies greatly depend on the mass of the A atom [17,70]. These external modes show very different pressure coefficients, since they involve different Gd-O bonds, with different compressibility, as seen in Section 2 [14]. Due to the marked differences between the pressure dependence of the different modes, see Table 6, crossing and anti-crossing phenomena are observed in this region. In particular, one of the modes most affected by compression is the A$_g$ mode at 264.93 cm^{-1}; The lowest-frequency B$_g$ mode is almost not affected by pressure. In previous studies of monazite chromates [37] and phosphates [17], softening of the lower Raman modes was related to a pressure-driven instability of the monazite structure, which could result in a phase transition above 30 GPa [17]. In the case of the monazite-type GdPO$_4$, the lower modes are little affected by compression. As commented in a previous section, the extension of the simulation up to 60 GPa does not show evidence of soft phonons mode at the Γ point. The pressure-driven instability seems to be related to mechanical instability, rather than dynamical instability.

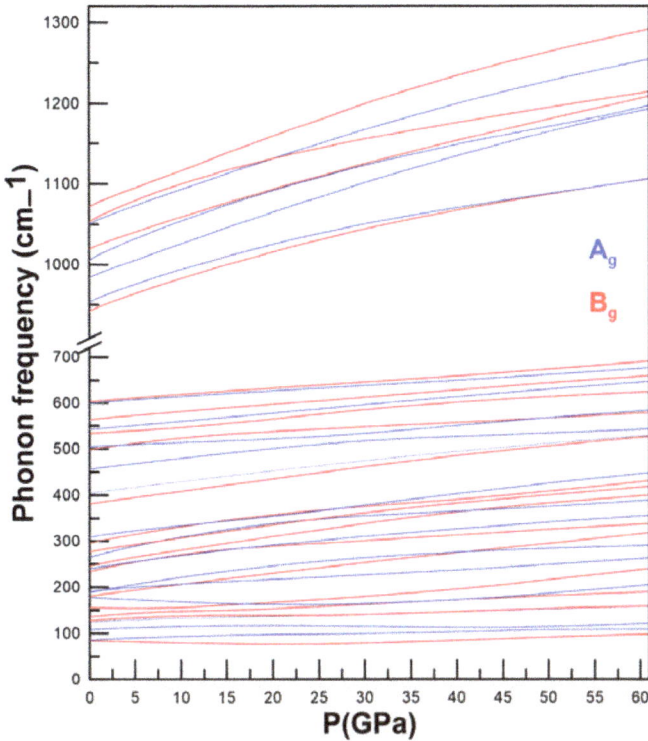

Figure 10. (**Color online**) Pressure dependence of the Raman modes of monazite-type GdPO$_4$.

Internal bending motions of the PO$_4$ tetrahedron (medium frequencies region) have smaller pressure coefficients. Two modes with frequencies A$_g$ = 506.02 and B$_g$ = 530 cm^{-1} at room pressure are the less affected by pressure, while the mode most sensitive to pressure in this region is the B$_g$ mode with a frequency 381.30 cm^{-1} at room pressure. The crossover of B$_g$ (499.32 cm^{-1}) and A$_g$ (506.02 cm^{-1}) modes is observed at ≈4 GPa in Figure 10, due to the different pressure dependences. This behavior could be related to the non-isotropic compression of monazite. This can cause the lower-frequency B$_g$ mode to move faster towards high frequencies than the higher-frequency mode.

The internal stretching modes (high frequency region) of the PO$_4$ tetrahedron have similar pressure coefficients, and they are among the modes whose frequency increases faster under compression.

Regarding the infrared modes of monazite GdPO$_4$, low-, medium-, and high frequencies modes have the same general behavior with pressure than the Raman modes have. This also happens for monazite chromates [34]. In particular, the modes less affected by pressure is the first A$_u$ mode (91.057 cm^{-1}); see Table 7 and Figure 11.

Table 7. Infrared frequencies at ambient pressure, their pressure coefficients, and Grüneisen parameters (γ). The pressure dependence of frequency has been fitted with a second order polynomial: $\omega = \omega_0 + \left(\frac{\partial \omega}{\partial P}\right)P + \left(\frac{\partial^2 \omega}{\partial P^2}\right)P^2$.

Mode	ω_0 (cm^{-1})	$\frac{\partial \omega}{\partial P}$ (cm^{-1}/GPa)	$\frac{\partial^2 \omega}{\partial P^2}$ (cm^{-1}/GPa^2)	γ
Au	91.05	0.219	0.007	0.332
Bu	102.50	1.442	0.005	1.946
Au	116.32	1.541	0.0125	1.832
Bu	168.18	0.722	0.013	0.594
Au	170.55	0.866	0.010	0.702
Bu	187.21	1.270	0.032	0.938
Au	189.765	2.761	0.041	2.012
Bu	205.99	2.328	0.021	1.563
Au	219.28	1.386	0.006	0.874
Bu	225.70	3.603	0.026	2.207
Au	245.96	3.397	0.032	1.910
Bu	254.49	3.153	0.013	1.713
Au	274.99	3.789	0.0231	1.905
Au	304.38	4.034	0.039	1.832
Bu	309.61	1.867	0.006	0.834
Bu	375.28	3.061	0.033	1.128
Au	388.32	2.186	0.0011	0.778
Au	466.32	1.013	0.0122	0.300
Bu	484.23	1.851	0.003	0.528
Au	507.73	1.501	0.0044	0.408
Bu	526.89	2.3165	0.0124	0.608
Au	533.09	1.864	0.0033	0.483
Bu	550.13	1.428	0.007	0.359
Bu	590.07	1.384	0.006	0.324
Au	613.26	1.765	0.008	0.398
Au	935.32	4.083	0.023	0.603
Bu	936.86	4.205	0.029	0.620
Au	974.23	3.643	0.017	0.517
Bu	986.50	3.483	0.014	0.488
Bu	1004.66	4.674	0.030	0.643
Au	1014.56	5.209	0.038	0.710
Bu	1066.28	4.956	0.024	0.642
Au	1082.61	4.763	0.024	0.608

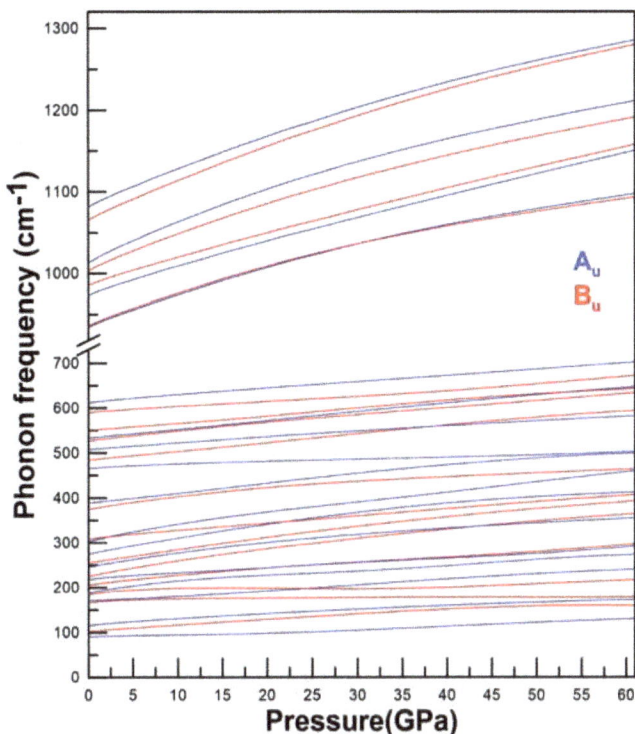

Figure 11. (Color online) Pressure dependence of the infrared modes of monazite-type GdPO$_4$.

The Grüneisen parameters in the high frequency region are very similar, while there are large differences between the Grüneisen parameters for the lower frequencies (Tables 6 and 7). Therefore, there are large differences between the restoring force on the atoms associated with the lowest modes.

4. Conclusions

The effects of high pressure on the structure of monazite-type GdPO$_4$ were studied from first principles, ab initio calculations. The simulations were extended up to 60 GPa, well above the maximum experimental pressure previously reported (30 GPa). The evolution of the lattice parameters under compression, the equations of state, and the axial and polyhedral compressibilities are reported. Moreover, the under-pressure distortion of the polyhedral units PO$_4$ and GdO$_9$ is shown. The compression of monazite-type GdPO$_4$ is not isotropic, and the GdO$_9$ polyhedra account for almost all the volume reduction under pressure.

The elastic constants and elastic stiffness coefficients were accurately determined, through which the elastic behavior of monazite-type GdPO$_4$ at high pressure was analyzed. The evolution with pressure of the B, G, and E elastic moduli, ν Poisson's ratio, and the B/G ratio was reported. At all pressures, this compound has shown to be ductile and more resistive to volume compression than to shear deformation ($B > G$). Furthermore, the high-pressure mechanical stability of this monoclinic compound was studied. It was found that monazite-type GdPO$_4$ becomes mechanically unstable above 54 GPa.

Finally, through the total and partial density of states, the contribution of each polynomial units to the vibrational modes was discussed. This contribution is similar to that found in other monazite phosphates. The evolution of the Raman- and infrared active modes of the monazite-type GdPO$_4$

as function of pressure was analyzed. No soft-modes were found up to 60 GPa at the center of the Brillouin zone.

A possible pressure-driven phase transition or amorphization might occur due to mechanical instability at the high pressure of 54 GPa. No dynamical instability was found up to this pressure.

The results of the presents study confirm the high stability under pressure of the monazite-type $GdPO_4$, which is very promising for practical applications.

Author Contributions: Both authors contributed equally to this work: simulations, analysis and interpretation of results and writing of the manuscript.

Acknowledgments: Research supported by the Spanish Ministerio de Economía y Competitividad, the Spanish Research Agency, and the European Fund for Regional Development under Grant Nos. MAT2016-75586-C4-3-P. We acknowledge the computer time provided by the RES (Red Española de Supercomputación) and the MALTA cluster.

Conflicts of Interest: The authors declare no conflict of interest.

References

1. Ni, Y.X.; Hughes, J.M.; Mariano, A.N. Crystal-chemistry of the monazite and xenotime structures. *Am. Mineral.* **1995**, *80*, 21–26. [CrossRef]

2. Morgan, P.E.D.; Marshall, D.B. Ceramic compounds of monazite and alumina. *J. Am. Ceram. Soc.* **1995**, *78*, 1553–1563. [CrossRef]

3. Kolitsch, U.; Holtstam, D. Crystal chemistry of REEXO$_4$ compounds (X = P, As, V). II. Review of REEXO$_4$ compounds and their stability fields. *Eur. J. Mineral.* **2004**, *16*, 117–126. [CrossRef]

4. Kaminskii, A.A.; Bettinelli, M.; Speghini, A.; Rhee, H.; Eichler, H.J.; Mariotto, G. Tetragonal YPO$_4$—A novel SRS-active crystal. *Laser Phys. Lett.* **2008**, *5*, 367–374. [CrossRef]

5. Clavier, N.; Podor, R.; Dacheux, N. Crystal chemistry of the monazite structure. *J. Eur. Ceram. Soc.* **2011**, *31*, 941–976. [CrossRef]

6. Errandonea, D. High-pressure phase transitions and properties of MTO$_4$ compounds with the monazite-type structure. *Phys. Status Solidi B Basic Solid State Phys.* **2017**, *254*, 1700016. [CrossRef]

7. Rubatto, D.; Hermann, J.; Buick, I.S. Temperature and bulk composition control on the growth of monazite and zircon during low-pressure anataxis (Mount Stafford, central Australia). *J. Petrol.* **2006**, *47*, 1973–1996. [CrossRef]

8. Grove, M.; Harrison, T.M. Monazite Th-Pb age depth profiling. *Geology* **1999**, *27*, 487–490. [CrossRef]

9. Meldrum, A.; Boatner, L.A.; Ewing, R.C. Displacive radiation effects in the monazite- and zircon-structure orthophosphates. *Phys. Rev. B* **1997**, *56*, 13805–13814. [CrossRef]

10. Ewing, R.C. The design and evaluation of nuclear-waste forms: Clues from mineralogy. *Can. Mineral.* **2001**, *39*, 697–715. [CrossRef]

11. Mullica, D.F.; Grossie, D.A.; Boatner, L.A. Coordination geometry and structural determinations of SmPO$_4$, EuPO$_4$ and GdPO$_4$. *Inorg. Chim. Acta F* **1985**, *109*, 105–110. [CrossRef]

12. Neumeier, S.; Arinicheva, Y.; Ji, Y.; Heuser, J.M.; Kowlaski, P.M.; Kegler, P.; Sclenz, H.; Bosbach, D.; Deissmann, G. New insights into phosphate based material for the immobilization of actinides. *Radiochim. Acta* **2017**, *105*, 961–984.

13. Lessing, P.A.; Erickson, A.W. Synthesis and characterization of gadolinium phosphate neutron absorber. *J. Eur. Ceram. Soc.* **2003**, *23*, 3049–3057. [CrossRef]

14. Lacomba-Perales, R.; Errandonea, D.; Meng, Y.; Bettinelli, M. High-pressure stability and compressibility of APO(4) (A = La, Nd, Eu, Gd, Er, and Y) orthophosphates: An x-ray diffraction study using synchrotron radiation. *Phys. Rev. B* **2010**, *81*, 064113. [CrossRef]

15. Errandonea, D.; Gomis, O.; Santamaria-Perez, D.; Garcia-Domene, B.; Munoz, A.; Rodriguez-Hernandez, P.; Achary, S.N.; Tyagi, A.K.; Popescu, C. Exploring the high-pressure behavior of the three known polymorphs of BiPO$_4$: Discovery of a new polymorph. *J. Appl. Phys.* **2015**, *117*, 105902. [CrossRef]

16. Achary, S.N.; Bevara, S.; Tyagi, A.K. Recent progress on synthesis and structural aspects of rare-earth phosphates. *Coord. Chem. Rev.* **2017**, *340*, 266–297. [CrossRef]

17. Errandonea, D.; Gomis, O.; Rodriguez-Hernandez, P.; Munoz, A.; Ruiz-Fuertes, J.; Gupta, M.; Achary, S.N.; Hirsch, A.; Manjon, F.J.; Peters, L.; et al. High-pressure structural and vibrational properties of monazite-type $BiPO_4$, $LaPO_4$, $CePO_4$, and $PrPO_4$. *J. Phys. Condens. Matter.* **2018**, *30*, 065401. [CrossRef] [PubMed]

18. Heffernan Karina, M.; Ross, N.L.; Spencer, E.C.; Boatner, L.A. The structural response of gadolinium phosphate to pressure. *J. Solid State Chem.* **2016**, *241*, 180–186. [CrossRef]

19. Wilkinson, T.M.; Wu, D.; Musselman, M.A.; Li, N.; Mara, N.; Packard, C.E. Mechanical behavior of rare-earth orthophosphates near the monazite/xenotime boundary characterized by nanoindentation. *Mater. Sci. Eng. A* **2017**, *691*, 203–210. [CrossRef]

20. Feng, J.; Xiao, B.; Zhou, R.; Pan, W. Anisotropy in elasticity and thermal conductivity of monazite-type $REPO_4$ (RE = La, Ce, Nd, Sm, Eu and Gd) from first-principles calculations. *Acta Mater.* **2013**, *61*, 7364–7383. [CrossRef]

21. Rustad, J.R. Density functional calculation of enthalpies of formation of rare-earth orthophosphates. *Am. Mineral.* **2012**, *97*, 791–799. [CrossRef]

22. Li, Y.; Kowalski, P.M.; Blanca-Romero, A.; Vinograd, V.; Bosbach, D. Ab initio calculation of excess properties of La_{1-x} (Ln, An)$_x$$PO_4$ solid solutions. *J. Sol. State Chem.* **2014**, *220*, 137–141. [CrossRef]

23. Kowalski, P.M.; Beridze, G.; Vinograd, V.L.; Bosbach, D. Heat capacities of lantanides and actinide monazite-type ceramics. *J. Nucl. Mater.* **2015**, *464*, 147–154. [CrossRef]

24. Kowalski, P.M.; Li, Y. Relationship between the thermodynamic excess properties of mixing and the elastic moduli in the monazite-type ceramics. *J. Eur. Ceram. Soc.* **2016**, *36*, 2093–2096. [CrossRef]

25. Gomis, O.; Lavina, B.; Rodriguez-Hernandez, P.; Munoz, A.; Errandonea, R.; Errandonea, D.; Bettinelli, M. High-pressure structural, elastic, and thermodynamic properties of zircon-type $HoPO_4$ and $TmPO_4$. *J. Phys. Condens. Matter.* **2017**, *29*, 095401. [CrossRef] [PubMed]

26. Errandonea, D.; Pellicer-Porres, J.; Martinez-Garcia, D.; Ruiz-Fuertes, J.; Friedrich, A.; Morgenroth, W.; Popescu, C.; Rodriguez-Hernandez, P.; Munoz, A.; Bettinelli, M. Phase Stability of Lanthanum Orthovanadate at High Pressure. *J. Phys. Chem. C* **2016**, *120*, 13749–13762. [CrossRef]

27. Errandonea, D.; Munoz, A.; Rodriguez-Hernandez, P.; Gomis, O.; Achary, S.N.; Popescu, C.; Patwe, S.J.; Tyagi, A.K. High-Pressure Crystal Structure, Lattice Vibrations, and Band Structure of $BiSbO_4$. *Inorg. Chem.* **2016**, *55*, 4958–4969. [CrossRef] [PubMed]

28. Errandonea, D.; Garg, A.B. Recent progress on the characterization of the high-pressure behaviour of AVO_4 orthovanadates. *Prog. Mater. Sci.* **2018**, *97*, 123–169. [CrossRef]

29. Achary, S.N.; Errandonea, D.; Munoz, A.; Rodriguez-Hernandez, P.; Manjon, F.J.; Krishna, P.S.R.; Patwe, S.J.; Grover, V.; Tyagi, A.K. Experimental and theoretical investigations on the polymorphism and metastability of $BiPO_4$. *Dalton Trans.* **2013**, *42*, 14999–15015. [CrossRef] [PubMed]

30. Errandonea, D.; Munoz, A.; Rodriguez-Hernandez, P.; Proctor, J.E.; Sapina, F.; Bettinelli, M. Theoretical and experimental study of the crystal structures, lattice vibrations, and band structures of monazite-type $PbCrO_4$, $PbSeO_4$, $SrCrO_4$, and $SrSeO_4$. *Inorg. Chem.* **2015**, *54*, 7524–7535. [CrossRef] [PubMed]

31. Hohenberg, P.; Kohn, W. Inhomogeneous electron gas. *Phys. Rev. B* **1964**, *136*, B864–B871. [CrossRef]

32. Kresse, G.; Furthmuller, J. Efficiency of ab-initio total energy calculations for metals and semiconductors using a plane-wave basis set. Comput. *Mater. Sci.* **1996**, *6*, 15–50.

33. Kresse, G.; Hafner, J. Ab-Initio molecular-dynamics simulation of the liquid-metal amorphous-semiconductor transition in germanium. *Phys. Rev. B* **1994**, *49*, 14251–14269. [CrossRef]

34. Kresse, G.; Hafner, J. Ab initio molecular-dynamics for liquid-metals. *Phys. Rev. B* **1993**, *47*, 558–561. [CrossRef]

35. Blöchl, P.E. Projector augmented-wave method. *Phys. Rev. B* **1994**, *50*, 17953–17979. [CrossRef]

36. Kresse, G.; Joubert, D. From ultrasoft pseudopotentials to the projector augmented-wave method. *Phys. Rev. B* **1999**, *59*, 1758–1775. [CrossRef]

37. Gleissner, J.; Errandonea, D.; Segura, A.; Pellicer-Porres, J.; Hakeem, M.A.; Proctor, J.E.; Raju, S.V.; Kumar, R.S.; Rodriguez-Hernandez, P.; Munoz, A.; et al. Monazite-type $SrCrO_4$ under compression. *Phys. Rev. B* **2016**, *94*, 134108. [CrossRef]

38. Perdew, J.P.; Ruzsinszky, A.; Csonka, G.I.; Vydrov, O.A.; Scuseria, G.E.; Constantin Lucian, A.; Zhou, X.; Burke, K. Restoring the density-gradient expansion for exchange in solids and surfaces. *Phys. Rev. Lett.* **2008**, *100*, 136406. [CrossRef] [PubMed]

39. Monkhorst, H.J.; Pack, J.D. Special points for Brillouin-zone integrations. *Phys. Rev. B* **1976**, *13*, 5188–5192. [CrossRef]
40. Mujica, A.; Rubio, A.; Munoz, A.; Needs, R.J. High-pressure phases of group-IV, III-V, and II-VI compounds. *Rev. Mod. Phys.* **2013**, *75*, 863–912. [CrossRef]
41. Nielsen, O.H.; Martin, R.M. Quantum-mechanical theory of stress and force. *Phys. Rev. B* **1985**, *32*, 3780–3791. [CrossRef]
42. Chetty, N.; Munoz, A.; Martin, R.M. 1st-principles calculation of the elastic-constants of AlAs. *Phys. Rev. B* **1989**, *40*, 11934–11936. [CrossRef]
43. Le Page, Y.; Saxe, P. Symmetry-general least-squares extraction of elastic data for strained materials from ab initio calculations of stress. *Phys. Rev. B* **2002**, *65*, 104104. [CrossRef]
44. Parlinski, K. Computer Code Phonon. Available online: http://www.computingformaterials.com/ (accessed on 14 July 2014).
45. Blanca-Romero, A.; Kowalski, P.M.; Beridze, G.; Schlenz, H.; Bosbach, D. Performance of DFT plus U Method for Prediction of Structural and Thermodynamic Parameters of Monazite-Type Ceramics. *J. Comput. Chem.* **2014**, *35*, 1339–1346. [CrossRef] [PubMed]
46. Birch, F. Finite elastic strain of cubic crystals. *Phys. Rev.* **1947**, *71*, 809–824. [CrossRef]
47. Du, A.; Wan, C.; Qu, Z.; Pan, W. Thermal Conductivity of Monazite-Type REPO$_4$ (RE = La, Ce, Nd, Sm, Eu, Gd). *J. Am. Ceram. Soc.* **2009**, *92*, 2687–2692. [CrossRef]
48. Li, H.; Zhang, S.; Zhou, S.; Cao, X. Bonding Characteristics, Thermal Expansibility, and Compressibility of RXO$_4$ (R = Rare Earths, X = P, As) within Monazite and Zircon Structures. *Inorg. Chem.* **2009**, *48*, 4542–4548. [CrossRef] [PubMed]
49. Momma, K.; Izumi, F. VESTA 3 for three-dimensional visualization of crystal, volumetric and morphology data. *J. Appl. Crystallogr.* **2011**, *44*, 1272–1276. [CrossRef]
50. Nye, J.F. *Physical Properties of Crystals. Their Representation by Tensor and Matrices*; Oxford University Press: Oxford, UK, 1957.
51. Wallace, D.C. *Thermodynamics of Crystals*; Dover Publications: New York, NY, USA, 1998.
52. Grimvall, G.; Magyari-Koepe, B.; Ozolins, V.; Persson, K.A. Lattice instabilities in metallic elements. *Rev. Mod. Phys.* **2012**, *84*, 945–986. [CrossRef]
53. Born, M.; Huang, K. *Dynamical Theory of Crystal Lattices*; Clarendon Press: London, UK, 1954.
54. Wallace, D.C. Thermoelasticity of stressed materials and comparison of various elastic constants. *Phys. Rev.* **1967**, *162*, 776–789. [CrossRef]
55. Voigt, W. *Lehrbuch der Kristallphysik (mit Ausschluss der Kristalloptik)*; B.G. Teubner: Leipzig/Berlin, Germany, 1928.
56. Reuss, A. Berechnung der Fließgrenze von Mischkristallen auf Grund der Plastizitätsbedingung für Einkristalle. *J. Appl. Math. Mech.* **1929**, *9*, 49–58. [CrossRef]
57. Hill, R. The elastic behaviour of a crystalline aggregate. *Proc. Phys. Soc. Lond.* **1952**, *65*, 349–355. [CrossRef]
58. Zhao, X.S.; Shang, S.L.; Liu, Z.K.; Shen, J.Y. Elastic properties of cubic, tetragonal and monoclinic ZrO$_2$ from first-principles calculations. *J. Nucl. Mater.* **2011**, *415*, 13–17. [CrossRef]
59. Caracas, R.; Ballaran, T.B. Elasticity of (K, Na)AlSi$_3$O$_8$ hollandite from lattice dynamic calculations. *Phys. Earth Planet. Inter.* **2010**, *181*, 21–26. [CrossRef]
60. Brazhkin, V.V.; Lyapin, A.G.; Hemley, R.J. Harder than diamond: Dreams and reality. *Philos. Mag. A* **2002**, *82*, 231–253. [CrossRef]
61. Greaves, G.N.; Greer, A.L.; Lakes, R.S.; Rouxel, T. Poisson's ratio and modern materials. *Nat. Mater.* **2011**, *10*, 823–837. [CrossRef] [PubMed]
62. Pugh, S.F. Relations between the elastic moduli and the plastic properties of polycrystalline pure metals. *Philos. Mag.* **1954**, *45*, 823–843. [CrossRef]
63. Silva, E.N.; Ayala, A.P.; Guedes, I.; Paschoal, C.W.A.; Moreira, R.L.; Loong, C.K.; Boatner, L.A. Vibrational spectra of monazite-type rare-earth orthophosphates. *Opt. Mater.* **2006**, *29*, 224–230. [CrossRef]
64. Begun, G.M.; Beall, G.W.; Boatner, L.A.; Gregor, W.J. Raman-spectra of the rare-earth ortho-phosphates. *J. RAMAN Spectrosc.* **1981**, *11*, 273–278. [CrossRef]
65. Ruschel, K.; Nasdala, L.; Kronz, A.; Hanchar, J.M.; Toebbens, D.M.; Skoda, R.; Finger, F.; Moeller, A. A Raman spectroscopic study on the structural disorder of monazite-(Ce). *Mineral. Petrol.* **2012**, *105*, 41–55. [CrossRef]

66. Huang, T.; Lee, J.S.; Kung, J.; Lin, C.M. Study of monazite under high pressure. *Solid State Commun.* **2010**, *150*, 1845–1850. [CrossRef]

67. Zhao, Z.; Zhang, X.; Zuo, J.A.; Ding, Z.J. Pressure effect on optical properties and structure stability of LaPO$_4$:Eu^{3+} Microspheres. *J. Nanosci. Nanotechnol.* **2010**, *10*, 7791–7794. [CrossRef] [PubMed]

68. Zhao, Z.; Zuo, J.A.; Ding, Z.J. Pressure effect on optical properties and structure stability of LaPO$_4$:Eu^{3+} hollow spheres. *J. Rare Earths* **2010**, *28*, 254–257. [CrossRef]

69. Grüneisen, E. Concerning the thermic expansion of metals. *Ann. Phys.* **1910**, *33*, 33–64. [CrossRef]

70. Ruiz-Fuertes, J.; Errandonea, D.; Lopez-Moreno, S.; Gonzalez, J.; Gomis, O.; Vilaplana, R.; Manjon, F.J.; Munoz, A.; Rodriguez-Hernandez, P.; Friedrich, A.; et al. High-pressure Raman spectroscopy and lattice-dynamics calculations on scintillating MgWO$_4$: Comparison with isomorphic compounds. *Phys. Rev. B* **2011**, *83*, 214112. [CrossRef]

![crystals](crystals logo)

Article

High Pressure Induced Insulator-to-Semimetal Transition through Intersite Charge Transfer in NaMn$_7$O$_{12}$

Davide Delmonte [1,*], Francesco Mezzadri [1,2], Fabio Orlandi [3], Gianluca Calestani [1,2], Yehezkel Amiel [4] and Edmondo Gilioli [1]

1   IMEM-CNR Parco Area delle Scienze 37/A, 43124 Parma, Italy; francesco.mezzadri@unipr.it (F.M.); gianluca.calestani@unipr.it (G.C.); edi@imem.cnr.it (E.G.)
2   Department of Chemistry, Life Sciences and Environmental Sustainability, University of Parma, Parco Area delle Scienze 17/A, 43124 Parma, Italy
3   ISIS Facility, STFC Rutherford Appleton Laboratory, Oxon OX11 0QX, UK; fabio.orlandi@stfc.ac.uk
4   School of Physics and Astronomy, Tel-Aviv University, Tel-Aviv 69978, Israel; hezy.amiel@gmail.com
*   Correspondence: davide.delmonte@imem.cnr.it

Received: 11 January 2018; Accepted: 31 January 2018; Published: 3 February 2018

Abstract: The pressure-dependent behaviour of NaMn$_7$O$_{12}$ (up to 40 GPa) is studied and discussed by means of single-crystal X-ray diffraction and resistance measurements carried out on powdered samples. A transition from thermally activated transport mechanism to semimetal takes place above 18 GPa, accompanied by a change in the compressibility of the system. On the other hand, the crystallographic determinations rule out a symmetry change to be at the origin of the transition, despite all the structural parameters pointing to a symmetrizing effect of pressure. Bond valence sum calculations indicate a charge transfer from the octahedrally coordinated manganese ions to the square planar ones, likely favouring the delocalization of the carriers.

Keywords: strongly correlated systems; high pressure; charge ordering; MIT

1. Introduction

Strongly correlated oxides can display a wide spectrum of intriguing properties, including magnetoresistance, half-metallicity, superconductivity, heavy fermion behaviour, and metal-insulator transition (MIT), with potential applications such as spintronics, superconductive devices or magnetoresistive memories. Often, different perturbations, like magnetic or electrical fields, temperature, doping or strain and pressure, can modify the material's electronic band structure [1]. Aside from the obvious structural variations (shortening of the bond distances, increased density, Jahn-Teller (JT) distortion), the application of external pressure to strongly correlated systems may lead to a wide range of phenomena, such as quenching of the orbital moment, spin crossover (high- to low-spin transition), metal-metal inter-valence charge transfer, MIT, etc. Mott insulators are likely to show metallization at high pressure due to competition between Coulomb repulsion and bandwidth. The latter is more susceptible to applied pressure [2], and therefore, experiments in which the bandwidth can be controlled with pressure represent an experimental proof for the existing theories on the behaviour of the strongly correlated electron systems, as reported in MnO [3], BaCoS$_2$ [4], Ba$_2$IrO$_4$ [5], Ca$_2$RuO$_4$ [6] and in the organic compound bis(ethylenedithio)tetrathiafulvalene [7]. Quite often, pressure-induced MIT is associated with symmetry changes, as for instance in SrTiO$_3$ [8] or BiNiO$_3$ [9,10], where the suppression of charge disproportionation involving the Bi^{3+}/Bi^{5+} ions is reported to bring from a triclinic insulating phase to higher-symmetry metallic phase. A similar case is PbCrO$_3$ [11], where at ambient pressure, Cr^{3+}/Cr^{6+} charge disproportionation occurs with the 6s-6p hybridization of the Pb^{2+} orbitals with the oxygen 2p ones accounting for the insulating properties, while high pressure induces a single Cr^{4+} state which gives rise to the metallic phase. In some cases, however, isostructural MIT can take place when the crystallographic modifications induced by applied pressure modulate the structural features without symmetry breaking, as in the case of $PrNiO_3$ [12], metal osmium [13], FeAs [14] and VO_2 [15]. In WSe_2, the MIT is caused by the anisotropic character of the compressibility, which induces a structural transition characterized by a sliding of the 2D layers [16]. In MoS_2. similar phenomena are observed, but accompanied by a discontinuous change in the lattice parameters variation with pressure [17]. Other mechanisms may be at the origin of high temperature metallization, as in the case of $LaMnO_3$, where the phenomenon is claimed to be caused in the paramagnetic phase by orbital splitting of the majority-spin e_g bands [18,19]. The effect of pressure on ordering phenomena is crucial, as in $Pr_{1-x}Ca_xMnO_3$ [20], where the suppression of charge order is identified as the origin of the MIT, while in $Pr_{0.6}Ca_{0.4}Mn_{0.96}Al_{0.04}O_3$ the quenching of JT distortion [21] is the driving force of metallization.

"Quadruple-perovskite" manganites, having general formula $A\,Mn_3Mn_4O_{12}$, exhibit strong electronic correlations [22]. These materials are characterized by A site cation ordering, where $\frac{3}{4}$ of the original perovskite A sites are occupied by Mn^{3+} ions in an uncommon square planar coordination (Figure 1), only stabilized by high-pressure/high-temperature (HP/HT) synthesis. The valence of the A ion tunes the Mn^{3+}/Mn^{4+} ratio on the B site, determining the crystal symmetry based on the specific charge and orbital ordering, as reported for compounds with A = Na [23], Ca [24], (Sr, Cd) [25], Pb [26], La [27,28], Bi [29,30], Y [31], and Pr [32]. The high electronic correlations, the significant lattice distortions, the full oxygen occupation (contrary to what usually happens in the conventional "simple" perovskites) and the chemical order (hetero-valent manganese atoms occupy different crystallographic sites), make this class of materials an intriguing playground for the study of possible instabilities induced by external stimuli on both the structural and physical properties.

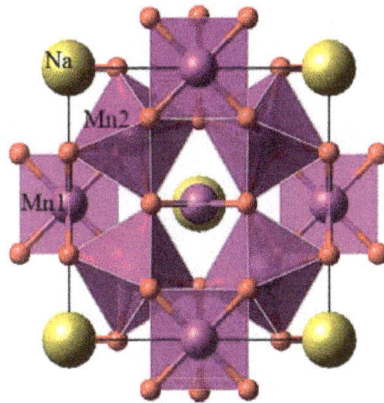

Figure 1. [100] projection of the quadruple perovskite structure having general formula $AA'_3B_4O_{12}$. In the present case, A, A' and B correspond to the Na1, Mn1 and Mn2 sites, respectively.

The sodium-substituted quadruple-perovskite manganite, $NaMn_7O_{12}$ ($NaMn_3Mn_4O_{12}$), shows features that make it particularly interesting for pressure-dependent crystallographic and physical studies. The presence of Na^+ at the A site induces a 1:1 ratio of Mn^{3+} and Mn^{4+} on the B site. At room temperature, the symmetry is cubic (*Im-3*) as a consequence of the delocalization of the Mn^{3+} e_g electrons [33], while below 175 K, the JT effect leads to a complete charge and orbital ordering, giving rise to superlattice reflections with propagation vector (1/2, 0, −1/2) observed by electron diffraction and powder synchrotron diffraction data [34]. The charge and orbital order scheme is then described by a monoclinic super-structure with *C2/m* symmetry, which, in agreement with band structure calculations [22], is characterized by a periodic scheme of JT apically elongated MnO_6 octahedra. The structural phase transition is accompanied by a change in both the resistivity and magnetic susceptibility and is clearly observed by diffraction techniques. The compound undergoes two different magnetic transitions at 125 K and 90 K related to the ordering of the Mn ions at the B and A' sites respectively. The weakly ferromagnetic transition observed below 125 K is related to spin canting of an E-type AFM structure with $(\frac{1}{2}\,0\,\frac{1}{2})$ propagation vector due to the large buckling of the MnO_6 octahedra. The 90 K transition is purely antiferromagnetic, and is ascribed to the ordering of the square planar Mn ions.

In the present work, the pressure dependent behaviour of $NaMn_7O_{12}$ is studied by means of single-crystal X-ray diffraction and resistance measurements up to about 40 GPa, showing the presence of an isostructural insulator-to-semimetal transition around 18 GPa, associated to an abrupt drop in the electrical resistance.

2. Experimental

$NaMn_7O_{12}$ samples were obtained by solid-state HP/HT reaction using a multi-anvil apparatus, as reported in Ref. [35]. The synthesis conditions were P = 6.0 GPa, T = 830 °C and reaction time 1 h. The $NaMn_7O_{12}$ crystals, having a typical size of 100 μm, were mechanically extracted by the bulk matrix. High-pressure electrical resistance measurements up to 40 GPa were performed on finely ground pure $NaMn_7O_{12}$ powder, with the van der Pauw method integrated in a Diamond Anvil Cell (DAC), using 300 μm culet anvils. A rhenium gasket was first filled with an insulator Al_2O_3-NaCl mixture (3:1 atomic ratio), which also acts as a pressure medium. The crystalline samples were placed inside a 100 μm cavity drilled within the pressed insulating layer. Six platinum thin strips were used as electrical probes for resistance measurements, as shown in Figure 2; the measurements were carried out in all cases using 4 of the 6 contacts in the Van der Pauw mode, the two "additional" Pt strips were placed to have backup contacts in case of breaking, which quite often happens as the pressure is increased. The Pt contacts were connected to copper leads, at the base of the diamond anvil, using a silver epoxy. At each pressure, under both compression and decompression cycles, the sample resistance was measured by the standard four-probe method as a function of temperature in a cryostat.

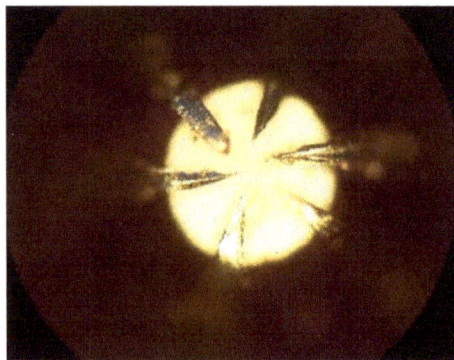

Figure 2. Diamond anvil cell's culet (300 μm diameter) with 6 Pt contacts.

Diffraction experiments were performed at the ID09 beam line of the ESRF (Grenoble, France) [36]. $NaMn_7O_{12}$ crystals were extracted from the HP/HT ceramic sample and introduced into a DAC with helium as a pressure medium to provide nearly hydrostatic conditions up to 40.8 GPa. For both resistivity and diffraction measurements, small rubies were inserted in the DAC, whose fluorescence was used as a pressure reference. The monochromatic X-ray radiation wavelength was set to λ = 0.41456 Å and diffraction data suitable for structure refinement were collected using a Mar555 flat panel detector in the range 2.5° < θ < 19° by 0.5° ω step in the angular interval ±30°. The structure refinement was carried out by the SHELXL software (SheldrickGoettingen, Germany) [37], making use of anisotropic atomic displacements parameters (a.d.p.) for all atoms.

3. Results and Discussion

3.1. Transport Analysis

In the "low-pressure" regime (P < 15 GPa), the DAC in situ electrical measurements performed on $NaMn_7O_{12}$ are in agreement with previous observations [33]. Electrical resistance vs. T show a semiconductor-like thermal activated transport mechanism related to the manganese e_g electrons delocalized by kT energy (Figure 3a). By increasing pressure, a sharp resistance drop is observed above 18 GPa, yielding a semimetallic behaviour (Figure 3b), with almost temperature invariance of the resistance vs. temperature. By lowering T, no decreasing trend of resistance is detected up to the maximum value of the applied pressure, so that the high-pressure state of $NaMn_7O_{12}$ behaves like a semimetal rather than a conventional metal.

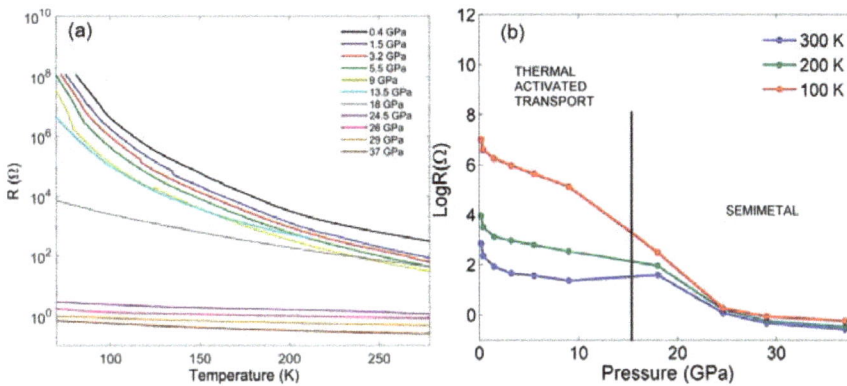

Figure 3. Electrical resistance vs. temperature of $NaMn_7O_{12}$ for different applied pressures (**a**) and electrical resistance vs. the applied pressure at different temperatures (**b**).

The Arrhenius plot derived from the R vs. T curves in the semiconducting regime shows the two critical temperatures of $NaMn_7O_{12}$, namely the charge ordering temperature T_{CO} = 176 K and the T_N = 90 K, distinguishable as deviations from the linear trend expected for a pure semiconductor, more evident at lower applied pressure (Figure 4). The E_A measured at the lowest reachable pressure (110 meV) is different from the value previously reported at ambient pressure (47 meV, [33]); measurements conducted on bulk pellets or ground powders are likely to give slightly different results, since the extrinsic phenomena and the occurrence of different percolative paths significantly affect the electrical transport mechanism. Moreover, the employed technique allows effective sample-wire contact only after the application of a (small) pressure; as a consequence, it was not possible to measure the electrical resistance below 0.4 GPa.

As the pressure increases, the charge-order transition anomaly becomes weaker and broader, in agreement with the symmetrizing effect of hydrostatic pressure, and is expected to finally lead to the melting of the charge and orbital orders. On the other hand, the anomaly observed at the T_N seems to move towards lower temperatures while getting more pronounced. By increasing pressure, the slope of the curves decreases, in accordance with the expected reduction of the thermally activated transport mechanism activation energy E_A, driven by the enhancement of the mechanical energy reducing the separation between the conduction and valence bands. Consequently, the E_A terms (obtained by the linear fit of the Arrhenius curves in the intermediate regime $T_N < T < T_{CO}$) plotted vs. the applied pressure, shows a linear decrease from 110 to 87 meV (Figure 5).

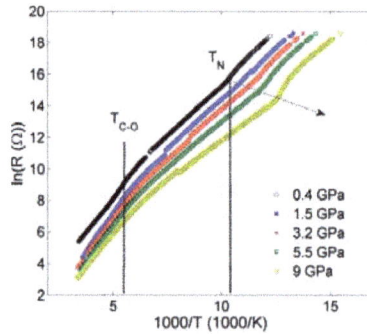

Figure 4. Arrhenius plot elaborated for the curves reported in Figure 2, limited to the thermal activated regime of $NaMn_7O_{12}$ (P < 15 GPa).

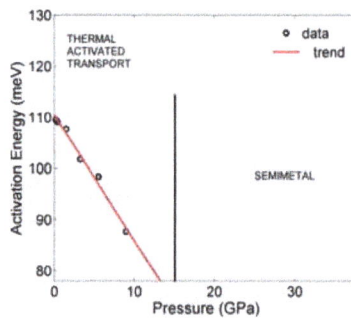

Figure 5. Calculated activation energy E_A vs. P in the "low pressure" thermal activated regime of $NaMn_7O_{12}$ (P < 15 GPa).

3.2. Structural Analysis

Single-crystal XRD experiments were carried out at room temperature in the 0.1–40.1 GPa range. Although the use of the DAC did not permit the collection of a significant portion of the Ewald sphere, the high symmetry of the analysed phase provides a sufficient number of equivalencies. The structure refined against the 0.3 GPa data is in excellent agreement with the previously reported crystallographic information: *Im-3* symmetry (S.G. number 204), compatible with the absence of charge and orbital ordering [33]. The Na ion lies at the *2a* site, while Mn1 and Mn2 are located at the *6b* and *8c* positions, respectively. Mn2 displays an octahedral coordination, while Mn1 a strongly distorted dodecahedral environment, often reported as square planar due to the large elongation of eight Mn-O bonds, four of which exceed 2.5 Å and four of which are above 3 Å (Figure 1).

It should be noted that all the atoms are located at high-symmetry positions with fixed x, y, z coordinates, except the sole oxygen atom, which is found at a low-symmetry site: 24 g with (0, y, z) coordinates.

While the *Im-3* symmetry and the crystal structure are retained through the entire investigated pressure range, a very small change in the compressibility coefficient is detectable around the electrical transition, as shown in Figure 6. The refined lattice parameters, relevant bond distances and Mn-O-Mn tilt angle are reported in Table 1. All the parameters show monotone variations in the investigated pressure range, with the cation-oxygen bond lengths decreasing with increasing pressure.

On the other hand, at about 18 GPa (matching the insulator to semimetal transition), the compressibility of the manganese-oxygen bond system abruptly decreases, with the Mn1 atom in square planar coordination displaying the most relevant variation (Figure 7).

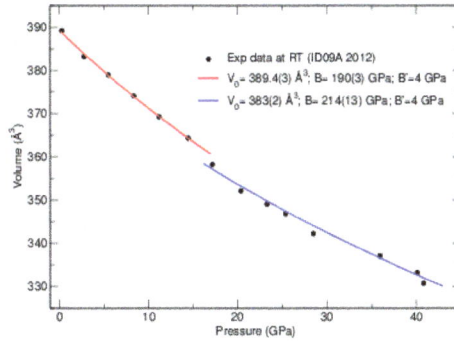

Figure 6. $NaMn_7O_{12}$ cell volume vs. pressure. Fittings were performed using the Birch–Murnaghan equation. B is the bulk modulus, and B′ is set to 4. A change in the compressibility is observed at about 16 GPa. Error bars are smaller than the corresponding symbols.

Table 1. Lattice parameter, cell volume and selected bond lengths of $NaMn_7O_{12}$ at different applied pressures.

P (GPa)	a (Å)	V (Å²)	Na1-O1 (Å)	Mn1-O1 (Å)	Mn1-O1 (Å)	Avg. Mn1-O1	Mn2-O1 (Å)	Mn2-O1-Mn2 (°)
0.3	7.301(2)	389.2(9)	2.644(7)	1.943(2)	2.6867(6)	2.3149(10)	1.916(7)	139.9(4)
2.8	7.264(2)	383.2(9)	2.629(4)	1.9320(15)	2.6734(6)	2.3027(8)	1.909(4)	140.1(2)
5.5	7.237(2)	379.0(9)	2.618(7)	1.924(3)	2.6648(6)	2.2944(15)	1.904(7)	140.2(4)
8.4	7.205(2)	374.0(9)	2.605(8)	1.918(3)	2.6610(12)	2.2895(16)	1.901(8)	140.3(4)
11.2	7.174(2)	369.3(9)	2.593(4)	1.9063(14)	2.6411(6)	2.2737(8)	1.891(4)	140.4(2)
14.5	7.142(2)	364.3(9)	2.582(3)	1.8974(11)	2.629(6)	2.263(3)	1.884(3)	140.45(16)
17.2	7.102(2)	358.3(9)	2.566(3)	1.8853(10)	2.6128(6)	2.2491(6)	1.878(3)	140.70(15)
20.4	7.062(2)	352.2(9)	2.546(4)	1.8745(13)	2.6042(6)	2.2394(7)	1.868(4)	140.7(2)
23.3	7.041(2)	349.1(9)	2.538(7)	1.867(2)	2.5932(6)	2.2301(10)	1.8682(7)	141.1(4)
25.4	7.026(2)	346.9(9)	2.534(4)	1.8637(14)	2.5881(6)	2.2259(8)	1.863(4)	142.0(2)
28.5	6.995(2)	342.3(9)	2.537(11)	1.857(10)	2.5606(6)	2.2088(6)	1.854(3)	141.2(6)
35.9	6.959(2)	333.2(9)	2.516(4)	1.8510(13)	2.5763(6)	2.2135(7)	1.859(4)	141.8(2)
40.1	6.933(2)	330.7(9)	2.490(6)	1.834(2)	2.5566(6)	2.1953(10)	1.853(6)	141.8(4)

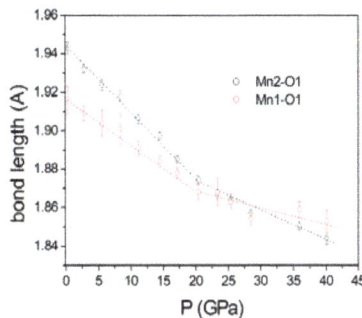

Figure 7. Pressure dependence of the manganese-oxygen bond lengths. A slope change is observed at 18 GPa, where the Mn1-O1 $\cong$ Mn2-O1 bonds.

Interestingly, the Mn-O-Mn bond angles (Figure 8) retain a linear trend over the whole pressure range, showing an opposite slope for the octahedral tilt angle (Mn2-O1-Mn2), which increases with pressure, contrary to (Mn1-O1-Mn2). This behaviour is a consequence of the symmetrizing effect of pressure, which forces the framework towards the undistorted simple perovskite geometry, where the Mn2-O1-Mn2 and Mn1-O1-Mn2 are 180° and 90°, respectively. In order to qualitatively evaluate the valence variations on the two symmetry-independent manganese ions induced by pressure, bond valence sum (BVS) calculations were performed by using the refined crystal data [38]. To this end, the Mn-O bond distances were normalized following the formula: $d_{norm} = d_P \cdot a_{P0}/a_P$, where d_P is the bond distance at a chosen pressure, a_P is the lattice parameter at the same pressure and a_{P0} is the $NaMn_7O_{12}$ lattice parameter at room pressure. This operation is required to eliminate from the calculation the simple effects of the compressibility, since the reference bond length used in BVS is valid at room pressure.

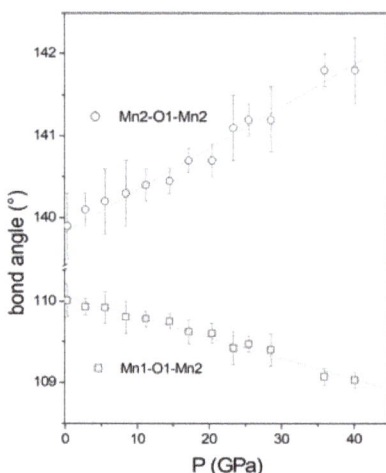

Figure 8. Mn-O-Mn bond angles in the 0–40 GPa pressure range. Mn2-O1-Mn2 monotonically evolves towards the 180° value, while Mn1-O1-Mn2 approaches 90°.

The obtained d_{norm} values and the corresponding valence of the manganese ions are reported in Figure 9. It should be noted that d_{norm} only depends on the oxygen atom displacement within the unit cell, given that the lattice parameter contribution to compressibility has been stripped out through normalization. Consequently, the linear trend displayed by the normalized values of Mn1-O1 and Mn2-O1 distances over the whole investigated pressure range, indicates that the anomaly previously observed in the refined bond length trends is directly related to the change in the lattice compressibility observed at the insulator-to-semimetal transition.

Moreover, the normalization procedure reveals a completely different behaviour for the Mn1 and Mn2 ions with respect to the unprocessed data. Indeed, the Mn-O bond distance in the square planar coordination increases by increasing the pressure, contrary to the octahedral site. The BVS values at the lowest reachable pressure (0.3 GPa), being 3.03 and 3.67 for the square planar and octahedral coordination, respectively, are quite close to the values expected by the nominal formula $NaMn^{III}_3(Mn^{III}_2Mn^{IV}_2)O_{12}$ (3 and 3.5). It is interesting to note that, when the average manganese valence is taken into account, the calculated value is not only in good agreement with the nominal one (3.29, given by 5 Mn^{3+} and 2 Mn^{4+} in the structure), but also remains practically constant, ranging from 3.39 to 3.36, over the whole pressure range. This indicates the substantial accuracy of the normalization applied to the bond distances and of the used model. On the other hand, it turns out that the system experiences, with increasing pressure, a relevant charge transfer from the Mn2 to the Mn1 sites. The BVS of Mn2 undergoes a 3% increase from 3.67 to 3.78, whereas those of Mn1 decreases from 3.03 to 2.79 (8%), causing the observed decrease of the normalized volume of the MnO_6 octahedra accompanied by the elongation of the normalized square planar bonds (Figure 9). On the basis of these results, the increase in pressure progressively extends the delocalization, even to the A' site electrons, playing an important role in determining the electric transition. Similar phenomena are reported for $LaCu_3Fe_4O_{12}$ [39], where intersite charge transfer at room temperature occurs by applying pressure between the A-site Cu and B-site Fe, leading to a first-order transition from $LaCu^{3+}_3Fe^{3+}_4O_{12}$ (at low pressure) to $LaCu^{2+}_3Fe^{3.75+}_4O_{12}$ (high pressure). The transition is accompanied by a structural phase transition related to the different charge ordering scheme yielding significant reduction of unit-cell volume and a change from antiferromagnetic-insulating to paramagnetic-metallic state. No symmetry change is observed in the present case, being $NaMn_7O_{12}$ not charge and orbital ordered at ambient conditions. As a consequence, the role of pressure in modifying the electronic band structure cannot be ruled out, but the charge transfer between octahedral to square planar sites probably plays the central role in inducing the semimetallic behaviour at high pressures instead of the pure metallic character observed in $LaCu_3Fe_4O_{12}$.

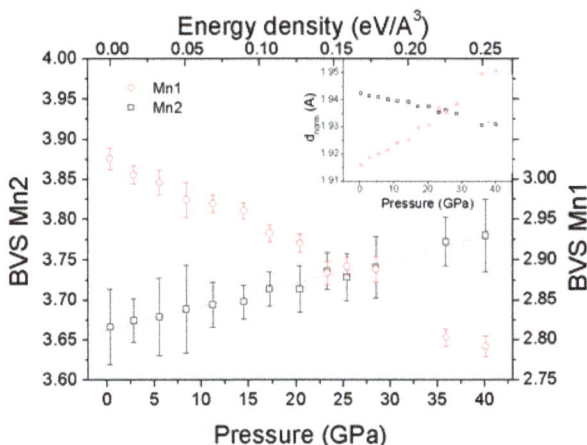

Figure 9. Bond Valence Sum and d_{norm} values (inset) as a function of pressure for the Mn1 and Mn2 atoms. The two sites display an opposite trend, pointing to an intersite charge transfer taking place from the octahedral to the square planar coordinated site.

4. Conclusions

The in situ, high-pressure powder electrical resistance measurements demonstrate that $NaMn_7O_{12}$, a metastable, charge-ordered manganite with quadruple-perovskite structure, undergoes an insulator-to-semimetal transition at about 18 GPa.

Across such transition, as refined by high-pressure single-crystal X-ray diffraction, $NaMn_7O_{12}$ retains the cubic (*Im-3*) structure with no symmetry breaking and non-relevant anomalies, or jump in the unit cell volume vs. P. On the contrary, charge transfer from the Mn2 (on the *B* site) to the Mn1 ions (on the A' site) is detected as the pressure increases, explaining the transformation of the system from an insulating to a semimetallic state. In this framework, the drop of the electric resistance can be interpreted as a significant electronic charge delocalization at the A' site favoured by the narrowing of the band gap.

Acknowledgments: The authors gratefully thank Andrea Prodi, Massimo Marezio, Andrea Gauzzi, Moshe Pasternak and Gregory Kh. Rozenberg for fruitful discussion. Edmondo Gilioli thanks the "SCENET exchange visits" program, funded by the NoE (Network of Excellence) SCENET (the European Network for Superconductivity, project No. BRRT CT98 5059). ESRF is acknowledged for provision of beamtime and Michael Hanfland and Marco Merlini for their assistance at the ID09A beamline.

Author Contributions: Edmondo Gilioli and Gianluca Calestani conceived the experiments; Yehezkel Amiel performed the DAC experiments; Davide Delmonte, Francesco Mezzadri, Fabio Orlandi, Gianluca Calestani and Edmondo Gilioli analyzed the data and contributed to the text; Davide Delmonte wrote, edited and submitted the paper.

Conflicts of Interest: The authors declare no conflict of interest.

References

1. Mizokawa, T. Metal-insulator transitions: Orbital control. *Nat. Phys.* **2013**, *9*, 612–613. [CrossRef]
2. Cui, C.; Tyson, T.A. Correlations between pressure and bandwidth effects in metal-insulator transitions in manganites. *Appl. Phys. Lett.* **2004**, *84*, 942–944. [CrossRef]
3. Yoo, C.S.; Maddox, B.; Klepeis, J.-H.P.; Iota, V.; Evans, W.; McMahan, A.; Hu, M.Y.; Chow, P.; Somayazulu, M.; Hausermann, D.; et al. First-Order isostructural mott transition in highly compressed MnO. *Phys. Rev. Lett.* **2005**, *94*, 115502. [CrossRef] [PubMed]
4. Yasui, Y.; Sasaki, H.; Sato, M.; Ohashi, M.; Sekine, Y.; Murayama, C.; Môri, N. Studies of pressure-induced mott metal-insulator transition of $BaCoS_2$. *J. Phys. Soc. Jpn.* **1999**, *68*, 1313–1320. [CrossRef]
5. Orii, D.; Sakata, M.; Miyake, A.; Shimizu, K.; Okabe, H.; Isobe, M.; Takayama-Muromachi, E.; Akimitsu, J. Pressure-induced metal-insulator transition of the Mott insulator Ba_2IrO_4. *J. Korean Phys. Soc.* **2013**, *63*, 349–351. [CrossRef]
6. Nakamura, F.; Goko, T.; Ito, M.; Fujita, T.; Nakatsuji, S.; Fukazawa, H.; Maeno, Y.; Alireza, P.; Forsythe, D.; Julian, S.R. From Mott insulator to ferromagnetic metal: A pressure study of Ca_2RuO_4. *Phys. Rev. B* **2002**, *65*, 220402. [CrossRef]
7. Oike, H.; Miyagawa, K.; Taniguchi, H.; Kanoda, K. Pressure-Induced mott transition in an organic superconductor with a finite doping level. *Phys. Rev. Lett.* **2015**, *114*, 067002. [CrossRef] [PubMed]
8. Dai, L.D.; Wu, L.; Li, H.P.; Hu, H.Y.; Zhuang, Y.K.; Liu, K.X. Evidence of the pressure-induced conductivity switching of yttrium-doped $SrTiO_3$. *J. Phys. Condens. Matter* **2016**, *28*, 475501. [CrossRef] [PubMed]
9. Ishiwata, S.; Azuma, M.; Takano, M. Pressure-induced metal-insulator transition in $BiNiO_3$. *Solid State Ion.* **2004**, *172*, 569–571. [CrossRef]
10. Cai, M.Q.; Yang, G.W.; Tan, X.; Cao, Y.L.; Wang, L.L.; Hu, W.Y.; Wang, G. Vacancy-driven ferromagnetism in ferroelectric $PbTiO_3$. *Appl. Phys. Lett.* **2007**, *91*, 101901. [CrossRef]
11. Cheng, J.; Kweon, K.E.; Larregola, S.A.; Ding, Y.; Shirako, Y.; Marshall, L.G.; Li, Z.-Y.; Li, X.; dos Santos, A.M.; Suchomel, M.R.; et al. Charge disproportionation and the pressure-induced insulator-metal transition in cubic perovskite $PbCrO_3$. *Proc. Natl. Acad. Sci. USA* **2015**, *112*, 1670–1674. [CrossRef] [PubMed]
12. Medarde, M.; Mesot, J.; Lacorre, P.; Rosenkranz, S.; Fischer, P.; Gobrecht, K. High-pressure neutron-diffraction study of the metallization process in $PrNiO_3$. *Phys. Rev. B* **1995**, *52*, 9248. [CrossRef]
13. Occelli, F.; Farber, D.L.; Badro, J.; Aracne, C.M.; Teter, D.M.; Hanfland, M.; Canny, B.; Couzinet, B. Experimental evidence for a high-pressure isostructural phase transition in osmium. *Phys Rev. Lett.* **2004**, *93*, 095502. [CrossRef] [PubMed]
14. Liu, Q.; Yu, X.; Wang, X.; Deng, Z.; Lv, Y.; Zhu, J.; Zhang, S.; Liu, H.; Yang, W.; Wang, L.; et al. Pressure-Induced isostructural phase transition and correlation of FeAs coordination with the superconducting properties of 111-Type Na1-xFeAs. *J. Am. Chem. Soc.* **2011**, *133*, 7892–7896. [CrossRef] [PubMed]
15. Arcangeletti, E.; Baldassarre, L.; Di Castro, D.; Lupi, S.; Malavasi, L.; Marini, C.; Perucchi, A.; Postorino, P. Evidence of a pressure-induced metallization process in monoclinic VO_2. *Phys. Rev. Lett.* **2007**, *98*, 196406. [CrossRef] [PubMed]
16. Wang, X.; Chen, X.; Zhou, Y.; Park, C.; An, C.; Zhou, Y.; Zhang, R.; Gu, C.; Yang, W.; Yang, Z. Pressure-induced iso-structural phase transition and metallization in WSe_2. *Sci. Rep.* **2017**, *7*, 46694. [CrossRef] [PubMed]

17. Chi, Z.H.; Zhao, X.M.; Zhang, H.; Goncharov, A.F.; Lobanov, S.S.; Kagayama, T.; Sakata, M.; Chen, X.J. Pressure-induced metallization of molybdenum disulfide. *Phys. Rev. Lett.* **2014**, *113*, 036802. [CrossRef] [PubMed]

18. Fuhr, J.D.; Avignon, M.; Alascio, B. Pressure-Induced Insulator-Metal Transition in LaMnO$_3$: A Slave-Boson Approach. *Phys. Rev. Lett.* **2008**, *100*, 216402. [CrossRef] [PubMed]

19. Loa, I.; Adler, P.; Grzechnik, A.; Syassen, K.; Schwarz, U.; Hanfland, M.; Rozenberg, G.K.; Gorodetsky, P.; Pasternak, M.P. Pressure-Induced quenching of the jahn-teller distortion and insulator-to-metal transition in LaMnO$_3$. *Phys. Rev. Lett.* **2001**, *87*, 125501. [CrossRef] [PubMed]

20. Moritomo, Y.; Kuwahara, H.; Tomioka, Y. Pressure effects on charge-ordering transitions in perovskite manganites. *Phys. Rev. B* **1997**, *55*, 7549–7556. [CrossRef]

21. Arumugam, S.; Thiyagarajan, R.; Kalaiselvan, G.; Sivaprakash, P. Pressure induced insulator-metal transition and giant negative piezoresistance in Pr$_{0.6}$Ca$_{0.4}$Mn$_{0.96}$Al$_{0.04}$O$_3$ polycrystal. *J. Magn. Magn. Mater.* **2016**, *417*, 69–74. [CrossRef]

22. Streltsov, S.V.; Khomskii, D.I. Jahn-Teller distortion and charge, orbital, and magnetic order in NaMn$_7$O$_{12}$. *Phys. Rev. B* **2014**, *89*, 201115. [CrossRef]

23. Marezio, M.; Dernier, P.D.; Chenavas, J.; Joubert, J.C. High pressure synthesis and crystal structure of NaMn$_7$O$_{12}$. *J. Solid State Chem.* **1973**, *6*, 16–20. [CrossRef]

24. Bochu, B.; Buevoz, J.L.; Chenavas, J.; Collomb, A.; Joubert, J.C.; Marezio, M. Bond lengths in 'CaMn$_3$' (Mn$_4$)O$_{12}$: A new Jahn-Teller distortion of Mn^{3+} octahedra. *Solid State Commun.* **1980**, *36*, 133–138. [CrossRef]

25. Belik, A.; Glazkova, Y.K.; Katsuya, Y.; Tanaka, M.; Sobolev, A.V.; Presniakov, A. Low-Temperature structural modulations in CdMn$_7$O$_{12}$, CaMn$_7$O$_{12}$, SrMn$_7$O$_{12}$, and PbMn$_7$O$_{12}$ perovskites studied by synchrotron X-ray powder diffraction and mössbauer spectroscopy. *J. Phys. Chem. C* **2016**, *120*, 8278–8288. [CrossRef]

26. Locherer, T.; Dinnebier, R.; Kremer, R.K.; Greenblatt, M.; Jansen, M. Synthesis and properties of a new quadruple perovskite: A-site ordered PbMn$_3$Mn$_4$O$_{12}$. *J. Solid State Chem.* **2012**, *190*, 277–284. [CrossRef]

27. Prodi, A.; Gilioli, E.; Cabassi, R.; Bolzoni, F.; Licci, F.; Huang, Q.; Lynn, J.W.; Affronte, M.; Gauzzi, A.; Marezio, M. Magnetic structure of the high-density Single-Valent e_g Jahn-Teller system LaMn$_7$O$_{12}$. *Phys. Rev. B* **2009**, *79*, 085105. [CrossRef]

28. Liu, X.J.; Lv, S.H.; Pan, E.; Meng, J.; Albrecht, J.D. First-principles study of crystal structural stability and electronic and magnetic properties in LaMn$_7$O$_{12}$. *J. Phys. Condens. Matter* **2010**, *22*, 246001. [CrossRef] [PubMed]

29. Mezzadri, F.; Calestani, G.; Calicchio, M.; Gilioli, E.; Bolzoni, F.; Cabassi, R.; Marezio, M.; Migliori, A. Synthesis and characterization of multiferroic BiMn$_7$O$_{12}$. *Phys. Rev. B* **2009**, *79*, 100106. [CrossRef]

30. Mezzadri, F.; Buzzi, M.; Pernechele, C.; Calestani, G.; Solzi, M.; Migliori, A.; Gilioli, E. Polymorphism and Multiferroicity in Bi$_{1-x/3}$(Mn$^{III}_3$)(Mn$^{III}_{4-x}$Mn$^{IV}_x$)O$_{12}$. *Chem. Mater.* **2011**, *23*, 3628–3635. [CrossRef]

31. Verseils, M.; Mezzadri, F.; Delmonte, D.; Baptiste, B.; Klein, Y.; Shcheka, S.; Chapon, L.C.; Hansen, T.; Gilioli, E.; Gauzzi, A. Effect of chemical pressure induced by La^{3+}/Y^{3+} substitution on the magnetic ordering of (AMn$_3$)Mn$_4$O$_{12}$ quadruple perovskite. *Phys. Rev. Mater.* **2017**, *1*, 064407. [CrossRef]

32. Mezzadri, F.; Calicchio, M.; Gilioli, E.; Cabassi, R.; Bolzoni, F.; Calestani, G.; Bissoli, F. High-pressure synthesis and characterization of PrMn$_7$O$_{12}$ polymorphs. *Phys. Rev. B* **2009**, *79*, 014420. [CrossRef]

33. Prodi, A.; Gilioli, E.; Gauzzi, A.; Licci, F.; Marezio, M.; Bolzoni, F.; Huang, Q.; Santoro, A.; Lynn, J. Charge, orbital and spin ordering phenomena in the mixed valence manganite (NaMn$^{3+}_3$)(Mn$^{3+}_2$Mn$^{4+}_2$)O$_{12}$. *Nat. Mater.* **2004**, *3*, 48–52.

34. Prodi, A.; Daoud-Aladine, A.; Gozzo, F.; Schmitt, B.; Lebedev, O.; van Tendeloo, G.; Gilioli, E.; Bolzoni, F.; Aruga-Katori, H.; Takagi, H.; et al. Commensurate structural modulation in the charge- and orbitally ordered phase of the quadruple perovskite (NaMn$_3$)Mn$_4$O$_{12}$. *Phys. Rev. B* **2014**, *90*, 180101. [CrossRef]

35. Gilioli, E.; Calestani, G.; Licci, F.; Gauzzi, A.; Bolzoni, F.; Prodi, A.; Marezio, M. P–T phase diagram and single crystal structural refinement of NaMn$_7$O$_{12}$. *Solid State Sci.* **2005**, *7*, 746–752. [CrossRef]

36. Merlini, M.; Hanfland, M. Single-crystal diffraction at megabar conditions by synchrotron radiation. *High Press. Res.* **2013**, *33*, 511–522. [CrossRef]

37. Sheldrick, G.M. Crystal structure refinement with SHELXL. *Acta Crystallogr. C* **2015**, *71*, 3–8. [CrossRef] [PubMed]

38. Brown, I.D.; Altermatt, D. Bond-valence parameters obtained from a systematic analysis of the Inorganic Crystal Structure Database. *Acta Crystallogr. B* **1985**, *41*, 244–247. [CrossRef]
39. Long, Y.; Kawakami, T.; Chen, W.; Saito, T.; Watanuk, T.; Nakakura, Y.; Liu, Q.; Jin, C.; Shimakawa, Y. Pressure effect on intersite charge transfer in a-site-ordered double-perovskite-structure oxide. *Chem. Mater.* **2012**, *24*, 2235–2239. [CrossRef]

crystals

MDPI

Review

Copper Delafossites under High Pressure—A Brief Review of XRD and Raman Spectroscopic Studies

Alka B. Garg [1,2,*] and Rekha Rao [2,3]

1 High Pressure and Synchrotron Radiation Physics Division, Bhabha Atomic Research Centre, Mumbai 400085, India
2 Homi Bhabha National Institute, Anushaktinagar, Mumbai 400094, India
3 Solid State Physics Division, Bhabha Atomic Research Centre, Mumbai 400085, India; rekhar@barc.gov.in
* Correspondence: alkagarg@barc.gov.in

Received: 10 May 2018; Accepted: 16 June 2018; Published: 19 June 2018

Abstract: Delafossites, with a unique combination of electrical conductivity and optical transparency constitute an important class of materials with their wide range of applications in different fields. In this article, we review the high pressure studies on copper based semiconducting delafossites with special emphasis on their structural and vibrational properties by synchrotron based powder X-ray diffraction and Raman spectroscopic measurements. Though all the investigated compounds undergo pressure induced structural phase transition, the structure of high pressure phase has been reported only for $CuFeO_2$. Based on X-ray diffraction data, one of the common features observed in all the studied compounds is the anisotropic compression of cell parameters in ambient rhombohedral structure. Ambient pressure bulk modulus obtained by fitting the pressure volume data lies between 135 to 200 GPa. Two allowed Raman mode frequencies E_g and A_{1g} are observed in all the compounds in ambient phase with splitting of E_g mode at the transition except for $CuCrO_2$ where along with splitting of E_g mode, A_{1g} mode disappears and a strong mode appears which softens with pressure. Observed transition pressure scales exponentially with radii of trivalent cation being lowest for $CuLaO_2$ and highest for $CuAlO_2$. The present review will help materials researchers to have an overview of the subject and reviewed results are relevant for fundamental science as well as possessing potential technological applications in synthesis of new materials with tailored physical properties.

Keywords: delafossites; high pressure; X-ray diffraction; phase transition; Raman spectroscopy

1. Introduction

Delafossites are ternary metal oxides belonging to a large family of compounds with the general formula $A^{+1}B^{+3}O^{2-}$, (A is monovalent cation Pt, Pd, Ag or Cu; B is trivalent transition metal). The primary member of the series is the mineral $CuFeO_2$ and the compounds adopting structure of $CuFeO_2$ are clubbed together as delafossites [1]. Scientific interest in this class of compounds is due to the diversity of the physical properties exhibited by them. They show a wide range of conductivity from insulating to metallic [2–4]. Most of the Cu and Ag based delafossites are semiconductors whereas Pt and Pd based compounds exhibit good metallic conductivity; with their room temperature in-plane conductivity reaching about a few $\mu\Omega$ cm, which is comparable to that of metallic elemental copper [5]. Along with good electrical conductivity, many of the delafossites show good transparency to optical photons and the compounds exhibiting the combination of these two properties are termed as transparent conducting oxides (TCO) [6–8]. Furthermore, depending on the donor or acceptor level in the band gap, they can show *p*-type or *n*-type conductivity. The origin of *p*-type conductivity in un-doped delafossites is either due to excess oxygen in the interstitials or copper vacancies [9]. Due to the unique combination of optical transparency and electrical conductivity, delafossites have been proposed to be useful in many areas, including in the solar energy industry [10], for liquid crystal displays [11], and in electro-chromatic materials for smart windows [12]. Thin films of a few delafossites find their uses as photocathodes to produce hydrogen by water splitting [13]. The catalytic activity of copper delafossites also finds applications in hydrogen production by decomposition of toxic H_2S gas [14]. Doping of a few delafossites also increases the *p*-type conductivity by an order of three [15,16]. Mineral $CuFeO_2$ finds its uses in medicine as a novel antimicrobial material [17], as an anode in lithium ion batteries [18], and as a gas sensor [19]. $CuCrO_2$ is being used as a catalyst for the production of chlorine [20]. Interestingly a few of these compounds also exhibit negative thermal expansion (NTE) behavior [21], which is attributed to the anharmonicity of linear O-Cu-O bond along the *c*-axis. A few members of the delafossite family where *B* atom is magnetic have attracted interest due to the multiferroic properties, wherein ferroelectricity is induced by magnetic ordering [22]. Low temperature investigations of these materials led to interesting magnetic behavior along with spin lattice coupling [23]. Various synthesis routes are also being employed to engineer different polymorphs of these semiconductors for applications as an absorber in solar cells by manipulating their band gaps. In fact, synthesis of tetrahedral structured wurtzite analogues with the same general formula ABO_2 has been successfully achieved [24]. These are direct narrow band gap semiconductors, unlike the rhombohedral modification where the band gap is indirect. Synthesis of delafossites in nanophase such as nanoparticles, nanowires, nanoplates and investigating the particle size effect on the properties of these materials is another interesting field which is currently being explored. Eu^{3+} doped $CuAlO_2$ single phase nanofibers prepared via an electrospinning technique show strong photoluminescence spectra with emission bands at 405 and 610 nm due to the intrinsic near-band-edge transition of $CuAlO_2$ and the f-f transition of the Eu^{3+} activator respectively [25]. Nanoplates of $CuGaO_2$ synthesized via low temperature hydrothermal method exhibited a blue emission at room temperature and free exciton emission at low temperature. P-N junction fabricated by these nanoparticles of *p*-type $CuGaO_2$ and *n*-type nano ZnO exhibited enhanced photocatalytic activity and light absorption properties [26]. Investigation of structural and vibrational behavior of $CuAlO_2$ as a function of particle size shows the expansion of lattice parameters and the cell volume with the reduction of particle size. Raman spectra shows large red shifts ($\sim$60 cm^{-1}) and line broadening ($\sim$50 cm^{-1}) as the particle size becomes of the order of 13 nm [27]. The electrochemical performance of nano $CuAlO_2$ with an average particle size of $\approx$20 nm demonstrated 12 times more catalytic activity in the electrolysis of water to the hydrogen evolution reactions and oxygen evolution reactions compared to bulk $CuAlO_2$ [28]. Research on thin films of delafossite is another interesting and widely investigated area for their potential uses in optoelectronic device fabrication [29–32]. As has been well established by the high pressure scientific community, compression is yet another way to engineer the crystal/electronic structure of materials, producing compounds with entirely different set of physical properties without altering their chemical composition [33,34]. The focus of the present article is to comprehend the high pressure studies on delafossite structured compounds till date in general, with special emphasis on their X-ray diffraction and Raman spectroscopic investigations. With the exception of one preliminary study on $PdCoO_2$ [35], most of the high pressure work has been focused on copper based delafossites, probably because of the difficulty in preparing the Ag/Pt/Pd based delafossites. Due to low decomposition temperature of their respective oxides, special synthesis techniques are required to synthesize Ag/Pt/Pd based delafossites [36,37]. It is to be noted here, that since it is easier to synthesize the rhombohedral polytype of delafossites, most of the experimental high pressure studies carried out are on rhombohedral structured delafossites [38].

2. Ambient Crystal Structure and Vibrational Properties of ABO_2

As shown in Figure 1, delafossite crystallizes in layered structure with hexagonal symmetry and can exist in two structural polymorphs, namely rhombohedral 3R with space group: *R-3m*, or hexagonal 2H with space group: *P6$_3$/mmc*. The building block of both the polytypes is layers of edge-connected BO_6 octahedra and triangular metallic planes of monovalent element *A* which are stacked along the c-axis. A cation is linearly coordinated with oxygen of upper and lower BO_6 layers. The difference in the two structures is the different orientation of the triangular metallic plane along the c-axis stacking. When two consecutive A layers are stacked with each layer rotated by 180° with respect to one another, the structure is hexagonal. Alternatively, when the layers are stacked in the same direction with respect to one another, resultant structure is rhombohedral. In the primitive rhombohedral cell, there are only four atoms: one A, one B and two oxygen atoms. However, in the triple hexagonal cell which is conventionally used to describe this structure, A and B cations occupy 3a (0,0,0) and 3b (0,0,0.5) Wyckoff positions respectively. The O atoms are situated at 6c (0,0,u) positions [39,40]. Each element in the delafossite structure forms the triangular lattice and stacks along the c-axis in the sequence B^{3+}-O^{2-}-A^+-O^{2-}-B^{3+}. One of the key features delafossite structure has in common with other layered structured oxides is its extremely accommodating nature and flexibility to host many different elements. Another structure commonly adopted by the ABO_2 compounds and related to delafossite is ordered rock salt. This structure too has a similar triangular lattice with same space group (R-3m), the only difference is in the stacking pattern of O^{2-}-A^+-O^{2-}layers. While the delafossite structure has a straight stacking, the ordered rock salt structure has a zigzag one (Figure 1). In both cases the rhombohedral (ABCABC...) stacking is realized among B^{3+} layers, although the distance between them is much shorter in the latter case.

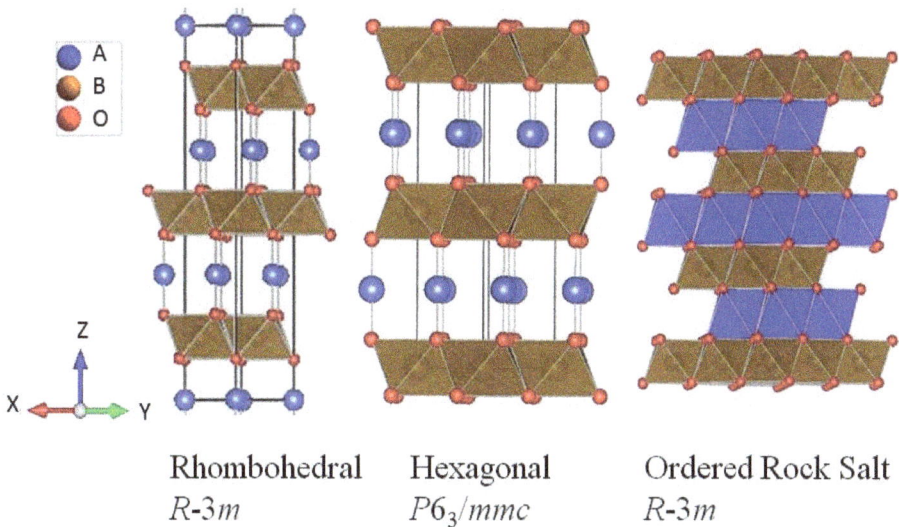

| Rhombohedral | Hexagonal | Ordered Rock Salt |
| R-3m | P6$_3$/mmc | R-3m |

Figure 1. Various crystal structures adopted by ABO_2 layered compounds. Rhombohedral and hexagonal polytypes adopted by delafossites have linear A-O bonding while in ordered rock salt structure, octahedra is formed around both A and B atoms.

As has been mentioned earlier, ABO_2 delafossites with rhombohedral structure (space group R-3m), consist of one formula unit of ABO_2 per primitive cell with four atoms, resulting in twelve vibrational modes at the zone center, which transform as $\Gamma = A_{1g} + E_g + 3A_{2u} + 3E_u$. Modes denoted by subscript g are Raman active modes and modes denoted by subscript u are infrared active which also includes acoustic modes $A_{2u}+E_u$. In Figure 2, we show the pictorial representation of the two Raman active modes. Delafossites with hexagonal structure consist of two formula units of ABO_2 per primitive cell with A at 2c, B at 2a and O at 4f position resulting in 24 modes of vibration at the zone center. Based on the structural information, using the nuclear site method [41], these vibrational modes are found to transform as $\Gamma = A_{1g} + E_{1g} + 2E_{2g} + 2B_{1g} + 3A_{2u} + 3E_{1u} + 2E_{2u} + 2B_{2u}$. Out of these A_{1g}, E_{1g}, E_{2g} and B_{1g} are Raman active modes while A_{2u}, E_{1u}, E_{2u} and B_{2u} are infrared active modes.

So, at ambient conditions, six distinct Raman modes are expected in hexagonal structure. Finally, in the ordered rock salt as the atoms occupy the same sites as in delafossite rhombohedral structure, the expected Raman modes are same in both structures. Table 1 shows the structural and vibrational details of the three structures.

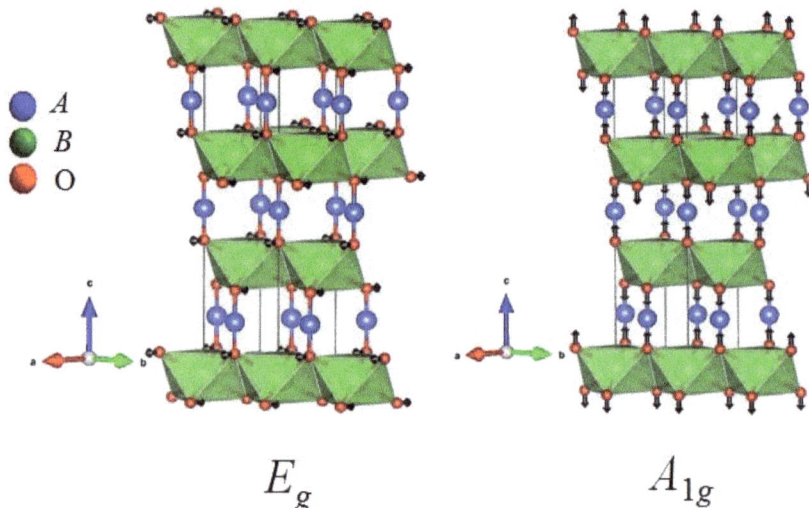

Figure 2. Eigen vectors for E_g and A_{1g} modes of vibration in ABO_2 delafossite compounds.

Table 1. Structural and vibrational details of ABO_2 type compounds.

Hexagonal $P6_3/mmc$, $Z = 2$	Atomic coordinates			Vibrations at the zone centre							
Wyckoff position	x	y	z	A_{1g}	E_{1g}	E_{2g}	B_{1g}	A_{2u}	E_{1u}	E_{2u}	B_{2u}
Monovalent cation A at 2c	1/3	2/3	1/4	-	-	1	1	1	1	-	-
Trivalent cation B at 2a	0	0	0	-	-	-	-	1	1	1	1
Oxygen at 4f	1/3	2/3	0.0892	1	1	1	1	1	1	1	1
Rhombohedral R-$3m$, $Z = 1$	Atomic co-ordinate			Vibrations at the zone centre							
Wyckoff position	x	y	z	A_{1g}	E_g	A_{2u}	E_u				
Monovalent cation A at 3b	0	0	0	-	-	1	1				
Trivalent cation B at 3a	0	0	1/2	-	-	1	1				
Oxygen at 6c	0	0	0.108	1	1	1	1				
Ordered rock salt R-$3m$, $Z = 1$	Atomic co-ordinate			Vibrations at the zone centre							
Wyckoff position	x	y	z	A_{1g}	E_g	A_{2u}	E_u				
Monovalent cation A at 3a	0	0	1/2	-	-	1	1				
Trivalent cation B at 3b	0	0	0	-	-	1	1				
Oxygen at 6c	0	0	0.743	1	1	1	1				

3. High Pressure Studies

The first high pressure in-situ X-ray diffraction (HP-XRD) measurement was reported on copper iron oxide, the representative of the series by Zhao et al. [42] in the year 1996. The measurement was carried out on powdered sample up to 10 GPa. Authors found an increase in c/a ratio indicating lattice anisotropic compression; however the ambient phase was found to be stable in the studied pressure range. On fitting the pressure volume data to 3rd order Birch-Murnaghan equation of state (BM-EOS) [43], ambient pressure bulk modulus, B_0, of the compound was reported to be 156 GPa, with its pressure derivative $B_0' = 2.6$. Nearly seven years later, Hasegawa et al. reported the HP-XRD measurements on metallic delafossite $PdCoO_2$ up to 10 GPa [35]. Structurally this compound was also found to be stable with anisotropic compression of lattice -parameters and increase in *c/a* ratio. The reported value of bulk modulus for the compound based on 3rd order BM-EOS fitting of experimental pressure volume data is 224(2) GPa with pressure derivative of bulk modulus $B_0' = 0.7(0.5)$. High value of bulk modulus and low value of B_0' indicates highly incompressible nature of the compound [43]. In both these measurements, methanol:ethanol in 4:1 ratio was used as pressure transmitting medium. Subsequently, with the availability of bright synchrotron sources, evolution in the diamond anvil cell (DAC) and detector technology, many of the delafossites have been investigated under high pressure with increased pressure range. Revisiting the compression behavior of $CuFeO_2$ up to 30 GPa using XRD along with ^{57}Fe Mössbauer and Fe & Cu K-edge X-ray absorption spectroscopy methods, reveal a sequence of electronic-magnetic pressure-induced transitions along with structural transition to more isotropic $C2/c$ structure with onset of long range antiferromagnetic order at 18 GPa. Beyond 23 GPa, interionic valence exchange between Cu and Fe leads to a four-fold coordinated Cu, resulting in another crystallographic structure with space group P-$3m$. All the observed transitions are reversible with minimal hysteresis [44]. However, a neutron diffraction experiment on isotropically compressed $CuFeO_2$ indicated suppression of long range magnetic ordering at around 7.9 GPa [45]. X-ray diffraction data on $CuGaO_2$, collected up to 28.1 GPa, at two different temperatures, indicated pressure induced phase transition in the compound beyond 24 GPa [46]. As observed in $CuFeO_2$ and $PdCoO_2$, anisotropy in the axial compressibility was also observed in this compound. Though the transition was found to be irreversible, no details of the high pressure phase were provided in this article. A report on vibrational behavior up to 33.3 GPa on the same compound followed in the year 2005, which happens to be the first delafossite whose vibrational properties were investigated experimentally under high pressure [47]. Based on splitting of the E_g mode, authors reported a structural phase transition in the compound beyond 26 GPa. Raman measurements on single crystal of $CuAlO_2$ up to 48 GPa indicated a pressure driven phase transition at around 34 ($\pm$2) GPa, which is completed by 37 ($\pm$2) GPa [48]. Raman data on the pressure cycled sample showed the presence of two modes as observed in the ambient sample, indicating the reversibility of the phase transition. Based on density functional theory, the phase transition is related to the dynamic instability in the compound [49]. High quality X-ray diffraction and X-ray absorption measurements on $CuAlO_2$, also indicated the presence of phase transition around 35($\pm$2) GPa [50]. However, first-principles calculations on $CuAlO_2$ under high pressure showed transformation to a leaning delafossite structure at 60 GPa with an increased energy gap due to the enhanced covalency of Cu 3d and O 2p states [51]. Optical absorption measurements on thin films of $CuAlO_2$ (indirect band gap) and $CuScO_2$ (direct band gap) up to 20 GPa indicated two phase transitions in $CuScO_2$ at 13 and 18 GPa [52], however the structures of high pressure phases have not been identified. High pressure behavior of $CuInO_2$, which is the only copper based delafossite that can be doped with both *n*- and *p*-type ion [53], has not been investigated experimentally, however, its structural, elastic, mechanical and optical properties have been reported by first-principles density-functional theory [54]. The two polytypes of the compound with 3R and 2H phases become unstable beyond 9.3 and 8.7 GPa with the value of bulk modulus as 121 and 117 GPa which are nearly 20% less than the earlier reported values of 156 and 146 GPa respectively [55]. The dielectric, ferroelectric and *ac* calorimetric measurements on $CuCrO_2$ have revealed the increase in magnetic transition temperature T_N remarkably on pressurization. However, the magnitude of the dielectric anomaly at T_N is suppressed by applying pressure and the magnitude of the spontaneous polarization below T_N is abruptly suppressed at around 8 GPa [56]. We have investigated the high pressure behavior of $CuCrO_2$ and $CuLaO_2$ using synchrotron based X-ray diffraction and Raman spectroscopic technique on polycrystalline samples followed by Raman studies on single crystal of $CuFeO_2$. For all these XRD measurements, Mao-Bell type of diamond anvil cell with stainless steel gasket pre-indented to a thickness of 40–80 μm and central hole of 100–200 μm was employed as the sample chamber. Methanol:ethanol in 4:1 ratio by volume was used as pressure transmitting medium and ruby fluorescence technique was employed for in-situ pressure calibration [57]. In X-ray diffraction measurements, equation of state data of standard like gold/copper was used for pressure calibration [58]. Rietveld/Lebail analysis of the XRD data was carried out using

GSAS software [59]. All Raman spectroscopic measurements were carried out using a 532 nm laser in back scattering geometry. Polycrystalline samples of all these compounds were synthesized using conventional solid state route. Single crystals of $CuFeO_2$ used in the present work were grown by the floating zone technique [60]. Readers can refer to earlier publications to get more detail about the sample synthesis and experimental details.

In Figure 3, we show the refined ambient XRD data on $CuCrO_2$, $CuFeO_2$ and $CuLaO_2$ along with residuals while their Raman spectra are depicted in Figure 4. All the observed diffraction peaks could be fitted with rhombohedral symmetry, indicating the single phase formation of these compounds. Raman data of all three compounds have two prominent Raman modes along with a few disorder induced non-zone center modes. While the vibrations in the direction of Cu-O bonds along the c-axis are represented by A_{1g} modes, vibrations in the direction perpendicular to c-axis correspond to E_g modes (Figure 2). As seen in Figure 4, both the frequencies in all three compounds shift to higher values as the ionic radii of trivalent cation decreases from La^{3+} to Cr^{3+} which is a consequence of lattice contraction due to decrease in B^{3+} ionic radii. Higher frequency modes are identified to be A_{1g} and the lower frequency mode as E_g from *ab-initio* calculations [48] as well as polarized Raman measurements on single crystals [61]. In Table 2, we give ionic radii of various trivalent cations [62] along with lattice parameters and Raman frequencies of copper delafossites in rhombohedral symmetry [63–67]. There are various efforts to substitute trivalent cation and investigate the effect of chemical doping on structural and vibrational properties on delafossite systems. Depending on the ionic radii, there is a contraction or expansion of the lattice which results in increase/decrease in the frequency of the Raman modes, particularly the E_g modes which are highly sensitive to the ionic radii of the trivalent cation. In doped $CuCrO_2$, the lattice parameters were found to vary according to Vegard's law with broadening in the reflection due to local lattice distortion as a result of difference in ionic radii between Cr^{3+} and trivalent dopants [68]. The effect of scandium doping in $CuCrO_2$ [69] and $CuFeO_2$ [70] indicated the softening of both the modes. Temperature dependence of the two modes of $CuFeO_2$ was found to decrease with increasing temperatures and the behavior was attributed to thermal expansion of the lattice and phonon–phonon interaction [71].

Table 2. Ionic radii, Raman mode frequencies, lattice parameters and bond-lengths for various copper delafossite compounds.

Delafossite	Ionic Radii of Trivalent Cation (Å) [62]	Raman Mode Frequency		Lattice Parameter		Bond-Length		Ref.
		E_g (cm^{-1})	A_{1g} (cm^{-1})	a (Å)	c (Å)	Cu-O (Å)	M-O (Å)	
$CuLaO_2$	1.032	318	652	3.8326	17.092	1.760	2.466	[63]
$CuPrO_2$	0.99			3.7518	17.086	1.789	2.411	[63]
$CuNdO_2$	0.983			3.7119	17.085	1.836	2.370	[63]
$CuSmO_2$	0.958			3.6628	17.078	1.880	2.325	[63]
$CuEuO_2$	0.947			3.6316	17.074	1.895	2.302	[63]
$CuYO_2$	0.90			3.5330	17.136	1.827	2.285	[64]
$CuInO_2$	0.8	378	678	3.2922	17.388	1.845	2.172	[65]
$CuScO_2$	0.745			3.2204	17.099	1.831	2.121	[66]
$CuFeO_2$	0.645	352	692	3.0351	17.166	1.835	2.033	[61]
$CuGaO_2$	0.62	368	729	2.9770	17.171	1.848	1.996	[46]
$CuCrO_2$	0.615	454	703	2.9767	17.111	1.8455	1.989	[67]
$CuAlO_2$	0.535	418	767	2.8584	16.958	1.8617	1.912	[50]

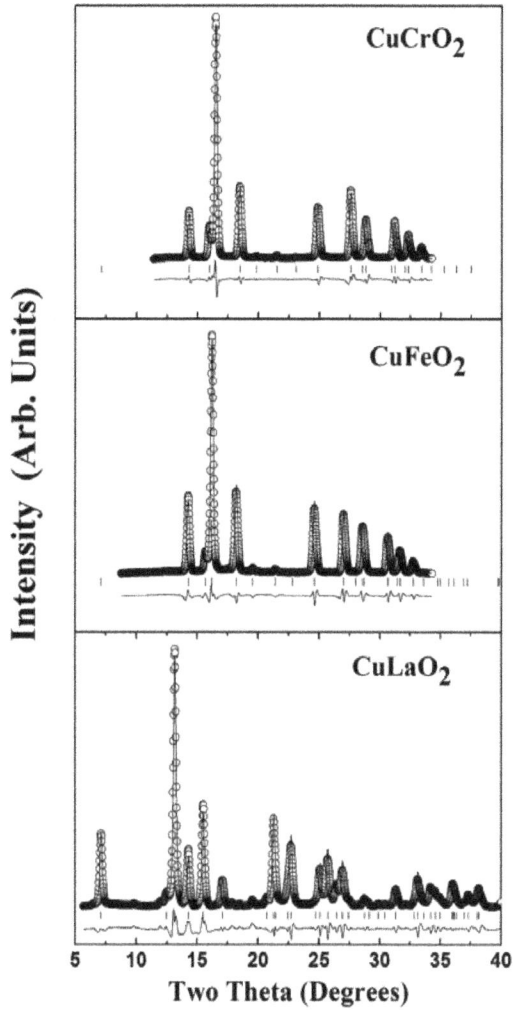

Figure 3. Rietveld refined ambient pressure and temperature X-ray diffraction patterns of as-synthesized CuLaO$_2$, CuFeO$_2$ and CuCrO$_2$ showing single phase formation of the compound in rhombohedral structure. Difference plot is also plotted. Vertical tick marks represent allowed reflection of delafossite structure with *R*-3*m* space group.

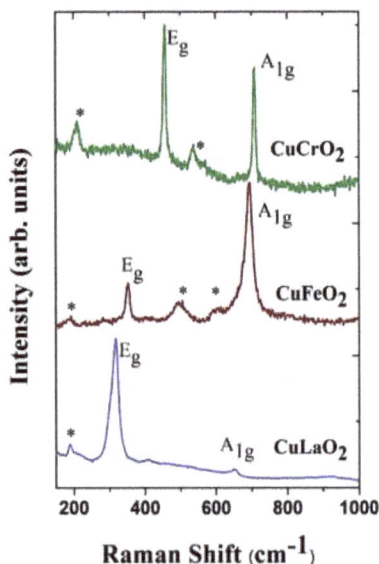

Figure 4. Raman spectra of as-synthesized $CuLaO_2$, $CuFeO_2$ and $CuCrO_2$, showing two allowed Raman modes, a few weak modes shown by asterisks are disorder induced non-zone centre modes.

Figure 5 shows the high pressure Raman data on $CuLaO_2$ which shows interesting sequence of phase transitions. Unlike other delafossites, structural transition in $CuLaO_2$ takes place at a relatively low pressure of 1.8 GPa. Appearance of several new modes beyond 1.8 GPa indicates lower symmetry of the high-pressure phase. The nature of changes observed in the Raman spectra at 1.8 GPa are similar to $CuAlO_2$ at 34 GPa [48]. Beyond 7 GPa, there is sudden loss of Raman intensity as the compound becomes opaque indicative of electronic/structure changes. The changes are irreversible from 8 GPa [72]. In Figure 6, pressure evolutions of XRD data of $CuLaO_2$ up to 36 GPa are shown. Data at 0.7 and 1.6 GPa could be fitted with the ambient structure. However, data collected at 4.2 GPa shows appearance of a few peaks at $2\theta = 6.6°$ and $13.8°$. On further pressurization, these two peaks build up in intensity while the peak intensity from ambient structure reduces. Data collected beyond 8 GPa, shows disappearance of the peaks corresponding to the first HP phase along with clusters of new peaks with broadening. This is the same pressure region where there is a complete loss of Raman intensity. Possible reasons for loss of Raman intensity could be that the second high pressure phase is Raman inactive or the reduction in band gap across the transition which results in increase in absorption. Indeed, our electrical resistance measurements under high pressure show a considerable drop in the resistance, indicating a reduction in band gap [72]. Compressibility was found to be highly anisotropic and further investigation to identify the high pressure structure is in progress. Pressure evolution of XRD data on $CuCrO_2$ (Figure 7) do not reveal any major changes in the data except for the shifting of diffraction peaks to higher angle, indicative of lattice compression. Refined pattern with residuals at two pressures are shown in Figure 8. Similar types of refinements were obtained for all the data points. High pressure Raman data on $CuCrO_2$ shown in Figure 9 indicates usual pressure hardening of both modes up to 24.5 GPa however major changes are observed beyond 24.5 GPa with splitting of E_g mode and appearance of a new mode which softens with pressure and grows in intensity with further pressurization. At 31 GPa, the delafossite modes completely disappear with an intense broad mode at lower frequency [73]. The features in the Raman spectra are similar to that seen in lithium intercalated compound $LiCoO_2$ [74] which crystallizes in closely related structure of layered rock salt. However, in our XRD measurements, we have not reached the pressures at which transition has been observed in Raman data. Experimental pressure volume data obtained by the Rietveld refinement [59] of XRD data for the low pressure phase when fitted to 3rd order Birch-Murnaghan equation of state results in ambient pressure bulk modulus as 154(25) and 156(2.8) GPa with their pressure derivative of bulk modulus as 4.8(0.5) and 5.3(0.5) for $CuLaO_2$ and $CuCrO_2$ respectively. It is to be noted here that bulk modulus and pressure derivative of bulk modulus are highly correlated [75]. In Figure 10 we show the normalized pressure volume data for $CuCrO_2$ and $CuLaO_2$. As one can see, the data of $CuLaO_2$ almost overlaps with that of $CuCrO_2$, indicating the similarity of bulk modulus of both compounds. Nearly same value of bulk modulus for the two compounds indicates that it is mainly the compression of Cu-O bonds which contribute to the overall compressibility of the compounds. Normalized a and c axis for $CuCrO_2$ and $CuLaO_2$ in the ambient pressure phase are plotted in Figure 11. One can clearly see an anisotropic compression of the axes in both the compounds. Interestingly, anisotropy in axial compressibility seems to be the only common feature of all the studied delafossites with a R-$3m$ structure, which results in the regularization of oxygen octahedra around B atom which is slightly distorted at ambient conditions [54]. High pressure Raman measurements on single crystals of $CuFeO_2$, shown in Figure 12a, could well reproduce the structural changes reported earlier [52]. While both the Raman modes harden under pressures up to 18 GPa, the first transition around 18 GPa to the reported monoclinic $C2/c$ phase is indicated by splitting of the E_g mode. This is accompanied by softening of A_g mode thereafter [76]. Figure 12b shows the pressure dependence of mode frequencies. Rapid softening of the high frequency mode is understood as due to the change in Fe-O bond-length which ultimately results in change in copper coordination leading to the second high pressure phase transition above 23 GPa. Unlike in $CuLaO_2$ and $CuAlO_2$, there are no additional modes in the low frequency region (around 100 cm^{-1}) in $CuFeO_2$ across the transition. In Figure 13, we have summarized the bulk modulus of various copper delafossites where except for $CuYO_2$ and $CuInO_2$, the experimental values have been plotted. The high value of bulk modulus for $CuAlO_2$ and $CuGaO_2$ are expected because of the covalent nature of B-O bond in these compounds, however we need more accurate measurements on a few other compounds to establish an empirical relation for bulk modulus in ABO_2 delafossites.

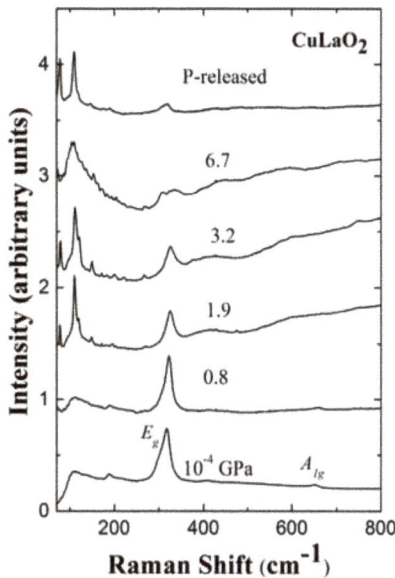

Figure 5. Raman spectra of $CuLaO_2$ at various pressures. Note the appearance of new modes above 2 GPa.

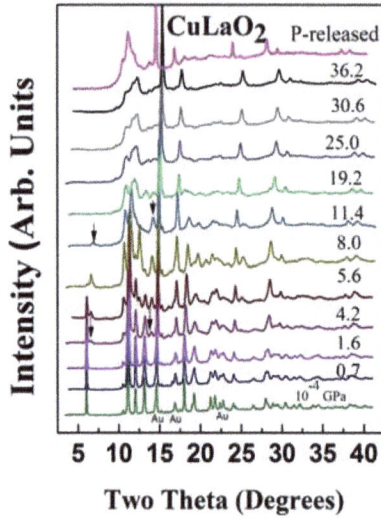

Figure 6. X-ray diffraction data for $CuLaO_2$ at a few selected pressures. Arrow indicates appearance of new peaks indicating instability in the ambient phase. Diffraction peaks from gold, used as insitupressure marker is indicated with Au.

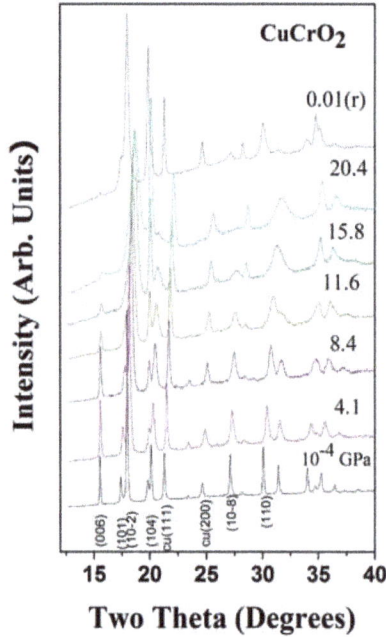

Figure 7. X-ray diffraction data at a few selected pressures for $CuCrO_2$ along with the released data. Diffraction peaks from in-situ pressure calibrant (Cu) are also indicated. Numbers denote the pressure in GPa.

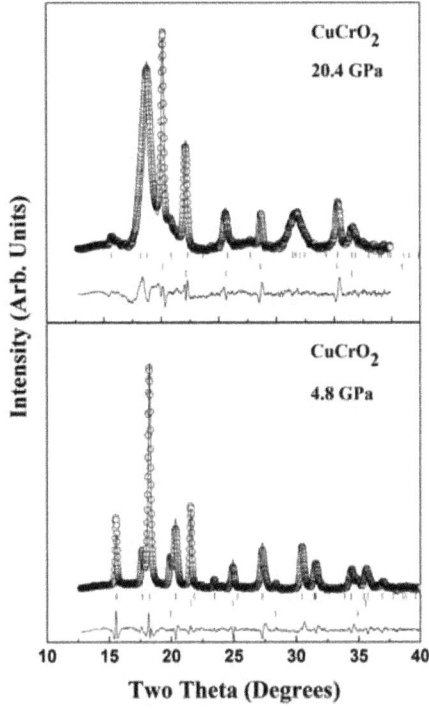

Figure 8. Observed, calculated and difference plot of X-ray powder patterns for CuCrO$_2$ at 4.8 GPa, at 20.4 GPa. Top, middle and bottom vertical marks indicate Bragg reflections from the sample, pressure calibrant (Cu) and sample chamber (W) respectively. The difference between fitted and observed data is also plotted.

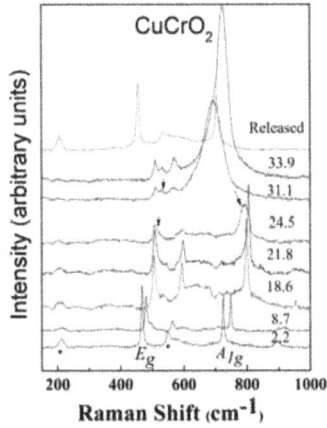

Figure 9. Pressure evolution of Raman spectra of CuCrO$_2$ at a few selected pressures. Arrow indicates the splitting of E_g mode and appearance of new modes at higher frequency which soften with pressure.

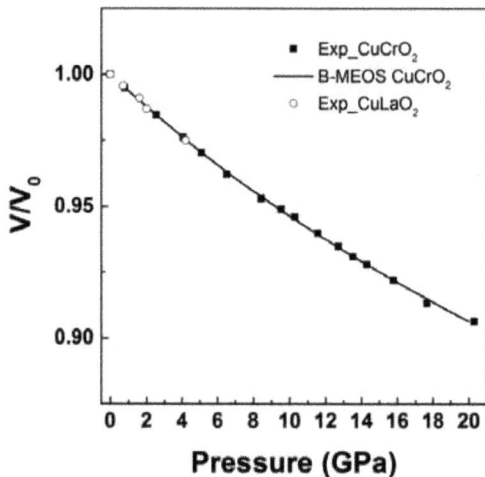

Figure 10. Normalized pressure–volume data for CuCrO$_2$ and CuLaO$_2$. Symbols are explained in the figure.

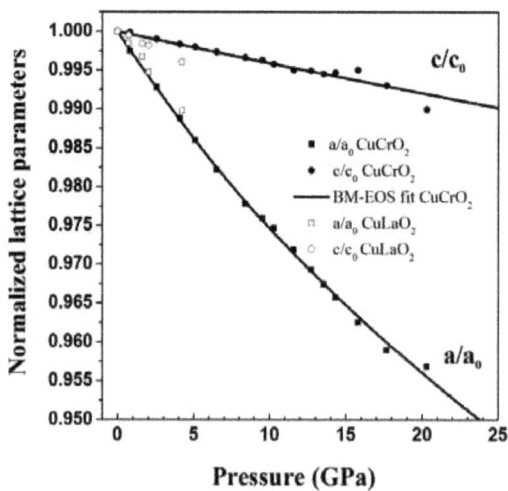

Figure 11. Normalized cell parameters of CuCrO$_2$ and CuLaO$_2$. Both the samples show anisotropic compression. Symbol are explained in the figure.

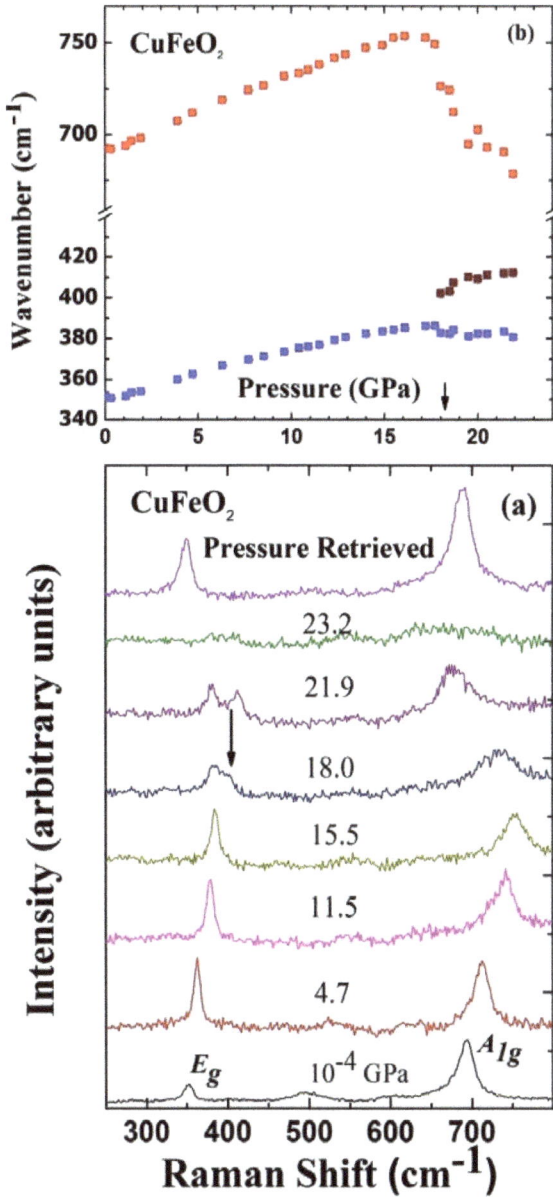

Figure 12. Evolution of Raman modes with pressure for CuFeO$_2$ (**a**), Mode frequency vs. Pressure (**b**). Arrow in (**a**) shows splitting of E_g mode indicating the transition. Figures after Reference [76].

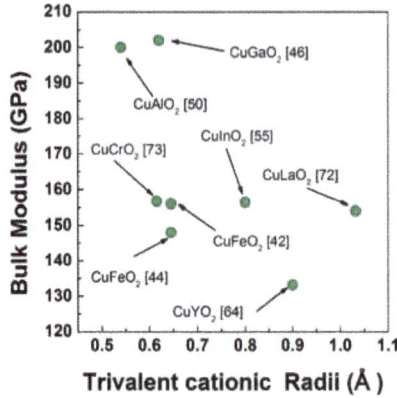

Figure 13. Landscape of bulk modulus of various copper delafossites as a function of trivalent cationic radii. Numbers in the bracket indicate the references.

In Figure 14, we have plotted the trivalent ionic radii vs. transition pressure for available copper delafossites. Based on this data, we found following empirical relation between transition pressure and cationic radii.

$$Y = Y_0 + A \times \exp(-R_0 \times X)$$

where Y = transition pressure in GPa; Y_0, A and R_0 are constants with their numerical values as = −1.74, 430.73 and 4.6 respectively. X = trivalent cationic radii in Å. In spite of various elements at *B* site (transition metal, rare earth, group III), the transition pressure scaling systematically indicates that the size of the trivalent cation plays an important role in phase transition pressure in delafossites. Empirical relations have been proposed earlier in general for compounds containing different rare-earth ions with widely spaced radii [77] to predict transition pressures. In the present article, from the available data for delafossites, we have found a non-linear dependence of transition pressure with cationic radii, which is unusual. Only future high pressure studies on other delafossites can validate this empirical relationship.

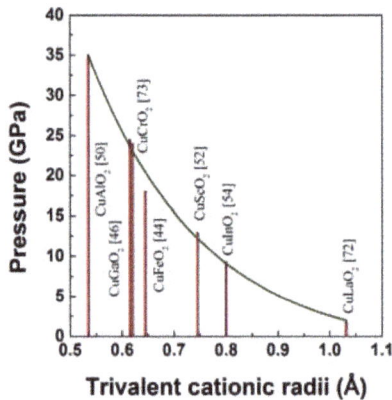

Figure 14. Transition pressure of various copper delafossites as a function of trivalent cationic radii. Green line is a fitted exponential function showing decrease in the transition pressure with the increase in trivalent cationic radii. Numbers in the bracket indicate the references.

4. Summary

To summarize, we have reviewed the structural and vibrational properties of copper delafossites under high pressure. While there is a similarity between the high pressure vibrational behavior of $CuGaO_2$, $CuAlO_2$, $CuLaO_2$ and $CuFeO_2$, the high pressure behavior of $CuCrO_2$ is quite different and unique, where an intense Raman mode appears in the high pressure phase which softens under pressure. Another feature common to all the delafossites investigated so far is the anisotropy in the compressibility of cell axes in the initial phase and the rapid increase in c/a ratio leading to structural phase transition. Surprisingly, in spite of having good quality synchrotron based XRD data, the structure of high pressure phase has only been reported for $CuFeO_2$. Probably high quality single crystal X-ray diffraction data under high pressure may be helpful to get the structure of high pressure phase. Till now the high pressure studies are focused on copper based delafossites with rhombohedral symmetry, however delafossites with hexagonal symmetry have not yet been investigated, and hence systematic investigations on these compounds is required. Another area of research under high pressure is the effect of non-hydrostatic stresses [78] on the crystal structure and band gap of these materials which may result in synthesis of new metastable polymorph with improved properties. Doped delafossites with reduced band gap and increased conductivity can also be investigated under high pressure to obtain compounds with tailor made properties for specific applications.

Conflicts of Interest: The authors declare no conflicts of interest.

References

1. Shannon, R.D.; Rogers, D.B.; Prewitt, C.T. Chemistry of noble metal oxides I. Syntheses and properties of ABO_2 delafossite compounds. *Inorg. Chem.* **1971**, *10*, 713–718. [CrossRef]
2. Dordor, P.; Chaminade, J.P.; Wichainchai, A.; Marquestaut, E.; Doumerc, J.P.; Pouchard, M.; Hagenmuller, P. Crystal growth and electrical properties of $CuFeO_2$ single crystals. *J. Solid State Chem.* **1988**, *75*, 105–112. [CrossRef]
3. Marquardt, M.A.; Ashmore, N.A.; Cann, D.P. Crystal chemistry and electrical properties of the delafossite structure. *Thin Solid Films* **2006**, *496*, 146–156. [CrossRef]
4. Kawazoe, H.; Yasukawa, M.; Hyodo, H.; Kurita, M.; Yanagi, H.; Hosono, H. P-type electrical conduction in transparent thin films of $CuAlO_2$. *Nature* **1997**, *389*, 939–942. [CrossRef]
5. Mackenzie, A.P. The Properties of ultrapure delafossite metals. *Rep. Prog. Phys.* **2017**, *80*, 032501–032519. [CrossRef] [PubMed]
6. Banerjee, A.N.; Chattopadhyay, K.K. Recent developments in the emerging field of crystalline p-type transparent conducting oxide thin films. *Prog. Cryst. Growth Charact. Mater.* **2005**, *50*, 52–105. [CrossRef]
7. Walsh, A.; Da Silva, J.L.F.; Wei, S.-H. Multi-component transparent conducting oxides: Progress in materials modelling. *J. Phys. Condens. Matter* **2011**, *23*, 334210. [CrossRef] [PubMed]
8. Sheng, S.; Fang, G.; Li, C.; Xu, S.; Zhao, X. P-type transparent conducting oxides. *Phys. Status Solidi A* **2006**, *203*, 1891–1900. [CrossRef]
9. King, P.D.C.; Veal, T.D. Conductivity in transparent oxide semiconductors. *J. Phys. Condens. Matter* **2011**, *23*, 334214. [CrossRef] [PubMed]
10. Yu, M.; Natu, G.; Ji, Z.; Wu, Y. P-type dye-sensitized solar cells based on delafossite $CuGaO_2$ nanoplates with saturation photovoltages exceeding 460 mv. *J. Phys. Chem. Lett.* **2012**, *3*, 1074–1078. [CrossRef] [PubMed]
11. Chae, G.J. A modified transparent conducting oxide for flat panel displays only. *J. Appl. Phys.* **2001**, *40*, 1282–1286. [CrossRef]
12. Granqvist, C.G.; Azens, A.; Hjelm, A.; Kullman, L.; Niklasson, G.A.; Ronnow, D.; Mattsson, M.S.; Veszele, M.; Vaiva, G. Recent advances in electrochromics for smart windows applications. *Sol. Energy* **1998**, *63*, 199–276. [CrossRef]
13. Diaz-Garcia, A.K.; Lana-Villarreal, T.; Gomez, R. Sol–gel copper chromium delafossite thin films as stable oxide photocathodes for water splitting. *J. Mater. Chem. A* **2015**, *3*, 19683–19687. [CrossRef]

14. Gurunathan, K.; Baeg, J.O.; Lee, S.M.; Subramanian, E.; Moon, S.J.; Kong, K.J. Visible light assisted highly efficient hydrogen production from H_2S decomposition by $CuGaO_2$ and $CuGa_{1-x}In_xO_2$ delafossite oxides bearing nanostructured co-catalysts. *Catal. Commun.* **2008**, *9*, 395–402. [CrossRef]

15. Kykyneshi, R.; Nielsen, B.C.; Tate, J.; Li, J.; Sleight, A.W. Structural and transport properties of $CuSc_{1-x}Mg_xO_{2+y}$ delafossites. *J. Appl. Phys.* **2004**, *96*, 6188–6194. [CrossRef]

16. Mazumder, N.; Sen, D.; Ghorai, U.K.; Roy, R.; Saha, S.; Das, N.S.; Chattopadhyay, K.K. Realizing direct gap, polytype, group IIIA delafossite: Ab initio forecast and experimental validation considering prototype $CuAlO_2$. *J. Phys. Chem. Lett.* **2013**, *4*, 3539–3543. [CrossRef]

17. Qiu, X.; Liu, M.; Sunada, K.; Miyauchi, M.; Hashimoto, K. A facile one-step hydrothermal synthesis of rhombohedral $CuFeO_2$ crystals with antivirus property. *Chem. Commun.* **2012**, *48*, 7365–7367. [CrossRef] [PubMed]

18. Dong, Y.; Cao, C.; Chui, Y.S.; Zapien, J.A. Facile hydrothermal synthesis of $CuFeO_2$ hexagonal platelets/rings and graphene composites as anode materials for lithium ion batteries. *Chem. Commun.* **2014**, *50*, 10151–10154. [CrossRef] [PubMed]

19. Patzsch, J.; Balog, I.; Krau, P.; Lehmann, C.W.; Schneider, J.J. Synthesis, characterization and p–n type gas sensing behaviour of $CuFeO_2$ delafossite type inorganic wires using Fe and Cu complexes as single source molecular precursors. *RSC Adv.* **2014**, *4*, 15348–15355. [CrossRef]

20. Amrute, A.P.; Larrazabal, G.O.; Mondelli, C.; Perez-Ramirez, J. $CuCrO_2$ Delafossite: A stable copper catalyst for chlorine production. *Angew. Chem. Int. Ed.* **2013**, *52*, 9772–9775. [CrossRef] [PubMed]

21. Ahmed, S.I.; Dalba, G.; Fornasini, P.; Vaccari, M.; Rocca, F.; Sanson, A.; Li, J.; Sleight, A.W. Negative thermal expansion in crystals with the delafossite structure: An extended X-ray absorption fine structure study of $CuScO_2$ and $CuLaO_2$. *Phys. Rev. B* **2009**, *79*, 104302. [CrossRef]

22. Seki, S.; Onose, Y.; Tokura, Y. Spin-driven ferroelectricity in triangular lattice antiferromagnets $ACrO_2$ (A = Cu, Ag, Li, or Na). *Phys. Rev. Lett.* **2008**, *101*, 067204. [CrossRef] [PubMed]

23. Zhong, C.; Cao, H.; Fang, J.; Jiang, X.; Ji, X.; Dong, Z. Spin-lattice coupling and helical-spin driven ferroelectric polarization in multiferroic $CuFeO_2$. *Appl. Phys. Lett.* **2010**, *97*, 094103. [CrossRef]

24. Omata, T.; Nagatani, H.; Suzuki, I.; Kita, M.; Yanagi, H.; Ohashi, N. A new direct and narrow band gap oxide semiconductor applicable as a solar cell absorber. *J. Am. Chem. Soc.* **2014**, *136*, 3378–3381. [CrossRef] [PubMed]

25. Liu, Y.; Gong, Y.; Mellott, N.P.; Wang, B.; Ye, H.; Wu, Y. Luminescence of delafossite-type $CuAlO_2$ fibers with Eu substitution for Al cations. *Sci. Technol. Adv. Mater.* **2016**, *17*, 200–209. [CrossRef] [PubMed]

26. Shi, L.; Wang, F.; Wang, Y.; Wang, D.; Zhao, B.; Zhang, L.; Zhao, D.; Shen, D. Photoluminescence and photocatalytic properties of rhombohedral $CuGaO_2$ nanoplates. *Sci. Rep.* **2016**, *6*, 21135. [CrossRef] [PubMed]

27. Yassin, O.A.; Alamri, S.N.; Joraid, A.A. Effect of particle size and laser power on the Raman spectra of $CuAlO_2$ delafossite nanoparticles. *J. Phys. D Appl. Phys.* **2013**, *46*, 235301. [CrossRef]

28. Ahmed, J.; Mao, Y. Delafossite $CuAlO_2$ nanoparticles with electrocatalytic activity toward oxygen and hydrogen evolution reactions. In *Nanomaterials for Sustainable Energy*; ACS Symposium Series; American Chemical Society: Washington, DC, USA, 2015; Chapter 4; Volume 1213, pp. 57–72.

29. Harada, T.; Fujiwara, K.; Tsukazaki, A. Highly conductive $PdCoO_2$ ultrathin films for transparent electrodes. *APL Mater.* **2018**, *6*, 046107. [CrossRef]

30. Deng, Z.; Fang, X.; Wu, S.; Dong, W.; Shao, J.; Wang, S.; Lei, M. The morphologies and optoelectronic properties of delafossite $CuFeO_2$ thin films prepared by PEG assisted. *J. Sol-Gel Sci. Technol.* **2014**, *71*, 297–302. [CrossRef]

31. Sinnarasa, I.; Thimont, Y.; Presmanes, L.; Barnabé, A.; Tailhades, P. Thermoelectric and transport properties of delafossite $CuCrO_2$:Mg thin films prepared by RF magnetron sputtering. *Nanomaterials* **2017**, *7*, 157. [CrossRef] [PubMed]

32. Barnabe, A.; Thimont, Y.; Lalanne, M.; Presmanes, L.; Tailhades, P. P-type conducting transparent characteristics of delafossite Mg-doped $CuCrO_2$ thin films prepared by RF-sputtering. *J. Mater. Chem. C* **2015**, *3*, 6012–6024. [CrossRef]

33. Errandonea, D. Exploring the properties of MTO_4 compounds using high-pressure powder X-ray diffraction. *Cryst. Res. Technol.* **2015**, *50*, 729–736. [CrossRef]

34. Errandonea, D.; Ruiz-Fuertes, A. A Brief review of the effects of pressure on wolframite-type oxides. *Crystals* **2018**, *8*, 71. [CrossRef]

35. Hasegawa, M.; Tanaka, M.; Yagi, T.; Takei, H.; Inoue, A. Compression behavior of the delafossite-type metallic oxide PdCoO$_2$ below 10 GPa. *Solid State Commun.* **2003**, *128*, 303–307. [CrossRef]

36. Sheets, W.C.; Mugnier, E.; Barnabe, A.; Marks, T.J.; Poeppelmeier, K.R. Hydrothermal synthesis of delafossite-type oxides. *Chem. Mater.* **2006**, *18*, 7–20. [CrossRef]

37. Kumar, S.; Miclau, M.; Christine, M. Hydrothermal synthesis of AgCrO$_2$ delafossite in supercritical water: A new single-step process. *Chem. Mater.* **2013**, *25*, 2083–2088. [CrossRef]

38. Jin, Y.; Chuamanov, G. Solution synthesis of pure 2H CuFeO$_2$ at low temperatures. *RSC Adv.* **2016**, *6*, 26392–26397. [CrossRef]

39. Effenberger, H. Structure of Hexagonal Copper(I) Ferrite. *Acta Crystallogr. Sect. C Cryst. Struct. Commun.* **1991**, *47*, 2644–2646. [CrossRef]

40. Godinho, K.G.; Morgan, B.J.; Allen, J.P.; Scanlon, D.O.; Watson, G.W. Chemical bonding in copper-based transparent conducting oxides: CuMO$_2$ (M = In, Ga, Sc). *J. Phys. Condens. Matter* **2011**, *23*, 334201. [CrossRef] [PubMed]

41. Rousseau, D.L.; Bauman, R.P.; Porto, S.P.S. Normal mode determination in crystals. *J. Raman Spectrosc.* **1981**, *10*, 253–290. [CrossRef]

42. Zhao, T.R. X-ray diffraction study of copper iron oxide [CuFeO$_2$] under pressures up to 10 GPa. *Mater. Res. Bull.* **1997**, *32*, 151–157. [CrossRef]

43. Birch, F. Finite strain isotherm and velocities for single-crystal and polycrystalline NaCl at high pressures and 300° K. *J. Geophys. Res. Solid Earth* **1978**, *83*, 1257–1268. [CrossRef]

44. Xu, W.M.; Rozenberg, G.K.; Pasternak, M.P.; Kertzer, M.; Kurnosov, A.; Dubrovinsky, L.S.; Pascarelli, S.; Munoz, M.; Vaccari, M.; Hanfland, M.; et al. Pressure-induced Fe-Cu cationic valence exchange and its structural consequences: High-pressure studies of delafossite CuFeO$_2$. *Phys. Rev. B* **2010**, *81*, 104110. [CrossRef]

45. Terada, N.; Osakabe, T.; Kitazawa, H. High-pressure suppression of long range magnetic order in the triangular lattice antiferromagnet CuFeO$_2$. *Phys. Rev. B* **2010**, *83*, 020403. [CrossRef]

46. Pellicer-Porres, J.; Segura, A.; Ferrer-Roca, C.; MartiiNez-Garcii, A.D.; Sans, J.A.; MartiiNez, E.; Itie, J.P.; Polian, A.; Baudelet, F.; Munoz, A.; et al. Structural evolution of the CuGaO$_2$ delafossite under high pressure. *Phys. Rev. B* **2004**, *69*, 024109. [CrossRef]

47. Pellicer-Porres, J.; Segura, A.; Martínez, E.; Saitta, A.M.; Polian, A.; Chervin, J.C.; Canny, B. Vibrational properties of delafossite CuGaO$_2$ at ambient and high pressure. *Phys. Rev. B* **2005**, *72*, 064301. [CrossRef]

48. Pellicer-Porres, J.; Martínez-García, D.; Segura, A.; Rodríguez-Hernández, P.; Muñoz, A.; Chervin, J.C.; Garro, N.; Kim, D. Pressure and temperature dependence of the lattice dynamics of CuAlO$_2$ investigated by Raman scattering experiments and ab initiocalculations. *Phys. Rev. B* **2006**, *74*, 184301. [CrossRef]

49. Liu, Q.J.; Liu, Z.T.; Feng, L.P.; Tian, H.; Liu, W.T.; Yan, F. Density functional theory study of 3r–and 2h–CuAlO$_2$ under pressure. *Appl. Phys. Lett.* **2010**, *97*, 141917. [CrossRef]

50. Pellicer-Porres, J.; Segura, A.; Ferrer-Roca, C.; Polian, A.; Munsch, P.; Kim, D. XRD and XAS structural study of CuAlO$_2$ under high pressure. *J. Phys. Condens. Matter* **2013**, *25*, 115406. [CrossRef] [PubMed]

51. Nakanishi, A.; Katayama-Yoshida, H. Pressure-induced structural transition and enhancement of energy gap of CuAlO$_2$. *J. Phys. Soc. Jpn.* **2011**, *80*, 024706. [CrossRef]

52. Gilliland, S.; Pellicer-Porres, J.; Segura, A.; Muñoz, A.; Rodríguez-Hernández, P.; Kim, D.; Lee, M.S.; Kim, T.Y. Electronic structure of CuAlO$_2$ and CuScO$_2$delafossites under pressure. *Phys. Status Solidi B* **2007**, *244*, 309–314. [CrossRef]

53. Nie, X.; Su-Huai, W.; Zhang, S.B. Bipolar doping and band-gap anomalies in delafossite transparent conductive oxides. *Phys. Rev. Lett.* **2002**, *88*, 066405. [CrossRef] [PubMed]

54. Liu, W.; Liu, Q.; Liu, Z.-T. First principles studies of structural, mechanical, electronic, optical properties and pressure-induced phase transition of CuInO$_2$ polymorph. *Physica B* **2012**, *407*, 4665–4670. [CrossRef]

55. Jayalakshmi, V.; Murugan, R.; Palanivel, B. Electronic and structural properties of CuMO$_2$ (M = Al, Ga, In). *J. Alloy. Compd.* **2005**, *388*, 19–22. [CrossRef]

56. Aoyama, T.; Miyake, A.; Kagayama, T.; Shimizu, K.; Tsuyoshi, K. Pressure effects on the magnetoelectric properties of a multiferroic triangular-lattice antiferromagnet CuCrO$_2$. *Phys. Rev. B* **2013**, *87*, 094401. [CrossRef]

57. Piermarini, G.J.; Block, S.; Barnett, J.D. Hydrostatic limits in liquids and solids to 100 kbar. *J. Appl. Phys.* **1973**, *44*, 5377–5382. [CrossRef]

58. Carter, W.T.; Marsh, S.P.; Fritz, J.N.; McQueen, R.G. *Accurate Characterization of the High Pressure Environment*; Lloyd, E.C., Ed.; NBS Special Pub.: Washington, DC, USA, 1971; Volume 326, p. 147.

59. Larson, A.C.; Von Dreele, R.B. *GSAS: General Structure Analysis System*; Report LAUR 86-748; Los Alamos National Laboratory: Los Alamos, NM, USA, 2000.

60. Petrenko, O.A.; Balakrishnan, G.; Lees, M.R.; Paul, D.M.; Hoser, A. High-magnetic-field behavior of the triangular-lattice antiferromagnet $CuFeO_2$. *Phys. Rev. B* **2000**, *62*, 8983. [CrossRef]

61. Aktas, O.; Truong, K.D.; Otani, T.; Balakrishnan, G.; Clouter, M.J.; Kimura, T.; Quirion, G. Raman scattering study of delafossite magnetoelectric multiferroic compounds: $CuFeO_2$ and $CuCrO_2$. *J. Phys. Condens. Matter.* **2012**, *24*, 036003. [CrossRef] [PubMed]

62. Shannon, R.D. Revised effective ionic radii and systematic studies of interatomic distances in halides and chalcogenides. *Acta Cryst. A* **1976**, *32*, 751–767. [CrossRef]

63. Miyasaka, N.; Doi, Y.; Hinatsu, Y. Synthesis and magnetic properties of $ALnO_2$ (A =Cu or Ag; Ln = rare earths) with the delafossite structure. *J. Solid State Chem.* **2009**, *182*, 2104–2110. [CrossRef]

64. Cheng, C.; Lv, Z.L.; Cheng, Y.; Ji, G.F. Structural, elastic and electronic properties of $CuYO_2$ from first-principles study. *J. Alloy. Compd.* **2014**, *603*, 183–189. [CrossRef]

65. Shimode, M.; Sasaki, M.; Mukaida, K. Synthesis of the delafossite-type $CuInO_2$. *J. Solid State Chem.* **2000**, *151*, 16–20. [CrossRef]

66. Li, J.; Yokochi, A.F.T.; Sleight, A.W. Oxygen intercalation of two polymorphs of $CuScO_2$. *Solid State Sci.* **2004**, *6*, 831–839. [CrossRef]

67. Poienar, M.; Hardy, V.; Kundys, B.; Singh, K.; Maignan, A.; Damay, F.; Martin, C. Revisiting the properties of delafossite $CuCrO_2$: A single crystal study. *J. Solid State Chem.* **2012**, *185*, 56–61. [CrossRef]

68. Elkhouni, T.; Amami, M.; Hlil, E.K.; Salah, A.B. The structural, anisotropic magnetization, and spectroscopic study of delafossite $CuCr_{1-x}M_xO_2$ systems. *J. Supercond. Nov. Magn.* **2015**, *28*, 1895–1903. [CrossRef]

69. Elkhouni, T.; Amami, M.; Strobel, P.; Salah, A.B. Structural and magnetic properties of substituted delafossite-type oxides $CuCr_{1-x}Sc_xO_2$. *World J. Condens. Matter Phys.* **2013**, *3*, 1–8. [CrossRef]

70. Elkhoun, T.; Amami, M.; Hlil, E.K.; Salah, A.B. Effect of Spin dilution on the magnetic state of delafossite $CuFeO_2$ with an $S = 5/2$ antiferromagnetic triangular sublattice. *J. Supercond. Novel Magn.* **2015**, *28*, 1439–1447. [CrossRef]

71. Pavunny, S.P.; Kumar, A.; Katiyar, R.S. Raman spectroscopy and field emission characterization of delafossite $CuFeO_2$. *J. Appl. Phys.* **2010**, *107*, 013522. [CrossRef]

72. Salke, N.P.; Garg, A.B.; Rao, R.; Achary, S.N.; Gupta, M.K.; Mittal, R.; Tyagi, A.K. Phase transitions in delafossite $CuLaO_2$ at high pressures. *J. Appl. Phys.* **2014**, *115*, 133507. [CrossRef]

73. Garg, A.B.; Mishra, A.K.; Pandey, K.K.; Sharma, S.M. Multiferroic $CuCrO_2$ under high pressure: In situ X-ray diffraction and Raman spectroscopic studies. *J. Appl. Phys.* **2014**, *116*, 133514. [CrossRef]

74. Inaba, M.; Iriyama, Y.; Ogumi, Z.; Todzuka, Y.; Tasaka, A. Raman study of layered rock-salt $LiCoO_2$ and its electrochemical lithium deintercalation. *J. Raman Spectrosc.* **1997**, *28*, 613–617. [CrossRef]

75. Gomis, O.; Lavina, B.; Rodríguez-Hernández, P.; Muñoz, A.; Errandonea, R.; Errandonea, D.; Bettinelli, M. High-pressure structural, elastic, and thermodynamic properties of zircon-type $HoPO_4$ and $TmPO_4$. *J. Phys. Condens. Matter* **2017**, *29*, 095401. [CrossRef] [PubMed]

76. Salke, N.P.; Kamali, K.; Ravindran, T.R.; Balakrishnan, G.; Rao, R. Raman spectroscopic studies of $CuFeO_2$ at high pressures. *Vib. Spectrosc.* **2015**, *81*, 112–118. [CrossRef]

77. Mota, D.A.; Almeida, A.; Rodrigues, V.H.; Costa, M.M.R.; Tavares, P.; Bouvier, P.; Guennou, M.; Kreisel, J.; Moreira, J.A. Dynamic and structural properties of orthorhombic rare-earth manganites under high pressure. *Phys. Rev. B* **2014**, *90*, 054104. [CrossRef]

78. Garg, A.B.; Errandonea, D.; Rodríguez-Hernández, P.; Muñoz, A. $ScVO_4$ under non-hydrostatic compression: A new metastable polymorph. *J. Phys. Condens. Matter* **2017**, *29*, 055401. [CrossRef] [PubMed]

Review

A Brief Review of the Effects of Pressure on Wolframite-Type Oxides

Daniel Errandonea [1],* and **Javier Ruiz-Fuertes [1,2]**

[1]   Departament de Física Aplicada-ICMUV, MALTA Consolider Team, Universitat de València, 46100 Burjassot, Spain; ruizfuertesj@unican.es

[2]   DCITIMAC, MALTA Consolider Team, Universidad de Cantabria, 39005 Santander, Spain

*   Correspondence: daniel.errandonea@uv.es; Tel.: +34-96-354-4475

Received: 8 January 2018; Accepted: 29 January 2018; Published: 31 January 2018

Abstract: In this article, we review the advances that have been made on the understanding of the high-pressure (HP) structural, vibrational, and electronic properties of wolframite-type oxides since the first works in the early 1990s. Mainly tungstates, which are the best known wolframites, but also tantalates and niobates, with an isomorphic ambient-pressure wolframite structure, have been included in this review. Apart from estimating the bulk moduli of all known wolframites, the cation–oxygen bond distances and their change with pressure have been correlated with their compressibility. The composition variations of all wolframites have been employed to understand their different structural phase transitions to post-wolframite structures as a response to high pressure. The number of Raman modes and the changes in the band-gap energy have also been analyzed in the basis of these compositional differences. The reviewed results are relevant for both fundamental science and for the development of wolframites as scintillating detectors. The possible next research avenues of wolframites under compression have also been evaluated.

Keywords: wolframite; high pressure; phase transitions; crystal structure; phonons; band structure

1. Introduction

Wolframite is an iron manganese tungstate mineral. The name is normally used to denote the family of isomorphic compounds. The crystal structure of a wolframite was first solved for $MgWO_4$ by Broch [1] in 1929 with the exception of the oxygen positions and then completely determined for $NiWO_4$ by Keeling [2] in 1957. The structure of wolframite, monoclinic with space group $P2/c$, is adopted by all tungstates AWO_4 with the divalent cation A with an ionic radius in octahedral coordination $r_A < 0.9$ Å. In addition, $CdWO_4$, both M = In and Sc niobates $MNbO_4$ and tantalates $MTaO_4$, and a metastable high-pressure (HP) and high-temperature polymorph in molybdates $AMoO_4$ [3] have an isostructural crystal structure. Most wolframites, and in particular tungstates and molybdates with an A^{2+} cation with a completely empty or full outer d shell, have been extensively studied for their applications in scintillating detectors for X-ray tomography, high-energy particle physics, and dosimetry devices [4–6]. The reason behind this is their high light yield when hit by γ-particles or X-rays despite long scintillation times of a few μs [7]. In fact, the search for different polymorphs of the scintillating wolframites with enhanced properties, i.e., a faster scintillating response, has motivated the study of these materials under high pressure conditions. In addition, wolframites with transition metals with unfilled outer d-shells are magnetic. In particular, hübnerite $MnWO_4$, which shows three different antiferromagnetic phases below 13.7 K, is a type II multiferroic material [8,9], exhibiting ferroelectricity induced by helical magnetic ordering.

After the pioneering work of Young and Schwartz [3], dated from 1963, in which the synthesis of different wolframite-type molybdates was reported at ~0.6 GPa and 900 °C, only two works were published about the HP behavior of wolframites during the 1990s. A detailed structural study of the wolframite structure of $MgWO_4$, $MnWO_4$, and $CdWO_4$ was performed at room temperature by Macavei and Schultz [10] up to 9.3 GPa using single crystal X-ray diffraction (XRD). A Raman spectroscopy study of $CdWO_4$ at up to ~40 GPa was carried out by Jayaraman et al. [11] in which a structural phase transition of $CdWO_4$ was found at 20 GPa. However, during the last decade, the interest in the behavior of wolframites under high pressure has inspired many works that have contributed to our understanding of their structural [12–16], vibrational [15–21], and electronic [14,22] properties under compression.

In the following sections, we shall review first the main features of the behavior and trends of the structure of wolframites under high pressure and the advances on the structural determination of their HP phases. Finally, the effects of pressure on the vibrational, electronic, and optical properties shall be presented with a special emphasis on $CdWO_4$, whose HP phase has recently been fully solved.

2. The Wolframite Structure at High Pressure

The best description of the wolframite structure at ambient pressure was formulated by Kihlborg and Gebert [23] in 1969, unfolding the structure of the Jahn–Teller distorted $CuWO_4$ wolframite. The wolframite structure of an AXO_4 compound at ambient conditions (Figure 1a) can be described as a framework of oxygen atoms in approximately hexagonal close packing with the cations (A and $X = W$), octahedrally coordinated and occupying half of the octahedral sites. In this structure, the octahedral units of the same cations share edges forming alternating zig-zag chains and conferring the structure with a layer-like $AOXO$ configuration in the [100] direction. A particular case worth mentioning is $CuWO_4$, not included in the list of wolframites presented in the introduction. Although, according to the ionic radius of Cu, it should also crystallize in the same structure as $NiWO_4$ and $ZnWO_4$, it does it in a distorted version of the wolframite structure. In $CuWO_4$, as a consequence of the Jahn–Teller effect of Cu^{2+} in octahedral coordination [23,24], the Cu^{2+} ion requires a distortion that is achieved by a shear parallel to the b-axis along each copper plane. This has as a consequence a displacement of oxygen layers, destroying the twofold symmetry and lowering the space group from $P2/c$ to $P\bar{1}$.

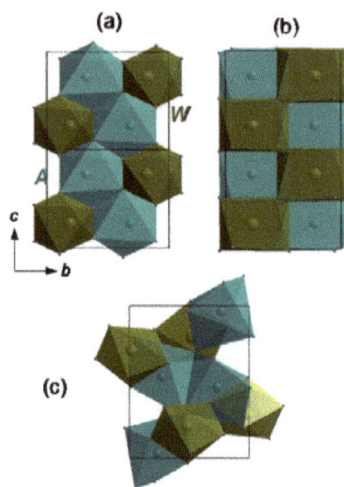

Figure 1. (a) General ambient-pressure wolframite-type structure of AXO_4 ($X = W$) compounds and high-pressure (HP) structure of (b) $InTaO_4$ and (c) $CdWO_4$. In (a), the coordination polyhedra of the divalent cation A (W) are shown in green (turquoise). In (b), the same color code is used for In and Ta polyhedra, respectively. In (c), Cd (W) polyhedra are shown in green (turquoise).

   In wolframites, the monoclinic *a* and *c* unit-cell lattice parameters are similar, although the *c*-axis is slightly larger than the *a*-axis. However, the *b*-axis is, in general, around 15% larger than the other axes. The monoclinic angles are always very close to 90°. As was first shown by Macavei and Shultz [10] in MgWO$_4$, MnWO$_4$, and CdWO$_4$, this causes the wolframite structure to suffer an anisotropic contraction under pressure (Figure 2). For instance, in MnWO$_4$ [13], the axial compressibility (defined as $k_x = (-1/x) \times \partial x/\partial P$) of the *b*-axis ($k_b = 3.3(1) \times 10^{-3}$ GPa^{-1}) almost doubles the axis compressibility of *a* and *c* ($k_a = 2.0(1) \times 10^{-3}$ GPa^{-1} and $k_c = 1.6(1) \times 10^{-3}$ GPa^{-1}). The same anisotropic compression has been found in CdWO$_4$, MgWO$_4$, and other wolframites [10–14]. This fact, together with a continuous increase in the monoclinic β angle, indicates that, under compression, the monoclinic structure tends to distort, at least in tungstates. We shall show in the next section that this continuous distortion causes a phase transition to a structure with a lower symmetry in most wolframite-type tungstates. The only exception is CdWO$_4$, which increases its space-group symmetry at the phase transition. In the structure of wolframite, the octahedra of different cations only share corners between them, while octahedra of the same cation share both corners and edges. Hence, we can isolate the AO_6 and XO_6 polyhedral units as independent blocks in the structure. In this picture, each type of polyhedra would respond differently to the effect of pressure. In particular, one would expect the XO_6, with high valence (6$^+$ in W or Mo, and 5$^+$ in Ta or Nb) to be much less compressible than the *A* cations with 2$^+$ or 3$^+$ valences. Such an approach has often been used to study the bulk compressibility of scheelites, a structure adopted by tungstates with *A* cations with a large ionic radius. However, differently from wolframites, in scheelites, the also-isolated *X* cations are tetrahedrally coordinated. Therefore, the XO_4 polyhedra are almost pressure-independent, and the bulk modulus is very well predicted by the empirical equation $B_0 = N \cdot Z/d_{A-O}{}^3$, first proposed by Hazen and Finger [25], which assumes that the bulk modulus is proportional (N being the proportion constant) to the formal charge of the *A* cation, *Z*, and inversely proportional to the cation–anion distance to the third power $d_{A-O}{}^3$. In the case of scheelites, this proportion constant is N = 610 [26]. The question is how such a model would work for wolframites. In Figure 3, we show the estimated and experimentally available bulk modulus of all known wolframites. The agreement is excellent considering N = 661 for wolframite-type tungstates and molybdates and N = 610 for tantalates and niobates, indicating that assuming that the XO_6 in wolframites is a rigid and almost pressure-incompressible unit is justified in terms of the bulk modulus. In Section 4, we shall study how this approximation works in terms of describing the Grüneisen parameters of the Raman-active modes. The experimental bulk modulus values are shown in Table 1. Figure 3 also shows that the bulk moduli of tantalates and niobates is around 30% larger than the bulk modulus of tungstates. This fact, a result of a higher valence of the *A* cation (3$^+$ instead of 2$^+$), reinforces the negligible effect of pressure on the XO_6 polyhedra since the empirical formula is also valid despite the valence of Ta and Nb is 5$^+$ instead of 6$^+$.

Table 1. Experimental bulk modulus (B_0) of different wolframites. Results were obtained from [10] and [12–16].

Compounds	B_0 (GPa)	Reference
MgWO$_4$	144–160	[10,12]
MnWO$_4$	131–145	[10,13]
ZnWO$_4$	145	[12]
CdWO$_4$	136–123	[10,14]
InTaO$_4$	179	[15]
InNbO$_4$	179	[16]

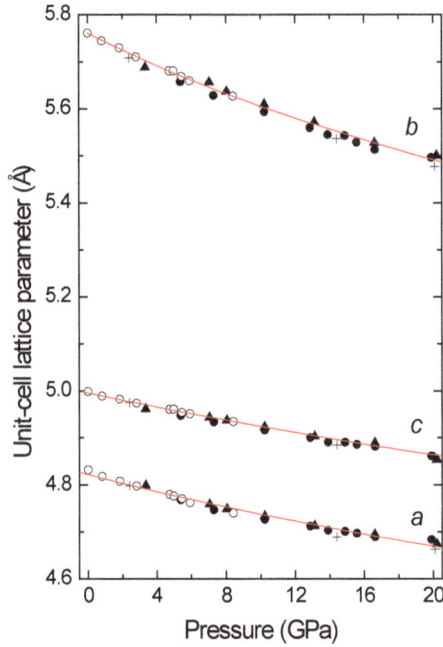

Figure 2. Pressure dependence of the unit-cell lattice parameters of $MnWO_4$. Empty circles represent Macavei's and Schulz's data [10], full circles and triangles represent data obtained by powder X-ray diffraction, and the crosses represent data obtained by single crystal X-ray diffraction [14]. The continuous lines are fits to a third-order Birch–Murnaghan equation of state.

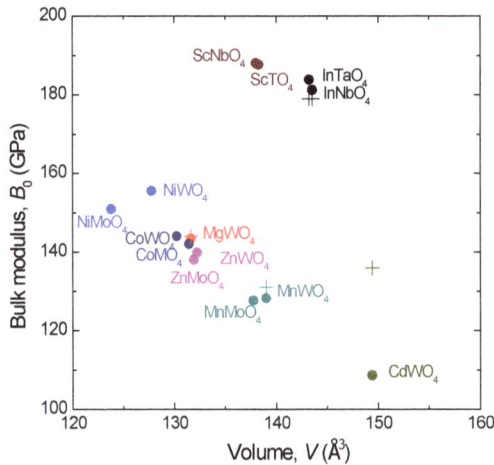

Figure 3. Volume change of the bulk modulus B_0 of all known compounds with a wolframite-type structure. Crosses (+) represent experimental results. Circles are the estimated B_0 considering the WO_6, MoO_6, TaO_6, and NbO_6 as rigid units and the following dependence with the A–O distances: $B_0 = 610 \cdot Z/d_{A-O}^3$ in the case of tantalates and niobates and $B_0 = 661 \cdot Z/d_{A-O}^3$ in the case of tungstates and molybdates. Different colors are used for different compounds, the same color being used for the compound name.

Regarding the atomic positions under compression in wolframites, while the oxygen atoms barely change their positions, the cations tend to shift along the high symmetry *b*-axis [10]; the *A* cations either shift down or up depending on the compound, but the W cations largely shift down along the *b*-axis. For tantalates and niobates, no reliable atomic positions exist under pressure.

3. Phase Transitions

Under compression, most wolframites undergo a phase transformation to a different polymorph, with tungstates transiting at around 20 GPa, and $InTaO_4$ and $ScNbO_4$ doing so a few GPa below. However, the post-wolframite phase depends on the compound. Thus, while tungstates, except $CdWO_4$, apparently transform to a triclinic version of wolframites, similar to that of $CuWO_4$, with space group $P\bar{1}$ and a similar unit-cell, tantalates and niobates also do it but to another distorted version of wolframite that keeps the same space group $P2/c$. $CuWO_4$ and $CdWO_4$ are the only (pseudo)wolframites that, under pressure, increase their space-group symmetry, with $CuWO_4$ transforming to a normal wolframite structure with space group $P2/c$, despite keeping the Jahn–Teller distortion, and $CdWO_4$ introducing a screw axis in space group $P2_1/c$ and doubling the unit cell. The solution of the HP phase of the tungstates has been approached unsuccessfully with powder X-ray diffraction in $ZnWO_4$ and $MgWO_4$ [12] and with single crystal X-ray diffraction in $MnWO_4$ [13]. However, based on the careful indexation of the observed reflections, the study of the systematic extinctions of the HP phase, and the number of active Raman modes observed in the HP phase, which we shall show in the next section, a triclinic structure is proposed as the post-wolframite structure of normal wolframites. This proposed post-wolframite structure would be very similar to the low-pressure phase, but would be described as being in space group $P\bar{1}$. Unfortunately, in the particular case of $MnWO_4$, the single crystal, despite of a small volume collapse of only 1% in the phase transition, dramatically deteriorates (Figure 4) with the appearance of more than two triclinic HP domains during the phase transition coexisting with the monoclinic low-pressure phase. This fact prevents a correct integration of the reflection intensities and therefore an accurate determination of the atomic positions in the HP phase. Since the experiments in $MnWO_4$ were carried out using Ne as a pressure-transmitting medium, the phase coexistence observed is likely inherent to the properties of the sample and not caused by non-hydrostatic effects. The study of this issue deserves future efforts.

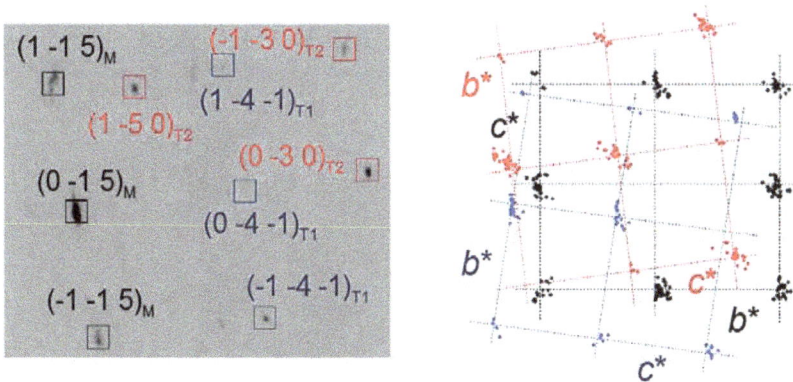

Figure 4. Section of one single-crystal X-ray diffraction frame of $MnWO_4$ (**left**) showing the existence of the reflections of two HP triclinic domains (T1, T2) along the monoclinic (M) reflections. The corresponding projection of the reciprocal space on the (*b**; *c**) plane (**right**). Dots represent the location of the measured reflections projected along the *a**-axis. The axes of the unit cells are shown as dashed lines. The monoclinic reflections are in black, while the triclinic reflections of the two domains are in blue (T1) and red (T2).

In the previous section, we mentioned that $CdWO_4$, with Cd having an ionic radius in octahedral coordination above 0.9 Å, is just above the limit to be a wolframite at ambient pressure. This implies that the Cd–O distances are larger than the A–O distances in the remaining wolframites and that its bulk modulus is therefore the lowest one in the series (Figure 3). In the extreme case of $MgWO_4$, the Cd–O distances are around 12% longer than the Mg–O distances and the bulk modulus of $MgWO_4$ is 25% larger than that of $CdWO_4$. These differences have an influence in the phase transition of $CdWO_4$ that emerge for instance in the Raman spectrum of the HP phase of $CdWO_4$, presenting 36 instead of 18 modes observed in the post-wolframite phase of normal wolframites (Section 4). In fact, such an increase in modes relates to the doubling of the unit cell that the post-wolframite of $CdWO_4$ presents above 20 GPa. The structure (Figure 1c) was solved with single-crystal X-ray diffraction at 20 GPa [14]. According to Macavei and Shultz [10], the y coordinate of Cd moves fast under pressure. In the phase transition, the a-axis of the HP unit cell remains in the [100] direction of the low-pressure cell, increasing its length, while the new [010] and [001] directions form from the [0$\bar{1}$1] and [0$\bar{1}$1] directions of the low-pressure wolframite cell, respectively. Such a transformation, which can be described by the following transformation matrix [1 0 0, 0 $\bar{1}$ 1, 0 $\bar{1}$ $\bar{1}$] implies a doubling of the unit cell from $Z = 2$ in the wolframite structure to the $Z = 4$ of the post-wolframite of $CdWO_4$. Such a phase transition gives rise to the formation of a screw axis in the [010] direction of the HP phase and therefore a space-group symmetry increase from $P2/c$ to $P2_1/c$. Regarding the coordination of the tungsten and cadmium ions, it is increased to 7-fold and 6+1-fold, respectively, when the phase transition occurs (Figure 1c). This coordination increase contrasts with the phase transition undergone by normal wolframites, which keep the same octahedral coordination for both cations up to the highest pressure reached according to Raman spectroscopy in the case of the other tungstates. According to powder X-ray diffraction, a coordination increase occurs at the phase transition in tantalates and niobates as well [15,16]. Considering that the Cd–O distances are abnormally large in the wolframite-type structure, the coordination increase associated with the phase transition relates well with $CdWO_4$, which has a more compact and stable structure that differs more from the wolframite-type structure than the HP phase of the wolframite-type tungstates that keep the cationic coordination. This coordination change is directly related to a band-gap energy drop associated with the phase transition as we shall show in Section 4 [14].

In the case of $InTaO_4$ and $InNbO_4$, both compounds undergo phase transitions at 13 and 10.8 GPa, respectively. These are pressures around 10 GPa below the phase transition of wolframite-type tungstates. This indicates that, when the X cation is substituted by Ta or Nb, the wolframite structure becomes less robust, probably, as the result of the lowering of the nominal charge of the ion and therefore weakening the X–O bond. Under high pressure, $InTaO_4$ and $InNbO_4$ undergo the same phase transition to a structure (Figure 1b) solved with powder X-ray diffraction [15,16]. Their post-wolframite structure consists on a packing of the wolframite-type structure with the X cations increasing their coordination from 6 to 8-fold filling the channels that the alternating $AOXO$ zigzag chains create in the [001] direction in the wolframite structure. This alternating pattern is kept in the HP phase but with the A and X cations having an eight-fold coordination in the HP phase. The coordination increase in both kind of cations and the packing increase generate a different interaction between the AO_8 and XO_8 polyhedra that now share also edges in addition to corners between them. This implies that they cannot longer be considered as separated blocks.

In order to conclude this section, we shall comment on the case of $CuWO_4$, which, though it presents a distorted triclinic version of the monoclinic wolframite structure at ambient pressure, also transforms to the monoclinic wolframite structure above ~9 GPa [23]. As we explained in the previous section, the low-pressure structure of $CuWO_4$ is also formed by CuO_6 and WO_6 polyhedra arranged in a very similar way to wolframite. However, due to the degeneracy breaking of the $3d^9$ orbitals of Cu^{2+} into five electronic levels (two singlets and one doublet) as a result of the Jahn–Teller effect, two of the six Cu–O distances are longer than the other four, thus lowering its local structure from quasi O_h to quasi D_{4h} symmetry. Under high pressure, the two longest Cu–O distances reduce; however, before quenching the Jahn–Teller distortion, the system finds an easy distortion direction, and the elongated axes of the CuO_6 polyhedra shift from the $[11\bar{1}]$ in the low-pressure phase to the $[101]$ in the HP phase. Thus, the structure transforms to a wolframite structure in space group $P2/c$ that still accommodates the Jahn–Teller distortion of Cu^{2+}.

4. Raman Spectroscopy

Raman spectroscopy is a technique used to observe vibrational modes in a solid and is one of the most informative probes for studies of material properties under HP [24]. Since the study performed two decades ago by Fomichev et al. in $ZnWO_4$ and $CdWO_4$ [25], the Raman spectra of wolframite-type tungstates have been extensively characterized. Studies have been carried out for synthetic crystals [17–20] and minerals [26] and have been also performed for $InTaO_4$ [15] and $InNbO_4$ [16]; however, the characterization of the Raman-active vibrations in wolframite-type molybdates is missing. Indeed, the Raman spectra of $CoMoO_4$, $MnMoO_4$, and $MgMoO_4$ have been characterized [27–29], but for polymorphs different from wolframite.

According to group-theory analysis, a crystal structure isomorphic to wolframite has 36 vibrational modes at the Γ point of the Brillouin zone: $\Gamma = 8A_g + 10B_g + 8A_u + 10B_u$. Three of these vibrations correspond to acoustic modes ($A_u + 2B_u$) and the rest are optical modes, 18 of which are Raman active ($8A_g + 10B_g$) and 15 of which are infrared active ($7A_u + 8B_u$). Typical Raman spectra, taken from the literature [17–20] of wolframite-type tungstates are shown in Figure 5. In the figure, it can be seen that the frequency distribution (and intensity) of the Raman-active modes is qualitatively similar in $CdWO_4$, $ZnWO_4$, $MnWO_4$, and $MgWO_4$. Since Raman-active vibrations correspond either to A_g or B_g modes, polarized Raman scattering and selection rules can be combined to identify the symmetry of modes [30,31]. The expected 18 Raman modes have been measured for $CdWO_4$, $ZnWO_4$, $MnWO_4$, and $MgWO_4$, and 15 modes for $CoWO_4$, $FeWO_4$, and $NiWO_4$. The frequencies of the modes are summarized in Table 2. As we commented above, the frequency distribution of modes is similar in all tungstates, with 4 high-frequency modes, two isolated modes around 500–550 cm^{-1}, and the remaining 12 modes at frequencies below 450 cm^{-1} in all of the compounds.

Table 2. Raman frequencies, ω (cm^{-1}), measured in different wolframite-type tungstates [17–20]. The symmetry of the different modes is given. The pressure coefficients, $d\omega/dP$ (cm^{-1}/GPa), are included in parenthesis for those compounds that are available. The asterisks identify the internal modes of the WO_6 octahedron.

Mode	$MgWO_4$ ω (dω/dP)	$MnWO_4$ ω (dω/dP)	$ZnWO_4$ ω (dω/dP)	$CdWO_4$ ω (dω/dP)	$FeWO_4$ ω	$CoWO_4$ ω	$NiWO_4$ ω
B_g	97.4 (0.69)	89 (0.73)	91.5 (0.95)	78 (0.52)	86	88	91
A_g	155.9 (0.26)	129 (0.02)	123.1 (0.65)	100 (0.69)	124	125	141
B_g	185.1 (0.51)	160 (0.22)	145.8 (1.20)	118 (1.02)	154	154	165
B_g	215.0 (0.63)	166 (0.78)	164.1 (0.72)	134 (0.82)	174	182	190
B_g	266.7 (1.01)	177 (1.03)	189.6 (0.67)	148 (1.51)		199	201
A_g	277.1 (0.55)	206 (2.01)	196.1 (2.25)	177 (0.71)	208		
B_g	313.9 (1.99)	272 (2.03)	267.1 (1.32)	249 (2.14)	266	271	298
A_g	294.1 (1.92)	258 (0.30)	276.1 (0.87)	229 (0.29)			298
B_g	384.8 (4.95)	294 (2.02)	313.1 (1.74)	269 (1.41)	299	315	
A_g	351.9 (3.52)	327 (1.50)	342.1 (1.74)	306 (0.04)	330	332	354
B_g	405.2 (1.47)	356 (4.09)	354.1 (3.87)	352 (4.55)			
A_g*	420.4 (1.59)	397 (1.69)	407 (1.65)	388 (2.33)	401	403	412
B_g	518.1 (3.30)	512 (2.86)	514.5 (3.18)	514 (3.86)	500	496	505
A_g*	551.6 (3.00)	545 (2.39)	545.5 (3.00)	546 (2.32)	534	530	537
B_g*	683.9 (4.09)	674 (4.20)	677.8 (3.90)	688 (4.35)	653	657	663
A_g*	713.2 (3.35)	698 (3.08)	708.9 (3.30)	707 (3.92)	692	686	688
B_g*	808.5 (3.69)	774 (3.58)	786.1 (4.40)	771 (4.30)	777	765	765
A_g*	916.8 (3.19)	885 (1.63)	906.9 (3.70)	897 (3.66)	878	881	887

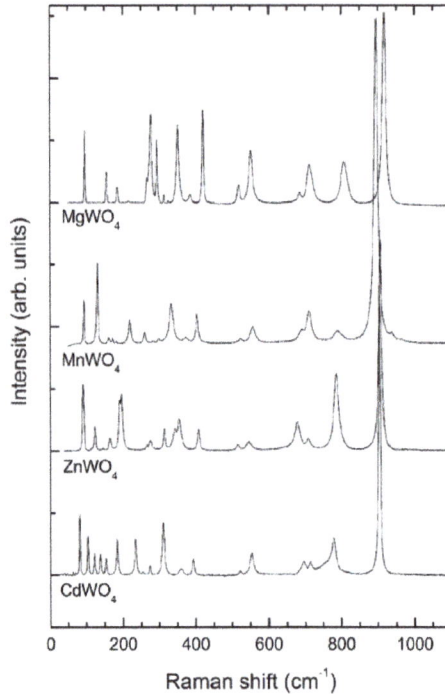

Figure 5. Raman spectra of CdWO$_4$, ZnWO$_4$, MnWO$_4$, and MgWO$_4$ at ambient conditions.

The similitude among the Raman spectra of wolframite-type tungstates can be explained, as a first approximation, by the fact that Raman modes can be classified as internal and external modes with respect to the WO$_6$ octahedron [19]. Six internal stretching modes are expected to arise from the WO$_6$ octahedron. Four of them should have A$_g$ symmetry and the other two B$_g$ symmetry. Since the W atom is heavier than any of the divalent cations in AWO$_4$ wolframites (e.g., Mg or Mn) and W–O covalent bonds are less compressible than A–O bonds, the internal stretching modes of WO$_6$ are the four modes in the high-frequency part of the Raman spectrum (2A$_g$ + 2B$_g$) plus one A$_g$ mode located near 550 cm^{-1} and one A$_g$ mode with a frequency near 400 cm^{-1}. To facilitate identification by the reader, the six internal modes are identified by an asterisk in Table 2. Notice that these six modes have pressure coefficients that do not change much from one compound to the other, and are among the largest pressure coefficients (dω/dP). These observations are consistent with the fact that these modes are associated with internal vibrations on the WO$_6$ octahedron and with the well-known incompressibility of it [10], explained in the previous section.

The twelve Raman modes not corresponding to internal stretching vibrations of the WO$_6$ octahedron imply either bending or motions of WO$_6$ units against the divalent atom A. These modes are in the low-frequency region and are of particular interest because they are very sensitive to structural symmetry changes. Most of these modes (except to B$_g$ modes) have smaller pressure coefficients than do the internal modes. As expected, all Raman-active modes harden under compression, which indicates that the phase transitions induced by pressure are not associated with soft-mode mechanisms. A similar behavior has been reported for wolframite-type MnWO$_4$–FeWO$_4$ solid-solutions [32].

Raman spectroscopy is very sensitive to pressure-driven phase transitions in wolframites. In order to illustrate this, we show in Figure 6 a selection of Raman spectra measured in MnWO$_4$ at different pressures. The experiments were carried out in a diamond-anvil cell using neon as a pressure medium to guarantee quasi-hydrostatic conditions [20]. Clear changes in the Raman spectrum are detected at 25.7 GPa. In particular, new peaks are detected. These changes are indicative of the onset of a structural phase transition. Above 25.7 GPa, gradual changes occur in the Raman spectrum—the Raman modes of the low-pressure wolframite phase only beyond 35 GPa disappear—which indicates the completion of the phase transition. In parallel, the Raman modes of the high-pressure (HP) phase steadily gain intensity from 25.7 to 35 GPa. In Figure 6, the Raman spectra reported at 37.4 and 39.3 GPa correspond to a single post-wolframite phase. In the same figure, it can be seen that the phase transition is reversible, the Raman spectrum collected at 0.5 GPa under decompression corresponding to the wolframite phase. The number of Raman-active modes detected for this HP phase of MnWO$_4$ is 18 (the same number as that for the wolframite-type phase). These modes have been assigned in accordance with the crystal structure proposed for the HP phase [13,31]. This structure is a triclinic distortion of wolframite, and the same number of Raman-active modes is expected; however, all of them have A$_g$ symmetry in the HP phase. Table 3 shows the frequencies of the Raman-active modes of the HP phase of MnWO$_4$ and its pressure coefficients. A detailed discussion of the Raman spectrum of the HP phase can be found elsewhere [20] and is beyond the scope of this article. Here, we will just mention its most relevant features: (1) The modes are no longer isolated in three groups, which suggests that vibrations cannot be explained with a model that assumes that the WO$_6$ octahedron is an isolated unit. (2) The most intense mode is the highest frequency mode (Figure 6), which indicates that it might be associated with a W–O stretching vibration. (3) Interestingly, this mode drops in frequency in comparison to the highest frequency mode in the low-pressure wolframite phase. This is indicative of an increase in W–O bond length evolving towards an effective increasing of the W–O coordination. (4) All the modes of the HP phase harden under compression, but the pressure coefficients of the different modes are smaller than those in the low-pressure phase, which can be explained by the decrease in the compressibility of MnWO$_4$.

Table 3. The Raman active modes of the HP phase of MnWO$_4$ at 34 GPa with their pressure coefficients.

Mode	ω (cm^{-1})	$d\omega/dP$ (cm^{-1} GPa^{-1})
A$_g$	146	0.8
A$_g$	186	0.09
A$_g$	196	1.78
A$_g$	217	1.73
A$_g$	242	1.58
A$_g$	292	0.78
A$_g$	314	1.34
A$_g$	370	2.02
A$_g$	388	2.6
A$_g$	446	2.46
A$_g$	495	1.5
A$_g$	511	1.34
A$_g$	586	1.19
A$_g$	676	1.86
A$_g$	710	1.46
A$_g$	784	1
A$_g$	810	3.26
A$_g$	871	0.69

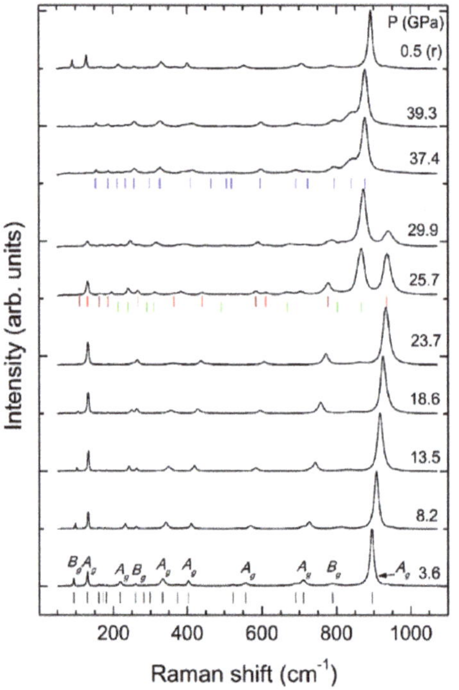

Figure 6. Raman spectra of $MnWO_4$ at selected pressures. Raman modes of the wolframite phase are identified by ticks (and labeled) in the lowest trace (3.6 GPa). At 25.7 GPa, the modes of the low- and high-pressure phase are identified by red and green ticks, respectively. At 37.4 GPa, the ticks identify the 18 Raman-active modes of the HP phase. Pressures (in GPa) are indicated in each Raman spectrum. The spectrum labeled as 0.5(r) was collected at 0.5 GPa after decompression.

Phase transitions are detected by Raman spectroscopy in other AWO_4 tungstates from 20 to 30 GPa [17–19]. In $CdWO_4$, $ZnWO_4$, and $MgWO_4$ phase coexistence is found between the low- and high-pressure phases in a pressure range of 10 GPa. The Raman spectra of the HP phase of all of the compounds studied up to now resemble very much those of the HP phase of $MnWO_4$. However, there are small discrepancies that suggest that, even though the HP phase is structurally related to wolframite, there are some differences in their HP phases. These can be clearly seen when comparing the Raman spectra published for the HP phases of $CdWO_4$ [18] and $MnWO_4$ [20]. In $CdWO_4$, the phase transition to the post-wolframite structure leads to a doubling of the unit-cell (hence, the formula unit increases from 2 to 4) [14], with the consequent increase in the number of Raman modes. They have been assigned to 19 A_g + 17 B_g modes. Out of them, only 26 modes have been observed [14,18]. The frequency of these modes and the pressure coefficients are summarized in Table 4. These values agree very well with theoretical predictions for the monoclinic HP phase of $CdWO_4$ [18]. The larger number of modes observed in the HP phase of $CdWO_4$ in comparison with $MnWO_4$ clearly indicates that the crystal structures of the two compounds are different. In spite of this fact, in the post-wolframite phase of $CdWO_4$, there is a substantial drop in the frequency of the highest frequency mode (the most intense one), as observed in $MnWO_4$ [20]. In the case of $CdWO_4$, a clear correlation can be established between the frequency drop in this W–O stretching mode and the increase in the tungsten–oxygen coordination number [13]. Regarding the pressure dependence of the Raman modes in post-wolframite $CdWO_4$, in Table 4, it can be seen that, as in $MnWO_4$, in the HP of $CdWO_4$ phase the vibrational modes are less affected by compression. In the HP phase of $CdWO_4$, there are two modes with negative pressure coefficients. The existence of modes with negative pressure slopes might be an indication of structural instability like that observed in related scheelite-type tungstates [33].

Table 4. The Raman-active modes of the HP phase of $CdWO_4$ at 26.9 GPa with their pressure coefficients. Twenty-six out of the 36 modes expected were measured.

Mode	ω (cm^{-1})	$d\omega/dP$ (cm^{-1} GPa^{-1})
A_g	69	1.96
B_g	88	1.94
A_g	99	0.09
B_g	130	0.38
A_g	146	1.35
A_g	155	0.97
B_g	165	0.19
A_g	185	1.26
B_g	209	1.26
A_g	243	−0.06
B_g	279	2.53
A_g	290	0.99
B_g	315	3.00
B_g	378	1.65
A_g	401	2.31
A_g	428	3.03
B_g	475	2.51
A_g	486	2.71
B_g	512	2.33
B_g	590	2.62
A_g	673	−0.82
A_g	688	2.81
B_g	710	1.60
A_g	766	2.12
A_g	824	2.23
A_g	864	2.04

To conclude the discussion on tungstates, we would like to state that, in $ZnWO_4$ and $MgWO_4$ [7,19], fewer than 18 Raman modes have been found for the HP phase. The frequency distribution of Raman modes more closely resembles that of $MnWO_4$ than that of $CdWO_4$. However, no definitive conclusion can be stated on the structure of the HP phase of $ZnWO_4$ or $MgWO_4$ only from Raman spectroscopy measurements. Indeed, the accurate assignment of the modes of the HP phases of $ZnWO_4$ and $MgWO_4$ requires HP single-crystal experiments similar to those already carried out with $MnWO_4$ and $CdWO_4$ [13,14].

HP Raman studies have been also carried out for wolframite-type $InTaO_4$ and $InNbO_4$ [15,16]. In both compounds, the Raman spectrum and its pressure evolution resemble those for the tungstates described above. Again, in the low-pressure phase, the modes that change more under compression are the highest frequency modes, which correspond to internal stretching vibrations of the TaO_6 or NbO_6 octahedron. In these compounds, the phase transition occurs around 15 GPa. At the transition, there is a redistribution of high-frequency modes (which seems to be a fingerprint of a transition to post-wolframite), which involves a drop in frequency in the highest frequency mode and other changes consistent with coordination changes determined by XRD experiments [15,16].

5. Electronic Structure and Band Gap

The knowledge of the electronic band structure of wolframite-type compounds is important for the development of the technological applications of these materials. The study of the pressure effects in the band gap has been proven to be an efficient tool for testing the electronic band structure of materials. The first efforts to accurately determine the band structure of wolframites were made by Abraham et al. [34]. These authors focused on $CdWO_4$, comparing it with the better known scheelite-type $CdMoO_4$. By means of density-functional theory calculations, they found that the lowermost conduction bands of $CdWO_4$ are controlled by the crystal-field-splitting of the $5d$ bands of tungsten (slightly hybridized with O $2p$ states). On the other hand, the upper part of the valence band is mainly constituted by O $2p$ states (slightly hybridized with W $5d$ states). As a consequence, $ZnWO_4$, $CdWO_4$, and $MgWO_4$ have very similar band-gap energy E_g. Accurate values of E_g have been determined by means of optical-absorption measurements, E_g being equal to 3.98, 4.02, and 4.06 eV in $ZnWO_4$, $CdWO_4$, and $MgWO_4$, respectively [22]. On the contrary, $MnWO_4$ is known to have a considerably smaller band gap, E_g = 2.37 eV [22]. This distinctive behavior is the consequence of the contribution of the Mn $3d^5$ orbitals to the states near the Fermi energy. Basically, Mg ($3s^2$), Zn ($3d^{10}$), or Cd ($4d^{10}$) have filled electronic shells and therefore do not contribute either to the top of the valence band or the bottom of the conduction band. However, in $MnWO_4$, Mn ($3d^5$) contributes to the top of the valence band and the bottom of the conduction band, reducing the band-gap energy in comparison to $ZnWO_4$, $CdWO_4$, and $MgWO_4$. A similar behavior to that of $MnWO_4$ is expected for $NiWO_4$, $CoWO_4$, and $FeWO_4$ due to the presence of Ni $3d^8$, Co $3d^7$, and Fe $3d^6$ states. Therefore, among the wide dispersion of values reported for $NiWO_4$ (2.28 eV < E_g < 4.5 eV) [35], those in the lowest limit appear to be the most realistic values. Notice that the above conclusions are in agreement with the fact that $CuWO_4$, a distorted wolframite, in which Cu $3d^9$ states contribute to the top of the conduction band, has a band-gap energy of 2.3 eV [36]. Another relevant difference between the band structure of the first and second group of compounds is that it necessarily implies that $ZnWO_4$, $CdWO_4$, and $MgWO_4$ are direct band gap materials, but the other wolframites are indirect band-gap materials [22].

An analogous behavior to that of the wolframite-tungstates is expected for isomorphic molybdates. In this case, the states near the Fermi level will be basically dominated by MoO_4^{2-} [37], so wolframite-type molybdates should have slightly smaller E_g than the tungsten-containing counterpart [38]. Unfortunately, less efforts have been dedicated to molybdates than to tungstates. However, the above stated hypothesis has been confirmed in the case of wolframite-type $ZnMoO_4$, which has E_g = 3.22 eV [39]. A similar value is expected for $MgMoO_4$, ruling out estimations that range from 4.5 to 5.5 eV [40]. On the other hand, the band-gap energy of 2.2 eV determined for $NiMoO_4$ [41] is fully consistent with the conclusions extracted from $MnWO_4$. In this case, the Ni $3d^8$ states will be responsible of the reduction of its E_g in comparison with $ZnMoO_4$.

$InTaO_4$ and $InNbO_4$ are promising materials for photocatalytic water splitting applications [42]. Both materials have been found to be indirect band-gap semiconductors, with E_g = 3.79 eV in $InTaO_4$ [16]. In this compound, the O $2p$ states, with a small amount of mixing of the In $4d$ states, dominate the upper part of the valence band, and Ta $5d$ states and In $5s$ dominate the lower conduction bands. An analogous situation is expected for $InNbO_4$, with the only difference being that Nb $4d$ contributes to the bottom of the conduction band and the Ta $5d$ states do not. This fact makes E_g slightly smaller in $InNbO_4$ (nearly 3.4 eV) than in $InTaO_4$ [43]. An explanation for this phenomenon comes from the larger Pauling's electronegativity of Nb (1.6) in comparison with Ta (1.5) [44]. Within a basic tight-binding approach, E_g is proportional to the overlap integral between the wave functions of atoms. Since Ta has a smaller Pauling's electronegativity than Nb, in a tantalate, the electron transfer from Ta to neighboring oxygen atoms is expected to be larger than the electron transfer from Nb to the neighboring oxygen atoms in a niobate. Therefore, the superposition of wave functions of Ta and O is larger than that of Nb and O, resulting in a larger E_g in $InTaO_4$ than in $InNbO_4$. The same argument can be used to justify the fact that, systematically, molybdates have a smaller E_g than tungstates, which has been stated in the previous paragraph.

We shall discuss now the influence of pressure on the band structure of wolframites. We will first focus on wolframite-type tungstates. The pressure dependence of E_g has been determined from optical-absorption experiments up to 10 GPa for $CdWO_4$, $MgWO_4$, $MnWO_4$, and $ZnWO_4$ [22]. The experiments were carried out on single-crystal samples, under quasi-hydrostatic conditions up to a maximum pressure which is far away the transition pressure. The obtained results are summarized in Figure 7. There it can be seen that $MnWO_4$ has a very distinctive behavior. For $MgWO_4$, $ZnWO_4$, and $CdWO_4$, the pressure dependence of E_g can be represented by a linear function with a positive slope close to $dE_g/dP = 13$ meV/GPa [22]. However, a different pressure dependence is followed by $MnWO_4$, in which E_g redshifts at -22 meV/GPa [22]. An explanation to the different behavior of E_g in $MnWO_4$ comes from the contribution of Mn $3d^5$ states to the bottom of the conduction band. Whereas for $MgWO_4$, $ZnWO_4$, and $CdWO_4$, under compression the bottom of the conduction band goes up in energy, the contribution of Mn states makes it to go down [22]. On the other hand, for the four compounds, the top of the valence band keeps the same energy. This is translated into an increase in E_g for $MgWO_4$, $ZnWO_4$, and $CdWO_4$ and a decrease in E_g for $MnWO_4$. The same behavior in $MnWO_4$ has been reported for triclinic wolframite-related $CuWO_4$ [35] and can be predicted for $CoWO_4$, $FeWO_4$, and $NiWO_4$.

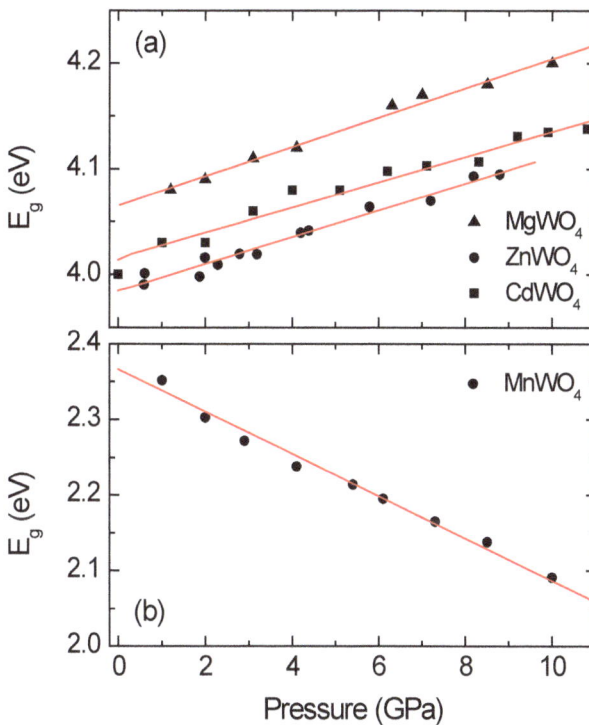

Figure 7. Pressure dependence of the band-gap energy in (**a**) wolframite-type $MgWO_4$, $ZnWO_4$, $CdWO_4$, and (**b**) $MnWO_4$.

The same arguments used to explain the HP behavior of the band-gap in wolframites are useful for understanding the related scheelite-type and monazite-type oxides [45]. An example of it are scheelite-type $CaMoO_4$ ($CaWO_4$) and $PbMoO_4$ ($PbWO_4$) [46,47]. Another example of it is monazite-type $SrCrO_4$ and $PbCrO_4$ [48,49]. In all these compounds, the Pb-containing compounds have a smaller E_g than their isomorphic compounds. This is due to the contribution of Pb $6s$ ($6p$) states to the top (bottom) of the valence (conduction) band. In addition, the Pb states make the band gap to close under compression. As a consequence, dE_g/dP is negative in the Pb containing compounds, but has the opposite sign in the other compounds.

Studies beyond the pressure range of stability have been carried out for $CdWO_4$ [14]. In this compound optical-absorption studies under quasi-hydrostatic conditions have been performed up to 23 GPa. The pressure dependence determined of E_g is shown in Figure 8. As described above, in the first compression steps, E_g increases with pressure. However, at 16.9 GPa, the behavior of E_g changes, moving to lower energies under compression up to 19.5 GPa. This change is caused by a band crossing of the direct and indirect band gaps [14], this result being consistent with Raman and XRD measurements. Beyond 19.5 GPa, a drastic color change occurs in the sample as the result of a sharp band gap reduction to ~3.5 eV [14]. The changes are triggered by the onset of the structural phase transition described in a previous section. In the HP phase of CdWO$_4$ E_g decreases with pressure. Regarding the slope change observed at 16.9 GPa, it is a consequence of a direct to indirect band-gap transition caused by the modification of the electronic band structure under compression [14,50]. Such band crossing has also been observed in other wolframites, such as InTaO$_4$ [15]. This phenomenon should influence not only the band gap but also other band-structure parameters, such as the effective masses, having a strong influence in transport properties [51], an issue that deserves to be explored in the future.

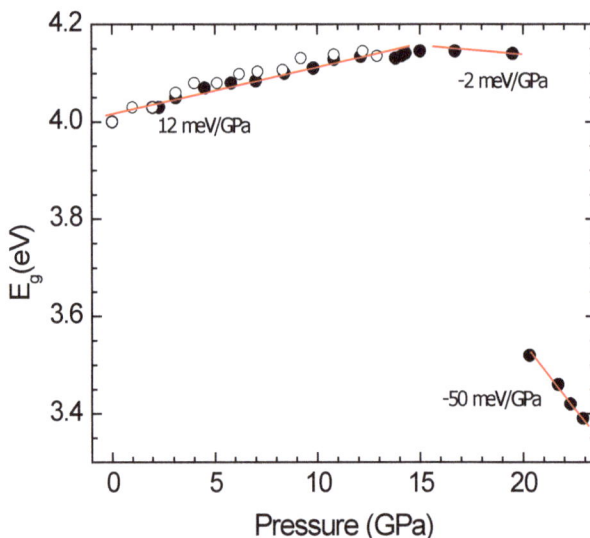

Figure 8. Pressure dependence of the band-gap energy for the low- and high-pressure phases of CdWO$_4$.

Let us discuss now the case of InTaO$_4$. In this compound, E_g has been experimentally determined up to 23 GPa [15]. Calculations have been also carried out [15]. The results are shown in Figure 9. Calculations underestimate the value of E_g, which is typical of density-functional theory; however, they nicely reproduce the pressure dependence of E_g. In the figure, it can be seen that, when pressure is increased, there is a blueshift of E_g with a change of the dE_g/dP around 5 GPa. As in CdWO$_4$, this singularity occurs due to a band crossing [15], which in the case of InTaO$_4$ is triggered by changes induced in the top of the valence band by pressure. In particular, at around 5 GPa, the maximum of the valence band changes from the Y point of the Brillouin zone to the Z point. On the other hand, when increasing the pressure beyond 13 GPa, InTaO$_4$ changes from colorless to yellow [15]. The change in color is correlated to a band-gap collapse and is associated with a structural phase transition found by Raman and XRD experiments [15]. The HP phase has been found to have a direct band gap (the low-pressure wolframite has an indirect gap). In contrast with the low-pressure phase, in the HP phase E_g redshifts with pressure. This is a consequence of the fact that, under compression, the valence band shifts slightly faster towards high energies than the conduction band.

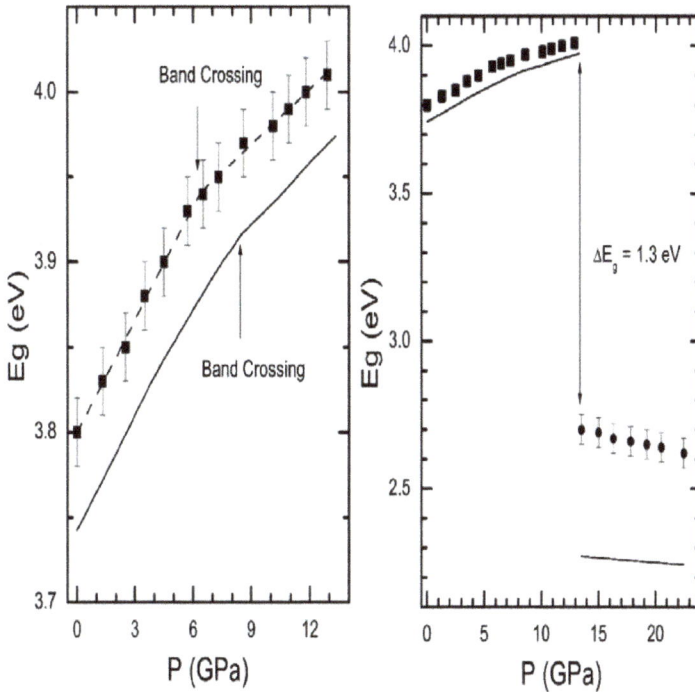

Figure 9. (**left**) Pressure dependence of E_g in the low-pressure phase of InTaO$_4$. The change in the pressure dependence caused by the band-crossing induced by pressure is indicated with arrows. (**right**) Pressure dependence of E_g in the low- and high-pressure phases of InTaO$_4$. The band-gap collapse associated with the transition is indicated (E_g = 1.3 eV). In both figures, symbols correspond to experiments and lines to calculations.

To close this section, we would like to comment on the possible pressure-induced metallization of wolframites. It has been suggested, based upon resistivity measurements, that ternary oxides related to wolframite might metallize through band overlapping at relative low pressures (12–30 GPa) [52,53]. So far, no evidence of metallization has been detected for all the studied wolframites, either in the pressure range of stability of the low-pressure phase or in the post-wolframite phases up to the maximum pressure achieved in experiments (45 GPa in ZnWO$_4$ and CdWO$_4$) [17,18]. In particular, the sample darkening associated with a semiconductor–metal transition has never been detected in any wolframite under compression. In addition, density-functional theory calculations also exclude the possibility of metallization in the wolframite and post-wolframite phases [14,15]. One of the reasons preventing metallization is the robustness of the WO$_6$ (MO$_6$, NbO$_6$, and TaO$_6$) octahedron, which is a less compressible polyhedral unit within the crystal structure. As a consequence, the application of high pressures is not enough to increase the overlap between the electronic wave-functions of transition metals (W, Mo, Nb, and Ta) and oxygen atoms to broaden the electronic bands and create the eventual delocalization of the electrons requested for metallization [54]. The fact that in distorted wolframite CuWO$_4$ a HP phase transition takes place without any significant reduction of the Jahn–Teller distortion [55] suggests that this compound is the best candidate for metallization driven by band overlap under compression.

6. Concluding Remarks

In the sections presented above, we have described the recent advances made on the understanding of the structural, vibrational, and electronic properties of wolframites under compression. The discussion has mainly been based on experimental results; however, *ab initio* calculations have been crucial for the interpretation of experiments [12–16]. Between the recent advances, the one that has likely had more influence in the field has been the structural solution of the different post-wolframite structures. The precise knowledge of the pressure structural stability of different wolframites and their HP structures has opened two new avenues to explore: (i) the study of the scintillating properties of the HP polymorphs by means of HP photoluminescence studies; (ii) the changes produced by this structural change on the magnetic properties of those wolframites with an open d outer shell like multiferroic $MnWO_4$; and (iii) the behavior under compression of wolframite alloys like $MnW_{1-x}Mo_xO_4$ [56]; in particular, $CdW_{1-x}Mo_xO_4$ whose end-members have either the wolframite or scheelite structure. For these biphasic alloy systems, their HP behavior is unpredictable [57].

In recent years, some works have appeared to deal with the pressure–temperature magnetic phase diagram of pure [9,58] or cobalt alloyed [59] $MnWO_4$. Those works have found that pressure is able to disrupt the fine equilibrium of the frustrated antiferromagnetic phase of $MnWO_4$ (AF1) but enhance the Néel temperature of the AF3 and AF4 magnetic phases of pure or lowly Co-doped $MnWO_4$ and highly Co-doped $MnWO_4$, respectively. Considering the direct effect that pressure has on the spin structure that is even able to cause a spin–flop transition for highly Co-doped $MnWO_4$ [59], one can expect a new and fascinating phase diagram in the HP phase of $MnWO_4$, where distortions are expected to be higher due to the lowering of symmetry from space group $P2/c$ to $P\bar{1}$. So far, the only study done in this direction has been done with $CuWO_4$; according to calculations in that study, the structural phase transition from $P2/c$ to $P\bar{1}$ also involves an antiferromagnetic to ferromagnetic order [60].

In summary, though some fundamental questions still remain to be completely solved, such as the crystallization of $CdWO_4$ in wolframite structures in spite of the size of Cd or the atomic coordinates of the post-wolframite phase of $MnWO_4$, this brief review shows (i) the great advances that have been done with respect to this family of compounds since the pioneering works of Macavei and Shultz [10] and Jayaraman et al. [11] and (ii) the avenues that have yet to be explored.

Acknowledgments: J. Ruiz-Fuertes thanks the Spanish Ministerio de Economía y Competitividad (MINECO) for the support through the Juan de la Cierva Program (IJCI-2014-20513). This work was supported by the Spanish MINECO, the Spanish Research Agency (AEI), and the European Fund for Regional Development (FEDER) under project number MAT2016-75586-C4-1-P. The authors are grateful to all of the collaborators who participated in the research reviewed here.

Conflicts of Interest: The authors declare no conflict of interest.

References

1. Broch, E.K. Skrifter Untersuchungen ueber Kristallstrukturen des Wloframittypus und des Scheelittypus Norske Videnskaps Akademi i Oslo. *Mat. Nat. Kl.* **1929**, *8*, 20.
2. Keeling, R.O., Jr. The structure of $NiWO_4$. *Acta Crystallogr.* **1957**, *10*, 209–213. [CrossRef]
3. Young, A.P.; Schwartz, C.M. High-pressure synthesis of molybdates with the wolframite structure. *Science* **1963**, *141*, 348. [CrossRef] [PubMed]
4. Rathee, S.; Tu, D.; Monajemi, T.T.; Rickey, D.W.; Fallone, B.G. A bench-top megavoltage fan-beam CT using $CdWO_4$-photodiode detectors. I. System description and detector characterization. *Med. Phys.* **2006**, *33*, 1078–1089. [CrossRef] [PubMed]
5. Mikhailik, V.B.; Kraus, H. Performance of scintillation materials at cryogenic temperatures. *Phys. Status Solidi B* **2010**, *247*, 1583–1599. [CrossRef]
6. Silva, M.M.; Novais, S.M.V.; Silva, E.S.S.; Schimitberger, T.; Macedo, Z.S.; Bianchi, R.F. $CdWO_4$-on-MEH-PPV:PS as a candidate for real-time dosimeters. *Mater. Chem. Phys.* **2012**, *136*, 317–319. [CrossRef]

7. Burachas, S.P.; Danevich, F.A.; Georgadze, A.S.; Klapdor-Kleingrothaus, H.V.; Kobychev, V.V.; Kropivyansky, B.N.; Kuts, V.N.; Muller, A.; Muzalevsky, V.V.; Nikolaiko, A.S.; et al. Large volume $CdWO_4$ crystal scintillators. *Nucl. Instrum. Methods Phys. Res. Sect. A* **1996**, *369*, 164–168. [CrossRef]

8. Lautenschläger, G.; Weitzel, H.; Vogt, T.; Hock, R.; Böhm, A.; Bonnet, M.; Fuess, H. Magnetic phase transitions of $MnWO_4$ studied by the use of neutron diffraction. *Phys. Rev. B* **1993**, *48*, 6087. [CrossRef]

9. Chaudhury, R.P.; Yen, F.; de la Cruz, C.R.; Lorenz, B.; Wang, Y.Q.; Sun, Y.Y.; Chu, C.W. Thermal expansion and pressure effect in $MnWO_4$. *Physica B* **2008**, *403*, 1428–1430. [CrossRef]

10. Macavei, J.; Schulz, H. The crystal structure of wolframite type tungstates at high pressure. *Z. Kristallogr.* **1993**, *207*, 193–208.

11. Jayaraman, A.; Wang, S.Y.; Sharma, S.K. New pressure-induced phase changes in $CdWO_4$ from Raman spectroscopic and optical microscopic studies. *Curr. Sci.* **1995**, *69*, 44–48.

12. Ruiz-Fuertes, J.; López-Moreno, S.; Errandonea, D.; Pellicer-Porres, J.; Lacomba-Perales, R.; Segura, A.; Rodríguez-Hernández, P.; Muñoz, A.; Romero, A.H.; González, J. High-pressure phase transitions and compressibility of wolframite-type tungstates. *J. Appl. Phys.* **2010**, *107*, 083506. [CrossRef]

13. Ruiz-Fuertes, J.; Friedrich, A.; Gomis, O.; Errandonea, D.; Morgenroth, W.; Sans, J.A.; Santamaría-Pérez, D. High-pressure structural phase transition in $MnWO_4$. *Phys. Rev. B* **2015**, *91*, 104109. [CrossRef]

14. Ruiz-Fuertes, J.; Friedrich, A.; Errandonea, D.; Segura, A.; Morgenroth, W.; Rodríguez-Hernández, P.; Muñoz, A.; Meng, Y. Optical and structural study of the pressure-induced phase transition of $CdWO_4$. *Phys. Rev. B* **2017**, *95*, 174105. [CrossRef]

15. Errandonea, D.; Popescu, C.; Garg, A.B.; Botella, P.; Martínez-García, D.; Pellicer-Porres, J.; Rodríguez-Hernández, P.; Muñoz, A.; Cuenca-Gotor, V.; Sans, J.A. Pressure-induced phase transition and band-gap collapse in the wide-band-gap semiconductor $InTaO_4$. *Phys. Rev. B* **2016**, *93*, 35204. [CrossRef]

16. Garg, A.B.; Errandonea, D.; Popescu, C.; Martínez-García, D.; Pellicer-Porres, J.; Rodríguez-Hernández, P.; Muñoz, A.; Botella, P.; Cuenca-Gotor, V.P.; Sans, J.A. Pressure-Driven Isostructural Phase Transition in $InNbO_4$: In Situ Experimental and Theoretical Investigations. *Inorg. Chem.* **2017**, *56*, 5420–5430. [CrossRef] [PubMed]

17. Errandonea, D.; Manjón, F.J.; Garro, N.; Rodríguez-Hernández, P.; Radescu, S.; Mujica, A.; Muñoz, A.; Tu, C.Y. Combined Raman scattering and ab initio investigation of pressure-induced structural phase transitions in the scintillator $ZnWO_4$. *Phys. Rev. B* **2008**, *78*, 54116. [CrossRef]

18. Lacomba-Perales, R.; Errandonea, D.; Martínez-García, D.; Rodríguez-Hernández, P.; Radescu, S.; Mújica, A.; Muñoz, A.; Chervin, J.C.; Polian, A. Phase transitions in wolframite-type $CdWO_4$ at high pressure studied by Raman spectroscopy and density-functional theory. *Phys. Rev. B* **2009**, *79*, 94105. [CrossRef]

19. Ruiz-Fuertes, J.; Errandonea, D.; López-Solano, S.; González, J.; Gomis, O.; Vilaplana, R.; Manjón, F.J.; Muñoz, A.; Rodríguez-Hernández, P.; Friedrich, A.; et al. High-pressure Raman spectroscopy and lattice-dynamics calculations on scintillating $MgWO_4$: Comparison with isomorphic compounds. *Phys. Rev. B* **2011**, *83*, 214112. [CrossRef]

20. Ruiz-Fuertes, J.; Gomis, O.; Errandonea, D.; Friedrich, A.; Manjón, F.J. Room-temperature vibrational properties of multiferroic $MnWO_4$ under quasi-hydrostatic compression up to 39 GPa. *J. Appl. Phys.* **2014**, *115*, 43510. [CrossRef]

21. Dai, R.C.; Ding, X.; Wang, Z.P.; Zhang, Z.M. Pressure and temperature dependence of Raman scattering of $MnWO_4$. *Chem. Phys. Lett.* **2013**, *586*, 76–80. [CrossRef]

22. Ruiz-Fuertes, J.; López-Moreno, S.; López-Solano, J.; Errandonea, D.; Segura, A.; Lacomba-Perales, R.; Muñoz, A.; Radescu, S.; Rodríguez-Hernández, P.; Gospodinov, M.; et al. Pressure effects on the electronic and optical properties of AWO_4 wolframites (A = Cd, Mg, Mn, and Zn): The distinctive behavior of multiferroic $MnWO_4$. *Phys. Rev. B* **2012**, *86*, 125202. [CrossRef]

23. Ruiz-Fuertes, J.; Friedrich, A.; Pellicer-Porres, J.; Erradonea, D.; Segura, A.; Haussühl, E.; Tu, C.-Y.; Polian, A. Structure Solution of the High-Pressure Phase of $CuWO_4$ and Evolution of the JahnTeller Distortion. *Chem. Mater.* **2011**, *23*, 4220–4226. [CrossRef]

24. Goncharov, A. Raman Spectroscopy at High Pressures. *Int. J. Spectrosc.* **2012**, *2012*, 617528. [CrossRef]

25. Fomichev, V.V.; Kondratov, O.I. Vibrational spectra of compounds with the wolframite structure. *Spectrochim. Acta A* **1994**, *50*, 1113–1120. [CrossRef]

26. Crane, M.; Frost, R.L.; Williams, P.A.; Kloprogge, J.T. Raman spectroscopy of the molybdate minerals chillagite (tungsteinian wulfenite-I4), stolzite, scheelite, wolframite and wulfenite. *J. Raman Spectrosc.* **2002**, *33*, 62–66. [CrossRef]

27. Kanesake, I.; Hashiba, H.; Matsumura, I. Polarized Raman spectrum and normal coordinate analysis of α-MnMoO₄. *J. Raman Spectrosc.* **1988**, *19*, 213–218. [CrossRef]

28. Coelho, M.N.; Freire, P.T.C.; Maczka, M.; Luz-Lima, C.; Saraiva, G.D.; Paraguassu, W.; Filho, A.G.S.; Pizani, P.S. High-pressure Raman scattering of MgMoO₄. *Vib. Spectrosc.* **2013**, *68*, 34–39. [CrossRef]

29. De Moura, A.P.; de Oliveira, L.H.; Pereira, P.F.S.; Rosa, I.L.V.; Li, M.S.; Longo, E.; Varela, J.A. Structural, Optical, and Magnetic Properties of NiMoO₄ Nanorods Prepared by Microwave Sintering. *Adv. Chem. Eng. Sci.* **2012**, *2*, 465.

30. Iliev, M.N.; Gospodinov, M.M.; Litvinchuk, A.P. Raman spectroscopy of MnWO₄. *Phys. Rev. B* **2009**, *80*, 212302. [CrossRef]

31. Errandonea, D.; Pellicer-Porres, J.; Martínez-García, D.; Ruiz-Fuertes, J.; Friedrich, A.; Morgenroth, W.; Popescu, C.; Rodríguez-Hernández, P.; Muñoz, A.; Bettinelli, M. Phase Stability of Lanthanum Orthovanadate at High Pressure. *J. Phys. Chem. C* **2016**, *120*, 13749. [CrossRef]

32. Maczka, M.; Ptak, M.; da Silva, K.P.; Freire, P.T.C.; Hanuza, J. High-pressure Raman scattering and an anharmonicity study of multiferroic wolframite-type Mn₀.₉₇Fe₀.₀₃WO₄. *J. Phys. Condens. Matter* **2012**, *4*, 345403. [CrossRef] [PubMed]

33. Errandonea, D.; Manjon, F.J. On the ferroelastic nature of the scheelite-to-fergusonite phase transition in orthotungstates and orthomolybdates. *Mater. Res. Bull.* **2009**, *44*, 807–811. [CrossRef]

34. Abraham, Y.; Holzwarth, N.A.W.; Williams, R.T. Electronic structure and optical properties of CdMoO₄ and CdWO₄. *Phys. Rev. B* **2000**, *62*, 1733. [CrossRef]

35. Zawawi, S.M.M.; Yahya, R.; Hassan, A.; Mahmud, H.N.M.E.; Daud, M.N. Structural and optical characterization of metal tungstates (MWO₄; M = Ni, Ba, Bi) synthesized by a sucrose-templated method. *Chem. Cent. J.* **2013**, *7*, 80. [CrossRef] [PubMed]

36. Ruiz-Fuertes, J.; Errandonea, D.; Segura, A.; Manjón, F.J.; Zhu, Z.; Tu, C.Y. Growth, characterization, and high-pressure optical studies of CuWO₄. *High Press. Res.* **2008**, *28*, 565–570. [CrossRef]

37. Errandonea, D. Comment on "Molten salt synthesis of barium molybdate and tungstate microcrystals". *Mater. Lett.* **2009**, *63*, 160–161. [CrossRef]

38. Lacomba-Perales, R.; Ruiz-Fuertes, J.; Errandonea, D.; Martınez-Garcıa, D.; Segura, A. Optical absorption of divalent metal tungstates: Correlation between the band-gap energy and the cation ionic radius. *EPL* **2008**, *83*, 37002. [CrossRef]

39. Cavalcante, L.S.; Moraes, E.; Almeida, M.A.P.; Dalmaschio, C.J.; Batista, N.C.; Varela, J.A.; Longo, E.; Li, M.S.; Andrés, J.; Beltrán, A. A combined theoretical and experimental study of electronic structure and optical properties of β-ZnMoO₄ microcrystals. *Polyhedron* **2013**, *54*, 13–25. [CrossRef]

40. Spasskii, D.A.; Kolobanov, V.N.; Mikhailin, V.V.; Berezovskaya, L.Y.; Ivleva, L.I.; Voronina, I.S. Luminescence peculiarities and optical properties of MgMoO₄ and MgMoO₄:Yb crystals. *Opt. Spectrosc.* **2009**, *106*, 556–563. [CrossRef]

41. De Moura, A.P.; de Oliveira, L.H.; Rosa, I.L.V.; Xavier, C.S.; Lisboa-Filho, P.N.; Li, M.S.; la Porta, F.A.; Longo, E.; Varela, J.A. Structural, Optical, and Magnetic Properties of NiMoO₄ Nanorods Prepared by Microwave Sintering. *Sci. World J.* **2015**, *2015*, 315084. [CrossRef] [PubMed]

42. Zou, Z.; Ye, J.; Sayana, K.; Arakawa, H. Direct splitting of water under visible light irradiation with an oxide semiconductor photocatalyst. *Nature* **2001**, *414*, 625. [CrossRef] [PubMed]

43. Li, G.L.; Yin, Z. Theoretical insight into the electronic, optical and photocatalytic properties of InMO₄ (M = V, Nb, Ta) photocatalysts. *Phys. Chem. Chem. Phys.* **2011**, *13*, 2824–2833. [CrossRef] [PubMed]

44. Pauling, L. The energy of single bonds and the relative electronegativity of atoms. *J. Am. Chem. Soc.* **1932**, *54*, 3570–3582. [CrossRef]

45. Errandonea, D. High-pressure phase transitions and properties of MTO₄ compounds with the monazite-type structure. *Phys. Status Solidi B* **2017**, *254*, 1700016. [CrossRef]

46. Panchal, V.; Garg, N.; Poswal, H.K.; Errandonea, D.; Rodríguez-Hernández, P.; Muñoz, A.; Cavalli, E. The high-pressure behavior of CaMoO₄. *Phys. Rev. Mater.* **2017**, *1*, 43605. [CrossRef]

47. Lacomba-Perales, R.; Errandonea, D.; Segura, A.; Ruiz-Fuertes, J.; Rodríguez-Hernández, P.; Radescu, S.; Lopez-Solano, J.; Mujica, A.; Muñoz, A. A combined high-pressure experimental and theoretical study of the electronic band-structure of scheelite-type AWO_4 (A = Ca, Sr, Ba, Pb) compounds. *J. Appl. Phys.* **2011**, *110*, 43703. [CrossRef]

48. Bandiello, E.; Errandonea, D.; Martinez-Garcia, D.; Santamaria-Perez, D.; Manjón, F.J. Effects of high-pressure on the structural, vibrational, and electronic properties of monazite-type $PbCrO_4$. *Phys. Rev. B* **2012**, *85*, 24108. [CrossRef]

49. Gleissner, J.; Errandonea, D.; Segura, A.; Pellicer-Porres, J.; Hakeem, M.A.; Proctor, J.E.; Raju, S.V.; Kumar, R.S.; Rodríguez-Hernández, P.; Muñoz, A.; et al. Monazite-type $SrCrO_4$ under compression. *Phys. Rev. B* **2016**, *94*, 134108. [CrossRef]

50. Baj, M.; Dmowski, L.H.; Slupinski, T. Direct proof of two-electron occupation of Ge-DX centers in GaAs codoped with Ge and Te. *Phys. Rev. Lett.* **1993**, *71*, 3529. [CrossRef] [PubMed]

51. Errandonea, D.; Segura, A.; Manjon, F.J.; Chevy, A. Transport measurements in InSe under high pressure and high temperature: Shallow-to-deep donor transformation of Sn related donor impurities. *Semicond. Sci. Technol.* **2003**, *18*, 241. [CrossRef]

52. Garg, A.B.; Shanavas, K.V.; Wani, B.N.; Sharma, S.M. Phase transition and possible metallization in $CeVO_4$ under pressure. *J. Solid State Chem.* **2013**, *203*, 273–280. [CrossRef]

53. Duclos, S.J.; Jayaraman, A.; Espinosa, G.P.; Oxper, A.S.; Maines, R.G. Raman and optical absorption studies of the pressure-induced zircon to scheelite structure transformation in $TbVO_4$ and $DyV0_4$. *J. Phys. Chem. Solids* **1989**, *8*, 769–775. [CrossRef]

54. Patterson, J.R.; Aracne, C.M.; Jackson, D.D.; Malba, V.; Weir, S.T. Pressure-induced metallization of the Mott insulator MnO. *Phys. Rev. B* **2004**, *69*, 220101(R). [CrossRef]

55. Ruiz-Fuertes, J.; Segura, A.; Rodríguez, F.; Errandonea, D.; Sanz-Ortiz, M.N. Anomalous High-Pressure Jahn-Teller Behavior in $CuWO_4$. *Phys. Rev. Lett.* **2012**, *108*, 166402. [CrossRef] [PubMed]

56. Blanco-Gutierrez, V.; Demourgues, A.; Lebreau, E.; Gaudon, M. You have full text access to this content Phase transitions in $Mn(Mo_{1-x}W_x)O_4$ oxides under the effect of high pressure and temperature. *Phys. Status Solids B* **2016**, *253*, 2043–2048. [CrossRef]

57. Taoufyq, A.; Guinneton, F.; Valmalette, J.C.; Arab, M.; Benlhachemi, A.; Bakiz, B.; Villain, S.; Lyoussi, A.; Nolibe, G.; Gavarri, J.R. Structural, vibrational and luminescence properties of the $(1-x)CaWO_4-xCdWO_4$ system. *J. Solid State Chem.* **2014**, *219*, 127–137. [CrossRef]

58. Hardy, V.; Payen, C.; Damay, F.; Meddar, L.; Josse, M.; Andre, G. Phase transitions and magnetic structures in $MnW_{1-x}Mo_xO_4$ compounds (x ≤ 0.2). *J. Phys. Condens. Matter* **2016**, *28*, 33600359. [CrossRef] [PubMed]

59. Wang, J.; Ye, F.; Chi, S.; Fernandez-Baca, J.A.; Cao, H.; Tian, W.; Gooch, M.; Poudel, N.; Wang, Y.; Lorenz, B.; et al. Pressure effects on magnetic ground states in cobalt-doped multiferroic $Mn_{1-x}Co_xWO_4$. *Phys. Rev. B* **2016**, *93*, 155164. [CrossRef]

60. Ruiz-Fuertes, J.; Errandonea, D.; Lacomba-Perales, R.; Segura, A.; González, J.; Rodríguez, F.; Manjón, F.J.; Ray, S.; Rodríguez-Hernández, P.; Muñoz, A.; et al. High-pressure structural phase transitions in $CuWO_4$. *Phys. Rev. B* **2010**, *81*, 224115. [CrossRef]

crystals

MDPI

Article

Unravelling the High-Pressure Behaviour of Dye-Zeolite L Hybrid Materials

Lara Gigli [1], Rossella Arletti [2,3,*], Ettore Fois [4], Gloria Tabacchi [4], Simona Quartieri [5], Vladimir Dmitriev [6] and Giovanna Vezzalini [7]

[1]   Elettra-Sincrotrone Trieste, Strada Statale 14-km 163,5 in AREA Science Park, 34149 Basovizza, Trieste, Italy; lara.gigli@elettra.eu
[2]   Dipartimento di Scienze della Terra, Università degli Studi di Torino, Via Valperga Caluso 35, 10125-Torino, Italy
[3]   Interdepartmental Centre "Nanostructure Interfaces and Surfaces NIS", Via Pietro Giuria 7, 10125-Torino, Italy
[4]   Dipartimento di Scienza e Alta Tecnologia and INSTM, Università degli Studi dell'Insubria, Via Valleggio 9, I-22100 Como, Italy; ettore.fois@uninsubria.it (E.F.); gloria.tabacchi@uninsubria.it (G.T.)
[5]   Dipartimento di Scienze Matematiche e Informatiche, Scienze Fisiche e Scienze della Terra, Università degli Studi di Messina, Viale Ferdinando Stagno d'Alcontres 31, 98166-Messina S.Agata, Italy; simona.quartieri@unime.it
[6]   Swiss-Norwegian Beam Line at ESRF, BP220, 38043 Grenoble CEDEX, France; dmitriev@esrf.fr
[7]   Dipartimento di Scienze Chimiche e Geologiche, Università degli Studi di Modena e Reggio Emilia, Via Giuseppe Campi 103, 41125-Modena, Italy; mariagiovanna.vezzalini@unimore.it
*   Correspondence: rossella.arletti@unito.it; Tel.: +39-011-670-5129

Received: 11 January 2018; Accepted: 29 January 2018; Published: 2 February 2018

Abstract: Self-assembly of chromophores nanoconfined in porous materials such as zeolite L has led to technologically relevant host-guest systems exploited in solar energy harvesting, photonics, nanodiagnostics and information technology. The response of these hybrid materials to compression, which would be crucial to enhance their application range, has never been explored to date. By a joint high-pressure in situ synchrotron X-ray powder diffraction and ab initio molecular dynamics approach, herein we unravel the high-pressure behaviour of hybrid composites of zeolite L with fluorenone dye. High-pressure experiments were performed up to 6 GPa using non-penetrating pressure transmitting media to study the effect of dye loading on the structural properties of the materials under compression. Computational modelling provided molecular-level insight on the response to compression of the confined dye assemblies, evidencing a pressure-induced strengthening of the interaction between the fluorenone carbonyl group and zeolite L potassium cations. Our results reveal an impressive stability of the fluorenone-zeolite L composites at GPa pressures. The remarkable resilience of the supramolecular organization of dye molecules hyperconfined in zeolite L channels may open the way to the realization of optical devices able to maintain their functionality under extreme conditions.

Keywords: high-pressure chemistry; nanomaterials; supramolecular chemistry; self-assembly; X-ray diffraction; zeolites; density functional calculations; ab initio molecular dynamics; structural refinements; artificial antenna systems; organic-inorganic hybrid materials

---

1. Introduction

Zeolites are crystalline natural or synthetic porous materials consisting of corner-sharing tetrahedral units, characterized by a regular arrangement of cages and channels of molecular size [1,2]. The nanometric-scale geometry of zeolite pore systems allows the intensive use of zeolites in several fields such as in molecular separation processes and in heterogeneous catalysis [3,4].

Besides being instrumental in traditional applications, the ordered arrangements of zeolitic cages has long been recognized as a route to create advanced materials based on confined, organized nanostructures, such as luminescent metal clusters [5–8], quantum dots/wires [9] or lanthanides [10]. Turning to more complex guests, the incorporation of fluorescent molecules in zeolite cages generally enhances their emission properties [11–13] because it allows to obtain high concentrations of chromophores while limiting the formation of aggregates (which negatively affect the emission intensity) [14]. Incorporated dyes exhibit a preferential orientation of the electronic transition dipole moments, which endows -dye-zeolite host–guest compounds with exceptional energy transfer properties, similar to antenna systems in natural photosynthetic organisms [11,15–25].

Zeolite L (Figure 1) features a unique one-dimensional 12-membered (12 MR) channel system running along the [001] direction. One of the greatest advantages of zeolite L is that high-quality crystals with a desired aspect ratio can be easily synthesized [26,27]. Also importantly, their surfaces can be modified selectively by the attachment of molecules carrying a specific chemical functionality [28,29], thus allowing zeolite L crystals to be sealed (to avoid leakage of dye molecules) [30–32], or interfaced to molecules [33], nanoparticles [34,35], living cells [36,37] and rigid supports [38]. Based on these features, zeolite L has been fully exploited as matrix for dye encapsulation and the molecules organization has been hierarchically enhanced up to the macroscopic scale, resulting into functional materials with remarkable electro-optical and energy-transfer performances, already exploited in devices [39]. Composites based on zeolite L as host matrix have reached an advanced stage of development [40] and are presently used as effect pigments [41], in solar energy harvesting [11], or in biomedical technology [42,43].

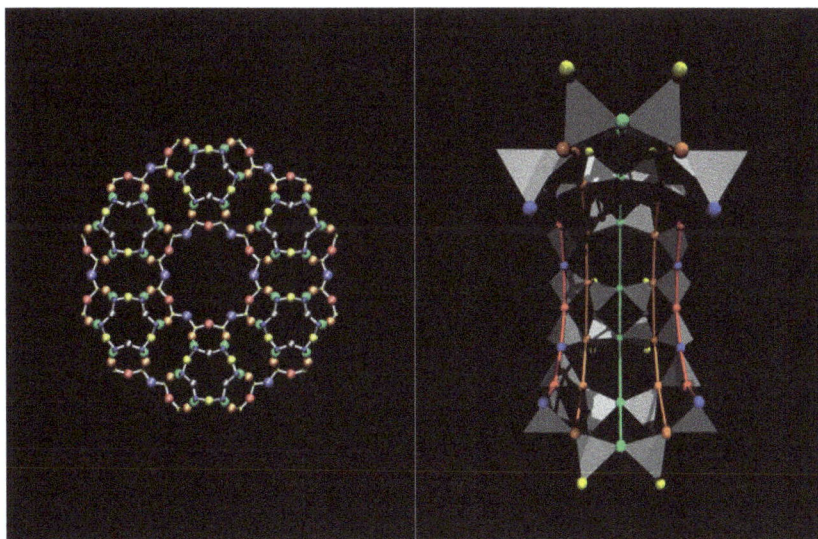

Figure 1. Zeolite L and its one-dimensional nanochannels. Left panel: View along the channel axis (parallel to [001]) highlighting the larger channel, delimited by a 12-membered ring (12 MR) and surrounded by smaller 8 MR channels. The tetrahedral atoms of the framework (T = Si, Al) are represented by small balls, O atoms by big balls. Colour codes: T1 = yellow, T2 = blue, O1 = blue (big), O2 = red, O3 = grey, O4 = orange, O5 = yellow (big), O6 = green; Right panel: Half-section of the 12 MR channel, viewed perpendicular to the 12 MR channel axis. T-atoms (Si, Al) are represented by grey tetrahedra, O atoms by big balls, with same colour codes as in the left panel. The coloured lines are a guide to the eye drawn to represent the different rings in the cage and to evidence the channel curvature under a perspective view.

Such ongoing progress in applications has been accompanied, in the latest years, by a deeper molecular-level understanding of the confined photoactive assemblies, which has been achieved through computational modelling [44–50], often combined with multi-technique experimental analyses [45,50–57]. The key role of water in tuning the organization of the confined chromophores has been revealed [45,46,58], as well as the stabilizing effect of potassium cations in composites with carbonyl dyes [44,57,59].

The high-pressure behaviour of zeolites has been intensively studied in the last 15 years, as documented in various reviews [50,60–66], both with pore penetrating and non-penetrating pressure transmitting media (PTM). One reason for this interest is that high pressures, combined with the confining environments of nanometric pores, may reveal unexpected chemical phenomena (e.g., pressure-induced hydration [67–82], ionic conductivity [83], guest exchange [84–88], or realization of new materials otherwise unattainable [89–97]). With this paper we would like to answer the question: how do confined supramolecular assemblies of water and chromophores respond to compression?

In this paper, it is reported the study of the high-pressure evolution of zeolite–dye hybrids. Specifically, we selected the zeolite L (ZL)–fluorenone (FL) hybrid (ZL/FL from now on), subject of previous investigations performed by our group [56]. We are interested in exploring whether the compression could favour a more ordered distribution of the dye molecules in the zeolite L channels and if this process leads to an improvement of the optical properties of the hybrid material. In fact, it has been shown that the properties of dye-zeolite systems depend on the molecular orientation, arrangement and packing inside the channel [14,15,20,44–46,56,58], which control the guest-guest and host-guest interactions. To unravel and understand at molecular level the compression behaviour of the inclusion composites we adopted an integrated experimental-theoretical approach, based on the use of high pressure (HP) in situ synchrotron X-ray powder diffraction (XRPD) and ab initio molecular dynamics [98] simulations, which already captured a nice example of pressure-induced supramolecular organization in zeolites [97]. Herein, high-pressure experiments were performed up to 6 GPa using non-penetrating pressure transmitting media on samples characterized by three different contents of fluorenone molecules per unit cell, in order to verify how the dye loading influences the structural properties of the hybrid material under compression.

2. Materials and Methods

2.1. Zeolite Dye Hybrids: Structural Details at Ambient Conditions

The LTL framework [1] is built from columns of cancrinite cages stacked with double six membered rings (D6R) along the *c* axis. These columns are connected to form larger circular 12-ring (12 MR) channels and smaller elliptical 8-ring (8 MR) channels both running along the *c* axis (8 MR ⊥ [001], channel ∥ [001]). The main channels—with dimensions ranging from of 7.1 to 12 Å—are connected to the parallel 8 MR channels by a non-planar boat shaped 8 MR (8 MR ∥ [001]). In the as-synthesized K-LTL [56,99] [$K_{8.46}(Al_{8.35} \cdot Si_{27.53})O_{72} \cdot 17.91H_2O$, *s.g.* P6/mmm, *a* = 18.3795, *c* = 7.5281], three positions of K cations were located: site KB in cancrinite cage, site KC in the 8 MR and site KD in the main 12 MR channel.

The ZL/FL composites here investigated are those synthesized and previously characterized at ambient pressure (P_{amb}) by Gigli et al. [56]. Specifically, three hybrids at different fluorenone loadings were investigated, containing 0.5, 1 and 1.5 molecules of colorant per unit cell (from now on ZL/0.5FL, ZL/1FL, ZL/1.5FL).

In the structures of ZL/0.5FL and ZL/1FL at P_{amb} [56] the fluorenone molecules are sited in the 12 MR on the mirror planes parallel to the *c* axis, statistically occupying only one of the six equivalent positions (Figure 2a,b). The oxygen of the carbonyl group of the FL molecule is strongly interacting with potassium atoms KD located along the walls of the main channel. Along with the dye, water molecules were also located in the 12 MR channel: specifically, 14.7 molecules/u.c. in the ZL/0.5FL composite and 9.7 molecules/u.c. in the ZL/1FL one. The structure of ZL/1.5FL resulted to be too complex to be unravelled by structural refinement (i.e., due to high symmetry constrains imposed and to the low electronic density of the atoms), thus a theoretical approach was used. Density functional calculations were indeed able to provide a reliable structural model of the packing of fluorenone inside the composite at room pressure conditions: more specifically, the supramolecular organization of the confined dyes consisted of pairs of molecules—positioned roughly on top of each other, with their long axes nearly parallel to the channel axis—alternated by a dye molecule oriented at about 45° with respect to the zeolite channel axis (Figure 3) [56]. Moreover, calculations evidenced that such a peculiar arrangement—originally called "dye-nanoladder"—was due not only to the structural constraints imposed by the ZL-nanochannel geometry but also to a complex network of intermolecular interactions, among which the dominant one was the coordination of the carbonyl oxygen to the K⁺ extra framework cations (KD)).

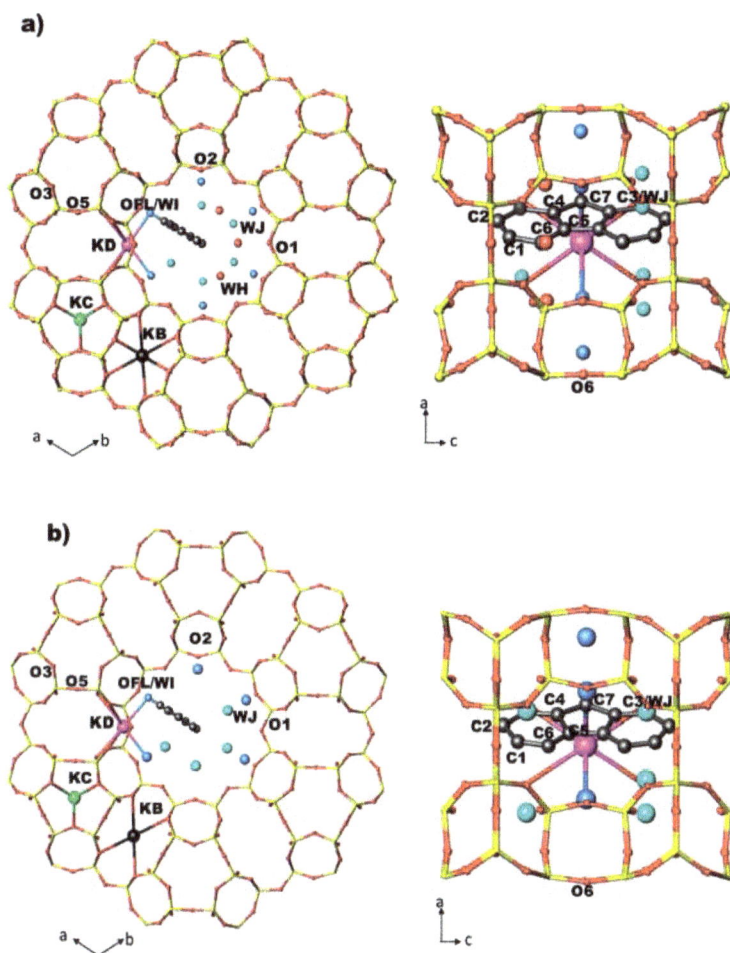

Figure 2. Arrangement of FL and water molecules in the 12 MR channel along [001] and along [010] directions in ZL/0.5FL (**a**) and ZL/1.0FL (**b**) at P_{amb} [56].

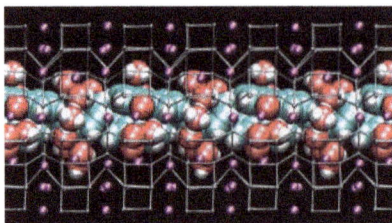

Figure 3. Structure of the hydrated ZL/1.5FL model at P_{amb}—(side view of the 12 MR channel ∥ [001] axis [56]). Tetrahedral atoms (T = Si, Al) of the zeolite L framework are represented as grey stick-and-ball. K^+ = purple spheres. Water and fluorenone atoms are depicted in van der Waals representation (cyan = C, red = O, white = H) to highlight the close-packing arrangement of guest species inside the zeolite L nanochannel.

2.2. Synchrotron X-ray Powder Diffraction Experiments

All the samples were studied with silicon oil (s.o.) as non-penetrating PTM. The in-situ HP XRPD experiments were performed at the SNBL1 (BM01a) beamline at ESRF (European Synchrotron Radiation Facility) with fixed wavelength of 0.72 Å, using a modified Merril-Basset Diamond Anvil Cell (DAC) [100,101]. The pressure was measured using the ruby fluorescence method on the non-linear hydrostatic pressure scale [102]. The estimated error in the pressure values is 0.05 GPa. Bidimensional diffraction patterns were recorded on a PILATUS2M-Series detector (commercialised by Dectris-Switzerland, Baden-Dättwil, Switzerland) (pixel dimension 172 μm) at a fixed distance of 195 mm from the sample; the exposure times were 300 s for each collected pattern. One dimensional diffraction patterns were obtained in the 2θ range 0°–43° by integrating the two dimensional images with the program Fit2D [103]. Some patterns were collected upon pressure release (labelled (rev) in Tables and Figures). Unit cell parameters determination through Rietveld refinements was possible for all the ramps (Table S1). The Rietveld structural refinements were performed on the ZL/0.5FL, ZL/1FL composites at 2.09 and 2.01 GPa, respectively (from now on will be both labelled 2 GPa (i.e., the highest pressures at which the quality of the diffraction patterns allowed the structural refinement) and upon complete decompression (P_{amb} (rev)). All the structural refinements were carried out in the space group P6/mmm starting from the atomic coordinates reported by Gigli et al. [56], using the GSAS package [104] with EXPGUI interface [105]. The background curves were fitted by a Chebyschev polynomial function with 24 coefficients for all the samples. The pseudo-Voight profile function proposed by Thomson et al. [106] was used with refined Gaussian (GW) e Lorentzian (LX) terms and a 0.1% cut-off was applied to the peak intensities. The scale factor and 2θ zero shift were accurately refined in all patterns of the data set. Soft constraints were imposed on tetrahedral bond lengths (Si–O = 1.63 Å) as well as on the C–C (in the range 1.39–1.48 Å) and C–O (1.19 Å) distances, with tolerance values of 0.03 Å. The isotropic displacement parameters were constrained in the following way: the same value for all tetrahedral cations, a second value for all framework oxygen atoms and a third value for water molecules. For both the ZL/FL systems, isotropic displacement parameters for FL atoms were kept equal to P_{amb} values. Occupancy factor for FL were not varied and kept equal to P_{amb} values for all the refinement [56] and a fourth for the oxygen atoms of the water molecules. Details of the structural refinements are reported in Table 1. Atomic coordinates, site occupancies and isotropic displacement parameters are reported in Table S2 while interatomic distances in Table S3.

2.3. Theoretical Modelling

The behaviour of the ZL/1.5FL sample was simulated both at ambient conditions and at compression corresponding to 1.95 GPa in s.o. adopting the experimental cell parameters. The simulation cell consisted of two crystallographic cells along the 12 MR channel (*c* axis). The corresponding stoichiometry of the simulated system is $K_{18}[Al_{18}Si_{54}O_{144}]13 \cdot (H_2O) \ 3 \cdot FL$, as determined in [56]. The systems were simulated, at room temperature, via first-principles molecular dynamics [98]. The PBE approximation to Density Functional Theory was adopted for the electron-electron interactions and augmented with dispersion corrections via the Grimme D2 approach [107] for the FL-FL interactions. Pseudopotentials and basis set expansion were the same adopted in [56]. Such a computational approach has demonstrated to provide a satisfactory description of diverse systems involving adsorbate-surfaces [108–122] or host-guest interactions [123–134] including also processes at zeolite interfaces [32,59] and high pressure conditions [97,135–140].

Equations of motion were integrated with the Car Parrinello Lagrangean [98,141] adopting a time step of 0.121 fs. The starting configuration for the P_{amb} simulation corresponded to the minimum energy structure determined in [56]. An equilibration trajectory of 5 ps was followed by 12 ps of production run. The final configuration obtained at P_{amb} conditions was used as starting configuration for the 1.95 GPa compression simulation. Also in this case, data were gathered from a 12 ps trajectory after a 5 ps equilibration run. In all cases, temperature was controlled via Nose-Hoover thermostats set at 300 K. In all the simulations, performed with the CPMD code [142], no constraints was imposed to the systems except for the experimentally determined cell parameters.

3. Results

3.1. Structure of the ZL-FL Composites at Different Pressure Conditions

Figure 4 shows selected powder patterns of the three ZL-FL composites compressed in s.o. as a function of pressure. With increasing pressure the peak intensities decrease and the peak profiles become broader. Notwithstanding this, complete X-ray amorphization is not achieved up to the highest investigated pressure (about 6 GPa). All the observed peaks are consistent with the P/6mmm *s.g.*, thus ruling out any *P*-induced phase transition. The patterns collected upon decompression demonstrate that the *P*-induced effects are almost completely reversible. In fact, the features of the ambient-pressure pattern (P_{amb}) and the unit cell parameters are rather well recovered upon P release. The decrease of the cell parameters for ZL/0.5FL, ZL/1FL, ZL/1.5FL samples up to 6 GPa are: Δa = −3%, −2.7%, −2.3%; Δc = −4.6%, −4.4%, −3.7% accounting for a ΔV = −10.1%, −9.5% and −8.0%, respectively (See Figure 5 and Table S1). These values indicate that the compressibility changes with the loading and that the FL molecules hosted in the channels stiffen the structure.

The hexagonal lattice undergoes a slight anisotropic compression (Figure 5), with *c* as the most compressible axis. This is probably due to the presence of bonds among the FL carbonyl groups and KD potassium cations, lying along the *a* direction, that stiffen the structure.

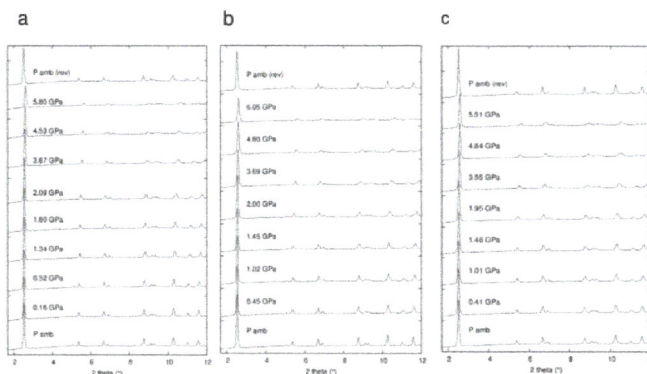

Figure 4. Selected integrated powder patterns (in the 2θ range 3°–12°) of ZL/0.5FL (**a**), ZL/1FL (**b**) and ZL/1.5FL (**c**) compressed in s.o., reported as a function of pressure. The powder patterns at the top (P_{amb} (rev)) is collected upon pressure release.

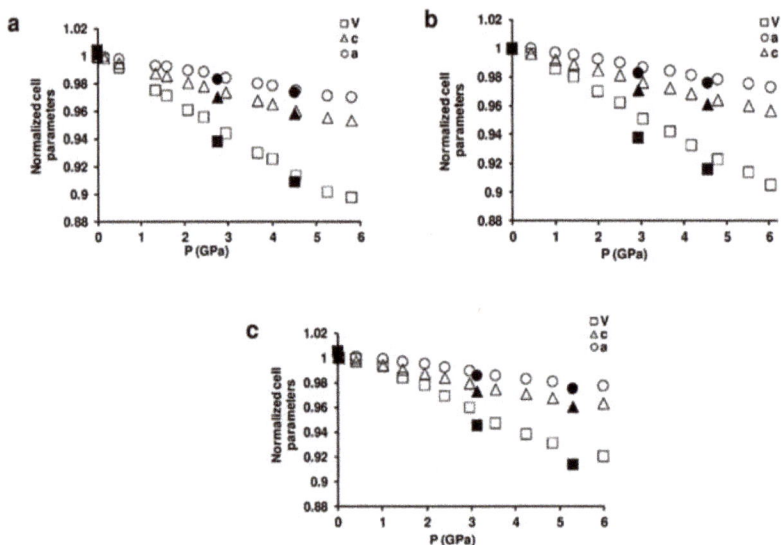

Figure 5. Variations of normalized lattice parameters as a function of pressure for: (**a**) ZL/0.5FL; (**b**) ZL/1FL; (**c**) ZL/1.5FL. The errors associated with the cell parameters are smaller than the symbols used. The full black symbols are associated to decompression ramp.

Table 1. Experimental and structural parameters for selected XRPD refinements performed on the ZL/0.5FL and ZL/1FL composites compressed in silicon oil (s.o.) at 2 GPa and upon decompression (P_{amb} (rev)). [1] From Ref. [56].

	ZL/0.5FL		
P (GPa)	P_{amb} [1]	**2 GPa**	P_{amb} (rev)
Space group	P6/mmm	P6/mmm	P6/mmm
a (Å)	18.3860 (4)	18.1788 (9)	18.4349 (8)
c (Å)	7.5228 (2)	7.3866 (4)	7.5498 (6)
V (Å)	2202.4 (1)	2114.0 (2)	2222.0 (2)
R F2 (%)	7.3	12.3	11.2
No. variables	73	82	82
No. obs.	1319	2474	2474
No. refl.	944	618	618
	ZL/1FL		
P (GPa)	P_{amb} [1]	**2 GPa**	P_{amb} (rev)
Space group	P6/mmm	P6/mmm	P6/mmm
a (Å)	18.3940 (6)	18.250 (1)	18.3962 (8)
c (Å)	7.5203 (3)	7.4063 (5)	7.5189 (4)
V (Å)	2203.5 (1)	2136.3 (2)	2203.6 (2)
R F2 (%)	7.8	11.9	13.7
No. variables	81	83	83
No. obs.	1319	2422	2404
No. refl.	946	679	639

Framework Modifications

The framework *P*-induced deformations can be summarized as follows (see Table 2 and Figure 6):

(i)   In the ZL/0.5FL at 2 GPa, both the diameters (O1–O1 and O2–O2) of the 12 MR shorten. The shortening of the 12 MR O1–O1 diameter is reflected in the lengthening of the O1–O1 diameter of 8 MR channel ($\parallel$ [001]), which becomes more elliptical. Upon pressure release, the original values of the 12 MR diameters are almost regained (remaining slightly smaller than those observed at P_{amb}), while the 8 MR ones are strongly lengthened with respect to the original values, in accordance with the increase of *a* parameter (see Table 1 and Table S1).

(ii) In the ZL/1FL sample, O2–O2 (12 MR) diameter decreases with pressure while O1–O1 increases—probably due to the presence of a larger numbers of FL molecules in the channels with respect to the ZL/0.5FL system. The O1–O1 (12 MR) increase is balanced by the decrease of O1–O1 diameter of the 8 MR channel running alongside ($\parallel$ [001]): as a consequence, 8 MR becomes less elliptical. Upon pressure release, the 12 MR opening remains slightly smaller than that observed at P_{amb}. The significant enlargement of the 8 MR leads to an overall value for *a* parameter comparable with that at P_{amb}.

(iii) The 8 MR window (8 MR $\parallel$ [001]) (O1–O1, O6–O6), parallel to the *c* axis, becomes more circular at 2 GPa in both the samples. Once the pressure is released the starting values are regained.

(iv) The O3–O5–O3 and O5–O3–O5 angle variations indicate that the D6R slightly increases its ditrigonal distortion in both samples. Upon pressure release the P_{amb} features are almost recovered.

Table 2. Experimental framework distances obtained from the refinements performed on the ZL/0.5FL and ZL/1.0FL composites at P_{amb} (from Ref. [56]), at 2 GPa and upon decompression (P_{amb} (rev)). [1] From ref. [56].

	ZL/0.5FL			ZL/1FL		
	P_{amb} [1]	2 GPa	P_{amb} (rev)	P_{amb} [1]	2 GPa	P_{amb} (rev)
	12 MR			12 MR		
O1–O1	10.10	9.76	9.95	10.14	10.30	10.00
O2–O2	10.52	10.43	10.56	10.46	10.12	10.45
	8 MR $\perp$ [001]			8 MR $\perp$ [001]		
O1–O1	8.29	8.41	8.48	8.26	7.95	8.39
O5–O5	4.63	4.54	4.75	4.60	4.70	5.22
* E	1.79	1.85	1.78	1.79	1.69	1.61
	8 MR $\parallel$ [001]			8 MR $\parallel$ [001]		
O1–O1	7.52	7.39	7.54	7.52	7.41	7.51
O6–O6	4.66	4.45	4.63	4.68	4.42	4.60
	D6R			D6R		
O5–O3–O5	147.67	151.82	145.63	149.18	152.6	143.00
O3–O5–O3	91.68	86.51	93.23	90.09	87.7	95.97
	12 MR maximum diameter					
O6–O6	15.60	15.44	15.69	15.67	15.67	15.91

* *E* = (Ellipticity) is the ratio between the largest and the smallest O–O diameters.

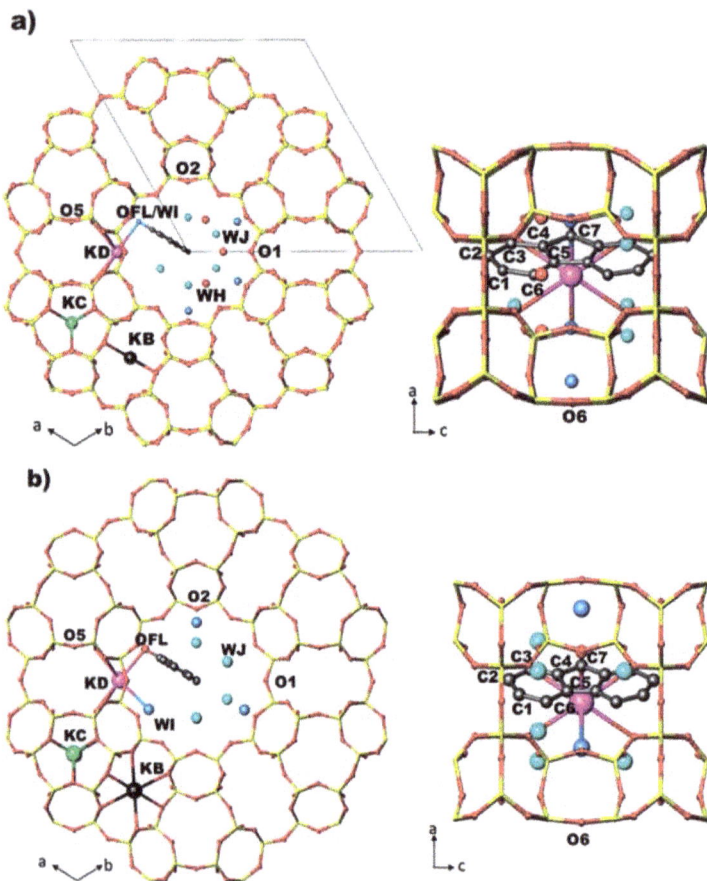

Figure 6. Arrangement of FL and water molecules in the 12 MR channel along [001] and along [010] directions in the ZL/0.5FL and ZL/1.0FL composites (**a**,**b**) at 2 GPa, as obtained by Rietveld refinements.

Extra Framework Species

The distribution of the K cations in both composites at P_{amb} can be described as follows (see Table S2 and S3):

(i)   site KB—in the centre of the cancrinite cage—is fully occupied and coordinated to six framework oxygen atoms O3;

(ii)   site KC—in the centre of the 8 MR channel—is fully occupied and coordinated to four oxygen atoms O5;

(iii)   site KD—near the wall of the main 12 MR channel—is partially occupied and coordinated to six oxygen atoms (O4, O6), two water molecules (WH and WI) and to the oxygen atom of FL molecule (OFL).

At P_{amb}, in ZL/0.5FL and ZL/1.0FL composites the water molecules (14.7 and 9.7 per unit cell, respectively) are distributed over three extra framework sites (WH, WI, WJ). All of them are located in the main channel. WH site is present only in ZL/0.5FL composite, the other two sites have the same positions of the oxygen atom and of the C3 carbon atom of FL molecule (labelled OFL/WI and C3/WJ, respectively, in Ref. [56]).

Upon compression at 2 GPa the following structural features are observed:

(i) the distances between KB, KC and the coordinating framework oxygen atoms (Table S3 and Ref. [56]) decrease as a consequence of the shape modifications of both 8 MR channel aperture and D6R. The distances between the cation in the main channel (KD) and O4 and O6 decrease as well. All these effects are more marked in the ZL/1FL system.

(ii) OFL–KD distance decreases in ZL/0.5FL and remains almost constant in ZL/1.0FL.

(iii) Compression induces the splitting of OFL/WI and C3/WJ sites, which in the P_{amb} structures of ZL/0.5FL and ZL/1FL samples [56], occupy single sites. After the splitting, WJ increases its distance from the framework O2 atom, approaching WI site (Table S3). After pressure release, the original positions are recovered in the ZL/0.5FL composite, while this does not happen in the ZL/1FL sample.

(iv) At 2 GPa, the shape, orientation and arrangement of the fluorenone molecules in the main channel do not change with respect to ambient pressure.

All these *P*-induced deformations are reversible and, once the pressure is released, the original features of the zeolite and the distances among the FL molecules are almost recovered (Tables 2, S2 and S3).

3.2. Structure of the ZL/1.5FL Composite from First-Principles Molecular Dynamics

As evidenced by the above-discussed data, XRPD refinements provided a satisfactory description of the P-induced structural modifications of the composites characterized by low and moderate dye content. On the other hand, the great number of low-occupancy sites found for water and fluorenone molecules and the high symmetry hindered the structural refinement of the sample containing the maximum amount of dye. Such a difficulty was previously encountered in the room pressure refinement of the ZL/1.5FL composite and it was overcome by integrating the experimental data on cell parameters with theoretical modelling for the atomic coordinates [56].

Hence, encouraged by this result, we exploited again theoretical modelling for achieving an atomistic structural description of the ZL/1.5FL composite at high-pressure conditions. Practically, we used the room-pressure coordinates as an initial guess to determine the composite structure at cell parameters corresponding to 1.95 GPa and we run first-principles molecular dynamics simulations for both $P = P_{amb}$ and $P = 1.95$ GPa in order to study the pressure-induced changes of the supramolecular organization inside the zeolite nanochannels at molecular-level detail.

The first remarkable observation is that the unique dye-architecture found at room pressure remains stable at high pressure conditions (Figure 7), with minimal alterations of its intermolecular distances and essentially without significant perturbation of the FL molecular geometry, apart from slight instantaneous distortions from the ideal gas-phase structure, mainly ascribable to thermal motion.

Significantly, the leading interaction stabilizing the confined fluorenone superstructure, i.e., the coordination to potassium cations, not only is maintained but it appears also to be, on average, strengthened upon compression. Indeed, by comparing the pair distribution functions relative to the K-OFL interaction, we deduce a significant shortening of the minimum coordination distance, which passes from 3.1 to 2.9 Å in going from room pressure to 1.95 GPa (Figure 8). Moreover, the splitting of the first maximum into two peaks becomes much more evident upon compression, indicating that the interaction of the carbonyl group and potassium may have different degrees of strength among the hyperconfined fluorenone molecules. Taken as a whole, these data indicate that the supramolecular architecture of dyes responds to pressure essentially by approaching with the carbonyl groups the potassium cations in the 12 MR channel, without undergoing appreciable modifications of both intra- and inter-molecular distances.

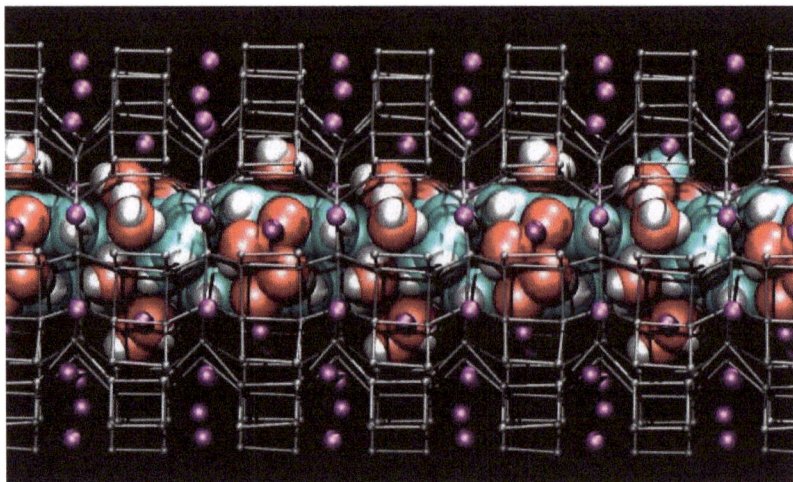

Figure 7. Snapshot from first-principles molecular dynamics (FPMD) simulation showing the structure of the hydrated ZL/1.5FL model with cell parameters corresponding to P = 1.95 GPa (side view of the 12 MR channel ∥ [001] axis). Colours are the same as reported in Figure 3.

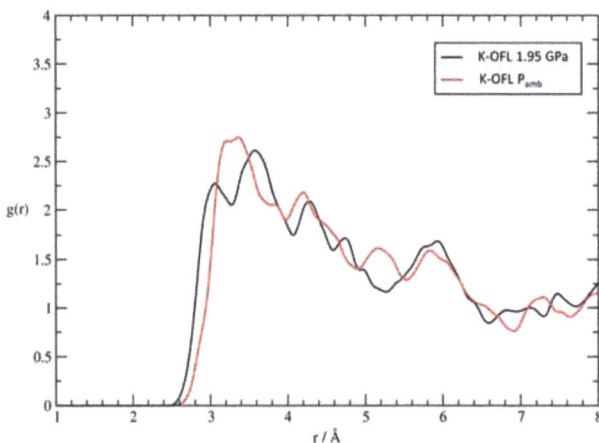

Figure 8. Pair distribution functions g(r) from the room-temperature FPMD simulations relative to P_{amb} (red) and P = 1.95 GPa (black), calculated for the potassium extra framework cations of zeolite L with the carbonyl oxygens of fluorenone.

By considering now water molecules, which share the channel space with fluorenone, a striking similarity emerges by comparing their pair distribution functions at room pressure and 1.95 GPa (Figure 9). Actually, for both water protons (Figure 9a) and water oxygens (Figure 9b) the two curves are almost superposable, showing also very close positions for first maximum peaks of water-water hydrogen bonding (corresponding, at room pressure, to H_w–O_w and O_w–O_w distances of 1.85 and 2.80 Å, respectively—see upper panels of Figure 9a,b). In particular, only the shoulder of the H_w–O_w peak—found at 1.78 Å at P_{amb}—appears to be very slightly displaced towards greater distances. Also the interaction of water with framework oxygens—which, as normally found for zeolitic water, is weaker than water–water hydrogen bonding [143–168]—appears to be nearly unperturbed upon compression and characterized by H_w–O_f and O_w–O_f distances of 1.95 and 2.90 Å, respectively (see centre panels of Figure 9a,b). Furthermore, the interaction of water with potassium cations is nearly unaffected by pressure at short range distances (with first maximum position at 2.90 Å, see Figure 9b, bottom panel), indicating that the coordination environment of the zeolite L extra framework cations is, on average, spectacularly insensitive to compression: no variation of the K coordination number occurs and distances from water oxygens remain basically unaltered. Finally, very minor changes are detected for the water-carbonyl interaction: the first peak position passes from 1.80 to 1.75 Å upon compression, indicating a slightly stronger interaction of water molecules with the carbonyl group of the dye (Figure 9a, bottom panel).

Interestingly, whereas in the ZL/0.5FL and ZL/1FL composites the O6–O6 distance (i.e., the maximum diameter of the 12 MR channel) either remains constant (ZL/1FL) or slightly shortens (ZL/0.5FL) with pressure (Table 2), such a distance increases in the ZL/1.5FL composite. Accordingly, the calculations also predict a decrease of the 12 MR window (i.e., the channel opening) of ZL/1.5FL with compression (see O1–O1 and O2–O2 distances in Table 3). This behaviour is due to the peculiar close-packing arrangement of the extra framework content at maximum dye loading (ZL/1.5FL). Specifically, as compression occurs mainly along the channel axis, the guest species—water molecules and dye nanoladder—can only respond by further clustering in the maximum-diameter region of the channel (Figure 7), thus explaining the O6–O6 lengthening and the 12 MR window narrowing. The different pressure response of the composites according to the dye loading is a consequence of the stiffening/template effects of the extra framework content, already observed in several high-pressure studies on zeolites and zeolite-based materials (see, e.g., Refs. [62,63,66,77,81,139,169–171]).

Taken together, besides the impressive stability of the fluorenone nanoladder, these results underline a quite surprising and important feature of the confined supramolecular system: even though water is smaller than the dye and hence in principle more easily displaceable upon compression, the arrangement of the water molecules in ZL/0.5FL and ZL/1FL is only slightly modified by the application of hydrostatic pressure—in particular, WJ increases its distance from the framework O2 atoms—and remains essentially unperturbed in ZL/1.5FL. This is ascribable to the fact that the P_{amb} structure of the ZL/1.5FL composite has already a close-packing arrangement of the extra framework species. In particular, all water molecules are fully stabilized by coordination to potassium cations and by the network of hydrogen-bond interactions with fluorenone carbonyl groups, water and framework oxygens [56].

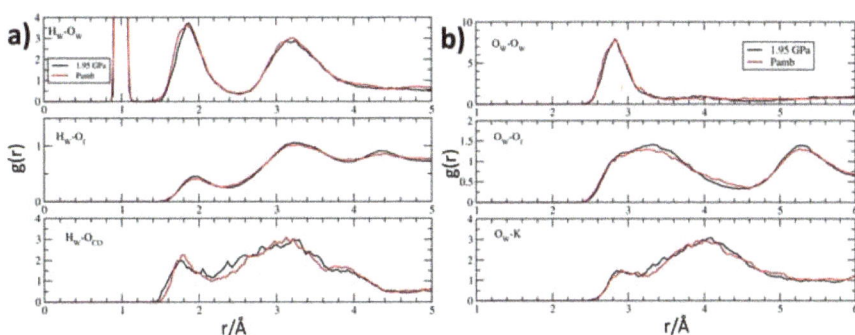

Figure 9. Pair distribution functions g(r) from the room-temperature FPMD simulations relative to P_{amb} (red) and P = 1.95 GPa (black), calculated for: (**a**) the water hydrogens H_w with water oxygens O_w (top), framework oxygens O_f (middle), carbonyl oxygens O_{FL} (bottom); (**b**) the water oxygens O_w with water oxygens (top), framework oxygens O_f (middle) and potassium cations (bottom).

Table 3. Calculated average diameters of 12 MR channel in Å from room-temperature FPMD simulations at different compression for the hydrated ZL/1.5FL system.

Distance	P_{amb}	1.95 GPa
O1–O1	9.934	9.878
O2–O2	10.431	10.409
O6–O6	15.520	15.614

4. Discussion

The main motivation of our study was to elucidate the behaviour of luminescent hybrid materials, obtained via the inclusion of the fluorenone dye in zeolite L microcrystals, under compression. To answer this question, a thorough study has been performed using non-penetrating pressure transmitting fluid (silicon oil) in the high pressure X-ray diffraction experiments and complementing those results by ab initio molecular dynamics simulations, which allow to explore the finite temperature behaviour of the system at atomistic level detail.

If we consider that the technological applications of these composites rely on the presence of confined regular arrays of photoactive species, it is important to understand how this supramolecular organization of dye molecules responds to non-ambient conditions, specifically an applied pressure. Remarkably, all of the experiments and simulations evidenced the approaching of the fluorenone carbonyl group to potassium—i.e., the strengthening of the main stabilizing interaction in zeolite L composites with carbonyl dyes. Also notably, the applied pressure did not cause appreciable distortions of the molecular geometry of the chromophore, because of the zeolite framework ability to withstand compression effects, as already demonstrated in other high pressure studies of zeolites with non-penetrating fluids [135,138–140,169–171].

Experiments performed on the samples with low and medium fluorenone contents showed that the arrangement and the distances of the dye molecules, upon increasing pressure, do not change and their behaviour is consistent with the observed shortening of the *c*-parameter of the zeolite host. This pressure-induced shortening of intermolecular distances between dye molecules is however reversible upon returning to P_{amb}, because it depends solely on non-covalent interactions. Also, such an effect is much less pronounced in the case of the high-concentrated composite (1.5FL/ZL), where the dye molecules are organized in a very close-packed arrangement already at room pressure conditions.

All these observations suggest that also the optical properties of the composites should not be significantly perturbed by external pressures of GPa-scale, an extremely important feature in the perspective of potential applications.

5. Conclusions

Overall, this study highlights the stability upon compression of the zeolite L/fluorenone adducts. This very important fact is of relevance in view of possible applications of such a kind of dye-zeolite composites at conditions different from the standard ones. We have also monitored an intriguing fluorenone-loading dependence of the system response to the applied pressure. However, some common features have been evidenced as well. In particular, compression brings about a strengthening of the fluorenone–zeolite L interactions, as shown by the shortening of the distance between the carbonyl oxygen atoms of the dye and the K^+ cations. Also, the molecular structure of the dye guests is barely influenced by compression. This is essentially due to the zeolite framework, which bears most of the compression effects—as evidenced by the (moderate) changes in the cell parameters and by the (moderate) changes in the 12 MR channel—that are relevant for the zeolite pores volume. Such a framework stability upon compression induces only a modest variation of the pores volumes, which affects only to a minor extent the molecular structure of the organic guests. In conclusion, we have shown for the first time the stability towards pressure of dye-zeolite L adducts, which suggests that also the technologically relevant photophysical properties of these appealing materials could be exploited at high pressure conditions.

Supplementary Materials: The following are available online at http://www.mdpi.com/2073-4352/8/2/79/s1, Table S1: Unit-cell parameters of ZL/0.5FL, ZL/1FL, ZL/1.5FL samples at the investigated pressures; Table S2: Refined atomic positions, occupancy factors and displacement parameters of ZL/0.5FL and ZL/1FL at 2 GPa and upon decompression (P_{amb} (rev)); Table S3: Extra framework bond distances <3.2 Å for the ZL/0.5FL, ZL/1.0FL composites at 2 GPa and upon decompression (P_{amb} (rev)).

Acknowledgments: BM01 beamline and the European Synchrotron Radiation Facility are acknowledged for the allocation of experimental beamtime. This work was supported by the Italian MIUR, within the frame of the following projects: PRIN2015 "ZAPPING" High-pressure nano-confinement in Zeolites: the Mineral Science know-how APPlied to engineerING of innovative materials for technological and environmental applications (2015HK93L7), ImPACT (FIRB RBFR12CLQD), and University of Insubria Far 2016.

Author Contributions: Rossella Arletti, Simona Quartieri, Giovanna Vezzalini, Gloria Tabacchi and Ettore Fois conceived the topic and designed the experiments; Lara Gigli, Rossella Arletti, Simona Quartieri, Giovanna Vezzalini, performed the experiments and analysed the data; Vladimir Dmitriev provided fundamental assistance during the experiments; Gloria Tabacchi and Ettore Fois performed the molecular dynamics simulations; all the authors contributed to write the paper.

Conflicts of Interest: The authors declare no conflict of interest. The founding sponsors had no role in the design of the study; in the collection, analyses, or interpretation of data; in the writing of the manuscript and in the decision to publish the results.

References

1. Baerlocher, C.; McCusker, L.B.; Olson, D.H. *Atlas of Zeolite Framework Types*; Elsevier: Amsterdam, The Netherland, 2007; ISBN 9780444530646.
2. Baur, W.; Fischer, R.X. *Microporous and Other Framework Materials with Zeolite-Type Structures*; Springer: Berlin, Germany, 2017; ISBN 9783662542514.
3. Čejka, J.; Centi, G.; Perez-Pariente, J.; Roth, W.J. Zeolite-based materials for novel catalytic applications: Opportunities, perspectives and open problems. *Catal. Today* **2012**, *179*, 2–15. [CrossRef]
4. Corma, A. Heterogeneous catalysis: Understanding for designing designing for applications. *Angew. Chem. Int. Ed.* **2016**, *55*, 6112–6113. [CrossRef] [PubMed]
5. Seifert, R.; Kunzmann, A.; Calzaferri, G. The yellow color of silver-containing zeolite A. *Angew. Chem. Int. Ed.* **1998**, *37*, 1521–1524. [CrossRef]
6. Baldansuren, A.; Roduner, E. EPR experiments of Ag species supported on NaA. *Chem. Phys. Lett.* **2009**, *473*, 135–137. [CrossRef]
7. Coutiño-Gonzalez, E.; Baekelant, W.; Steele, J.A.; Kim, C.W.; Roeffaers, M.B.J.; Hofkens, J. Silver clusters in zeolites: From self-assembly to ground-breaking luminescent properties. *Acc. Chem. Res.* **2017**, *50*, 2353–2361. [CrossRef] [PubMed]
8. Dong, B.; Retoux, R.; de Waele, V.; Chiodo, S.G.; Mineva, T.; Cardin, J.; Mintova, S. Sodalite cages of EMT zeolite confined neutral molecular-like silver clusters. *Microporous Mesoporous Mater.* **2017**, *244*, 74–82. [CrossRef]
9. Kim, H.S.; Yoon, K.B. Preparation and characterization of CdS and PbS quantum dots in zeolite Y and their applications for nonlinear optical materials and solar cell. *Coord. Chem. Rev.* **2014**, *263–264*, 239–256. [CrossRef]
10. Wang, Y.; Li, H. Luminescent materials of zeolite functionalized with lanthanides. *CrystEngComm* **2014**, *16*, 9764–9778. [CrossRef]
11. Calzaferri, G.; Huber, S.; Maas, H.; Minkowski, C. Host-guest antenna materials. *Angew. Chem. Int. Ed.* **2003**, *42*, 3732–3758. [CrossRef] [PubMed]
12. Hashimoto, S. Zeolite photochemistry: Impact of zeolites on photochemistry and feedback from photochemistry to zeolite science. *J. Photochem. Photobiol. C Photochem. Rev.* **2003**, *4*, 19–49. [CrossRef]
13. Alarcos, N.; Cohen, B.; Ziółek, M.; Douhal, A. Photochemistry and photophysics in silica-based materials: Ultrafast and Single molecule spectroscopy observation. *Chem. Rev.* **2017**, *117*, 13639–13720. [CrossRef] [PubMed]

14. Busby, M.; Devaux, A.; Blum, C.; Subramaniam, V.; Calzaferri, G.; De Cola, L. Interactions of perylene bisimide in the one-dimensional channels of zeolite L. *J. Phys. Chem. C* **2011**, *115*, 5974–5988. [CrossRef]

15. Calzaferri, G.; Méallet-Renault, R.; Brühwiler, D.; Pansu, R.; Dolamic, I.; Dienel, T.; Adler, P.; Li, H.; Kunzmann, A. Designing dye-nanochannel antenna hybrid materials for light harvesting, transport and trapping. *ChemPhysChem* **2011**, *12*, 580–594. [CrossRef] [PubMed]

16. Gartzia-Rivero, L.; Bañuelos, J.; López-Arbeloa, I. Photoactive nanomaterials inspired by nature: LTL zeolite doped with laser dyes as artificial light harvesting systems. *Materials (Basel)* **2017**, *10*, 495. [CrossRef] [PubMed]

17. Martínez-Martínez, V.; García, R.; Gómez-Hortigüela, L.; Sola Llano, R.; Pérez-Pariente, J.; López-Arbeloa, I. Highly luminescent and optically switchable hybrid material by one-pot encapsulation of dyes into MgAPO-11 unidirectional nanopores. *ACS Photonics* **2014**, *1*, 205–211. [CrossRef]

18. Sola-Llano, R.; Fujita, Y.; Gómez-Hortigüela, L.; Alfayate, A.; Ujii, H.; Fron, E.; Toyouchi, S.; Perez-Pariente, J.; Lopez-Arbeloa, I.; Martinez-Martinez, V. One-directional antenna systems: Energy transfer from monomers to J-Aggregates within 1D nanoporous aluminophosphates. *ACS Photonics* **2017**, *5*, 151–157. [CrossRef]

19. Ramamurthy, V.; Lakshminarasimhan, P.; Grey, C.P.; Johnston, L.J. Energy transfer, proton transfer and electron transfer reactions within zeolites. *Chem. Commun.* **1998**, 2411–2424. [CrossRef]

20. Calzaferri, G. Nanochannels: Hosts for the supramolecular organization of molecules and complexes. *Langmuir* **2012**, *28*, 6216–6231. [CrossRef] [PubMed]

21. Shim, T.; Lee, M.H.; Kim, D.; Kim, H.S.; Yoon, K.B. Fluorescence Properties of Hemicyanine in the Nanoporous Materials with Varying Pore Sizes. *J. Phys. Chem. B* **2009**, *113*, 966–969. [CrossRef] [PubMed]

22. Kim, D.; Kim, H.S. Enhancement of fluorescence from one- and two-photon absorption of hemicyanine dyes by confinement in silicalite-1 nanochannels. *Microporous Mesoporous Mater.* **2017**, *243*, 69–75. [CrossRef]

23. Doungmanee, S.; Siritanon, T.; Insuwan, W.; Jungsuttiwong, S.; Rangsriwatananon, K. Multi step energy transfer between three Si_LTL and SiGe_LTL zeolite-loaded dyes. *J. Porous Mater.* **2017**, 1–9. [CrossRef]

24. Hu, D.D.; Lin, J.; Zhang, Q.; Lu, J.N.; Wang, X.Y.; Wang, Y.W.; Bu, F.; Ding, L.F.; Wang, L.; Wu, T. Multi-Step Host-Guest Energy Transfer Between Inorganic Chalcogenide-Based Semiconductor Zeolite Material and Organic Dye Molecules. *Chem. Mater.* **2015**, *27*, 4099–4104. [CrossRef]

25. Noh, T.H.; Jang, J.; Hong, W.; Lee, H.; Jung, O.-S. Truncated trigonal prismatic tubular crystals consisting of a zeolite L-mimic metal-organic framework. *Chem. Commun.* **2014**, *50*, 7451–7454. [CrossRef] [PubMed]

26. Ruiz, A.Z.; Brühwiler, D.; Ban, T.; Calzaferri, G. Synthesis of zeolite L. Tuning size and morphology. *Monatshefte fur Chemie* **2005**, *136*, 77–89. [CrossRef]

27. Cho, H.S.; Hill, A.R.; Cho, M.; Miyasaka, K.; Jeong, K.; Anderson, M.W.; Kang, J.K.; Terasaki, O. Directing the distribution of potassium cations in zeolite-LTL through crown ether addition. *Cryst. Growth Des.* **2017**, *17*, 4516–4521. [CrossRef]

28. Maas, H.; Calzaferri, G. Trapping energy from and injecting energy into dye-zeolite nanoantennae. *Angew. Chem. Int. Ed.* **2002**, *41*, 2284–2288. [CrossRef]

29. Brühwiler, D.; Calzaferri, G. Selective functionalization of the external surface of zeolite L. *C. R. Chim.* **2005**, *8*, 391–398. [CrossRef]

30. Dieu, L.-Q.; Devaux, A.; López-Duarte, I.; Victoria Martínez-Díaz, M.; Brühwiler, D.; Calzaferri, G.; Torres, T. Novel phthalocyanine-based stopcock for zeolite L. *Chem. Commun.* **2008**, *0*, 1187–1189. [CrossRef] [PubMed]

31. Li, P.; Wang, Y.; Li, H.; Calzaferri, G. Luminescence enhancement after adding stoppers to europium(III) nanozeolite L. *Angew. Chem. Int. Ed.* **2014**, *53*, 2904–2909. [CrossRef] [PubMed]

32. Tabacchi, G.; Fois, E.; Calzaferri, G. Structure of Nanochannel Entrances in Stopcock-Functionalized Zeolite L Composites. *Angew. Chem. Int. Ed.* **2015**, *54*, 11112–11116. [CrossRef] [PubMed]

33. Albuquerque, R.Q.; Popović, Z.; De Cola, L.; Calzaferri, G. Luminescence quenching by O_2 of a Ru^{2+} complex attached to zeolite L. *ChemPhysChem* **2006**, *7*, 1050–1053. [CrossRef] [PubMed]

34. Beierle, J.M.; Roswanda, R.; Erne, P.M.; Coleman, A.C.; Browne, W.R.; Feringa, B.L. An improved method for site-specific end modification of zeolite L for the formation of zeolite L and gold nanoparticle self-assembled structures. *Part. Part. Syst. Charact.* **2013**, *30*, 273–279. [CrossRef]

35. Ramachandra, S.; Popović, Z.D.; Schuermann, K.C.; Cucinotta, F.; Calzaferri, G.; De Cola, L. Förster resonance energy transfer in quantum dot-dye-loaded zeolite L nanoassemblies. *Small* **2011**, *7*, 1488–1494. [CrossRef] [PubMed]

36. Popović, Z.; Otter, M.; Calzaferri, G.; De Cola, L. Self-Assembling living systems with functional nanomaterials. *Angew. Chem. Int. Ed.* **2007**, *46*, 6188–6191. [CrossRef] [PubMed]

37. Strassert, C.A.; Otter, M.; Albuquerque, R.Q.; Hone, A.; Vida, Y.; Maier, B.; De Cola, L. Photoactive hybrid nanomaterial for targeting, labeling killing antibiotic-resistant bacteria. *Angew. Chem. Int. Ed.* **2009**, *48*, 7928–7931. [CrossRef] [PubMed]

38. Wang, Y.; Li, H.; Feng, Y.; Zhang, H.; Calzaferri, G.; Ren, T. Orienting zeolite L microcrystals with a functional linker. *Angew. Chem. Int. Ed.* **2010**, *49*, 1434–1438. [CrossRef] [PubMed]

39. Devaux, A.; Calzaferri, G.; Belser, P.; Cao, P.; Brühwiler, D.; Kunzmann, A. Efficient and robust host-guest antenna composite for light harvesting. *Chem. Mater.* **2014**, *26*, 6878–6885. [CrossRef]

40. Brühwiler, D.; Calzaferri, G.; Torres, T.; Ramm, J.H.; Gartmann, N.; Dieu, L.-Q.; López-Duarte, I.; Martínez-Díaz, M.V. Nanochannels for supramolecular organization of luminescent guests. *J. Mater. Chem.* **2009**, *19*, 8040–8067. [CrossRef]

41. Woodtli, P.; Giger, S.; Müller, P.; Sägesser, L.; Zucchetto, N.; Reber, M.J.; Ecker, A.; Brühwiler, D. Indigo in the nanochannels of zeolite L: Towards a new type of colorant. *Dyes Pigments* **2017**, *149*, 456–461. [CrossRef]

42. El-Gindi, J.; Benson, K.; De Cola, L.; Galla, H.J.; Seda Kehr, N. Cell adhesion behavior on enantiomerically functionalized zeolite L monolayers. *Angew. Chem. Int. Ed.* **2012**, *51*, 3716–3720. [CrossRef] [PubMed]

43. Greco, A.; Maggini, L.; De Cola, L.; De Marco, R.; Gentilucci, L. Diagnostic implementation of fast and selective integrin-mediated adhesion of cancer cells on functionalized zeolite L monolayers. *Bioconjug. Chem.* **2015**, *26*, 1873–1878. [CrossRef] [PubMed]

44. Fois, E.; Tabacchi, G.; Calzaferri, G. Interactions, behavior stability of fluorenone inside zeolite nanochannels. *J. Phys. Chem. C* **2010**, *114*, 10572–10579. [CrossRef]

45. Zhou, X.; Wesolowski, T.A.; Tabacchi, G.; Fois, E.; Calzaferri, G.; Devaux, A. First-principles simulation of the absorption bands of fluorenone in zeolite L. *Phys. Chem. Chem. Phys.* **2013**, *15*, 159–167. [CrossRef] [PubMed]

46. Fois, E.; Tabacchi, G.; Calzaferri, G. Orientation and order of xanthene dyes in the one-dimensional channels of zeolite L: Bridging the gap between experimental data and molecular behavior. *J. Phys. Chem. C* **2012**, *116*, 16784–16799. [CrossRef]

47. Viani, L.; Minoia, A.; Cornil, J.; Beljonne, D.; Egelhaaf, H.J.; Gierschner, J. Resonant energy transport in dye-filled monolithic crystals of zeolite L: Modeling of inhomogeneity. *J. Phys. Chem. C* **2016**, *120*, 27192–27199. [CrossRef]

48. Van Speybroeck, V.; Hemelsoet, K.; Joos, L.; Waroquier, M.; Bell, R.G.; Catlow, C.R.A. Advances in theory and their application within the field of zeolite chemistry. *Chem. Soc. Rev.* **2015**, *44*, 7044–7111. [CrossRef] [PubMed]

49. Calzaferri, G. Entropy in multiple equilibria, theory and applications. *Phys. Chem. Chem. Phys.* **2017**, *19*, 10611–10621. [CrossRef] [PubMed]

50. Tabacchi, G. Supramolecular Organization in Confined Nanospaces. *ChemPhysChem* **2018**, in revision.

51. Manzano, H.; Gartzia-Rivero, L.; Bañuelos, J.; López-Arbeloa, I. Ultraviolet-visible dual absorption by single BODIPY dye confined in LTL zeolite nanochannels. *J. Phys. Chem. C* **2013**, *117*, 13331–13336. [CrossRef]

52. Cucinotta, F.; Guenet, A.; Bizzarri, C.; Mroz, W.; Botta, C.; Milian-Medina, B.; Gierschner, J.; De Cola, L. Energy transfer at the zeolite l boundaries: Towards photo- and electroresponsive materials. *Chempluschem* **2014**, *79*, 45–57. [CrossRef]

53. Fois, E.; Gamba, A.; Medici, C.; Tabacchi, G. Intermolecular electronic excitation transfer in a confined space: A first-principles study. *ChemPhysChem* **2005**, *6*, 1917–1922. [CrossRef] [PubMed]

54. Insuwan, W.; Rangsriwatananon, K.; Meeprasert, J.; Namuangruk, S.; Surakhot, Y.; Kungwan, N.; Jungsuttiwong, S. Combined experimental and theoretical investigation on photophysical properties of trans-azobenzene confined in LTL zeolite: Effect of cis-isomer forming. *Microporous Mesoporous Mater.* **2014**, *197*, 348–357. [CrossRef]

55. Insuwan, W.; Rangsriwatananon, K.; Meeprasert, J.; Namuangruk, S.; Surakhot, Y.; Kungwan, N.; Jungsuttiwong, S. Combined experimental and theoretical investigation on Fluorescence Resonance Energy Transfer of dye loaded on LTL zeolite. *Microporous Mesoporous Mater.* **2017**, *241*, 372–382. [CrossRef]

56. Gigli, L.; Arletti, R.; Tabacchi, G.; Fois, E.; Vitillo, J.G.; Martra, G.; Agostini, G.; Quartieri, S.; Vezzalini, G. Close-Packed dye molecules in zeolite channels self-assemble into supramolecular nanoladders. *J. Phys. Chem. C* **2014**, *118*, 15732–15743. [CrossRef]

57. Gigli, L.; Arletti, R.; Tabacchi, G.; Fabbiani, M.; Vitillo, J.G.; Martra, G.; Devaux, A.; Miletto, I.; Quartieri, S.; Calzaferri, G.; et al. Structure and host-guest interactions of perylene-diimide dyes in zeolite l nanochannels. *J. Phys. Chem. C* **2018**, accepted. [CrossRef]

58. Fois, E.; Tabacchi, G.; Devaux, A.; Belser, P.; Brühwiler, D.; Calzaferri, G. Host-guest interactions and orientation of dyes in the one-dimensional channels of zeolite L. *Langmuir* **2013**, *29*, 9188–9198. [CrossRef] [PubMed]

59. Tabacchi, G.; Calzaferri, G.; Fois, E. One-dimensional self-assembly of perylene-diimide dyes by unidirectional transit of zeolite channel openings. *Chem. Commun.* **2016**, *52*, 11195–11198. [CrossRef] [PubMed]

60. Arletti, R.; Ferro, O.; Quartieri, S.; Sani, A.; Tabacchi, G.; Vezzalini, G. Structural deformation mechanisms of zeolites under pressure. *Am. Mineral.* **2003**, *88*, 1416–1422. [CrossRef]

61. Gatta, G.D. A comparative study of fibrous zeolites under pressure. *Eur. J. Mineral.* **2005**, *17*, 411–422. [CrossRef]

62. Gatta, G.D. Does porous mean soft? On the elastic behaviour and structural evolution of zeolites under pressure. *Zeitschrift für Kristallographie Cryst. Mater.* **2008**, *223*, 160–170. [CrossRef]

63. Gatta, G.D.; Lee, Y. Zeolites at high pressure: A review. *Mineral. Mag.* **2014**, *78*, 267–291. [CrossRef]

64. Vezzalini, G.; Arletti, R.; Quartieri, S. High-pressure-induced structural changes, amorphization and molecule penetration in MFI microporous materials: A review. *Acta Crystallogr. Sect. B Struct. Sci. Cryst. Eng. Mater.* **2014**, *70*, 444–451. [CrossRef] [PubMed]

65. Fraux, G.; Coudert, F.-X.; Boutin, A.; Fuchs, A.H. Forced intrusion of water and aqueous solutions in microporous materials: From fundamental thermodynamics to energy storage devices. *Chem. Soc. Rev.* **2017**, *46*, 7421–7437. [CrossRef] [PubMed]

66. Gatta, G.D.; Lotti, P.; Tabacchi, G. The effect of pressure on open-framework silicates: Elastic behaviour and crystal–fluid interaction. *Phys. Chem. Miner.* **2017**, 1–24. [CrossRef]

67. Lee, Y.; Vogt, T.; Hriljac, J.A.; Parise, J.B.; Hanson, J.C.; Kim, S.J. Non-framework cation migration and irreversible pressure-induced hydration in a zeolite. *Nature* **2002**, *420*, 485–489. [CrossRef] [PubMed]

68. Lee, Y.; Hriljac, J.A.; Parise, J.B.; Vogt, T. Pressure-induced hydration in zeolite tetranatrolite. *Am. Mineral.* **2006**, *91*, 247–251. [CrossRef]

69. White, C.L.I.M.; Ruiz-Salvador, A.R.; Lewis, D.W. Pressure-Induced Hydration Effects in the Zeolite Laumontite. *Angew. Chem. Int. Ed.* **2004**, *43*, 469–472. [CrossRef] [PubMed]

70. Lee, Y.; Kao, C.C.; Kim, S.J.; Lee, H.H.; Lee, D.R.; Shin, T.J.; Choi, J.Y. Water nanostructures confined inside the quasi-one-dimensional channels of LTL zeolite. *Chem. Mater.* **2007**, *19*, 6252–6257. [CrossRef]

71. Lee, Y.; Kim, S.J.; Ahn, D.C.; Shin, N.S. Confined water clusters in a synthetic rubidium gallosilicate with zeolite LTL topology. *Chem. Mater.* **2007**, *19*, 2277–2282. [CrossRef]

72. Seoung, D.; Lee, Y.; Kao, C.-C.; Vogt, T.; Lee, Y. Two-Step pressure-induced superhydration in small pore natrolite with divalent extra-framework cations. *Chem. Mater.* **2015**, *27*, 3874–3880. [CrossRef]

73. Cailliez, F.; Trzpit, M.; Soulard, M.; Demachy, I.; Boutin, A.; Patarin, J.; Fuchs, A.H. Thermodynamics of water intrusion in nanoporous hydrophobic solids. *Phys. Chem. Chem. Phys.* **2008**, *10*, 4817. [CrossRef] [PubMed]

74. Likhacheva, A.Y.; Seryotkin, Y.V.; Manakov, A.Y.; Goryainov, S.V.; Ancharov, A.I.; Sheromov, M.A. Pressure-induced over-hydration of thomsonite: A synchrotron powder diffraction study. *Am. Mineral.* **2007**, *92*, 1610–1615. [CrossRef]

75. Likhacheva, A.; Seryotkin, Y.; Manakov, A.; Goryainov, S.; Ancharov, A.; Sheromov, M. Anomalous compression of scolecite and thomsonite in aqueous medium to 2 GPa. *High Press. Res.* **2006**, *26*, 449–453. [CrossRef]

76. Arletti, R.; Vezzalini, G.; Quartieri, S.; Di Renzo, F.; Dmitriev, V. Pressure-induced water intrusion in FER-type zeolites and the influence of extraframework species on structural deformations. *Microporous Mesoporous Mater.* **2014**, *191*, 27–37. [CrossRef]

77. Arletti, R.; Vezzalini, G.; Morsli, A.; Di Renzo, F.; Dmitriev, V.; Quartieri, S. Elastic behavior of MFI-type zeolites: 1-Compressibility of Na-ZSM-5 in penetrating and non-penetrating media. *Microporous Mesoporous Mater.* **2011**, *142*, 696–707. [CrossRef]

78. Lotti, P.; Gatta, G.D.; Merlini, M.; Liermann, H.-P. High-pressure behavior of synthetic mordenite-Na: An in situ single-crystal synchrotron X-ray diffraction study. *Zeitschrift für Kristallographie Cryst. Mater.* **2015**, *230*, 201–211. [CrossRef]

79. Comboni, D.; Gatta, G.D.; Lotti, P.; Merlini, M.; Hanfland, M. Crystal-fluid interactions in laumontite. *Microporous Mesoporous Mater.* **2018**, *263*, 86–95. [CrossRef]

80. Lotti, P.; Arletti, R.; Gatta, G.D.; Quartieri, S.; Vezzalini, G.; Merlini, M.; Dmitriev, V.; Hanfland, M. Compressibility and crystal–fluid interactions in all-silica ferrierite at high pressure. *Microporous Mesoporous Mater.* **2015**, *218*, 42–54. [CrossRef]

81. Lotti, P.; Gatta, G.D.; Comboni, D.; Merlini, M.; Pastero, L.; Hanfland, M. AlPO4-5 zeolite at high pressure: Crystal-fluid interaction and elastic behavior. *Microporous Mesoporous Mater.* **2016**, *228*, 158–167. [CrossRef]

82. Kim, Y.; Choi, J.; Vogt, T.; Lee, Y. Structuration under pressure: Spatial separation of inserted water during pressure-induced hydration in mesolite. *Am. Mineral.* **2018**, *103*, 175–178. [CrossRef]

83. Goryainov, S.V.; Secco, R.A.; Huang, Y.; Likhacheva, A.Y. Pressure-induced ionic conductivity of overhydrated zeolite NaA at different water/zeolite ratios. *Microporous Mesoporous Mater.* **2013**, *171*, 125–130. [CrossRef]

84. Lee, Y.; Lee, Y.; Seoung, D.; Im, J.H.; Hwang, H.J.; Kim, T.H.; Liu, D.; Liu, Z.; Lee, S.Y.; Kao, C.C.; et al. Immobilization of large, aliovalent cations in the small-pore zeolite K-natrolite by means of pressure. *Angew. Chem. Int. Ed.* **2012**, *51*, 4848–4851. [CrossRef] [PubMed]

85. Liu, D.; Chen, X.; Ma, Y.; Liu, Z.; Vogt, T.; Lee, Y. Spectroscopic and computational characterizations of alkaline-earth- and heavy-metal-exchanged natrolites. *ChemPlusChem* **2014**, *79*, 1096–1102. [CrossRef]

86. Im, J.; Seoung, D.; Hwang, G.C.; Jun, J.W.; Jhung, S.H.; Kao, C.-C.; Vogt, T.; Lee, Y. Pressure-Dependent structural and chemical changes in a metal–organic framework with one-dimensional pore structure. *Chem. Mater.* **2016**, *28*, 5336–5341. [CrossRef]

87. Kremleva, A.; Vogt, T.; Rösch, N. Monovalent cation-exchanged natrolites and their behavior under pressure. A computational study. *J. Phys. Chem. C* **2013**, *117*, 19020–19030. [CrossRef]

88. Kremleva, A.; Vogt, T.; Rösch, N. Potassium-exchanged natrolite under pressure. computational study vs experiment. *J. Phys. Chem. C* **2014**, *118*, 22030–22039. [CrossRef]

89. Santoro, M.; Gorelli, F.A.; Bini, R.; Haines, J.; van der Lee, A. High-pressure synthesis of a polyethylene/zeolite nano-composite material. *Nat. Commun.* **2013**, *4*, 1557. [CrossRef] [PubMed]

90. Santoro, M.; Dziubek, K.; Scelta, D.; Ceppatelli, M.; Gorelli, F.A.; Bini, R.; Thibaud, J.M.; Di Renzo, F.; Cambon, O.; Rouquette, J.; et al. High pressure synthesis of all-transoid polycarbonyl [-(C=O)-]n in a zeolite. *Chem. Mater.* **2015**, *27*, 6486–6489. [CrossRef]

91. Scelta, D.; Ceppatelli, M.; Santoro, M.; Bini, R.; Gorelli, F.A.; Perucchi, A.; Mezouar, M.; Van Der Lee, A.; Haines, J. High pressure polymerization in a confined space: Conjugated chain/zeolite nanocomposites. *Chem. Mater.* **2014**, *26*, 2249–2255. [CrossRef]

92. Santoro, M.; Scelta, D.; Dziubek, K.; Ceppatelli, M.; Gorelli, F.A.; Bini, R.; Garbarino, G.; Thibaud, J.M.; Di Renzo, F.; Cambon, O.; et al. Synthesis of 1D polymer/zeolite nanocomposites under high pressure. *Chem. Mater.* **2016**, *28*, 4065–4071. [CrossRef]

93. Jordá, J.L.; Rey, F.; Sastre, G.; Valencia, S.; Palomino, M.; Corma, A.; Segura, A.; Errandonea, D.; Lacomba, R.; Manjón, F.J.; et al. Synthesis of a novel zeolite through a pressure-induced reconstructive phase transition process. *Angew. Chem. Int. Ed.* **2013**, *52*, 10458–10462. [CrossRef] [PubMed]

94. Arletti, R.; Ronchi, L.; Quartieri, S.; Vezzalini, G.; Ryzhikov, A.; Nouali, H.; Daou, T.J.; Patarin, J. Intrusion-extrusion experiments of MgCl2 aqueous solution in pure silica ferrierite: Evidence of the nature of intruded liquid by in situ high pressure synchrotron X-ray powder diffraction. *Microporous Mesoporous Mater.* **2016**, *235*, 253–260. [CrossRef]

95. Arletti, R.; Leardini, L.; Vezzalini, G.; Quartieri, S.; Gigli, L.; Santoro, M.; Haines, J.; Rouquette, J.; Konczewicz, L. Pressure-induced penetration of guest molecules in high-silica zeolites: The case of mordenite. *Phys. Chem. Chem. Phys.* **2015**, *17*, 24262–24274. [CrossRef] [PubMed]

96. Santamaría-Pérez, D.; Marqueño, T.; MacLeod, S.; Ruiz-Fuertes, J.; Daisenberger, D.; Chuliá-Jordan, R.; Errandonea, D.; Jordá, J.L.; Rey, F.; McGuire, C.; et al. Structural evolution of CO_2-filled pure silica LTA zeolite under high-pressure high-temperature conditions. *Chem. Mater.* **2017**, *29*, 4502–4510. [CrossRef]

97. Arletti, R.; Fois, E.; Gigli, L.; Vezzalini, G.; Quartieri, S.; Tabacchi, G. Irreversible Conversion of a water–ethanol solution into an organized two-dimensional network of alternating supramolecular units in a hydrophobic zeolite under pressure. *Angew. Chem. Int. Ed.* **2017**, *56*, 2105–2109. [CrossRef] [PubMed]

98. Car, R.; Parrinello, M. Unified approach for molecular dynamics and density-functional theory. *Phys. Rev. Lett.* **1985**, *55*, 2471–2474. [CrossRef] [PubMed]

99. Gigli, L.; Arletti, R.; Quartieri, S.; Di Renzo, F.; Vezzalini, G. The high thermal stability of the synthetic zeolite K–L: Dehydration mechanism by in situ SR-XRPD experiments. *Microporous Mesoporous Mater.* **2013**, *177*, 8–16. [CrossRef]

100. Merrill, L.; Bassett, W.A. Miniature diamond anvil pressure cell for single crystal X-ray diffraction studies. *Rev. Sci. Instrum.* **1974**, *45*, 290–294. [CrossRef]

101. Miletich, R.; Allan, D.R.; Kuhs, W.F. High-Pressure single-crystal techniques. *Rev. Mineral. Geochem.* **2000**, *41*, 445–519. [CrossRef]

102. Angel, R.J.; Bujak, M.; Zhao, J.; Gatta, G.D.; Jacobsen, S.D. Effective hydrostatic limits of pressure media for high-pressure crystallographic studies. *J. Appl. Crystallogr.* **2007**, *40*, 26–32. [CrossRef]

103. Hammersley, A.P.; Svensson, S.O.; Hanfland, M.; Fitch, A.N.; Hausermann, D. Two-dimensional detector software: From real detector to idealised image or two-theta scan. *High Press. Res.* **1996**, *14*, 235–248. [CrossRef]

104. Larson, A.C.; Von Dreele, R.B. *General Structure Analysis System (GSAS) Program*; Los Alamos National Laboratory: Los Alamos, NM, USA, 1994.

105. Toby, B.H. EXPGUI, a graphical user interface for GSAS. *J. Appl. Crystallogr.* **2001**, *34*, 210–213. [CrossRef]

106. Thompson, P.; Cox, D.E.; Hastings, J.B. Rietveld refinement of Debye–Scherrer synchrotron X-ray data from Al_2O_3. *J. Appl. Crystallogr.* **1987**, *20*, 79–83. [CrossRef]

107. Grimme, S. Semiempirical GGA-type density functional constructed with a long-range dispersion correction. *J. Comput. Chem.* **2006**, *27*, 1787–1799. [CrossRef] [PubMed]

108. Martínez-Suarez, L.; Siemer, N.; Frenzel, J.; Marx, D. Reaction network of methanol synthesis over Cu/ZnO Nanocatalysts. *ACS Catal.* **2015**, *5*, 4201–4218. [CrossRef]

109. Bandoli, G.; Barreca, D.; Gasparotto, A.; Seraglia, R.; Tondello, E.; Devi, A.; Fischer, R.A.; Winter, M.; Fois, E.; Gamba, A.; et al. An integrated experimental and theoretical investigation on Cu(hfa)2·TMEDA: Structure, bonding and reactivity. *Phys. Chem. Chem. Phys.* **2009**, *11*, 5998–6007. [CrossRef] [PubMed]

110. Fois, E.; Tabacchi, G.; Barreca, D.; Gasparotto, A.; Tondello, E. "Hot" Surface activation of molecular complexes: Insight from modeling studies. *Angew. Chem. Int. Ed.* **2010**, *49*, 1944–1948. [CrossRef] [PubMed]

111. Martínez-Suárez, L.; Frenzel, J.; Marx, D. Cu/ZnO nanocatalysts in response to environmental conditions: Surface morphology, electronic structure, redox state and CO_2 activation. *Phys. Chem. Chem. Phys.* **2014**, *16*, 26119–26136. [CrossRef] [PubMed]

112. Barreca, D.; Fois, E.; Gasparotto, A.; Seraglia, R.; Tondello, E.; Tabacchi, G. How does CuII convert into CuI? An unexpected ring-mediated single-electron reduction. *Chem. Eur. J.* **2011**, *17*, 10864–10870. [CrossRef] [PubMed]

113. Tabacchi, G.; Fois, E.; Barreca, D.; Gasparotto, A. Opening the Pandora's jar of molecule-to-material conversion in chemical vapor deposition: Insights from theory. *Int. J. Quantum Chem.* **2014**, *114*, 1–7. [CrossRef]

114. Pietrucci, F.; Andreoni, W. Fate of a graphene flake: A new route toward fullerenes disclosed with ab initio simulations. *J. Chem. Theory Comput.* **2014**, *10*, 913–917. [CrossRef] [PubMed]

115. Fois, E.; Gamba, A.; Tabacchi, G.; Coluccia, S.; Martra, G. Ab initio study of defect sites at the inner surfaces of mesoporous silicas. *J. Phys. Chem. B* **2003**, *107*, 10767–10772. [CrossRef]

116. Tabacchi, G.; Fois, E.; Barreca, D.; Gasparotto, A. CVD precursors for transition metal oxide nanostructures: Molecular properties, surface behavior and temperature effects. *Phys. Status Solidi* **2014**, *211*, 251–259. [CrossRef]

117. Kraus, P.; Frank, I. On the dynamics of H_2 adsorption on the Pt(111) surface. *Int. J. Quantum Chem.* **2017**, *117*, e25407. [CrossRef]

118. Tabacchi, G.; Fois, E.; Barreca, D.; Carraro, G.; Gasparotto, A.; Maccato, C. *Advanced Processing and Manufacturing Technologies for Nanostructured and Multifunctional Materials II*; Ohji, T., Singh, M., Halbig, M., Eds.; Ceramic Engineering and Science Proceedings; John Wiley & Sons, Inc.: Hoboken, NJ, USA, 2015; ISBN 9781119211662.

119. Maccato, C.; Bigiani, L.; Carraro, G.; Gasparotto, A.; Seraglia, R.; Kim, J.; Devi, A.; Tabacchi, G.; Fois, E.; Pace, G.; et al. Molecular engineering of Mn II diamine diketonate precursors for the vapor deposition of manganese oxide nanostructures. *Chem. Eur. J.* **2017**, *23*, 17954–17963. [CrossRef] [PubMed]

120. Koizumi, K.; Nobusada, K.; Boero, M. Reducing the cost and preserving the reactivity in noble-metal-based catalysts: Oxidation of CO by Pt and Al-Pt alloy clusters supported on graphene. *Chem. Eur. J.* **2016**, *22*, 5181–5188. [CrossRef] [PubMed]

121. Deiana, C.; Tabacchi, G.; Maurino, V.; Coluccia, S.; Martra, G.; Fois, E. Surface features of TiO_2 nanoparticles: Combination modes of adsorbed CO probe the stepping of (101) facets. *Phys. Chem. Chem. Phys.* **2013**, *15*, 13391–13399. [CrossRef] [PubMed]

122. Deiana, C.; Fois, E.; Martra, G.; Narbey, S.; Pellegrino, F.; Tabacchi, G. On the simple complexity of carbon monoxide on oxide surfaces: Facet-Specific donation and backdonation effects revealed on TiO_2 anatase nanoparticles. *ChemPhysChem* **2016**, *17*, 1956–1960. [CrossRef] [PubMed]

123. Fois, E.; Gamba, A.; Tabacchi, G. Electronic spectra of Ti(IV) in zeolites: An ab initio approach. *ChemPhysChem* **2005**, *6*, 1237–1239. [CrossRef] [PubMed]

124. Fois, E.; Gamba, A.; Tabacchi, G. Structure and dynamics of a Brønsted acid site in a zeolite: An ab initio study of hydrogen sodalite. *J. Phys. Chem. B* **1998**, *102*, 3974–3979. [CrossRef]

125. Fois, E.; Gamba, A.; Tabacchi, G. First-principles simulation of the intracage oxidation of nitrite to nitrate sodalite. *Chem. Phys. Lett.* **2000**, *329*, 1–6. [CrossRef]

126. Tabacchi, G.; Silvi, S.; Venturi, M.; Credi, A.; Fois, E. Dethreading of a photoactive azobenzene-containing molecular axle from a crown ether ring: A computational investigation. *ChemPhysChem* **2016**, *17*, 1913–1919. [CrossRef] [PubMed]

127. Muñoz-Santiburcio, D.; Marx, D. Chemistry in nanoconfined water. *Chem. Sci.* **2017**, *8*, 3444–3452. [CrossRef] [PubMed]

128. Fois, E.; Gamba, A.; Tabacchi, G. Intracage chemistry: Nitrite to nitrate oxidation via molecular oxygen. A Car Parrinello study. *Stud. Surf. Sci. Catal.* **2001**, *140*, 251–268. [CrossRef]

129. Tabacchi, G.; Gianotti, E.; Fois, E.; Martra, G.; Marchese, L.; Coluccia, S.; Gamba, A. Understanding the vibrational and electronic features of Ti(IV) sites in mesoporous silicas by integrated ab initio and spectroscopic investigations. *J. Phys. Chem. C* **2007**, *111*, 4946–4955. [CrossRef]

130. Fois, E.; Gamba, A.; Spano, E.; Tabacchi, G. Rotation of molecules and ions in confined spaces: A first-principles simulation study. *J. Mol. Struct.* **2003**, *644*, 55–66. [CrossRef]

131. Spanó, E.; Tabacchi, G.; Gamba, A.; Fois, E. On the role of Ti(IV) as a Lewis acid in the chemistry of titanium zeolites: Formation, structure, reactivity aging of Ti-peroxo oxidizing intermediates. A first principles study. *J. Phys. Chem. B* **2006**, *110*, 21651–21661. [CrossRef] [PubMed]

132. Gamba, A.; Tabacchi, G.; Fois, E. TS-1 from first principles. *J. Phys. Chem. A* **2009**, *113*, 15006–15015. [CrossRef] [PubMed]

133. Tabacchi, G.; Vanoni, M.A.; Gamba, A.; Fois, E. Does negative hyperconjugation assist enzymatic dehydrogenations? *ChemPhysChem* **2007**, *8*, 1283–1288. [CrossRef] [PubMed]

134. Trudu, F.; Tabacchi, G.; Gamba, A.; Fois, E. First principles studies on boron sites in zeolites. *J. Phys. Chem. A* **2007**, *111*, 11626–11637. [CrossRef] [PubMed]

135. Fois, E.; Gamba, A.; Medici, C.; Tabacchi, G.; Quartieri, S.; Mazzucato, E.; Arletti, R.; Vezzalini, G.; Dmitriev, V. High pressure deformation mechanism of Li-ABW: Synchrotron XRPD study and ab initio molecular dynamics simulations. *Microporous Mesoporous Mater.* **2008**, *115*, 267–280. [CrossRef]

136. Gatta, G.D.; Tabacchi, G.; Fois, E.; Lee, Y. Behaviour at high pressure of $Rb_7NaGa_8Si_{12}O_{40} \cdot 3H_2O$ (a zeolite with EDI topology): A combined experimental–computational study. *Phys. Chem. Miner.* **2016**, *43*, 209–216. [CrossRef]

137. Arletti, R.; Fois, E.; Tabacchi, G.; Quartieri, S.; Vezzalini, G. Pressure-Induced penetration of water-ethanol mixtures in all-silica ferrierite. *Adv. Sci. Lett.* **2017**, *23*, 5966–5969. [CrossRef]

138. Fois, E.; Gamba, A.; Tabacchi, G.; Quartieri, S.; Arletti, R.; Vezzalini, G. High-pressure behaviour of yugawaralite at different water content: An ab initio study. *Stud. Surf. Sci. Catal.* **2005**, *155*, 271–280. [CrossRef]

139. Ferro, O.; Quartieri, S.; Vezzalini, G.; Fois, E.; Gamba, A.; Tabacchi, G. High-pressure behavior of bikitaite: An integrated theoretical and experimental approach. *Am. Mineral.* **2002**, *87*, 1415–1425. [CrossRef]

140. Fois, E.; Gamba, A.; Tabacchi, G.; Ferro, O.; Quartieri, S.; Vezzalini, G. A theoretical investigation on pressure-induced changes in the vibrational spectrum of zeolite bikitaite. *Stud. Surf. Sci. Catal.* **2002**, *142*, 1877–1884. [CrossRef]

141. Marx, D.; Hutter, J. *Ab Initio Molecular Dynamics: Basic Theory and Advanced Methods*; Cambridge University Press: Cambridge, UK, 2009; ISBN 9780511609633.

142. Copyright IBM Corp. 1990–2017; MPI für Festkörperforschung Stuttgart 1997–2001. *CPMD Code: Car Parrinello Molecular Dynamics*. Available online: http://www.cpmd.org/.

143. Bougeard, D.; Smirnov, K.S. Modelling studies of water in crystalline nanoporous aluminosilicates. *Phys. Chem. Chem. Phys.* **2007**, *9*, 226–245. [CrossRef] [PubMed]

144. Demontis, P.; Gulín-González, J.; Masia, M.; Suffritti, G.B. The behaviour of water confined in zeolites: Molecular dynamics simulations versus experiment. *J. Phys. Condens. Matter* **2010**, *22*, 284106. [CrossRef] [PubMed]

145. Fois, E.; Gamba, A.; Tabacchi, G.; Quartieri, S.; Vezzalini, G. On the collective properties of water molecules in one-dimensional zeolitic channels. *Phys. Chem. Chem. Phys.* **2001**, *3*, 4158–4163. [CrossRef]

146. Quartieri, S.; Sani, A.; Vezzalini, G.; Galli, E.; Fois, E.; Gamba, A.; Tabacchi, G. One-dimensional ice in bikitaite: Single-crystal X-ray diffraction, infra-red spectroscopy and ab initio molecular dynamics studies. *Microporous Mesoporous Mater.* **1999**, *30*, 77–87. [CrossRef]

147. Fois, E.; Tabacchi, G.; Quartieri, S.; Vezzalini, G. Dipolar host/guest interactions and geometrical confinement at the basis of the stability of one-dimensional ice in zeolite bikitaite. *J. Chem. Phys.* **1999**, *111*, 355–359. [CrossRef]

148. Godelitsas, A.; Armbruster, T. HEU-type zeolites modified by transition elements and lead. *Microporous Mesoporous Mater.* **2003**, *61*, 3–24. [CrossRef]

149. Martucci, A.; Alberti, A.; de Lourdes Guzman-Castillo, M.; Di Renzo, F.; Fajula, F. Crystal structure of zeolite omega, the synthetic counterpart of the natural zeolite mazzite. *Microporous Mesoporous Mater.* **2003**, *63*, 33–42. [CrossRef]

150. Fois, E.; Gamba, A.; Tabacchi, G.; Quartieri, S.; Vezzalini, G. Water molecules in single file: First-principles studies of one-dimensional water chains in zeolites. *J. Phys. Chem. B* **2001**, *105*, 3012–3016. [CrossRef]

151. Ceriani, C.; Fois, E.; Gamba, A.; Tabacchi, G.; Ferro, O.; Quartieri, S.; Vezzalini, G. Dehydration dynamics of bikitaite: Part II. Ab initio molecular dynamics study. *Am. Mineral.* **2004**, *89*, 102–109. [CrossRef]

152. Demontis, P.; Gulín-Gonzalez, J.; Suffritti, G.B. Molecular dynamics simulation study of superhydrated perdeuterated natrolite using a new interaction potential model. *J. Phys. Chem. B* **2006**, *110*, 7513–7518. [CrossRef] [PubMed]

153. Demontis, P.; Gulìn-Gonzàlez, J.; Jobic, H.; Masia, M.; Sale, R.; Suffritti, G.B. Dynamical properties of confined water nanoclusters: Simulation study of hydrated zeolite NaA: Structural and vibrational properties. *ACS Nano* **2008**, *2*, 1603–1614. [CrossRef] [PubMed]

154. Trudu, F.; Tabacchi, G.; Gamba, A.; Fois, E. Water in acid boralites: Hydration effects on framework B sites. *J. Phys. Chem. C* **2008**, *112*, 15394–15401. [CrossRef]

155. Fois, E.; Gamba, A.; Trudu, F.; Tabacchi, G. H_2O-induced trigonal-to-tetrahedral transition in boron zeolites. *Nuovo Cimento della Societa Italiana di Fisica. B Gen. Phys. Relativ. Astron. Math. Phys. Methods* **2008**, *123*, 1567–1574. [CrossRef]

156. Wang, C.H.; Bai, P.; Siepmann, J.I.; Clark, A.E. Deconstructing hydrogen-bond networks in confined nanoporous materials: Implications for alcohol-water separation. *J. Phys. Chem. C* **2014**, *118*, 19723–19732. [CrossRef]

157. Zhou, T.; Bai, P.; Siepmann, J.I.; Clark, A.E. Deconstructing the confinement effect upon the organization and dynamics of water in hydrophobic nanoporous materials: Lessons Learned from zeolites. *J. Phys. Chem. C* **2017**, *121*, 22015–22024. [CrossRef]

158. Fois, E.; Gamba, A.; Tabacchi, G. Bathochromic effects in electronic excitation spectra of hydrated Ti zeolites: A theoretical characterization. *ChemPhysChem* **2008**, *9*, 538–543. [CrossRef] [PubMed]

159. Balestra, S.R.G.; Hamad, S.; Ruiz-Salvador, A.R.; Domínguez−García, V.; Merkling, P.J.; Dubbeldam, D.; Calero, S. Understanding nanopore window distortions in the reversible molecular valve zeolite RHO. *Chem. Mater.* **2015**, *27*, 5657–5667. [CrossRef]

160. Fischer, M. Water adsorption in SAPO-34: Elucidating the role of local heterogeneities and defects using dispersion-corrected DFT calculations. *Phys. Chem. Chem. Phys.* **2015**, *17*, 25260–25271. [CrossRef] [PubMed]

161. Fischer, M. Structure and bonding of water molecules in zeolite hosts: Benchmarking plane-wave DFT against crystal structure data. *Zeitschrift für Kristallographie Cryst. Mater.* **2015**, *230*, 325–336. [CrossRef]

162. Fischer, M. Interaction of water with (silico)aluminophosphate zeotypes: A comparative investigation using dispersion-corrected DFT. *Phys. Chem. Chem. Phys.* **2016**, *18*, 15738–15750. [CrossRef] [PubMed]

163. Alabarse, F.G.; Haines, J.; Cambon, O.; Levelut, C.; Bourgogne, D.; Haidoux, A.; Granier, D.; Coasne, B. Freezing of water confined at the nanoscale. *Phys. Rev. Lett.* **2012**, *109*, 35701. [CrossRef] [PubMed]

164. Coudert, F.-X.; Vuilleumier, R.; Boutin, A. Dipole moment, hydrogen bonding and IR spectrum of confined water. *ChemPhysChem* **2006**, *7*, 2464–2467. [CrossRef] [PubMed]

165. Coudert, F.-X.; Cailliez, F.; Vuilleumier, R.; Fuchs, A.H.; Boutin, A. Water nanodroplets confined in zeolite pores. *Faraday Discuss.* **2009**, *141*, 377–398. [CrossRef] [PubMed]

166. Fischer, R.X.; Sehovic, M.; Baur, W.H.; Paulmann, C.; Gesing, T.M. Crystal structure and morphology of fully hydrated zeolite Na-A. *Zeitschrift für Kristallographie* **2012**, *227*, 438–445. [CrossRef]

167. Fois, E.; Gamba, A.; Tilocca, A. Structure and dynamics of the flexible triple helix of water inside VPI-5 molecular sieves. *J. Phys. Chem. B* **2002**, *106*, 4806–4812. [CrossRef]

168. Hernandez-Tamargo, C.E.; Roldan, A.; Ngoepe, P.E.; De Leeuw, N.H. Periodic modeling of zeolite Ti-LTA. *J. Chem. Phys.* **2017**, *147*, 74701. [CrossRef] [PubMed]

169. Fois, E.; Gamba, A.; Tabacchi, G.; Arletti, R.; Quartieri, S.; Vezzalini, G. The "template" effect of the extra-framework content on zeolite compression: The case of yugawaralite. *Am. Mineral.* **2005**, *90*, 28–35. [CrossRef]

170. Betti, C.; Fois, E.; Mazzucato, E.; Medici, C.; Quartieri, S.; Tabacchi, G.; Vezzalini, G.; Dmitriev, V. Gismondine under HP: Deformation mechanism and re-organization of the extra-framework species. *Microporous Mesoporous Mater.* **2007**, *103*, 190–209. [CrossRef]

171. Ballone, P.; Quartieri, S.; Sani, A.; Vezzalini, G. High-pressure deformation mechanism in scolecite: A combined computational-experimental study. *Am. Mineral.* **2002**, *87*, 1194–1206. [CrossRef]

crystals

MDPI

Review

Thermodynamic Picture of Dimer-Mott Organic Superconductors Revealed by Heat Capacity Measurements with External and Chemical Pressure Control

Yasuhiro Nakazawa *, Shusaku Imajo, Yuki Matsumura, Satoshi Yamashita and Hiroki Akutsu [iD]

Department of Chemistry, Graduate School of Science, Osaka University, Machikaneyama 1-1, Toyonaka, Osaka 560-0043, Japan; imajos12@chem.sci.osaka-u.ac.jp (S.I.); matsumuray16@chem.sci.osaka-u.ac.jp (Y.M.); sayamash@chem.sci.osaka-u.ac.jp (S.Y.); akutsu@chem.sci.osaka-u.ac.jp (H.A.)
* Correspondence: nakazawa@chem.sci.osaka-u.ac.jp; Tel.: +81-6-6850-5396

Received: 24 February 2018; Accepted: 19 March 2018; Published: 21 March 2018

Abstract: This article reviews and discusses the thermodynamic properties of dimer-Mott-type molecular superconductive compounds with $(BEDT\text{-}TTF)_2X$ composition, where BEDT-TTF is bis(ethylenedithio)tetrathiafulvalene and X denotes counter-anions, respectively. We focus mainly on the features occurring in the κ-type structure in which the d-wave superconductive phase appears depending on the Coulomb repulsion U and the bandwidth W, which is tunable by external and chemical pressures. First, we report the high-pressure ac (alternating current) calorimetry technique and experimental system constructed to measure single-crystal samples of molecule-based compounds to derive low-temperature thermodynamic parameters. Using extremely small resistance chips as a heater and a thermometer allows four-terminal detection of an accurate temperature and its oscillation in the sample part with sufficient sensitivity. From the analyses of the temperature dependence of the ac heat capacity of $\kappa\text{-}(BEDT\text{-}TTF)_2Cu(NCS)_2$ under external pressures, we discuss the changes in the peak shape of the thermal anomaly at the superconductive transition temperature T_c at various external pressures p. The rather sharp peak in $C_p T^{-1}$ at T_c = 9.1 K with a strong coupling character at ambient pressure is gradually reduced to weaker coupling as the pressure increases to 0.45 GPa concomitant with suppression of the transition temperature. This feature is compared with the systematic argument of the chemical–pressure effect on the basis of thermal anomalies around the superconductive transition of $\kappa\text{-}(BEDT\text{-}TTF)_2X$ compounds and other previously studied typical dimer-Mott 2:1 compounds. Finally, the discussion is extended to the chemical pressure effect on the normal state electronic heat capacity coefficient γ obtained by applying magnetic fields higher than H_{c2} and the residual γ^*, which remains in the superconductive state due to the induced electron density of states around the node structure. From the overall arguments with respect to both chemical and external pressures, we suggest that a crossover of the electronic state inside the superconductive phase occurs and the coupling strength of electron pairs varies from the electron correlation region near the metal-insulator boundary to the band picture region.

Keywords: organic superconductor; dimer-Mott system; heat capacity; electron correlations; d-wave; pressure

1. Introduction

Molecule-based superconductors are attracting much attention in condensed matter science because they provide a platform to discuss the physics related to the electron correlations in low-dimensional and relatively soft lattice systems [1–4]. Although there are several types of superconductive compounds in molecule-based systems, such as doped fullerene, graphite, polymers, and the recently observed picene [5–8], most are categorized as charge transfer complexes of organic donor/acceptor molecules with counter-ions. Electron pairs are formed by the attractive force in itinerant π-electrons released from frontier orbitals such as the HOMO (highest occupied molecular orbital) and LUMO (lowest unoccupied molecular orbital) to form electron bands. The band structure is determined by the molecular arrangement where the condensation energies are dominated by relatively weak van der Waals interactions.

Since a variety of molecular packings are possible in these charge transfer complexes, even in the same combination of donor/acceptor and counter-ions, the development of new superconductors and investigations of their physical properties have been extensively performed over the past four decades, as summarized in the literature [1–4,9–13]. To date, the most widely studied are the 2:1 complexes consisting of TMTSF (tetramethyltetraselenafulvalene), BEDT-TTF, BETS (bis(ethylenedithio)tetraselenafulvalene), and M(dmit)$_2$ (dmit is 1,3-dithiole-2-thione-4,5-dithiolate) with their counter-ions. Due to the segregated stacking of organic donor/acceptor molecules and counter-ion molecules, they show a layered structure of organic molecules and counter-ions to form quasi-one-dimensional and two-dimensional electronic systems [1–4,9–17]. Complexes of BEDT-TTF and BETS with monovalence counter-anions give numerous superconductive materials.

Among the various superconductive complexes, we have focused mainly on the κ-type dimer-Mott systems of BEDT-TTF and analogous molecules, which are extensively studied as a prototype of effective half-filling compounds. They sometimes show antiferromagnetic insulating states due to the electron correlation mechanism. The antiferromagnetic spin correlation due to the on-site Coulomb repulsion in the half-filled state plays a dominant role in determining the low-temperature electronic states. The electronic features of the dimer-Mott systems are summarized as a pressure–temperature phase diagram (Figure 1) where the ratio of the band energy W and the electron correlation in the dimer unit U are the essential parameters controlling the ground state features [14,15]. The antiferromagnetic Mott insulating phase changes the superconductive phase by applying pressure due to the decrease in the U/W ratio. Kanoda et al. mapped several compounds with rigidly dimerized BEDT-TTF compounds in the phase diagram and discussed the relation between the chemical and the external pressure effects in [11,14,15]. A superconductivity where T_c exceeds 10 K appears near the metal–insulator boundary, which is known as the Mott boundary. The diagram in Figure 1 is recognized as a typical bandwidth control of a two-dimensional electron correlation system in two-dimensional (2D) half-filling state [14,15,18–20]. It is known that the phase diagram dominated by the parameter V/t, where V is the intermolecular Coulomb and t the intermolecular transfer—discussed experimentally by Mori et al. [21] and theoretically by McKenzie et al. [22]—gives a pressure-sensitive feature. This phase diagram is realized in non-dimeric arrangements, such as θ-type BEDT-TTF systems. In the case of the dimer-Mott system, the pair symmetry of the superconductivity and the possibility of unconventional pairings with a nodal gap feature through NMR [23], STM [24], and thermodynamic [25,26] experiments have been discussed, although the electron–phonon interactions are considered to be relatively large in organic systems [27].

To investigate the relationship between superconductivity and electron correlation physics, which peculiarly appears in 2D dimer-Mott systems especially for κ-type compounds, and to discuss the mechanism of superconductivity, the characteristic features occurring inside the superconductive phase, as well as the features of the phase transition, must be pursued. Thermodynamic information from heat capacity measurements can provide quantitative information via the analysis of entropic information of the superconductivity [28–31]. By analyzing the features of the thermal anomalies at the transitions and low energy excitations simultaneously, it is possible to grasp systematic changes in the superconductive phase where the quantum mechanical crossover is observed. For this purpose, pressure-controlled thermodynamic measurements to tune the U/W ratio and systematic arguments with the parameter of this ratio are required [32,33]. In this article, we summarize the thermodynamic features in the superconductive phase of dimer-Mott compounds from heat capacity measurements under pressure. We also discuss the results by comparing with the features observed in the chemical pressure effects produced by changing the anions (X) in the κ-(BEDT-TTF)$_2$X system.

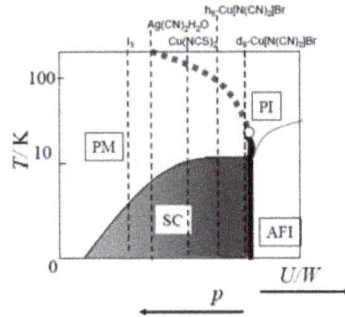

Figure 1. Electronic phase diagram of dimer-Mott compounds with a D_2X composition. Diagram is constructed based on the chemical and external pressure dependences of the physical properties of κ-(BEDT-TTF)$_2$X compounds by Kanoda in Refs. [14,15]. Position of several compounds with different counter-anions are shown by dashed lines. Horizontal axis is the U/W ratio, which is tunable by an external pressure.

2. Calorimetry Apparatus to Measure the Heat Capacity under Pressure

To measure the heat capacity of single-crystalline samples of organic superconductors at ambient pressure, we used the thermal relaxation calorimetry technique developed by Bachmann et al. [34,35]. We constructed and modified our original apparatus—that aimed to measure the heat capacity of molecule-based compounds—to realize accurate measurements for 80 µg–1 mg single crystals. Figure 2 shows a schematic drawing and photograph of the sample cell. Developments of the apparatus focusing on the technique to measure single crystals are reported in the literature [36–38].

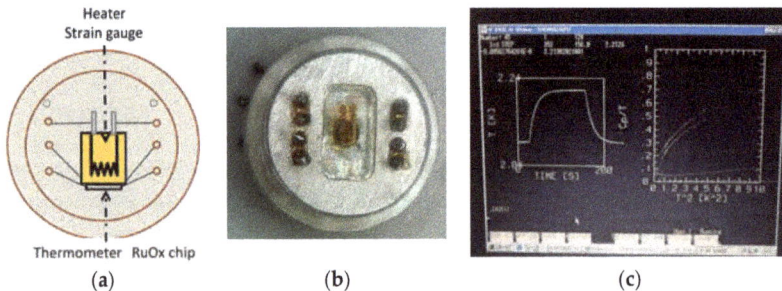

Figure 2. Schematic (**a**) and photograph (**b**) of the relaxation calorimetry cell used to measure the heat capacity of small amounts of samples of molecule-based compounds. Photograph (**c**) shows the relaxation curvature in the measurement.

The application of external pressure reduces the inter-atomic or inter-molecular distance in crystal lattices, inducing various structural transformations due to molecular or atomic arrangements, dielectric, magnetic, and transport properties, which are dominated by the orbital overlap between neighboring atoms and molecules. It also induces an increase in bandwidth W, which directly affects the U/W ratio. To measure the heat capacity while varying the external pressure, both the temperature and magnetic fields are important to realize new functionalities and to understand the mechanism of pressure-induced phase transitions. For heat capacity measurements of molecular compounds under pressure, we used the ac calorimetry technique. The adiabatic technique and the relaxation technique for heat capacity measurements are quite difficult because they usually require a substantial amount of sample and semi-adiabatic conditions around the sample from the surroundings. Such conditions cannot be realized when the crystal and the thermometer are in direct contact with the pressure medium. Therefore, the ac technique [39], which separates the difference of the temperature relaxation rates inside the sample parts from that of the surroundings, is the best way to detect precise thermodynamic information, as suggested by Eicher and Gey [40]. The detection of temperature oscillations using the frequencies in the range between $\tau_{ext}^{-1} < \omega < \tau_{int}^{-1}$, where τ_{ext} and τ_{int} are the internal and external temperature relaxation rates, respectively, can give an accurate heat capacity of the sample parts.

We constructed an ac heat capacity measurement apparatus for single-crystalline samples of molecular compounds. Measurements are available over a wide temperature range from a low temperature of about 0.7 K up to about 20 K while reducing the sample amount to 200 µg–2 mg [41–45]. The construction details are reported elsewhere [33,41,42]. The adoption of extremely small chips as the thermometer and the accurate detection using the four-terminal method have worked well to achieve high-resolution detection for organic superconductors, providing heat capacity measurements in the low-temperature region. Figure 3 shows a schematic view of the apparatus where the sample part and pressure cell are set with CuBe and NiCrAl. We utilized a ruthenium oxide tiny chip sensor with a room temperature resistance of 10^{2-4} Ω, dimensions of 0.6×0.3 mm^2, and a thickness of 0.2 mm. We also used another small-sized chip sensor with a resistance of 1 kΩ at room temperature. The GE (General Electric) vanish confirms good contact between the sample and the sensor and the sample parts were also coated by a small amount of epoxy (Stycast 1266, 2850FT, and 2850GT etc.) before being sealed inside a Teflon capsule with a pressure medium (Daphne 7373 oil, Idemitsu). The sample cell was set in the CuBe piston cylinder and clamped using CuBe screws on both sides.

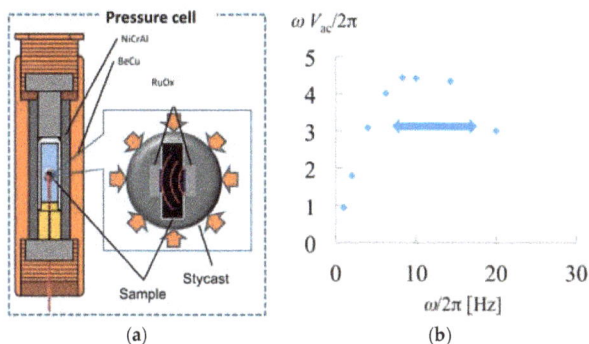

Figure 3. (a) Schematic of the ac calorimetry system in the hybrid pressure cell of CuBe and NiCrAl. Sample part consists of a single crystal sample and two chip sensors working as a thermometer and a heater; (b) a typical curve of the frequency dependence of the ac amplitude voltage to evaluate the measurement validity. In the plateau region, which is marked by the arrow, the oscillation frequency satisfies the appropriate conditions for the measurements.

The temperature modulation excited by the ON/OFF current of the heater was detected by the thermometer using an ac resistance bridge. The modulation occurs more rapidly than the relation of heat to the surroundings expressed by t_{ext}. Since the sensitivity of the ruthenium oxide chip increases in the low-temperature region, this technique is available even in the low-temperature region where thermocouples are not very sensitive. Furthermore, the ruthenium oxide chip sensor has a variable range hopping-type transport feature and can be utilized in extremely low-temperature regions with small magnet resistance.

The frequency dependence of the oscillation amplitude for typical organic crystals is shown in Figure 3b. Although this system is constructed for low-temperature experiments, the use of a tiny Pt chip thermometer and a Cernox thermometer make it possible to detect heat capacity measurements between 20 K and 300 K, as reported by Danda [45] and Konoike [46]. Danda et al. reported the Verwey transition of iron oxide (Fe$_3$O$_4$) in a higher temperature system [45]. The sharp peak due to magnetic and charge-ordering transition at 124 K is broadened by pressures up to 1 GPa without significantly changing the transition temperature. The most serious problem is to separate the background contribution since the extra contribution of the addenda, which consists of the chips, the lead wires, and the Stycast, is included.

3. Heat Capacity of κ-(BEDT-TTF)$_2$X under Pressure

In this section, we review the calorimetry results of dimer-Mott compounds under pressure. Figure 4 shows the temperature dependence of the heat capacities of four charge transfer complexes, including κ-(BEDT-TTF)$_2$Cu(NCS)$_2$, obtained by relaxation calorimetry. The details are reported in the published literature [47–51]. Compared with other compounds in the figure, κ-(BEDT-TTF)$_2$Cu(NCS)$_2$ exhibits a rather sharp peak as shown in Figure 4c [49,50]. The thermodynamics of this compound and κ-(BEDT-TTF)$_2$Cu[N(CN)$_2$]Br, which has a similar transition temperature, have already been thoroughly discussed. Several groups have analyzed the peak shape and the magnitude of the heat capacity jump at T_c for κ-(BEDT-TTF)$_2$Cu(NCS)$_2$ [49,50,52–57]. In the initial investigation stage of this material, the heat capacity jump is reported as evidence of the bulk nature of the superconductivity, as reported by Katsumoto et al. [52] and Andraka et al. [53]. Additionally, Graebner et al. [54] measured the ac heat capacity with an absolute precision around the peak and suggested, for the first time, a strong coupling nature with a larger condensation energy. More recently, Müller et al. [55] reported a detailed analysis of the peak shape. They showed the data over a wide temperature range to determine the α value of the coupling strength. The data are shown in Figure 4c where the magnitude of $CpT_c^{-1} \simeq 60$ mJ K^{-2} mol^{-1} is consistent with the strong coupling picture in Refs [26,49,50,54–56]. A similar thermodynamic feature has also been reported for κ-(BEDT-TTF)$_2$Cu[N(CN)$_2$]$_2$Br, which shows a stronger coupling peak shape [57].

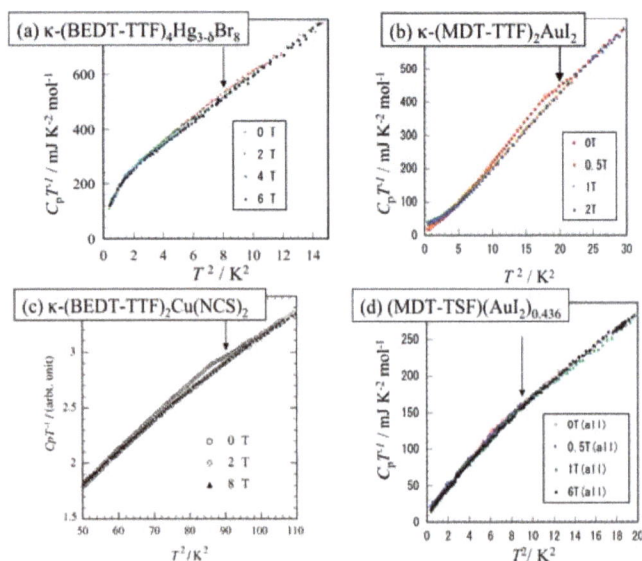

Figure 4. C_pT^{-1} vs. T^2 plot of superconductive compounds (**a**) κ-(BEDT-TTF)$_4$Hg$_{1-δ}$Br$_8$ (Refs. [47,48]); (**b**) κ-(MDT-TTF)$_2$AuI$_2$ (Ref. [49]); (**c**) κ-(BEDT-TTF)$_2$Cu(NCS)$_2$ (Refs. [49,50]); and (**d**) (MDT-TSF)(AuI$_2$)$_{0.436}$ (Ref. [51]) obtained under 0 T and with magnetic fields.

To confirm the thermodynamic peak in the ac heat capacity data obtained by the high-pressure calorimetric system, we have analyzed the data at ambient pressure and compared them with the relaxation calorimetry data. Figure 5a shows the ac heat capacity measurement results at ambient pressure obtained using a high-pressure calorimeter as a C_pT^{-1} vs. T^2 plot [32]. A thermal anomaly is detected as a hump structure around 9 K in this plot, although there is a rather large background due to the addenda heat capacity in the pressure cell. This anomaly is associated with the superconductive transition since the peak temperature and the peak shape resemble the preceding data obtained by the thermal relaxation technique in Figure 4c. Upon applying magnetic fields almost parallel to the plane in the present pressure calorimetry set up, the anomaly is suppressed through comparative analyses of the magnetic field dependence, as already reported [32]. The data obtained up to 7 T has a constant shift of C_pT^{-1} values at different magnetic fields (Figure 5a).

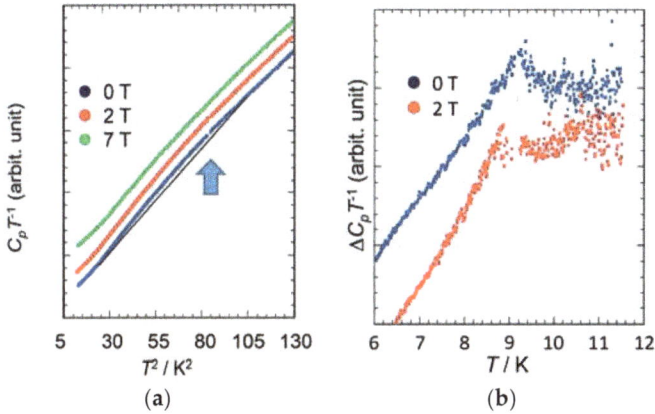

Figure 5. (a) C_pT^{-1} vs. T^2 plot of the ac heat capacity of κ-(BEDT-TTF)$_2$Cu(NCS)$_2$ at ambient pressure with magnetic fields of 0 T, 2 T, and 7 T applied parallel to the conducting layer. The values of C_pT^{-1} under magnetic fields are plotted with constant offsets in the vertical axis. The arrow shows the superconductive transition temperature. (**b**) Temperature dependence of the ΔC_pT^{-1} obtained by subtracting the 7-T data as the background ($\Delta C_pT^{-1}(H) = C_pT^{-1}(H) - C_pT^{-1}(7\ T)$) to evaluate the anomaly in the temperature dependence due to the electronic heat capacity around the transition (details are reported in Ref. [32]).

Usually, in quasi-2D superconductors, an increase in the magnetic field applied parallel to the plane drastically suppresses and broadens the thermal anomaly. To analyze the peak shape around T_c in detail, Figure 5b plots ΔC_pT^{-1} as the electronic heat capacity around the transition temperature. Here, ΔC_p is determined as the discrepancy of the heat capacity values in the 0-T and 7-T data assuming that the thermal anomaly due to superconductive transition is almost reduced in a magnetic field of 7 T. The analytic details were reported in Ref. [32]. The peak resembles the typical shape of the superconductive transition with a strong coupling feature. ΔC_p at 2 T is also evaluated and plotted as the discrepancy from the 7-T data. A slight downward shift in the transition temperature and suppression of the magnitude of heat capacity jump are observed.

Figure 6a shows the temperature dependence of the ac heat capacity at 0.15 GPa and Figure 6b shows the result of the similar analysis as was performed for ambient pressure data in Figure 5b. The transition temperature is reduced to about 5–6 K, which is consistent with the results of the proceeding work [44]. Subtracting the 7-T data also gives the relative change of the electronic heat capacity contribution ΔC_p for each field, which is shown in the ΔC_pT^{-1} vs. T plot in Figure 6b. It is difficult to see a distinct peak structure even for the data of 0 T; however, the magnetic field seems to suppress the peak and the transition temperature systematically. We can mention that the peak shape broadens compared with that at ambient pressure. The suppression of the peak structure becomes more remarkable in the case of 0.30 GPa and 0.45 GPa. Since the sensitivity of the resistance sensor increases in the lower temperature region, small anomalies, if any exist, can be more easily detected in the lower T_c cases. In fact, several compounds with a magnetic or superconductive transition around 2–4 K have been detected in a high-pressure calorimeter [41,42].

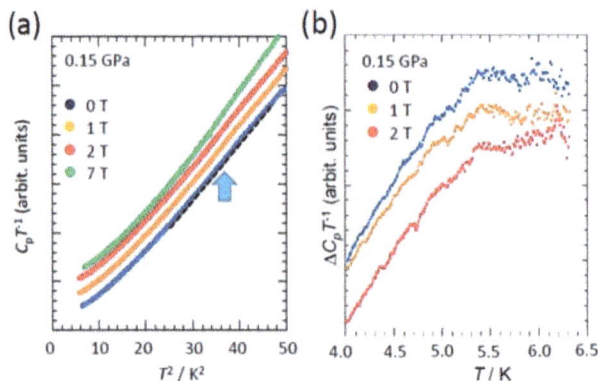

Figure 6. (a) $C_p T^{-1}$ vs. T^2 plot of the ac heat capacity of κ-(BEDT-TTF)$_2$Cu(NCS)$_2$ at 0.15 GPa with magnetic fields of 0 T, 1 T, 2 T, and 7 T. The values of $C_p T^{-1}$ under magnetic fields are plotted with constant offsets in the vertical axis. The arrow indicates the superconductive transition temperature at 0 T; (b) temperature dependence of the $\Delta C_p T^{-1}$ at 0.15 GPa obtained by subtracting the 7-T data as the background (details are reported in Ref. [32]).

To compare the relative change in the broadness and the magnitude of the heat capacity jump induced by the pressure in the same figure, Figure 7 plots the overall feature in the peak shape of $\Delta C_p T^{-1}$ obtained at ambient pressure, 0.15 GPa, and 0.30 GPa, where the temperature in the horizontal axis is the relative temperature normalized by the transition temperature (T_c) of each pressure. A qualitative comparison of the peak shape for the superconductive transition indicates that the reduction in the transition temperature is accompanied by a systematic change in the peak structure. As T_c decreases due to the increase in pressure, the magnitude of the peak of $\Delta C_p T^{-1}$ becomes smaller and the peak shape broadens gradually. This is explained by the continuous reduction in the coupling strength of the superconductivity in a dimer-Mott system. The sharp peak with a relatively high T_c means the coupling strength is strong and is almost as large as the condensation energy of the electron pairs. In contrast, broadening is considered to be the suppression of such pairing forces as the system becomes more metallic. κ-(BEDT-TTF)$_2$Cu[N(CN)$_2$]Br and κ-[(BEDT-TTF)$_{1-x}$(BEDSe-TTF)$_x$]$_2$Cu[N(CN)$_2$]Br compounds also show a broadening in the peak shape and a decrease in the heat capacity jump $\Delta C_p T^{-1}$ as the transition temperature is suppressed. However, the latter may have some extra effects related to the frustration due to the increase in the triangularity in the molecular arrangement of the dimer units. We have also reported the suppression of the peak by pressure in κ-(BEDT-TTF)$_2$Ag(CN)$_2$H$_2$O [44]. These features are consistent with κ-(BEDT-TTF)$_2$Cu(NCS)$_2$, as summarized in Figure 7.

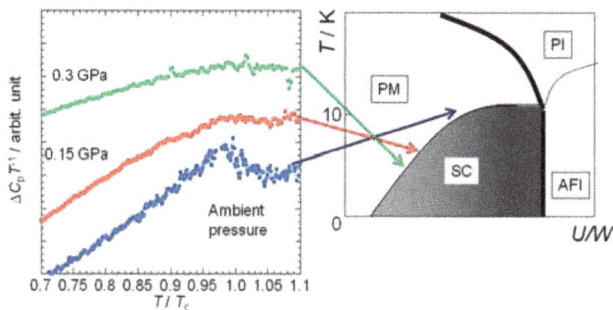

Figure 7. Systematic changes in the peak shape of κ-(BEDT-TTF)$_2$Cu(NCS)$_2$ as the pressure increases up to 0.30 GPa. (See details in Ref. [32].) Temperature in the horizontal axis is scaled by T_c to qualitatively evaluate the sharpness of the thermal anomaly. Correspondence of the data in each pressure is shown in the schematic phase diagram of the dimer-Mott system.

4. Chemical Pressure Effect on Dimer-Mott Organic Compounds

The systematic change in the peak shape in the κ-(BEDT-TTF)$_2$Cu(NCS)$_2$ and similar features observed in other κ-(BEDT-TTF)$_2$X compounds under pressures can be compared with the chemical pressure effects of the dimer-Mott system induced by varying the counter-anions. It is well known that the counter-anion size of organic charge transfer complexes induces a kind of pressure effect in the molecular arrangement, causing systematic changes in the physical properties. According to transport and NMR studies of the dimer-Mott phase diagram, the external pressure effects in 10 K class compounds with X = Cu(NCS)$_2$ and Cu[N(CN)$_2$]Br are consistent with the chemical pressure effects. Similar to the case where the external pressure increases, counter-ions with smaller volumes change the features from a strong correlation system to a normal metallic system. Many thermodynamic studies by heat capacity measurements of several κ-type compounds with different counter-ions have been performed up to now [49–60]. As mentioned in the previous section, the strong coupling nature of κ-(BEDT-TTF)$_2$Cu(NCS)$_2$ has been observed in 10 K class superconductors (Figures 4c and 8). The values of $\Delta C_p / \gamma T_c$ of two typical compounds κ-(BEDT-TTF)$_2$Cu(NCS)$_2$ and κ-(BEDT-TTF)$_2$Cu[N(CN)$_2$]Br—which were obtained using the normal state heat capacities obtained under strong magnetic fields above H_{c2} as the background heat capacity of exceed 2.0—are much larger than that of the BCS (Bardeen Cooper Schrieffer) weak coupling theory [54,55,57]. As compared with these higher T_c compounds, the heat capacity jump of middle-class (about 4–5 K) superconductors, such as κ-(BEDT-TTF)$_2$Ag(CN)$_2$H$_2$O (T_c = 5 K) [58], gives a $\Delta C_p / \gamma T_c$ value of 1.1. Although κ-(MDT-TTF)$_2$AuI$_2$ consists of a different asymmetric donor molecule, its T_c is 4.5 K [49] and it has a $\Delta C_p / \gamma T_c$ value of 1.4. These values are within the weak coupling region.

The plot in Figure 8 shows the systematic change in the peak shape for several superconductive compounds with the κ-type structure using the scaled temperature of $t = T / T_c$ for each compound. We also include the $\Delta C_p / T_c$ data of mixed crystals of κ-[(BEDT-TTF)$_{1-x}$(BEDSe-TTF)$_x$]$_2$Cu[N(CN)$_2$]Br compounds with x = 0.10. A clear tendency is observed; 10 K class superconductors have a sharper peak with a large $\Delta C_p T^{-1}$. This is typical for strong coupling systems, but 4–5 K class compounds have a smaller mean field type peak. The systematic change in the peak shape indicates a gradual crossover from a strong coupling region near the boundary to a weak coupling region in the superconductive phase, as shown schematically in Figure 8 (right) [32,33]. Although some exceptions exist, such as κ-(BEDT-TTF)$_2$I$_3$ being reported to have strong coupling features [59], the overall tendency seems to be consistent with the external pressure controlled effects. This feature is reasonably consistent with our observations of κ-(BEDT-TTF)$_2$Cu(NCS)$_2$ under pressure, as summarized in Figure 7.

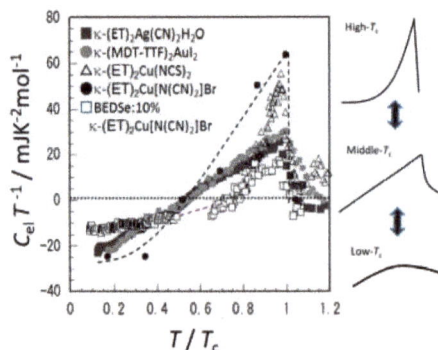

Figure 8. Shape of the thermal anomalies at the superconductive transitions of several compounds with the κ-D$_2$X composition. ET is an abbreviation of BEDT-TTF molecules. BEDSe-TTF10% means that a solid solution system of BEDT-TTF(90%) and BEDSe-TTF(10%) molecules. C_{el} in this plot is determined as the difference of the 0 T data and the heat capacity data obtained under magnetic fields higher than H_{c2} for each compound. (The data are from Ref. [33].) The dashed curves in the figure are a guide for the eyes to see temperature dependencies. The temperature in the horizontal axis is scaled by the transition temperature of each compound to compare the peak shape changes. The right part of the figure shows qualitatively the variation in the peak shapes in the dimer-Mott phase diagram.

The similar tendency observed between the data of the external and the chemical pressure-controlled experiments demonstrates that a crossover occurs inside the superconductive phase due to variations in the coupling strength of the superconductive electrons. A similar situation may occur in high T_c cuprates or heavy electron systems in which the quantum mechanical features produce a crossover in the ground state, giving complicated phase diagrams.

It is important to mention that although the peak shape of $\Delta C_p/T_c$ varies with the decrease in U/W, the temperature dependence of the heat capacity in the low-temperature region gives a quadratic temperature-dependent term, as suggested by the nodal feature of the d-wave formation [60]. Figure 9a,b compares the temperature dependence of the low-temperature heat capacity of two κ-type compounds with a high T_c (10 K) class and a middle T_c class. The existence of the quadratic temperature dependence in the electronic heat capacity clearly shows the difference in the BCS characters, even though the peak shapes are quite different, as we discussed in Figure 8. Since the large lattice contribution in the heat capacity creates ambiguity in background subtraction in such organic systems, the recovery of the γ by magnetic fields of both compounds also shows an $(H/H_{c2})^{1/2}$-dependence characteristic of d-wave pairings. The change in the nodal direction, even in the same κ-type structure, has been suggested theoretically, depending on the balance of the transfer integral inside the dimer and the ratio between the transfer integrals in the rectangle and the diagonal directions of the superconductivity. It is emphasized, however, that the electron correlations still seem to be an important mechanism for realizing superconductivity in this system. The existence of the nodes is also consistent with other experiments, such as the thermal conductivity, NMR, and STM experiments [23–26].

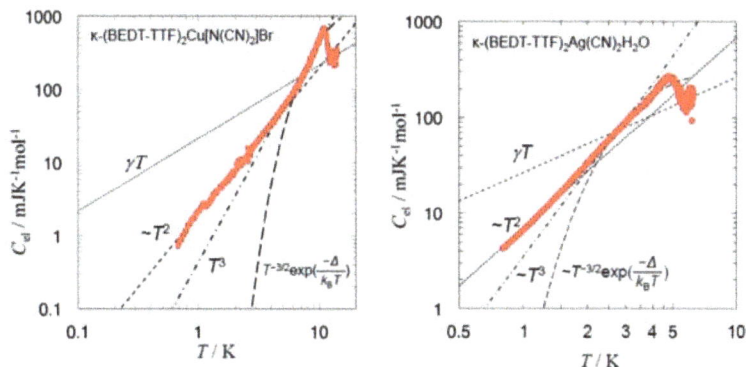

Figure 9. Temperature dependence of the electronic heat capacity determined as the difference of the 0-T data and the normal state data obtained under magnetic fields higher than H_{c2} for (a) κ-(BEDT-TTF)$_2$Cu[N(CN)$_2$]Br and (b) κ-(BEDT-TTF)$_2$Ag(CN)$_2$H$_2$O below the transition temperatures. (Ref. [60].) The lines and the curve shown in the figure denote T linear, T^2, T^3, and $T^{-3/2}\exp(-\Delta/k_BT)$ temperature dependence.

5. Normal State Electronic Heat Capacity Coefficient γ and the Residual γ*

A quantitative comparison and the overall discussion of the coupling strength observed in the peak shape of the heat capacity and magnitude of the electronic heat capacity coefficient γ of the normal state should reveal a more profound understanding of the superconductive state of the dimer-Mott system. For this purpose, low-temperature heat capacity data should be discussed systematically throughout the superconductive phase. In this section, we survey the normal state γ and the residual γ* in the superconductive state observed from the data of the low-temperature heat capacity, which reflects the low energy excitations in the superconductive phase. We discuss the peculiar feature and stability of the superconductive states in terms of the competition between the stability of the Fermi liquid nature and the magnitude of the electron correlations. Since the ground state of the κ-(BEDT-TTF)$_2$X compounds in question is superconductive with electron pairs, the heat capacity measurements with magnetic fields higher than H_{c2} should be performed to evaluate the electron density of state of the Fermi surface. The electronic heat capacity coefficient obtained under a magnetic field is defined as the normal state γ and is proportional to the electron density of states $D(\varepsilon_F)$ in the band theory.

To discuss the systematic changes of γ in the dimer-Mott-type superconductive phase, Figure 10b plots the γ values of several compounds with the conceptual phase diagram (Figure 10a). We use the values of the dimerized compounds of κ-(BEDT-TTF)$_2$X and some other compounds also considered as dimer-based systems. The data are taken from Table 9 in Ref. [28] and the low-temperature heat capacity results reported in Refs. [42,59–63]. In this plot, we also included β-(BEDT-TTF)$_2$I$_3$ in a well-annealed case according to the proposed phase diagram in Refs. [11,14,15]. The two arrows show that the systematic tendency occurs in the superconductive phase. Complexes of partially deuterated BEDT-TTF molecules with X = Cu[N(CN)$_2$]Br were used to systematically investigate the change in the normal state γ near the boundary region [62,63]. The normal state γ drastically decreases around the boundary despite the fact that T_c maintains a relatively high value around 10 K, indicating that the Mott boundary occurs as a first-order transition in the dimer-Mott system. The change in γ occurs in a region far away from the boundary, suggesting that the band-like feature reflecting in the electronic density of states systematically varies. In the discussion on this region, we include data of solid solution compounds of BEDT-TTF and BEDSe-TTF with X = Cu[N(CN)$_2$]Br since light doping (less than 10%) of BEDSe-TTF tends to reduce the T_c by keeping the bulk superconductive feature [33]. The arrow on the left side of the figure clearly indicates that the γ values increase as the U/W ratio increases. This is considered as a kind of Brinkman–Rice enhancement. It predicts that the increase in the electron correlation of the parameter U/W leads to an increase in the density of states similar to that observed in correlated metallic systems with a distinct band structure [63–65]. The normal state γ value gives a maximum around κ-(BEDT-TTF)$_2$Cu(NCS)$_2$, which shows a stable bulk superconductivity with a transition temperature of 9.1 K. In this compound, the cooling rate does not affect the transition temperature located close to the boundary, which differs from the case of X = Cu[N(CN)$_2$]Br (both deuterated and non-deuterated cases).

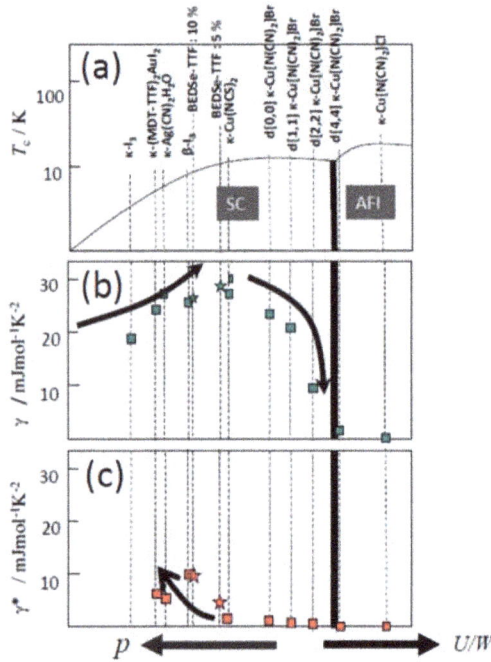

Figure 10. Systematic variations of the low-temperature thermodynamic parameters in the compounds mapped in the dimer-Mott phase diagram (**a**). Changes in (**b**) the normal states electronic heat capacity coefficient γ obtained by heat capacity data above H_{c2} and (**c**) the residual electronic heat capacity coefficient γ^* in the superconductive state at 0 T are plotted in accordance with the position in the phase diagram. The position of each compound is determined in reference to Refs [11,14,15]. The data are taken from Table 9 in Ref. [28] and the reported results data in Refs [42,59–63]. Systematic change of the thermodynamic parameters occurring with the change in chemical pressure, namely the change in the parameter of the U/W ratio, suggests that a kind of crossover exists inside the superconductive phase. Compounds shown as the name of the counter-anions are BEDT-TTF(ET) complexes. BEDSe-TTF 5%, 10% mean the solid solution systems of BEDT-TTF and BEDSe-TTF. (See text.) d[n,n]-Cu[N(CN)$_2$]Br denotes a Cu[N(CN)$_2$]Br compound with BEDT-TTF of which ethylene groups are partially deuterated (Ref. [61,62]) to tune the chemical pressure. Arrows show the characteristic tendency reflecting the physical properties.

The Brinkman–Rice-type enhancement is due to the increase in the enhanced correlation U in the region where the Fermi liquid picture holds. The increase in the correlation influences the effective mass of the normal electrons within the band picture and, therefore, it leads to an increase in γ values. The compounds in the left arrow region are considered to have a Fermi liquid character with a well-defined electron band. However, increasing U further leads the system to an anomalous metallic state where the picture with a gap-like structure between the upper and the lower Hubbard bands become dominant if the Mott boundary is approached. The relatively smaller normal state γ value despite the higher transition temperature and the stronger coupling character in the superconductivity in κ-(BEDT-TTF)$_2$Cu[(N(CN)$_2$]Br than Cu(NCS)$_2$ can be explained from this perspective. The decrease in γ in the partially deuterated compound denoted by d[1,1]-Cu[N(CN)$_2$]Br is considered as evidence of further opening of the Hubbard gap, leading to a decreased electron density of state as a band picture. On the other hand, the smaller γ values in the middle-class T_c compounds of κ-(BEDT-TTF)$_2$Ag(CN)$_2$H$_2$O and κ-(MDT-TTF)$_2$AuI$_2$ located in the band region can be explained in the frame of the Brinkman–Rice picture. Recently, Imajo et al. studied λ-(BETS)$_2$GaCl$_4$, which is also classified as a dimer-Mott-type superconductor and has a T_c of about 5.2 K. It shows a similar normal state $\gamma = 22.9$ mJK^{-2}mol^{-1} [66]. This result is consistent with this picture. We also include the γ values of κ-(BEDT-TTF)$_2$I$_3$ [59] and β-(BEDT-TTF)$_2$X (X = I$_3$, AuI$_2$) [67,68], which are also consistent with the above feature. Not only is the coupling strength reflected in the heat capacity jump but the feature of the normal state γ value also suggests that the electron correlations become stronger by approaching the boundary and are important for stabilizing the bulk superconductivity. When the transition temperature decreases with a decrease in the U/W ratio, the Mott–Hubbard picture is gradually suppressed and the Fermi liquid nature is enhanced. The change from strong coupling to weak coupling observed by pressure calorimetry is related to this crossover from the electron correlation region to the band-like region.

The crossover in the coupling nature originating from the magnitude of the electron correlation is also reflected in the low-temperature residual γ^* in the superconductive state (Figure 10c). The superconductivity of the dimer-Mott system is a d-wave with a nodal gap with four-fold symmetry around the two-dimensional Fermi surface, as detected by the magnetic angle-resolved heat capacity measurements. In such nodal gap superconductors, normal electrons inevitably remain due to the residual disorder in the crystals even without an external magnetic field. The effects of disorder and impurities can induce normal electrons around the nodal position with a very small gap. Figure 10c also shows the data for γ^*. As previously reported, the γ^* value shows a sample dependence but the value is about several percent of the normal state γ and less than 1–2 mJK^{-2}mol^{-1} in the cases of κ-(BEDT-TTF)$_2$Cu[N(CN)$_2$]Br and κ-(BEDT-TTF)$_2$Cu(NCS)$_2$. However, those with lower T_c compounds tend to have larger γ^* values despite the heat capacity giving a distinct peak at the transition. In the cases of κ-(BEDT-TTF)$_2$Ag(CN)$_2$H$_2$O and κ-(MDT-TTF)$_2$AuI$_2$, the reported γ^* values are 5.1 mJK^{-2}mol^{-1} and 5.9 mJK^{-2}mol^{-1}, respectively [33,49]. A more recent experiment on κ-(BEDT-TTF)$_2$Ag(CN)$_2$H$_2$O claims a smaller γ^* value (less than 2 mJK^{-2}mol^{-1}) using the fitting of the d-wave model [60]. This seems to be larger than those of κ-(BEDT-TTF)$_2$Cu[N(CN)$_2$]Br, and κ-(BEDT-TTF)$_2$Cu(NCS)$_2$. λ-(BETS)$_2$GaCl$_4$ also gives a relatively large value of 3.2 mJK^{-2}mol^{-1} [66].

Although the bulk nature of the superconductivity is retained in these compounds, pair breaking due to suppression of the electron correlation and the increase in the stability of Fermi liquid nature occurs in the superconductive phase. The competition of the band picture, which just enhances the electron density of state and the Mott–Hubbard picture, should exist in the superconductive phase; however, a gradual crossover occurs inside the superconductive phase. These characters can be considered to have similar physics in the over-doped region of high T_c cuprates. The heat capacity peak around the transition becomes much broader and the bulk nature of the superconductivity is suppressed in the over-doped region due to the Fermi liquid character, which stabilizes the band nature. This character is in contrast with the under-doped and optimal-doped regions [69]. Considering the overall thermodynamic features of dimer-Mott systems, the mechanism to produce a d-wave type superconductor is quite reasonably related to the electron correlations U, and the superconductivity is stabilized in the anomalous metallic region near the Mott boundary. Here, we discussed mainly κ-type compounds with a dimer lattice. More detailed thermodynamic information for other types of dimer-Mott structures with β-, β'-, and λ-type compounds is necessary in the future.

6. Conclusions

We systematically discuss the thermodynamic parameters to quantitatively evaluate the superconductive characters of the dimer-Mott system, which appear as tuning of the U/W ratio. The calorimetric information of κ-(BEDT-TTF)$_2$Cu(NCS)$_2$ and other counter-anion compounds reveals that the magnitude of the coupling strength is suppressed from the high T_c region to lower T_c region. The chemical pressures also confirm the changes in the peak shape of this thermal anomaly. A comparative discussion of the electronic heat capacity coefficients in the low-temperature heat capacity in both the superconductive state at 0 T and the normal state demonstrates that a crossover from the band character region to the strong electron correlation region exists in the superconductive phase. Such a crossover may support the unconventional nature of the superconductivity in dimer-Mott superconductors since the superconducting character near the strong correlation region provides stability compared with the lower T_c region where the band character becomes dominant. This feature resembles cuprate superconductors in which the electron correlations and band character compete with each other, yielding complicated electronic features in the optimal doping region. More systematic thermodynamic experiments with a parameter of pressure are required to further investigate the system.

Acknowledgments: This work is partly supported by CREST, JST Program in the area of "Establishment of Molecular Technology towards the Creation of New Functions".

Author Contributions: Yasuhiro Nakazawa conceived and designed the experiments; Hiroki Akutsu performed crystal syntheses and structure analyses; Shusaku Imajo, Yuki Matsumura, and Satoshi Yamashita performed the experiments and analyzed the data.

Conflicts of Interest: The authors declare no conflict of interests.

References

1. Ishiguro, T.; Yamaji, K.; Saito, G. *Organic Superconductors*; Springer: Heidelberg, Germany, 1998; pp. 1–245.
2. Williams, J.M.; Ferraro, J.R.; Thorn, R.J.; Carlson, K.D.; Geiser, U.; Wang, H.-H.; Kini, A.M.; Whangbo, M.-H. *Organic Superconductors*; Prentice-Hall, Inc.: Upper Saddle River, NJ, USA, 1992; pp. 65–179.
3. Lebet, A. *The Physics of Organic Superconductors and Conductors*; Springer: Berlin, Germany, 2008; pp. 1–704.
4. Uji, S.; Mori, T.; Takahashi, T. Focus on Organic Conductors. *Sci. Technol. Adv. Mater.* **2009**, *10*, 020301. [CrossRef] [PubMed]
5. Haddon, R.C.; Hebard, A.F.; Rosseinsky, M.J.; Murphy, D.W. Conducting films of C$_{60}$ and C$_{70}$ by alkali-metal doping. *Nature* **1991**, *350*, 320–322. [CrossRef]
6. Palstra, T.T.M.; Zhou, O.; Iwasa, Y.; Sulewski, P.E.; Flemming, R.M.; Zegarski, B.R. Superconductivity at 40K in cesium doped C$_{60}$. *Solid State Commun.* **1995**, *93*, 327–330. [CrossRef]
7. Greene, R.L.; Street, G.B.; Suter, L.J. Superconductivity in Polysulfur Nitride (SN)x. *Phys. Rev. Lett.* **1975**, *34*, 577–579. [CrossRef]
8. Kubozono, Y.; Mitamura, H.; Lee, X.; He, X.; Yamanari, Y.; Takahashi, Y.; Suzuki, Y.; Kaji, Y.; Eguchi, R.; Akaike, K.; et al. Metal-intercalated aromatic hydrocarbons: A new class of carbon-based superconductors. *Phys. Chem. Chem. Phys.* **2011**, *13*, 16476–16493. [CrossRef] [PubMed]
9. Kagoshima, S.; Kato, R.; Fukuyama, H.; Seo, H.; Kino, H. Interplay of Structural and Electronic Properties. In *Advances in Synthetic Metals Twenty Years of Progress in Science and Technology*; Chapter 4; Elsevier: Lausanne, Switzerland, 1999; pp. 262–316.
10. Mori, H. Materials Viewpoint of Organic Superconductors. *J. Phys. Soc. Jpn.* **2006**, *75*, 051003. [CrossRef]
11. Kanoda, K. Metal-Insulator Transition in κ-(ET)$_2$X and (DCNQI)$_2$M: Two Contrasting Manifestation of Electron Correlation. *J. Phys. Soc. Jpn.* **2006**, *75*, 051007. [CrossRef]
12. Kato, R. Development of π-Electron Systems Based on [M(dmit)$_2$] (M = Ni and Pd; dmit: 1,3-dithiole-2-thione-4,5-dithiolate) Anion Radicals. *Bull. Chem. Soc. Jpn.* **2014**, *87*, 355–374. [CrossRef]
13. Kino, H.; Fukuyama, H. Phase Diagram of Two-Dimensional Organic Conductors: (BEDT-TTF)$_2$X. *J. Phys. Soc. Jpn.* **1996**, *65*, 2158–2169. [CrossRef]
14. Kanoda, K. Recent progress in NMR studies on organic conductors. *Hyperfine Interact.* **1997**, *104*, 235–249. [CrossRef]

15. Kanoda, K. Electron correlation, metal-insulator transition and superconductivity in quasi-2D organic systems, (ET)$_2$X. *Phys. C Superconduct.* **1997**, *287*, 299–302. [CrossRef]

16. Kobayashi, H.; Cui, H.; Kobayashi, A. Organic metals and superconductors based on BETS (BETS = bis(ethylenedithio)tetraselenafulvalene). *Chem. Rev.* **2004**, *104*, 5265–5288. [CrossRef] [PubMed]

17. Fujiwara, H.; Kobayashi, H.; Fujiwara, E.; Kobayashi, A. An Indication of Magnetic-Field-Induced Superconductivity in a Bifunctional Layered Organic Conductor, κ-(BETS)$_2$FeBr$_4$. *J. Am. Chem. Soc.* **2002**, *124*, 6816–6817. [CrossRef] [PubMed]

18. Kagawa, F.; Miyagawa, K.; Kanoda, K. Unconventional critical behaviour in a quasi-two-dimensional organic conductor. *Nature* **2005**, *436*, 534–537. [CrossRef] [PubMed]

19. Gati, E.; Garst, M.; Manna, R.S.; Tutsch, U.; Wolf, B.; Bartosch, L.; Schubert, H.; Sasaki, T.; Schlueter, J.A.; Lang, M. Breakdown of Hooke's law of elasticity at the Mott critical endpoint in an organic conductor. *Sci. Adv.* **2016**, *2*, e1601646. [CrossRef] [PubMed]

20. Matsumura, Y.; Imajo, S.; Yamashita, S.; Akutsu, H.; Nakazawa, Y. Thermodynamic Investigation by Heat Capacity Measurements of κ-type Dimer-Mott Organic Compounds with Chemical Pressure Tuning. *Int. J. Mod. Phys. B* **2018**, *32*, 1840024. [CrossRef]

21. Mori, H.; Tanaka, S.; Mori, T. Systematic study of the electronic state in θ-type BEDT-TTF organic conductors by changing the electronic correlation. *Phys. Rev. B* **1998**, *57*, 12023–12029. [CrossRef]

22. McKenzie, R.H.; Merino, J.; Marston, J.B.; Sushkov, O.P. Charge ordering and antiferromagnetic exchange in layered molecular crystals of the θ type. *Phys. Rev. B* **2001**, *64*, 085109. [CrossRef]

23. Miyagawa, K.; Kanoda, K.; Kawamoto, A. NMR Studies on Two-Dimensional Molecular Conductors and Superconductors: Mott Transition in κ-(BEDT-TTF)$_2$X. *Chem. Rev.* **2004**, *104*, 5635–5654. [CrossRef] [PubMed]

24. Arai, T.; Ichimura, K.; Nomura, K.; Takasaki, S.; Yamada, J.; Nakatsuji, S.; Anzai, H. Tunneling spectroscopy on the organic superconductor κ-(BEDT-TTF)$_2$Cu(NCS)$_2$ using STM. *Phys. Rev. B* **2001**, *63*, 104518. [CrossRef]

25. Nakazawa, Y.; Kanoda, K. Low-temperature specific heat of κ-(BEDT-TTF)$_2$Cu[N(CN)$_2$]Br in the superconducting state. *Phys. Rev. B* **1997**, *55*, R8670–R8673. [CrossRef]

26. Taylor, O.J.; Carrington, A.; Schlueter, J.A. Specific-Heat Measurements of the Gap Structure of the Organic Superconductors κ-(ET)$_2$Cu[N(CN)$_2$]Br and κ-(ET)$_2$Cu(NCS)$_2$. *Phys. Rev. Lett.* **2007**, *99*, 057001. [CrossRef] [PubMed]

27. Girlando, A.; Masino, M.; Brillante, A.; Valle, R.G.D.; Venuti, E. BEDT-TTF organic superconductors: The role of phonons. *Phys. Rev. B* **2002**, *66*, 100507. [CrossRef]

28. Sorai, M.; Nakazawa, Y.; Nakano, M.; Miyazaki, Y. Calorimetric Investigation of Phase Transitions Occurring in Molecule-Based Magnets. *Chem. Rev.* **2013**, *113*, PR41-122. [CrossRef] [PubMed]

29. Gopal, E.S.R. *Specific Heats at Low Temperature*; Heywood Books: London, UK, 1966.

30. Wosnitza, J. Quasi-Two-Dimensional Organic Superconductors. *J. Low Temp. Phys.* **2007**, *146*, 641–667. [CrossRef]

31. Wosnitza, J. Superconductivity in Layered Organic Metals. *Crystals* **2012**, *2*, 248–265. [CrossRef]

32. Muraoka, Y.; Imajo, S.; Yamashita, S.; Akutsu, H.; Nakazawa, Y. Thermal Anomaly around the Superconductive Transition of κ-(BEDT-TTF)$_2$Cu(NCS)$_2$ with External Pressure and Magnetic Field Control. *J. Therm. Anal. Calorim.* **2016**, *123*, 1891–1897. [CrossRef]

33. Nakazawa, Y.; Yoshimoto, R.; Fukuoka, S.; Yamashita, S. Investigation on Electronic States of Molecule-Based Compounds by High-Pressure AC Calorimetry. *Curr. Inorg. Chem.* **2014**, *4*, 122–134. [CrossRef]

34. Bachmann, R.; DiSalvo, F.J., Jr.; Geballe, T.H.; Greene, R.L.; Howard, R.E.; King, C.N.; Kirsch, H.C.; Lee, K.N.; Schwall, R.E.; Thomas, H.-U.; et al. Heat Capacity Measurements on Small Samples at Low Temperatures. *Rev. Sci. Instrum.* **1972**, *43*, 205–214. [CrossRef]

35. Stewart, G.R. Measurement of low-temperature specific heat. *Rev. Sci. Instrum.* **1983**, *54*, 1–11. [CrossRef]

36. Sorai, M. *Comprehensive Handbook of Calorimetry and Thermal Analysis*; Wiley: New York, NY, USA, 2004.

37. Imajo, S.; Fukuoka, S.; Yamashita, S.; Nakazawa, Y. Construction of Relaxation Calorimetry for 10^{1-2} Micro-gram Samples and Heat Capacity Measurements of Organic Complexes. *J. Therm. Anal. Calorim.* **2016**, *123*, 1871–1876. [CrossRef]

38. Fukuoka, S.; Horie, Y.; Yamashita, S.; Nakazawa, Y. Development of Heat Capacity Measurement System for Single Crystals of Molecule-Based Compounds. *J. Therm. Anal. Calorim.* **2013**, *113*, 1303–1308. [CrossRef]

39. Sullivan, P.F.; Seidel, G. Steady-State, ac-Temperature Calorimetry. *Phys. Rev.* **1968**, *173*, 679–685. [CrossRef]

40. Eichler, A.; Gey, W. Method for the determination of the specific heat of metals at low temperatures under high pressures. *Rev. Sci. Instrum.* **1979**, *50*, 1445–1452. [CrossRef] [PubMed]
41. Kubota, O.; Nakazawa, Y. Construction of a low-temperature thermodynamic measurement system. *Rev. Sci. Instrum.* **2008**, *79*, 053901. [CrossRef] [PubMed]
42. Kubota, O.; Fukuoka, S.; Nakazawa, Y.; Nakata, K.; Yamashita, S.; Miyasaka, H. Thermodynamic Investigation of Coordination-Networked Systems of [Mn$_4$] Single-Molecule Magnets under Pressure. *J. Phys. Condens. Matter* **2010**, *22*, 026007. [CrossRef] [PubMed]
43. Tokoro, N.; Kubota, O.; Yamashita, S.; Kawamoto, A.; Nakazawa, Y. Thermodynamic Study of κ-(BEDT-TTF)$_2$Ag(CN)$_2$H$_2$O under Pressures and with Magnetic Fields. *J. Phys. Conf. Ser.* **2008**, *132*, 012010. [CrossRef]
44. Tokoro, N.; Fukuoka, S.; Kubota, O.; Nakazawa, Y. Low-temperature heat capacity measurements of κ-type organic superconductors under pressure. *Phys. B Condens. Matter* **2010**, *405*, S273–S276. [CrossRef]
45. Danda, M.; Muraoka, Y.; Yamamoto, T.; Nakazawa, Y. High-Pressure AC Calorimetry System Using Pt Chip Thermometer. *Netsu Sokutei* **2012**, *W39*, 29–32.
46. Konoike, T.; Uchida, K.; Osada, T. Specific Heat of the Multilayered Massless Dirac Fermion System. *J. Phys. Soc. Jpn.* **2013**, *81*, 043601. [CrossRef]
47. Naito, A.; Nakazawa, Y.; Saito, K.; Taniguchi, H.; Kanoda, K.; Sorai, M. Anomalous enhancement of electronic heat capacity in the organic conductors κ-(BEDT-TTF)$_4$Hg$_{3-\delta}$X$_8$ (X = Br, Cl). *Phys. Rev. B* **2005**, *71*, 054514. [CrossRef]
48. Yamashita, S.; Naito, A.; Nakazawa, Y.; Saito, K.; Taniguchi, H.; Kanoda, K.; Oguni, M. Drastic cooling rate dependence of thermal anomaly associated with the superconducting transition in κ-(BEDT-TTF)$_4$Hg$_{2.89}$Br$_8$. *J. Therm. Anal. Calorim.* **2005**, *81*, 591–594. [CrossRef]
49. Nakazawa, Y.; Yamashita, S. Thermodynamic Properties of κ-(BEDT-TTF)$_2$X Salts: Electron Correlations and Superconductivity. *Crystals* **2012**, *2*, 741–761. [CrossRef]
50. Yamashita, S.; Ishikawa, T.; Fujisaki, T.; Naito, A.; Nakazawa, Y.; Oguni, M. Thermodynamic behavior of the 10 K class organic superconductor κ-(BEDT-TTF)$_2$Cu(NCS)$_2$ studied by relaxation calorimetry. *Thermochim. Acta* **2005**, *431*, 123–126. [CrossRef]
51. Ishikawa, T.; Nakazawa, Y.; Ymashita, S.; Oguni, M.; Saito, K.; Takimiya, K.; Otsubo, T. Thermodynamic Study of an Incommensurate Organic Superconductor. *(MDT-TSF)*(AuI$_2$)$_{0.436}$. *J. Phys. Soc. Jpn.* **2006**, *75*, 074606. [CrossRef]
52. Katsumoto, S.; Kobayashi, S.; Urayama, H.; Yamochi, H.; Saito, G. Low-Temperature Specific Heat of Organic Superconductor κ-(BEDT-TTF)$_2$Cu(NCS)$_2$. *J. Phys. Soc. Jpn.* **1988**, *57*, 3672–3673. [CrossRef]
53. Andraka, B.; Kim, J.S.; Stewart, G.R.; Calson, K.D.; Wang, H.H.; Williams, J.M. Specific heat in high magnetic field of κ-di[bis(ethylenedithio)tetrathiafulvalene]-di(thiocyano)cuprate [κ-(ET)$_2$Cu(NCS)$_2$]: Evidence for strong-coupling superconductivity. *Phys. Rev. B* **1989**, *40*, 11345–11347. [CrossRef]
54. Graebner, J.E.; Haddon, R.C.; Chichester, S.V.; Glarum, S.H. Specific heat of superconducting κ-(BEDT-TTF)$_2$Cu(NCS)$_2$ near T_c [where BEDT-TTF is bis(ethylenedithio)tetrathiafulvalene]. *Phys. Rev. B* **1990**, *41*, 4808–4810. [CrossRef]
55. Müller, J.; Lang, M.; Helfrich, R.; Steglich, F.; Sasaki, T. High-resolution ac-calorimetry studies of the quasi-two-dimensional organic superconductor κ-(BEDT-TTF)$_2$Cu(NCS)$_2$. *Phys. Rev. B* **2002**, *65*, 140509. [CrossRef]
56. Lortz, R.; Wang, Y.; Demuer, A.; Böttger, P.H.M.; Bergk, B.; Zwicknagl, G.; Nakazawa, Y.; Wosnitza, J. Calorimetric Evidence for a Fulde-Ferrell-Larkin-Ovchinnikov Superconducting State in the Layered Organic Superconductor κ-(BEDT-TTF)$_2$Cu(NCS)$_2$. *Phys. Rev. Lett.* **2007**, *99*, 187002. [CrossRef] [PubMed]
57. Elsinger, H.; Wosnizta, J.; Wanka, S.; Hagel, J.; Schweitzer, D.; Strunz, W. κ-(BEDT-TTF)$_2$Cu[N(CN)$_2$]Br: A Fully Gapped Strong-Coupling Superconductor. *Phys. Rev. Lett.* **2000**, *84*, 6098–6101. [CrossRef] [PubMed]
58. Ishikawa, T.; Yamashita, S.; Nakazawa, Y.; Kawamoto, A.; Oguni, M. Calorimetric study of molecular superconductor κ-(BEDT-TTF)$_2$Ag(CN)$_2$H$_2$O which contains water in the anion layers. *J. Therm. Anal. Calorim.* **2008**, *92*, 435–438. [CrossRef]
59. Wosnitza, J.; Liu, X.; Schweitzer, D.; Keller, H.J. Specific heat of the organic superconductor κ-(BEDT-TTF)$_2$I$_3$. *Phys. Rev. B* **1994**, *50*, 12747–12751. [CrossRef]
60. Imajo, S.; Yamashita, S.; Akutsu, H.; Nakazawa, Y. Quadratic Temperature Dependence of Electronic Heat Capacities in the κ-Type Organic Superconductors. *Int. J. Mod. Phys. B* **2016**, *30*, 1642014. [CrossRef]

61. Nakazawa, Y.; Taniguchi, H.; Kawamoto, A.; Kanoda, K. Electronic specific heat at the boundary region of the metal-insulator transition in the two-dimensional electronic system of $\kappa-$(BEDT-TTF)$_2$Cu[N(CN)$_2$]Br. *Phys. Rev. B* **2000**, *61*, R16295–R16298. [CrossRef]

62. Nakazawa, Y.; Taniguchi, H.; Kawamoto, A.; Kanoda, K. Electronic specific heat of BEDT-TTF-based organic conductors. *Phys. B Condens. Matter* **2000**, *281–282*, 899–900. [CrossRef]

63. Brinkman, W.F.; Rice, T.M. Application of Gutzwiller's Variational Method to the Metal-Insulator Transition. *Phys. Rev. B* **1970**, *2*, 4302–4304. [CrossRef]

64. McWhan, D.B.; Remeika, J.P.; Rice, T.M.; Brinkman, W.F.; Maita, J.P.; Menth, A. Electronic Specific Heat of Metallic Ti-Doped V$_2$O$_3$. *Phys. Rev. Lett.* **1971**, *27*, 941–943. [CrossRef]

65. Tokura, Y.; Taguchi, Y.; Okada, Y.; Fujishima, Y.; Arima, A.; Kumagai, K.; Iye, Y. Filling dependence of electronic properties on the verge of metal–Mott-insulator transition in Sr$_{1-x}$La$_x$TiO$_3$. *Phys. Rev. Lett.* **1993**, *70*, 2126–2129. [CrossRef] [PubMed]

66. Imajo, S.; Kanda, N.; Yamashita, S.; Akutsu, H.; Nakazawa, Y.; Kumagai, H.; Kobayashi, T.; Kawamoto, A. Thermodynamic Evidence of *d*-Wave Superconductivity of the Organic Superconductor λ-(BETS)$_2$GaCl$_4$. *J. Phys. Soc. Jpn.* **2016**, *85*, 043705. [CrossRef]

67. Stewart, G.R.; Williams, J.M.; Wang, H.H.; Hall, L.N.; Perozzo, M.T.; Carlson, K.D. Bulk superconducting specific-heat anomaly in β-di[bis (ethylenedithio) tetrathiafulvalene] diiodoaurate [β-(ET)$_2$AuI$_2$]. *Phys. Rev. B* **1986**, *34*, 6509–6510. [CrossRef]

68. Andres, K.; Schwenk, H.; Veith, H. Peculiarities of Organic Superconductors of the (BEDT-TTF)$_2$X Family. *Phys. B+C* **1986**, *143*, 334–337. [CrossRef]

69. Matsuzaki, T.; Momono, N.; Oda, M.; Ido, M. Electronic specific heat of La$_{2-x}$Sr$_x$CuO4: Pseudogap formation and reduction of the superconducting condensation energy. *J. Phys. Soc. Jpn.* **2004**, *73*, 2232–2238. [CrossRef]

crystals

MDPI

Review

Effects of Carrier Doping on the Transport in the Dirac Electron System α-(BEDT-TTF)$_2$I$_3$ under High Pressure

Naoya Tajima

Department of Physics, Toho University, Miyama 2-2-1, Funabashi-shi, Chiba 274-8510, Japan;
naoya.tajima@sci.toho-u.ac.jp; Tel.: +81-47-472-6990

Received: 31 January 2018; Accepted: 6 March 2018; Published: 8 March 2018

Abstract: A zero-gap state with a Dirac cone type energy dispersion was discovered in an organic conductor α-(BEDT-TTF)$_2$I$_3$ under high hydrostatic pressures. This is the first two-dimensional (2D) zero-gap state discovered in bulk crystals with a layered structure. Moreover, the Dirac cones are highly tilted in a k-space. This system, thus, provides a testing ground for the investigation of physical phenomena in the multilayered, massless Dirac electron system with anisotropic Fermi velocity. Recently, the carrier injection into this system has been succeeded. Thus, the investigations in this system have expanded. The recent developments are remarkable. This effect exhibits peculiar (quantum) transport phenomena characteristic of electrons on the Dirac cone type energy structure.

Keywords: α-(BEDT-TTF)$_2$I$_3$; Dirac electron system; transport phenomena; inter-band effects of the magnetic field; carrier doping; quantum Hall effect

---

1. Introduction

The realization of the graphene opened the physics of the Dirac electron in a solid [1,2]. A rich variety of material with Dirac electrons has been discovered, and the recent progress of the physics for Dirac electrons has been brilliant [3–16]. However, the zero-gap material with the Fermi level at the Dirac point is limited. The physics at the vicinity of the Dirac point is the most significant. Among them, we have discovered the zero-gap material with Dirac-type energy dispersion in α-(BEDT-TTF)$_2$I$_3$ (BEDT-TTF = bis(ethylenedithio) tetrathiafulvalene) [17] (Figure 1) at high pressure. This is the first bulk (multilayered) 2D zero-gap system with Dirac electrons. Thus, this material has led the studies of the specific heat [18] and the nuclear magnetic resonance (NMR) [19] for the Dirac electron system. Another significant feature is that the Dirac cones are highly tilted as shown in Figure 1d,e [13–16]. Thus, this system has led to the peculiar transport characteristic of the electrons on the Dirac cone type energy structure [8–12,16,20–23]. This Dirac electron system, on the other hand, is next to the charge-ordered insulator phase in the temperature-pressure phase diagram. Therefore, strongly correlated Dirac electrons constitute one of the interesting recent studies [19,24,25].

The recent success of carrier injection helped the development of physics with regard to this system. In this review, the effects of carrier doping on the peculiar (quantum) transport phenomena are described. In the following, the electronic structure of α-(BEDT-TTF)$_2$I$_3$ and the experimental evidence of massless Dirac electron systems are briefly mentioned in this section. The methods of the carrier doping in Section 2 and those effects on the transport phenomena in Sections 3–5 are revealed. In Section 6, the Dirac type energy structure of this system is corrected.

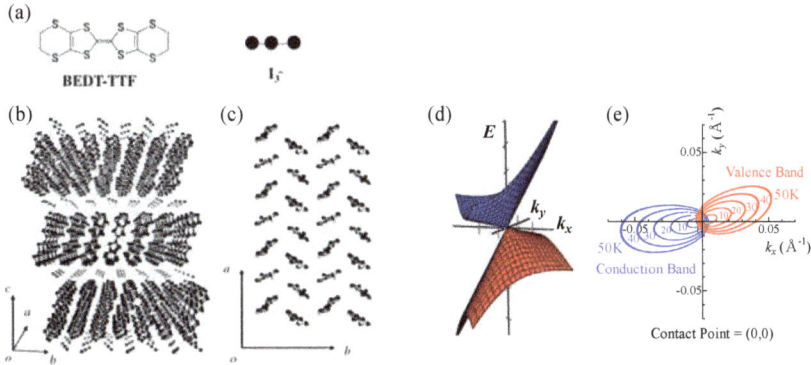

Figure 1. (**a**) BEDT-TTF molecule and $I_3{}^-$ anion, crystal structure of α-(BEDT-TTF)$_2$I$_3$ viewed from (**b**) *a*-axis and (**c**) *c*-axis, (**d**) band structure, and (**e**) energy contours near a Dirac point. Note that we take the origin to the position of the Dirac point.

1.1. Electronic Structure of α-(BEDT-TTF)$_2$I$_3$

α-(BEDT-TTF)$_2$I$_3$ is a member of the (BEDT-TTF)$_2$I$_3$ family [17]. The crystals consist of conductive layers of BEDT-TTF molecules and insulating layers of $I_3{}^-$ anions as shown in Figure 1a–c [26–28]. The difference of the arrangement of BEDT-TTF molecules gives rise to variations in the transport phenomena. Most are 2D metals with large Fermi surfaces, and some of them show a superconducting transition [26–28]. α-(BEDT-TTF)$_2$I$_3$, however, is different from other members. The band calculation indicated a semimetal with small Fermi pockets of the electron character and the hole character [8].

This material behaves as a metal down to 135 K, where it undergoes a phase transition to an insulator [17]. At temperatures below 135 K, an abrupt drop of the magnetic susceptibility suggests that a nonmagnetic state with a spin gap is realized [29]. The theory by Kino and Fukuyama [30] and Seo [31], the NMR study by Takano et al. [32], the Raman study by Wojciechowski et al. [33], and the spectroscopy study by Moldenhauer et al. [34] indicated that the origin of this transition was due to the charge disproportionation. Each BEDT-TTF molecule with approximately $0.5e$ has formed the horizontal charge stripe patterns for $+1e$ and 0 at temperatures below 135 K [33,35]. This phase transition is suppressed by the pressure above 1.5 GPa at room temperature [8,9,16].

The resistivity at high pressure is very peculiar. It is almost constant over the whole temperature range like dirty metals. The carrier mobility in dirty metals should be low, because the impurity scattering dominates the conduction. Thus, the resistance in dirty metals is temperature independent. However, the present situation is different. The large magnetoresistance at low temperatures indicated that the carrier mobility was extremely high. It was estimated to be approximately 10^5 cm/V·s at low temperatures [9,16,35–37]. The high carrier mobility led to the observation that the magnetic field warped the path of the electric currents [8]. So, this system is clean.

This is the motive with which this study has started. To clarify this mechanism, the Hall effect was investigated. Surprisingly, in the region from 300 to 2 K, the carrier (hole) density and the mobility change by approximately six orders of magnitude as shown in Figure 2. At low temperatures, the state of extremely low density of approximately 8×10^{14} cm^{-3} and extremely high mobility of approximately 3×10^5 cm^2/V·s is realized [10,16]. The independent resistance is due to the effects of changes in the density, and the mobility just cancels out [8–10,16].

According to the band calculations by Kobayashi et al. and first-principles band calculations by Kino and Miyazaki, this material under high pressures is in the zero-gap state of which the bottom of the conduction band and the top of the valence band touch each other at two points (we call these "Dirac points") in the first Brillouin zone [13–15]. The Fermi energy is located exactly on the Dirac point.

In the picture of 2D zero-gap energy structure with a linear dispersion, the peculiar transport phenomena were naturally understood. The carrier density proportional to the temperature squared is explained. When the Fermi energy E_F located at the Dirac point is temperature-independent, it is written as $n = \int D(E)f(E)dE \propto T^2$, in which $D(E)$ is the density of state and $f(E)$ is the Fermi distribution function. According to Mott's argument [38], on the other hand, the mean free path l of a carrier can never be shorter than the wavelength λ of the carrier, so $l \geq \lambda$. For the cases of high density of scattering centers, $l \sim \lambda$ ($lk \sim 1$). As the temperature is decreased, l becomes long, because λ becomes long (k becomes small) with the decreasing energy of the carriers. The Boltzmann transport equation gives the temperature-independent quantum conductivity as $\sigma_{xx} = 8e^2 \int v_x^2 \tau(-\partial f/\partial E)dk = 2e^2/h$, in which v_x is the velocity of Dirac electrons when the electric field along x-axis is applied and τ is the lifetime. The constant sheet resistance (resistivity per layer) R_s close to the value of the quantum resistance, $h/e^2 = 25.8$ kΩ, is derived. Many realistic theories for the sheet resistance in the zero-gap system give $R_s = gh/e^2$, in which g is a parameter of order unity [39–41]. Combining the temperature dependences of the carrier density and the resistivity with $\sigma = ne\mu$, on the other hand, the temperature dependence of the carrier mobility $\mu \propto T^{-2}$ was led as shown in Figure 2.

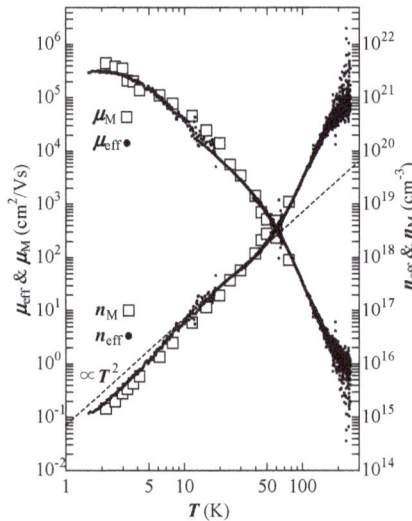

Figure 2. The carrier density and the mobility under the pressure of $p = 1.8$ GPa against the temperature. Close circles show the effective carrier density n_{eff} and the mobility μ_{eff} estimated from the Hall coefficient ($R_H = 1/ne$) and the conductivity ($\mu = \sigma/ne$). The magnetoresistance mobility μ_M and the density n_M, on the other hand, is shown by open square. The carrier density obeys $n \propto T^2$ from 10 K to 50 K (indicated by broken lines). Reproduced with permission from [9].

1.2. Experimental Evidence of Massless Dirac Electron System

We can see the remarkable characteristic transport of 2D Dirac electron systems in the magnetic field normal to the 2D plane. In the 2D massless Dirac electron system, the particles obeyed to the Weyl equation $H = v_F \boldsymbol{\sigma} \cdot \boldsymbol{p}$, in which $\boldsymbol{\sigma}$ is the Pauli matrix and $\boldsymbol{p}$ is the momentum. In the magnetic field, the gauge transformation from $\boldsymbol{p}$ into $\boldsymbol{p} + e\boldsymbol{A}$ derives the energy of Landau levels E_N as

$$E_N = \pm v_F \sqrt{2e\hbar|N||B|} \tag{1}$$

in which A is the vector potential, v_F is the Fermi velocity, N is the Landau index, and B is the magnetic field strength. This energy depends on the square root of B and N, which is different from that of the conventional conductors. At the Dirac point ($E = 0$), $N = 0$, and the Landau level, called the zero-mode, always appears [42].

For $k_B T < E_{\pm 1}$, the system is in the state of the quantum limit so that the zero-mode carriers dominate the conduction. In this situation, the carrier density per spin and per valley is given by $D(B) = B/2\phi_0$, in which $\phi_0 = h/e$ is the quantum flux and the Fermi distribution function at E_F is $1/2$. Strong magnetic fields induced the zero-mode carrier with high density.

Tajima et al. succeeded in detecting this effect in the longitudinal inter-layer magnetoresistance R_{zz} [11]. In this experiment, the Lorentz force is weak, because the electrical current and the magnetic field are parallel to each other. Thus, the effect of the magnetic field only gives rise to the change in the zero-mode carrier density. It leads to the remarkable negative interlayer magnetoresistance in the magnetic field above 0.2 T, as shown in Figure 3 [11]. An analytical formula for R_{zz} by Osada reproduced well the field and the angle dependences of R_{zz} [43].

Since each Landau level is broadened by the scattering of carriers and/or thermal energy, the zero-mode is sure to overlap with the other Landau levels at a low magnetic field. In such a region, the relationship of $R_{zz} \propto B^{-1}$ loses its validity. We can recognize this region in Figure 3, in which a positive magnetoresistance is observed. At $T_p = 4$ K, for example, this critical field B_p is approximately 0.2 T. At the magnetic field above B_p, the overlap between the zero-mode and the $N = \pm 1$ Landau levels E_1 will be sufficiently small so that the negative magnetoresistance is observed there. Thus, $E_1 \sim k_B T_p$ at B_p. The Fermi velocity v_F was estimated to be approximately 4×10^4 m/s [44,45].

Thus, the detection of zero-mode has demonstrated that this material under high pressure was composed of truly massless Dirac electron systems. The recent progress of this system has been remarkable. One example of this is the success of the carrier doping to this system. In the next section, the effects of the carrier doping on the transport phenomena are described.

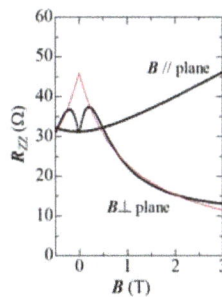

Figure 3. Field dependence of the interlayer magnetoresistance for $p = 1.7$ GPa at 4 K [11]. Remarkable negative magnetoresistance is observed at $B > 0.2$ T. Fitting curve (red line) is an Equation $R_{zz} \propto (|B| + B_0)^{-1}$ in the negative magnetoresistance region, in which B_0 is a fitting parameter that depends on the purity of a crystal [36].

2. Methods of Carrier Doping

In the Dirac electron systems, the Berry phase, which is fundamental concept for the geometry of the Bloch particles, plays an important role in quantum transport. In 2D massless Dirac electron system, the particles obey the Weyl equation $H = v_F \sigma \cdot p$ has π Berry phase. π Berry phase has yielded the new type of quantum Hall effect (half integer quantum Hall effect), which was first discovered in graphene. The success of the carrier doping (control of E_F) by the field effect transistor (FET) method led to this discovery.

Here, the detection of the Berry phase is briefly mentioned. The semiclassical quantization condition for a cyclotron orbit is written as $S_N = 2\pi(N + \gamma)/l_B^2$, in which S_N is the cross-section area of the Nth Landau level, $l_B = \sqrt{\hbar/eB}$ is the magnetic length, and γ $(0 \leq \gamma < 1)$ is the Onsager phase factor that is related to the Berry phase ϕ_B as $\gamma = 1/2 - \phi_B/2\pi$. In a conventional electron system, ϕ_B and γ are expected to be 0 and 1/2. Dirac particles, on the contrary, prefer $\phi_B = \pi$ and $\gamma = 0$. Dirac particles in a solid thus had been identified from the phase analysis of quantum oscillation (Shubnikov-de Haas oscillations: SdH) in a magnetic field.

In order to detect the quantum Hall effect (QHE) characterized by the electrons on the Dirac cones in α-(BEDT-TTF)$_2$I$_3$ under high pressure, the Fermi level should be moved from the Dirac point. However, control of the Fermi level by the field effect transistor (FET) method is much more difficult, because this crystal has a multilayered structure with high conductivity. Moreover, the conductivity in each layer is high.

Cannot we inject carriers to α-(BEDT-TTF)$_2$I$_3$ under high pressure? Cannot we detect the QHE experimentally in this system? The answer is "NO". One of the breakthroughs for the carrier injection to α-(BEDT-TTF)$_2$I$_3$ was suggested. Important results are that the SdH oscillations and the QHE associated with the special Landau level structure of Equation (1) were detected at low temperature [16,46]. In this section, we mention some unique methods of carrier doping in this system.

2.1. Effects of Dopant

Because of the characteristic energy spectrum, slight dopant brought a strong effect on the transport phenomena. The instability of I$_3^-$ anions is the main origin of the dopant. Depending on the dopant, there are two types of samples according to which the electrons or holes were doped. Moreover, sample (dopant) dependence of the resistivity and the Hall coefficient is strong at low temperatures. This characteristic feature provided experimentally an anomalous Hall conductivity caused by the inter-band effects of the magnetic field [12]. In the magnetic field, the vector potential plays an important role in the inter-band excitation of electrons [47]. In this situation, large diamagnetism and the anomalous Hall conductivity is derived by the orbital motion of virtual electron-hole pairs. This is the inter-band effect of the magnetic field. This effect is strongest when the chemical potential is located at the Dirac point. In Section 3, inter-band effects of magnetic field on the transport properties are described.

2.2. Electron Doping by the Annealing

Annealing of the crystals in a vacuum at high temperature gives rise to the lack of I$_3^-$ anions [48,49]. It yields mobile electrons. Annealing time and temperature are the control parameters of the density [50]. Recently, Tisserond et al. succeeded in injecting electrons and observed SdH oscillation at low temperature [51]. Effects of electron doping by the annealing on the transport properties are roughly mentioned in Section 4.

2.3. Hole Doping by Contact Electrification

The carrier density per layer of α-(BEDT-TTF)$_2$I$_3$ under a high pressure at low temperatures is estimated to be approximately 10^8 cm^{-2} [10]. Thus, the effects of hole doping can be detected on the transport phenomena by fixing a crystal onto a substrate weakly that is negatively charged. This is called the contact electrification method. The effects of hole doping on the quantum transport phenomena were detected by fixing a thin crystal onto a poly (ethylene naphthalate) (PEN) substrate (Figure 4) [46]. Positively charged substrate, on the other hand, dopes electrons. Effects of hole doping by the contact electrification on the transport phenomena are described in Section 5.

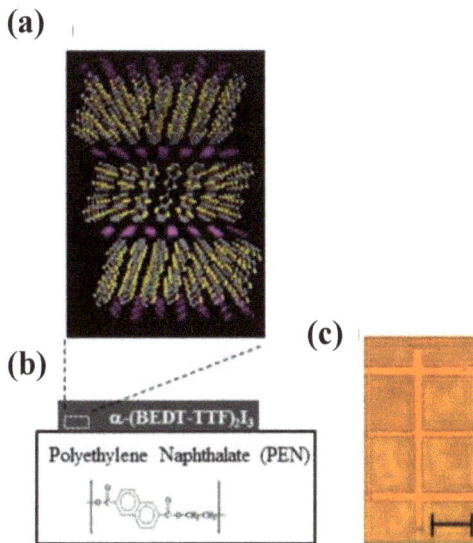

Figure 4. (**a**) Crystal structure of α-(BEDT-TTF)$_2$I$_3$ viewed from the *a*-axis; (**b**) schematic diagram of this system. The thickness of the crystal measured with a step profiler was approximately 100 nm; (**c**) optical image of a single crystal on a PEN substrate in the processed form. The crystal was cut using a pulsed laser beam with a wavelength of 532 nm. The scale bar is approximately 0.2 mm.

2.4. Experiments of the Transport Phenomena in α-(BEDT-TTF)$_2$I$_3$ Under High Pressure

A sample on which six to eight gold wire with a diameter of 15 μm is attached by the carbon paste and is put in a Teflon capsule filled with the pressure medium (DN-oil 7373, Idemitsu, Tokyo, Japan), and then the capsule is set in a clamp-type pressure cell made of hard alloy MP35N cell or BeCu/NiCrAl dual-structure cell. The pressure was examined by change in the resistance of Manganin wire at room temperature. Resistance of a crystal was measured by a conventional dc method with six to eight probes. An electrical current between 0.1 μA and 10 μA was applied in the 2D plane. The magneto transport phenomena were investigated at temperatures from 0.1 K to 300 K in the magnetic field up to 12 T.

3. Effects of Dopant: Inter-Band Effects of Magnetic Field

The sample (dopant) dependence of the resistivity and the Hall coefficient is strong at low temperatures, as shown in Figure 5, because this system has the characteristic energy spectrum. This fact is very important, because the dopant will throw light on the structure and the characteristic properties in the vicinity of Dirac point. In this section, anomalous Hall conductivity that originated from the inter-band effects of the magnetic field at the vicinity of Dirac point is examined [12]. Moreover, electron-hole symmetry of this system is revealed.

As mentioned in Section 1, R_H is proportional to T^2 at temperatures below 50 K. It was explained based on a single Dirac cone type energy structure. In the present system, however, two cones that touched at a Dirac point exclude such a simple situation. In the case that the two Dirac cones are strictly symmetric, the Hall coefficient will be zero, because the signals of the Hall effect due to carriers on the top and bottom of the Dirac cones cancel out. To detect the signal, the energy structure should be asymmetrical. The measurements of R_H should detect the signal that depends on the strength of asymmetry.

Tajima et al. examined the temperature dependence of R_H for seven samples (n1–n7) and found that the samples were classified into two groups. First group (n4 and n6) is that R_H is positive over the whole temperature range, as shown in Figure 5b. In another group (n1, n2, n3, n5, and n7), on the other hand, the polarity of R_H changes at a certain temperatures below 10 K, as shown in Figure 5c. The dopant with a low level of density in the sample leads to this difference. It plays an essential role in the polarity of R_H at low temperatures. The density and the kind of dopants were determined by R_H at the lowest temperature, in which the thermal excitation of carriers between the cones is negligible. The kind of dopant of the first group (n4 and n6) is hole. On the other hand, electron is the character of dopant in another group (n1, n2, n3, n5, and n7). Here, we have a deep interest in the change in the polarity of R_H in the second group (electron doped samples: n1, n2, n3, n5, and n7), as shown in Figure 5c. We find strong sample dependence of the temperature in which the polarity changes. Note that above 10 K, the curve of R_H for all samples is a single. It indicates that doping does not affect the electron energy structure. Thus, the effect of the dopant results in changes in E_F. The change in the polarity of R_H is understood as follows.

Strong temperature dependence of R_H indicates that the symmetry of the present Dirac cones is low. In this situation, the chemical potential μ should vary with the temperature. At the temperature at which $\mu = 0$, the polarity of R_H changes [12,20]. This fact is very important, because the temperature dependence of the transport phenomena provides its chemical potential dependence in the vicinity of the Dirac point. Here, let us detect the inter-band effects of the magnetic field in the Hall conductivity. The inter-band effect is strongest at $\mu = 0$ [47]. In the following, we examine the inter-band effect in the vicinity of $R_H = 0$ ($\mu = 0$).

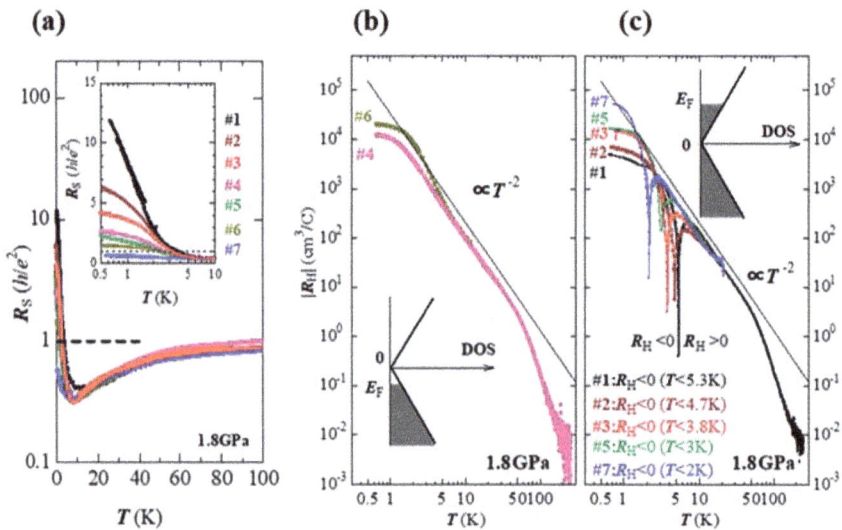

Figure 5. (**a**) Temperature dependence of R_s for seven samples under pressure of 1.8 GPa. R_s at temperature below 10 K is shown in the inset; (**b,c**) is the temperature dependence of R_H for hole-doped-type and electron-doped-type samples. Note that the absolute value of R_H is plotted. Thus, the dips in (**c**) indicate a change in the polarity. The inset of (**b,c**) shows the schematic illustration of the Fermi levels.

First, let us express μ as a function of T. The dopant density n_s was determined from $n_s = 1/R_H e$ at low temperature. Thus, E_F was calculated from the relationship $E_F = \hbar v_F \sqrt{n_s/\pi}$ with $v_F = 3.5 \times 10^4$ m/s. Note that the value of v_F for all samples is almost the same, because sample dependence of R_H is very weak at temperatures of above 7 K. For example, E_F/k_B for samples n1 and n7 are estimated to be 1.35 and 0.35 K, respectively. Kobayashi et al. theoretically demonstrated that μ is varied with the temperature as approximately $\mu/k_B = E_F/k_B = E_F/k_B - AT$ at low temperatures, in which a parameter A depends on the symmetry of the Dirac cone (v_F^h/v_F^e: v_F^h and v_F^e are the Fermi velocities for lower and upper Dirac cones) and is independent of E_F. At $R_H(T) = 0$, $\mu/k_B = 0$. Thus, $A = E_F/k_B T_0$ is estimated to be approximately 0.24 at $R_H(T = 0) = 0$. This experimental formula is consistent with the theoretical curve well [20]. Hence, v_F^h/v_F^e is estimated to be approximately 1.2 [12]. This is the electron-hole symmetry of this system. The detail for asymmetric Dirac cones in this system is described in Section 6.

The calculation of the Hall conductivity $\sigma_{xy} = \rho_{yx}/(\rho_{xx}^2 + \rho_{yx}^2)$ is the second step. The temperature dependence of σ_{xy} for samples n1 and n7 is shown in Figure 6a. In this calculation, it was assumed as $\rho_{xx} = \rho_{yy}$. The last step is that σ_{xy} is drawn in Figure 6b as a function for μ by replacing T by $T(\mu)$. We should compare this with the theoretical curve $\sigma_{xy}^{\text{theory}}$. Experimental data roughly reproduce the relation $\sigma_{xy} \sim g\sigma_{xy}^{\text{theory}}$, in which g is a parameter that depends on the temperature. It is significant that there is a peak and a dip structure in each curve at the vicinity of the point at which $\sigma_{xy} = 0$. In the magnetic field, the orbital motion of virtual electron-hole pairs by the vector potential plays an important role in σ_{xy}. The peak structure of σ_{xy} shown in Figure 6 is the characteristic feature due to the inter-band effects of the magnetic field [20]. The energy between two peaks is the damping, which depends on the density of scattering centers in the crystal. The intensity of the peak, on the other hand, depends on the damping and the tilt of the Dirac cones [20]. Note that σ_{xy} without the inter-band effects, on the other hand, has no peak structure.

Another important finding is a smooth change in the polarity of σ_{xy}. This is evidence that this material has an intrinsic zero-gap energy structure.

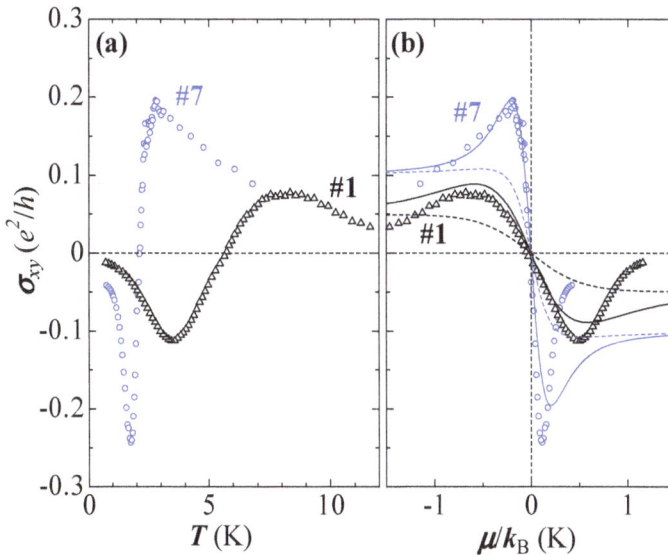

Figure 6. (a) Temperature dependence of the Hall conductivity for n1 and n7 in Figure 5; (b) chemical-potential dependence of the Hall conductivity for n1 and n7. Solid lines and dashed lines are the theoretical curves with and without the inter-band effects of the magnetic field by Kobayashi et al., respectively [20]. Reproduced with permission from [11].

4. Electron Doping by the Annealing

As mentioned in the previous section, the dopant with the density of ppm order gave rise to the intense effects on the transport phenomena. The origin of the dopant is the instability of I_3^- anions in a crystal. In a sense, this is the natural doping. Annealing in a vacuum at high temperature, on the other hand, enhances the lack of I_3^- anions so that the mobile electrons are yielded. In this section, the effects of annealing on the transport phenomena in α-(BEDT-TTF)$_2$I$_3$ under the pressure are mentioned.

The density of the doped electron by the annealing depends on the parameters of its time and temperature. The annealing at temperature above 80 °C, however, changes the crystal structure from α- to β-types [48]. Thus, Miura et al. investigated the resistivity of a crystal with a parameter of the time with 10 min steps of annealing in a vacuum of 10^{-3} Pa at 70 °C [50]. The effects of anneal (doping) were clearly seen in the value of resistivity and its temperature dependence as shown in Figure 7a. The effects of annealing lead this crystal to a metal. At 4.2 K, for example, the resistivity is decreased by about two orders of magnitude in duration of 40 min annealing. It is expected that electrons were doped to this system.

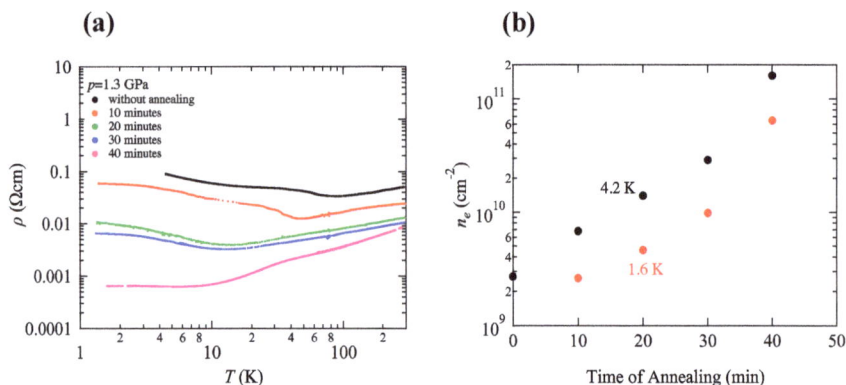

(a)

(b)

Figure 7. (a) Effects of the annealing on the resistivity in α-(BEDT-TTF)$_2$I$_3$ under pressure of 1.3 GPa; (b) annealing time dependence of electron density at 4.2 K and 1.6 K. Reproduced with permission from [50,51].

Here we note that this experiment was performed by a single crystal. First step was to investigate the resistivity of α-(BEDT-TTF)$_2$I$_3$ under the pressure of 1.3 GPa. As the second step, the crystal was annealed in the duration of 10 min in a vacuum about 10^{-3} Pa at 70 °C after pressure was removed. Then, the resistivity of this crystal under the pressure of 1.3 GPa was investigated again. For the after steps, the doping of electrons by the annealing was repeated in the same way.

The polarity of the Hall coefficient indicated the electron doping successes. At low temperature, the variation in the electron density estimated from the Hall coefficient is almost same as that in the resistivity. It is increased by about two orders of magnitude in a 40 min of anneal at 4.2 K as shown in Figure 7b.

Here, let us return to Figure 7a. The resistivity per layer in the case without annealing expressed as $R_s \sim h/e^2$ in a wide temperature region is the characteristic transport in the massless Dirac electron system in which E_F is located close to Dirac point as mentioned in Section 1. However, the system with E_F far from the Dirac point lacks validity of this law. The resistivity of the case with annealed duration of 40 min shows the metallic behavior.

Figure 8 is the recent highlight. Tisserond et al. succeeded in detecting SdH oscillation in a thick crystal under pressure of 2.2 GPa at 200 mK. [51]. Note that it is doped as a side effect of the elaboration of the gold electrical contacts deposited by Joule evaporation, with an unintentional annealing. Careful analysis of SdH oscillations, however, conduced the correction of the Dirac cones in this system. In Section 6, this will be disclosed.

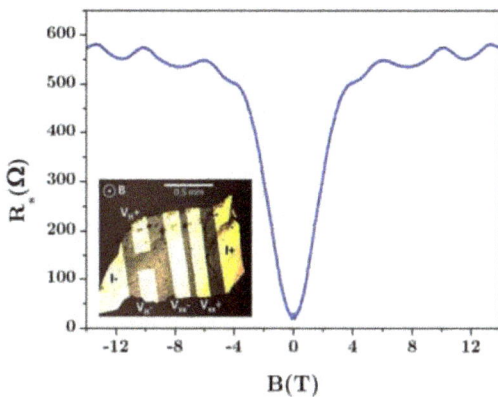

Figure 8. Field dependence of the magnetoresistance in α-(BEDT-TTF)$_2$I$_3$ under pressure of 2.2 GPa at 200 mK. Inset shows the photograph of sample. Reproduced with permission from [51].

5. Hole Doping by Contact Electrification

By fixing a thin crystal onto a substrate weakly negatively charged as shown in Figure 4, the effects of carrier doping by contact electrification can be detected in the transport. The polarity of Hall resistance R_{xy} indicates that the hole doping is successful. Note that holes should be injected into a few layers (pairs of BEDT-TTF molecular layers and I$_3^-$ anion layers). Main is the first from the interface.

5.1. Resistivity of α-(BEDT-TTF)$_2$I$_3$ on PEN Substrate under High Pressure

Figure 9 shows the temperature dependence of the resistivity of α-(BEDT-TTF)$_2$I$_3$ on PEN substrate under pressure of 1.7 GPa. We see clear effects of the hole-doping in the resistivity. First, the value of resistivity of the thin crystal on the PEN substrate is lower than those of the usual thick crystal. Most noticeable difference is seen at temperatures below 2 K. The resistivity for the thin crystal on the PEN substrate behaves as $\rho \propto T^2$. This is the characteristic transport in the Fermi liquid state.

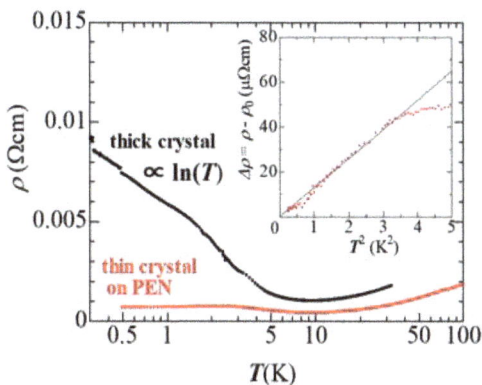

Figure 9. Temperature dependence of resistivity of α-(BEDT-TTF)$_2$I$_3$ on PEN substrate under pressure of 1.7 GPa. Inset shows $\Delta\rho = \rho - \rho_0$ against T^2, in which ρ_0 is the resistivity at a limit of $T = 0$.

Thus, carrier injection into α-(BEDT-TTF)$_2$I$_3$ with a layered structure was successful. Note that the resistivity depends on the crystal thickness, because the number of carrier-doped layers is small, as mentioned before.

5.2. Observations of SdH Oscillations and Quantized Hall Resistance

The signature of the Dirac electrons is seen in the quantum transport. The magnetic field dependence of resistance R_{xx} and the Hall resistance R_{xy} in the thin crystal on the PEN substrate under pressure of approximately 1.7 GPa were investigated at 0.5 K. We find clear oscillation of R_{xx} as shown in Figure 10a. The oscillation as a function of B^{-1} indicates the SdH signal with a frequency of $B_f \sim 9.18$ T. The second-order differential of R_{xx} clearly depicts the oscillation in Figure 10b.

Here, we regard the origin of the SdH oscillation is 2D massless Dirac electrons. In the Dirac electron systems, the circular orbit around the Dirac point in the magnetic field would yield Berry phase π, as mentioned before. The effect of the phase of the SdH oscillation is further probed in the semi-classical magneto-oscillation description. In general, the component of the SdH oscillation written by

$$\Delta R_{xx} = A(B) \cos\left[2\pi\left(B_f/B + \gamma\right)\right] \tag{2}$$

acquires the phase factor $\gamma = 1/2$ or 0 for normal electrons with $\phi_B = 0$ and Dirac particles with $\phi_B = \pi$. Here, $A(B)$ is the amplitude of the oscillation. We obtain phase factor γ by plotting the values of B^{-1} at the oscillation minima of ΔR_{xx} as a function of their number, as shown in Figure 10c. The linear extrapolated values of approximately $1/2$ at $B^{-1} = 0$ determine the phase factors γ, which are approximately 0. Thus, we identify the Dirac particles.

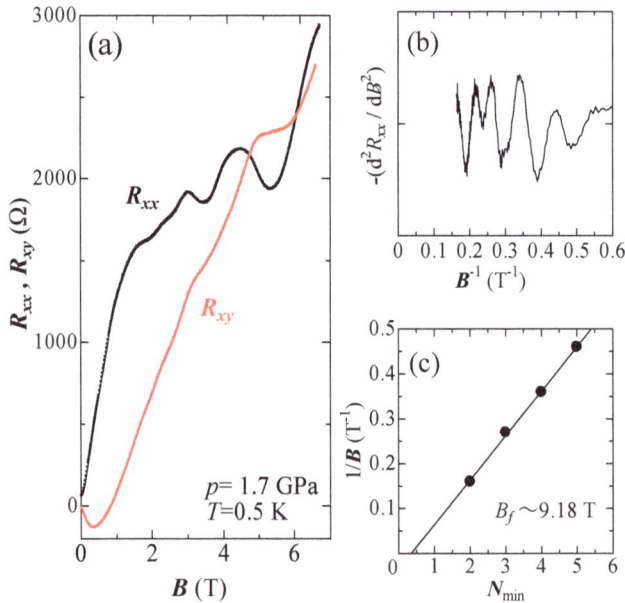

Figure 10. (a) Magnetic field dependence of R_{xx}, R_{xy}, and (b) SdH oscillation $(-d^2R_{xx}/dB^2)$ under the pressure of 1.7 GPa at 0.5 K; (c) value of B^{-1} for the SdH oscillation minima. In the Dirac electron systems, the linear extrapolation of the data to $B^{-1} = 0$ should be $1/2$. For normal electrons, on the other hand, it is shifted to 0.

Here, we notice the split of the SdH oscillation at $B^{-1} \sim 0.25$ T^{-1} in Figure 10b. The SdH oscillation represents the density of states of the Landau levels. Thus, it indicates the Zeeman splitting of $N = -2$ Landau level. The Landau level structure including its Zeeman splitting is written as

$$E_N = \pm v_F \sqrt{2e\hbar|N||B|} \pm \frac{g\mu_B B}{2} \tag{3}$$

in which g is the g factor and μ_B is the Bohr magneton. Combined this relation of $N = 2$ and $g = 2$ with $E_2/v_F = \hbar k = \hbar\sqrt{4\pi B_f/\phi_0} \sim 1.7 \times 1.0^{-26}$ Jm^{-1}s, the Fermi velocity is estimated to be $v_F \sim 4.3$ m/s, in which $g = 2$ and $\phi_0 = 4.14 \times 10^{-15}$ Tm^{-2} is the quantum flux.

The most significant finding is the quantum Hall state in this system. Two obvious R_{xy} plateaus are observed at magnetic fields of approximately 3.5 and 5.5 T, which show R_{xx} minima (Figure 10a). Based on the conventional 2D Dirac electron systems, the R_{xy} plateau is interpreted as follows.

R_{xy} quantization is in accordance with $1/R_{xy} = v \cdot e^2/h$, in which $v = \pm s(n + 1/2)$ is the quantized filling factor, and $s = 4$ is the fourfold spin/valley degeneracy. An outstanding effect on the Dirac electron system is that the factor of half-integer exists. Thus, probes of the quantum Hall plateaux for $|v| = 2,\ 6,\ 10,\ 14,\ \cdots$ are expected. In the data in Figure 10a, based on this step rule, $v = -6, -10, -14$ for the first layer at R_{xy} plateau, or anomalies are required from SdH oscillations against the Landau index shown in Figure 11.

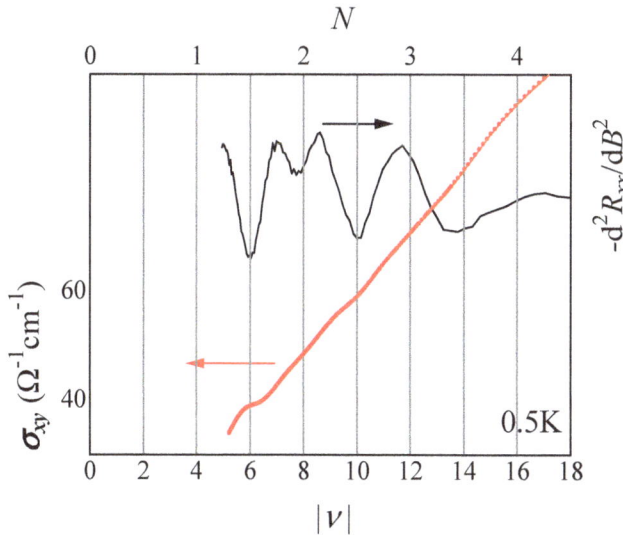

Figure 11. Landau index N or filling factor for first layer v dependence of $-d^2 R_{xx}/dB^2$ and σ_{xy}. Reproduced with permission from [45].

The estimation of v from the values of R_{xy} plateau in this system, however, is a serious problem. The multilayered structure with a few hole-doped layers (mostly single layer) gives rise to the lack of the validity of the estimation of v from the values of R_{xy} plateau because it depends on the thickness. On the other hand, the conductivities for the many undoped layers are finite. Hence, R_{xx} is not zero but shows minima at R_{xy} plateaux. At magnetic field, for example, we can see this effect on R_{xy}. In this magnetic field region, the polarity of R_{xy} is negative and yet holes are injected, as shown in Figure 10a. This is a frequently observed behavior of the present material induced by electron-type dopant (Section 3). However, detailed examination of the thickness (number of layer) dependence of the Hall conductivity revealed ν. For example, $\nu \sim -6$ at R_{xy} plateau of 5.5 T in Figure 10a [46]. This is expected ν shown in Figure 11.

6. Correction of Dirac Cones

In Section 5, the detection of SdH oscillations, whose phase was modified by π Berry phase, is evidence that this system is a 2D Dirac electron system. This measurement was done in the magnetic field below 7 T at 0.5 K. However, recent examination of the SdH oscillations at the magnetic field up to 12 T by Tisserond et al. suggested the correction of the Dirac cones [51]. In this section, the SdH signals in the high magnetic field are interpreted within the model of distorted Dirac cones.

The detections of the SdH oscillations were done by two carrier doping methods: anneal (Sample A: Figure 8) and contact electrification (Sample B). Both the oscillations with the $1/B$ periodicities and the phase factors of the Berry phase π indicate that the Dirac carriers are involved at low magnetic field. At high magnetic field, however, both the SdH signals are very peculiar. Both oscillations lost $1/B$ oscillations periodicity as shown in Figure 12 [51]. The detail structure of Dirac cones shows up the non-periodic SdH oscillations of the high magnetic field [52].

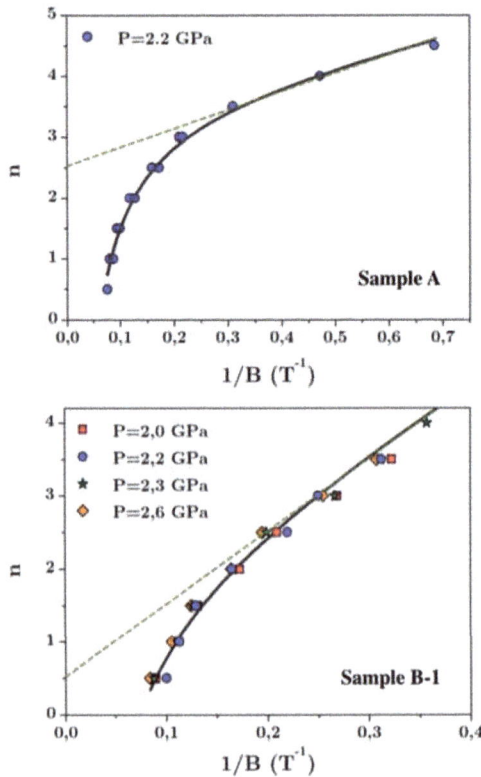

Figure 12. Construction of Landau plots from the analysis of the SdH oscillations. At low magnetic fields, the oscillations are $1/B$ periodic; they are SdH oscillations. The determination of their phase offset, connected to the Berry phase, indicates that the Dirac charge carriers are involved in the measured oscillations. At higher magnetic fields, the $1/B$ oscillations periodicity loss. Reproduced with permission from [51].

Tisserond et al. interpreted the SdH signals shown in Figure 12 based on the picture of distorted Dirac cones, of which the Hamiltonian is written as $H = v_F \sigma \cdot p + p^2/2m$, in which mass m is the curvature term. Figure 13 is the sketch of distorted Dirac cones. In the magnetic field, by the gauge transformation from p into $p + eA$, the Landau level structure without Zeeman effect term is calculated as

$$E_N = \hbar\omega_m N \pm v_F \sqrt{2e\hbar v_F^2 |N||B| + \left(\frac{\hbar\omega_m}{2}\right)^2} \tag{4}$$

in which $\omega_m = eB/m$ and $|N| \geq 1$. The positive part of this equation corresponds to the conduction band contribution and the negative part to that of the valence band. The curves (solid lines) of Landau index N against B^{-1} reproduced well the experimental curve as shown in Figure 13. The best fit mass parameter for Sample A is $|m| \sim 0.03m_0$. For Sample B, on the other hand, it is to be $|m| \sim 0.022m_0$. Note that others possible causes, such as a cone tilting or a Zeeman effect, are carefully ruled out in the Reference [46] and its Supplementary Materials.

The proposed interpretation corrects the band structure from (a) massless Dirac cone to distorted Dirac cone with (b) $m > 0$ or (c) $m < 0$ shown in Figure 13. This electron-hole asymmetry explains the temperature dependence of the chemical potential mentioned in Section 3 [12,20]. It suggests distorted Dirac cone shown in Figure 13b. Moreover, recent realistic calculations of the energy band and Landau level structures in this system by Kishigi and Hasegawa support our experimental results and those interpreted qualitatively [53].

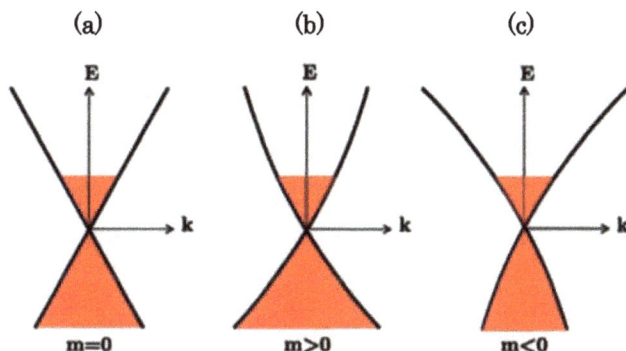

Figure 13. Distortion of the Dirac cone by a curvature term with a mass m. (**a**) massless Dirac cone; (**b**) distorted Dirac cone ($m > 0$); (**c**) distorted Dirac cone ($m < 0$). Reproduced with permission from [51].

7. Summary

We summarize this review as follows.

The effects of carrier doping on the peculiar (quantum) transport phenomena in α-(BEDT-TTF)$_2$I$_3$ under high pressure were described. We presented three unique methods for the carrier doping of this system. First, unstable I$_3^-$ anions yield the dopant in the crystal. This effect led to the detection of anomalous Hall conductivity that originated from the inter-band effects of the magnetic field; second, subjects that were annealed in a vacuum at high temperature enhanced the lack of I$_3^-$ anions so that the mobile electrons were yielded. The effects of annealing on the transport phenomena in α-(BEDT-TTF)$_2$I$_3$ under pressure were mentioned. Tisserond et al. succeeded in detecting SdH oscillation in a thick crystal under pressure of 2.2 GPa at 200 mK [51]. Lastly, we presented the effects of the carrier doping by the contact electrification on the transport phenomena. Only by fixing a thin crystal onto a substrate weakly negatively charged by contact electrification, the effects of carrier doping can be detected in the transport. A significant finding is the SdH oscillation and the quantum

Hall effect characterized by the Dirac type energy structure. Moreover, the distortion of the band structure Dirac cones was revealed.

The effects of carrier doping became more important for the organic Dirac electron system. However, control of the Fermi energy has not been achieved yet. Control of the Fermi energy and further investigations will lead us to interesting phenomena.

Acknowledgments: We thank M. Monteverde, Miura, T. Yamauchi, T. Yamaguchi, Y. Kawasugi, M. Suda, H. M. Yamamoto, R. Kato, Y. Nishio, K. Kajita, T. Morinari, A. Kobayashi, Y. Suzumura, and H. Fukuyama for valuable discussions. This work was supported by a Grant-in-Aid for Scientific Research (A) (No. 15H02108), (S) (No. 16H06346), (B) (No. 25287089), from the Ministry of Education, Culture, Sports, Science, and Technology, Japan.

Conflicts of Interest: The author declares no conflict of interest.

References

1. Novoselov, K.S.; Geim, A.K.; Morozov, S.V.; Jiang, D.; Katsnelson, M.I.; Grigorieva, I.V.; Dubonos, S.V.; Firsov, A.A. Two-dimensional gas of massless Dirac fermions in grapheme. *Nature* **2005**, *438*, 197–200. [CrossRef] [PubMed]

2. Zhang, Y.; Tan, Y.W.; Stormer, H.; Kim, P. Experimental observation of the quantum Hall effect and Berry's phase in grapheme. *Nature* **2005**, *438*, 201–204. [CrossRef] [PubMed]

3. Kempa, H.; Esquinazi, P.; Kopelevich, Y. Field-induced metal-insulator transition in the *c*-axis resistivity of graphite. *Phys. Rev. B* **2002**, *65*, 241101. [CrossRef]

4. Luk'yanchuk, I.A.; Kopelevich, Y. Phase Analysis of Quantum Oscillations in Graphite. *Phys. Rev. Lett.* **2004**, *93*, 166402. [CrossRef] [PubMed]

5. Fuseya, Y.; Ogata, M.; Fukuyama, H. Interband Contributions from the Magnetic Field on Hall Effects for Dirac Electrons in Bismuth. *Phys. Rev. Lett.* **2009**, *102*, 066601. [CrossRef] [PubMed]

6. Richard, P.; Nakayama, K.; Sato, T.; Neupane, M.; Xu, Y.-M.; Bowen, J.H.; Chen, G.F.; Luo, J.L.; Wang, N.L.; Dai, X.; et al. Observation of Dirac Cone Electronic Dispersion in $BaFe_2As_2$. *Phys. Rev. Lett.* **2010**, *104*, 137001. [CrossRef] [PubMed]

7. Zhang, H.; Liu, C.-X.; Qi, X.-L.; Dai, X.; Fang, Z.; Zhang, S.-C. Topological insulators in Bi_2Se_3, Bi_2Te_3 and Sb_2Te_3 with a single Dirac cone on the surface. *Nat. Phys.* **2009**, *5*, 438–442. [CrossRef]

8. Tajima, N.; Tamura, M.; Nishio, Y.; Kajita, K.; Iye, Y. Transport property of an organic conductor α-(BEDT-TTF)$_2$I$_3$ under high pressure: Discovery of a novel type of conductor. *J. Phys. Soc. Jpn.* **2000**, *69*, 543–551. [CrossRef]

9. Tajima, N.; Sugawara, S.; Tamura, M.; Kato, R.; Nishio, Y.; Kajita, K. Electronic phases in an organic conductor α-(BEDT-TTF)$_2$I$_3$: Ultra narrow gap semiconductor, superconductor, metal, and charge-ordered insulator. *J. Phys. Soc. Jpn.* **2006**, *75*, 051010. [CrossRef]

10. Tajima, N.; Sugawara, S.; Tamura, M.; Kato, R.; Nishio, Y.; Kajita, K. Transport properties of massless Dirac fermions in an organic conductor α-(BEDT-TTF)$_2$I$_3$ under pressure. *Europhys. Lett.* **2007**, *80*, 47002. [CrossRef]

11. Tajima, N.; Sugawara, S.; Kato, R.; Nishio, Y.; Kajita, K. Effect of the zero-mode Landau level on interlayer magnetoresistance in multilayer massless dirac fermion systems. *Phys. Rev. Lett.* **2009**, *102*, 176403. [CrossRef] [PubMed]

12. Tajima, N.; Kato, R.; Sugawara, S.; Nishio, Y.; Kajita, K. Inter-band effects of magnetic field on Hall conductivity in the multilayered massless Dirac fermion system α-(BEDT-TTF)$_2$I$_3$. *Phys. Rev. B* **2012**, *85*, 033401. [CrossRef]

13. Kobayashi, A.; Katayama, S.; Noguchi, K.; Suzumura, Y. Superconductivity in Charge Ordered Organic Conductor –α-(ET)$_2$I$_3$ Salt–. *J. Phys. Soc. Jpn.* **2004**, *73*, 3135–3148. [CrossRef]

14. Katayama, S.; Kobayashi, A.; Suzumura, Y. Pressure-induced zero-gap semiconducting state in organic conductor α-(BEDT-TTF)$_2$I$_3$ Salt. *J. Phys. Soc. Jpn.* **2006**, *75*, 054705. [CrossRef]

15. Kino, H.; Miyazaki, T. First-principles study of electronic structure in α-(BEDT-TTF)$_2$I$_3$ at ambient pressure and with uniaxial strain. *J. Phys. Soc. Jpn.* **2006**, *75*, 034704. [CrossRef]

16. Kajita, K.; Nishio, Y.; Tajima, N.; Suzumura, Y.; Kobayashi, A. Molecular Dirac Fermion Systems—Theoretical and Experimental Approaches—. *J. Phys. Soc. Jpn.* **2014**, *83*, 072002. [CrossRef]

17. Bender, K.; Hennig, I.; Schweitzer, D.; Dietz, K.; Endres, H.; Keller, H.J. Synthesis, structure and physical properties of a two-dimensional organic metal, di[bis(ethylenedithiolo)tetrathiofulva lene] triiodide, (BEDT-TTF)$^+_2$I$^-_3$. *Mol. Cryst. Liq. Cryst.* **1984**, *108*, 359–371. [CrossRef]

18. Konoike, T.; Uchida, T.; Osada, T. Specific heat of the multilayered massless Dirac fermion system. *J. Phys. Soc. Jpn.* **2012**, *81*, 043601. [CrossRef]

19. Hirata, M.; Ishikawa, K.; Miyagawa, K.; Tamura, M.; Berthier, C.; Basko, D.; Kobayashi, A.; Matsuno, G.; Kanoda, K. Observation of an anisotropic Dirac cone reshaping and ferrimagnetic spin polarization in an organic conductor. *Nat. Commun.* **2016**, *7*, 12666. [CrossRef] [PubMed]

20. Kobayashi, A.; Suzumura, Y.; Fukuyama, H. Hall Effect and Orbital Diamagnetism in Zerogap State of Molecular Conductor α-(BEDT-TTF)$_2$I$_3$. *J. Phys. Soc. Jpn.* **2008**, *77*, 064718. [CrossRef]

21. Kobayashi, A.; Suzumura, Y.; Fukuyama, H.; Goerbig, O. Tilted-cone-induced easy-plane pseudo-spin ferromagnet and Kosterlitz–Thouless transition in massless Dirac fermions. *J. Phys. Soc. Jpn.* **2009**, *78*, 114711. [CrossRef]

22. Proskrin, I.; Ogata, M.; Suzumura, Y. Longitudinal conductivity of massless fermions with tilted Dirac cone in magnetic field. *Phys. Rev. B* **2015**, *91*, 195413. [CrossRef]

23. Suzumura, Y.; Proskrin, I.; Ogata, M. Effect of tilting on the in-plane conductivity of Dirac electrons in organic conductor. *J. Phys. Soc. Jpn.* **2014**, *83*, 023701. [CrossRef]

24. Liu, D.; Ishikawa, K.; Takehara, R.; Miyagawa, K.; Tamura, M.; Kanoda, K. Insulating Nature of Strongly Correlated Massless Dirac Fermions in an Organic Crystal. *Phys. Rev. Lett.* **2016**, *116*, 226401. [CrossRef] [PubMed]

25. Hirara, M.; Ishikawa, K.; Matsuno, G.; Kobayashi, A.; Miyagawa, K.; Tamura, M.; Berthier, C.; Kanoda, K. Anomalous spin correlations and excitonic instability of interacting 2D Weyl fermions. *Science* **2017**, *358*, 1403–1406. [CrossRef] [PubMed]

26. Shibaeva, R.P.; Kaminskii, V.F.; Yagubskii, E.B. Crystal structures of organic metals and superconductors of (BEDT-TTP)-I system. *Mol. Cryst. Liq. Cryst.* **1985**, *119*, 361–373. [CrossRef]

27. Kobayashi, H.; Kato, R.; Kobayashi, A.; Nishio, Y.; Kajita, K.; Sasaki, W. Crystal and electronic structures of layered molecular superconductor, θ-(BEDT-TTF)$_2$(I$_3$)$_{1−x}$(AuI$_2$)$_x$. *Chem. Lett.* **1986**, *15*, 833–836. [CrossRef]

28. Kobayashi, H.; Kato, R.; Kobayashi, A.; Nishio, Y.; Kajita, K.; Sasaki, W. A new molecular superconductor, (BEDT-TTF)$_2$(I$_3$)$_{1−x}$(AuI$_2$)$_x$ (x < 0.02). *Chem. Lett.* **1986**, *15*, 789–792.

29. Rothaemel, B.; Forro, L.; Cooper, J.R.; Schilling, J.S.; Weger, M.; Bele, P.; Brunner, H.; Schweitzer, D.; Keller, H.J. Magnetic susceptibility of α and β phases of di[bis(ethylenediothiolo) tetrathiafulvalene] tri-iodide [(BEDT-TTF)$_2$I$_3$] under pressure. *Phys. Rev. B* **1986**, *34*, 704–712. [CrossRef]

30. Kino, H.; Fukuyama, H. On the phase transition of α-(ET)$_2$I$_3$. *J. Phys. Soc. Jpn.* **1995**, *64*, 1877–1880. [CrossRef]

31. Seo, H. Charge ordering in organic ET compounds. *J. Phys. Soc. Jpn.* **2000**, *69*, 805–820. [CrossRef]

32. Takano, Y.; Hiraki, K.; Yamamoto, H.M.; Nakamura, T.; Takahashi, T. Charge disproportionation in the organic conductor, α-(BEDT-TTF)$_2$I$_3$. *J. Phys. Chem. Solids* **2001**, *62*, 393–395. [CrossRef]

33. Wojciechowski, R.; Yamamoto, K.; Yakushi, K.; Inokuchi, M.; Kawamoto, A. High-pressure Raman study of the charge ordering in α-(BEDT-TTF)$_2$I$_3$. *Phys. Rev. B* **2003**, *67*, 224105. [CrossRef]

34. Moldenhauer, J.; Horn, C.H.; Pokhodnia, K.I.; Schweitzer, D.; Heinen, I.; Keller, H.J. FT-IR absorption spectroscopy of BEDT-TTF radical salts: Charge transfer and donor-anion interaction. *Synth. Met.* **1993**, *60*, 31–38. [CrossRef]

35. Kakiuchi1, T.; Wakabayashi, Y.; Sawa, H.; Takahashi, T.; Nakamura, T. Charge ordering in α-(BEDT-TTF)$_2$I$_3$ by synchrotron X-ray diffraction. *J. Phys. Soc. Jpn.* **2007**, *76*, 113702. [CrossRef]

36. Ojiro, T.; Kajita, K.; Nishio, Y.; Kobayashi, H.; Kobayashi, A.; Kato, R.; Iye, Y. A new magneto-pressure phase in α-(BEDT-TTF)$_2$I$_3$. *Synth. Met.* **1993**, *56*, 2268–2273. [CrossRef]

37. Kajita, K.; Ojiro, T.; Fujii, H.; Nishio, Y.; Kobayashi, H.; Kobayashi, A.; Kato, R. Magnetotransport Phenomena of α-Type (BEDT-TTF)$_2$I$_3$ under High Pressures. *J. Phys. Soc. Jpn.* **1993**, *61*, 23–26. [CrossRef]

38. Mott, N.F.; Davis, E.A. *Elecron Processes in Non-Crystalline Materials*; Clarendon: Oxford, UK, 1979.

39. Shon, N.H.; Ando, T. Quantum transport in Two-dimensional graphite system. *J. Phys. Soc. Jpn.* **1998**, *67*, 2421–2429. [CrossRef]

40. Ziegler, K. Delocalization of 2D dirac fermions: The role of a broken supersymmetry. *Phys. Rev. Lett.* **1998**, *80*, 3113–3116. [CrossRef]

41. Nomura, K.; MacDonald, A.H. Quantum transport of massless dirac fermions. *Phys. Rev. Lett.* **2007**, *98*, 076602. [CrossRef] [PubMed]

42. Ando, T. Theory of electronic states and transport in carbon nanotubes. *J. Phys. Soc. Jpn.* **2005**, *74*, 777–817. [CrossRef]

43. Osada, T. Negative interlayer magnetoresistance and Zero-Mode landau level in multilayer Dirac electron systems. *J. Phys. Soc. Jpn.* **2008**, *77*, 084711. [CrossRef]

44. Sugawara, S.; Tamura, M.; Tajima, N.; Sato, M.; Nishio, Y.; Kajita, K.; Kato, R. Temperature dependence of inter-layer longitudinal magnetoresistance in α-(BEDT-TTF)$_2$I$_3$: Positive versus negative contributions in a tilted Dirac cone system. *J. Phys. Soc. Jpn.* **2010**, *79*, 113704. [CrossRef]

45. Tajima, N.; Sato, M.; Sugawara, S.; Kato, R.; Nishio, Y.; Kajita, K. Spin and valley splittings in multilayered massless Dirac fermion system. *Phys. Rev. B* **2010**, *82*, 121420. [CrossRef]

46. Tajima, N.; Yamauchi, T.; Yamaguchi, T.; Suda, M.; Kawasugi, Y.; Yamamoto, H.M.; Kato, R.; Nishio, Y.; Kajita, K. Quantum Hall effect in multilayered massless Dirac fermion systems with tilted cones. *Phys. Rev. B* **2013**, *88*, 075315. [CrossRef]

47. Fukuyama, H. Anomalous orbital magnetism and hall effect of massless fermions in two dimension. *J. Phys. Soc. Jpn.* **2007**, *76*, 043711. [CrossRef]

48. Pokhodnia, K.I.; Graja, A.; Weger, M.; Schweitzer, D. Resonant Raman scattering from superconducting single crystals of (BEDT-TTF)$_2$I$_3$. *Z. Phys. B* **1993**, *90*, 127–133. [CrossRef]

49. Yoshimura, M.; Shigekawa, H.; Kawabata, K.; Saito, Y.; Kawazu, A. STM study of thin films of BEDT-TTF iodide. *Appl. Surf. Sci.* **1992**, *61*, 317–320. [CrossRef]

50. Miura, K. Effects of Annealing on the Transport in an Organic Zero-Gap Conductor α-(BEDT-TTF)$_2$I$_3$. Master's Thesis, Toho University, Chiba, Japan, 2011. (Supplementary material, 2017).

51. Tisserond, E.; Fuchs, J.N.; Goerbig, M.O.; Auban-Senzier, P.; Meziere, C.; Batail, P.; Kawasugi, Y.; Suda, M.; Yamamoto, H.M.; Kato, R.; et al. Aperiodic quantum oscillations of particle-hole asymmetric Dirac cones. *Europhys. Lett.* **2017**, *119*, 67001. [CrossRef]

52. Fortin, J.-Y.; Audouard, A. Effect of electronic band dispersion curvature on de Haas-van Alphen oscillations. *Eur. Phys. J. B* **2015**, *88*, 225. [CrossRef]

53. Kishigi, K.; Hasegawa, Y. Three-quarter Dirac points, Landau levels, and magnetization in α-(BEDT-TTF)$_2$I$_3$. *Phys. Rev. B* **2017**, *96*, 085430. [CrossRef]

crystals

MDPI

Article

Structural and Electronic Properties of (TMTTF)$_2X$ Salts with Tetrahedral Anions

Roland Rösslhuber †, Eva Rose †, Tomislav Ivek ‡, Andrej Pustogow, Thomas Breier, Michael Geiger, Karl Schrem, Gabriele Untereiner and Martin Dressel *

1. Physikalisches Institut, Universität Stuttgart, Pfaffenwaldring 57, D-70569 Stuttgart, Germany; roland.roesslhuber@pi1.physik.uni-stuttgart.de (R.R.); eva.rose@pi1.physik.uni-stuttgart.de (E.R.); tivek@ifs.hr (T.I.); andrej.pustogow@pi1.physik.uni-stuttgart.de (A.P.); acs@pi1.physik.uni-stuttgart.de (T.B.); m.geiger@fkf.mpg.de (M.G.); cienkowska-schmidt@pi1.physik.uni-stuttgart.de (K.S.); gabriele.untereiner@pi1.physik.uni-stuttgart.de (G.U.)
* Correspondence: dressel@pi1.physik.uni-stuttgart.de; Tel.: +49-711-685-64946
† These authors contributed equally to the work.
‡ Permanent address: Institut za fiziku, Bijenička 46, HR-10000 Zagreb, Croatia.

Academic Editor: Helmut Cölfen
Received: 14 February 2018; Accepted: 28 February 2018; Published: 4 March 2018

Abstract: Comprehensive measurements of the pressure- and temperature-dependent dc-transport are combined with dielectric spectroscopy and structural considerations in order to elucidate the charge and anion orderings in the quasi-one-dimensional charge-transfer salts (TMTTF)$_2X$ with non-centrosymmetric anions X = BF$_4$, ClO$_4$ and ReO$_4$. Upon applying hydrostatic pressure, the charge-order transition is suppressed in all three compounds, whereas the influence on the anion order clearly depends on the particular compound. A review of the structural properties paves the way for understanding the effect of the anions in their methyl cavities on the ordering. By determining the complex dielectric constant $\hat{\epsilon}(\omega, T)$ in different directions we obtain valuable information on the contribution of the anions to the dielectric properties. For (TMTTF)$_2$ClO$_4$ and (TMTTF)$_2$ReO$_4$, $\epsilon_{b'}$ exhibits an activated behavior of the relaxation time with activation energies similar to the gap measured in transport, indicating that the relaxation dynamics are determined by free charge carriers.

Keywords: Fabre salts; one-dimensional conductors; charge order; anion order; electronic transport; pressure dependence; dielectric spectroscopy

PACS: 71.30.+h, 74.70.Kn, 75.25.Dk, 71.27.+a, 72.15.-v, 77.22.Gm

---

1. Introduction

Electronic correlations play a decisive role in the quasi-one-dimensional molecular charge-transfer salts (TMTTF)$_2X$, where TMTTF means tetramethyltetrathiafulvalene. As a result of the 2:1 stoichiometry with positively charged (TMTTF)$^{\rho_0}$, ρ_0 = 0.5, and monovalent anions X^-, the (TMTTF)$_2X$ salts should have a 3/4-filled conduction band. The slight dimerization of the molecular stacks leads to a splitting into a completely filled lower and a half-filled upper band; accordingly, metallic properties are expected. This behavior is observed in the sibling TMTSF compounds, known as Bechgaard salts, here TMTSF stands for tetramethyltetraselenafulvalene. Due to the reduced dimension, Fermi liquid theory breaks down separating spin and charge degrees of freedom [1,2].

In the case of the Fabre-salts TMTTF, however, a minimum in resistivity $\rho(T)$ is observed at $T_\rho \approx 250$ K, followed by a strong increase at T_{CO} [3–5]. The first phase below T_ρ is referred to as charge localization or dimer Mott insulator, whereas the charge-order (CO) transition occurs at T_{CO}, usually somewhere below 160 K. Both states arise due to strong electronic correlations which are well described by the extended Hubbard model taking into account on-site U and nearest-neighbor Coulomb interactions V [6]. Intense research over the last decades revealed a colorful interplay of charge, spin and lattice degrees of freedom and established a generic phase diagram. The temperature-pressure diagram is shown in Figure 1 and includes (TMTTF)$_2$X and (TMTSF)$_2$X salts with centrosymmetric anions X = Br, PF$_6$, AsF$_6$, SbF$_6$, TaF$_6$, wherein the position can be tuned either by external pressure or by substituting the anions, causing chemical pressure. In the phase diagram Figure 1, the compounds with non-centrosymmetric anions, such as SCN, FSO$_3$, NO$_3$, BF$_4$, ClO$_4$ and ReO$_4$ are missing, as they exhibit ordering of the anions upon cooling.

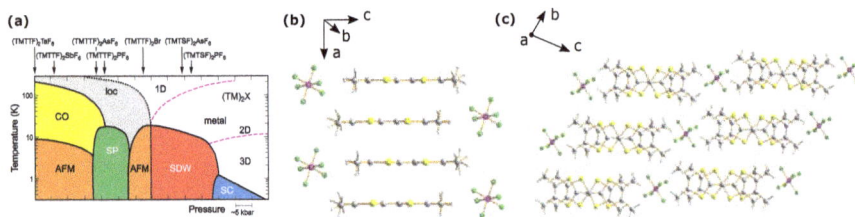

Figure 1. (a) The generic phase diagram of the tetramethyltetrathiafulvalene (TMTTF) and tetramethyltetraselenafulvalene (TMTSF) salts with centrosymmetric anions, as first suggested by Jérome [7] and further refined by other groups [2]. The position can be tuned either by external physical pressure or by chemical pressure via substituting the anion. The arrows indicate the ambient pressure position of the particular compounds with different anions. Decreasing the anion size corresponds to increasing pressure upon which the properties become more metallic and less one-dimensional. The full and dashed lines indicate phase transitions and crossovers, respectively. The various terms are shortened as follows: CO charge order, loc charge localization, SP spin-Peierls, AFM antiferromagnet, SDW spin density wave and SC superconductivity; (b) Crystal structure of (TMTTF)$_2$PF$_6$ [8] illustrating the stacking of the TMTTF molecules and the anions along the a-direction and, (c) in the bc-plane showing the separation of the TMTTF stacks by anions.

2. Structural Considerations

Before we report our experimental studies on the electronic properties of (TMTTF)$_2$X and discuss the temperature and pressure development, let us summarize the structural aspects known for this class of materials. As the best studied example, the crystal structure of (TMTTF)$_2$PF$_6$ under ambient conditions [8] is exemplarily shown in Figure 1b and c for the (TMTTF)$_2$X salts. Commonly they exhibit a triclinic space group P$\bar{1}$ with inversion centers located in between the TMTTF molecules and on the anions [9–14]. Most importantly, the planar TMTTF molecules are arranged in zig-zag stacks along the a-direction (Figure 1b) giving rise to the quasi-one-dimensional physics in the Fabre salts [3]. The anions separate the TMTTF stacks along the $b + c$ direction (Figure 1c) and mediate the inter-stack coupling. Moreover, due to the presence of the anions, a slight dimerization is introduced on the TMTTF stacks which decreases upon cooling [15]. This already indicates that the counter ions play a more complex role then just acting as spacers between the TMTTF stacks and that their symmetry affects the electronic properties [4,16–18].

The anions are located in cavities, which are constituted by the methyl end groups of the six nearest neighbor TMTTF molecules [19]. Depending on their size, symmetry, and polarizability, the anions and the methyl groups are slightly deformed, leading to a more or less snug fit in the cavities [14,16,17,20,21]. For that reason, the anions possess some limited rotational and translational degrees of freedom that are activated at higher temperatures [22,23]. The methyl groups basically form a centrosymmetric environment with symmetry axes close to the ones of the octahedral anions. This implies that for non-centrosymmetric entities, i.e., linear, triangular, tetrahedral anions, orientational disorder becomes an issue [17,24].

At lower temperatures—when thermal disorder is reduced—weak hydrogen-bonds are formed between the ligands of the counter ions and the closest methyl groups [25]. Moreover, there is a link between the S-atom of the nearest TMTTF molecule and the anion ligand O or F-atom [26], evidenced by an anion dependence of the sulfur to anion-ligand distance d_{AL-S} [4,18] (Table 1). It is important to emphasizes, that via the methyl groups any structural or charge modification in the TMTTF molecules influences the position of the anions and vice versa [17].

At elevated temperatures, the symmetry of the anions does not play a crucial role, for these entities are subject to rotation and disorder. Upon cooling, however, the motion is hindered and the anions get locked in certain positions, where the symmetry of the anions becomes an important aspect. The alternating orientation of the tetrahedra along and perpendicular to the stacks leads to a doubling of the lattice periodicity; hence the anion order is a structural transition with wave vector $q_{AO} = (\frac{1}{2}, \frac{1}{2}, \frac{1}{2})$. X-ray absorption near edge structure (XANES or NEXAFS) [27] and infrared optical spectroscopy [23] measurements unanimously identify a tetramerization of the TMTTF stacks at the AO transition T_{AO}, the latter also providing evidence for a 0110 charge pattern. A scheme of the corresponding modification of the structure and the charge pattern is shown in Figure 2e.

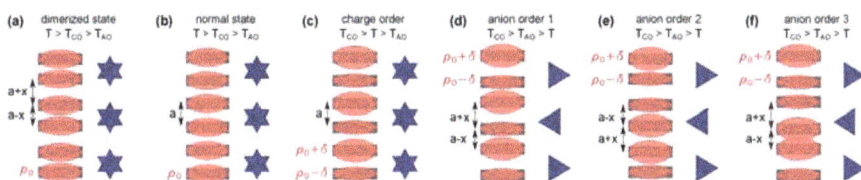

Figure 2. Sketch of the molecular and charge arrangements for different states of (TMTTF)$_2X$ salts. The organic molecules are depicted as rectangles; the amount of charge ρ_0, $\rho_0 \pm \delta$ is illustrated by red ellipses of different size. The anions are shown as blue hexagons or triangles, depending on the centrosymmetry. (**a**) The TMTTF stacks are dimerized at room temperature, $a \pm x$. The degree of dimerization decreases upon cooling, thus equalizing the TMTTF sites towards the uniform state in (**b**). Panel (**c**) illustrates how charge disproportionation develops as the charge-ordered state is entered with a 1010 charge pattern along the stacking direction. A priori, three different charge configurations are possible in the anion-ordered phase: (**d**) one with a 1010 pattern and (**e**,**f**) two with a 1100 pattern. Our optical studies [23] have revealed that the tetramerized state (**e**) forms due to the anion order, with an intra-dimer molecular distance $a - x$, and the dimers separated by $a + x'$ and $a + x''$.

In this respect, charge order as well as anion order in the (TMTTF)$_2X$ salts still bear some mystery. Among the remaining questions are (i) about the exact mechanism for CO involving the anions, hence (ii) about the nature of interplay, if there is any, between CO and AO and (iii) about the role of polarizability of the anions for CO and AO.

2.1. Charge Order

Charge order is caused by redistribution of electronic charge within the TMTTF stack resulting in an alternating arrangement of differently charged molecules (TMTTF)$^{+0.5+\delta}$ and (TMTTF)$^{+0.5-\delta}$ along the a-direction. This charge disproportionation is observed in all TMTTF salts considered here, except (TMTTF)$_2$ClO$_4$. Charge order lifts the inversion symmetry and conceptually doubles the unit cell. Halving the Brillouin zone leads to the opening of a gap in the electronic density of state [3]. The latter is observed as a strong increase of resistivity at T_{CO} [4,5] whereas the charge disproportionation is evidenced by nuclear magnetic resonance (NMR) [28,29], dielectric [30–36] and optical spectroscopy [23,37,38], as well as by synchrotron X-ray diffraction measurements [39]; the latter results evidence a 1010 charge pattern. The electrical transition proceeds continuously, wherein the charge disproportionation and the energy gap exhibit a temperature dependence that resembles the mean-field behavior of a second-order phase transition [4,23]. An overview of the transition temperatures and structural parameters determined at room temperature is given in Table 1.

Figure 2 illustrates the molecular and charge arrangements for the different states realized in the $(TMTTF)_2X$ salts. Panel a corresponds to the state at high temperatures, which exhibits a finite dimerization of the organic molecules. Upon cooling, the dimerization diminishes [15], approaching the arrangement shown in panel b. The charge disproportionation starts to develop upon entering the charge-ordered state with a 1010 pattern along the stacking direction as sketched in panel c.

2.2. Anion Contribution to Charge Order

From a general point of view, the $(TMTTF)_2X$ salts are stabilized by the charge transfer of half an electron from the TMTTF molecules to the anions. Therefore it seems reasonable that charge redistribution at the CO transition introduces structural modifications [17]. Nevertheless, the associated alternation in the crystal structure could not be resolved for a long time [15,40], and as a result, the charge order transition was first considered as a "structureless transition" of unknown origin [41]. Only recent ^{19}F-NMR studies [42], measurements of the thermal expansion [43,44], neutron scattering [45] and optical spectroscopy [38,46,47] provided the necessary evidence that CO is accompanied by lattice effects, attributed to a collective shift of the anions [17].

Pouget proposed [17] two different directions for the anion shift. First, the anion moves along the c-direction towards the methyl groups, as indicated by black arrows in Figure 3a,b. The resulting deformation of the methyl end-groups polarizes the hydrogen-bond network leading to a displacement of charge in the σ-bonds connected to the H-bonds. This indirectly stabilizes the excess of π-holes on the corresponding TMTTF molecule as indicated by a red ellipsoid, whereas blue ellipsoids mark charge-poor molecules (Figure 3).

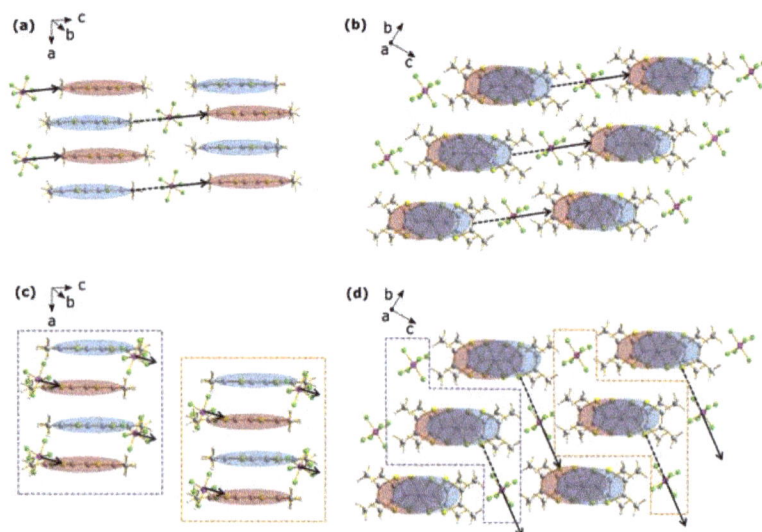

Figure 3. The anions shift along the $b + c$ direction towards the methyl groups and towards the sulfur atoms, as suggested by Pouget [17]. Charge-rich TMTTF molecules are indicated by a red ellipsoids whereas blue ellipsoids mark charge-poor molecules. (**a**) View in the ac-plane and (**b**) along the a-axis onto the cb-plane. The shift of the anions towards the S-atoms of the nearest TMTTF molecule depicted (**c**) in the ac-plane and (**d**) in the cb-plane. The polygons formed by the purple and orange dashed lines in panel d indicate which anion stacks are shown in the previous frame c.

Several observations support this suggestion of an anion shift with a strong contribution along the c-direction. First of all, thermal expansion measurements exhibit a clear anomaly at T_{CO}, which is most pronounced along the c-axis [43]. Secondly, the coupling between the methyl groups and the anions is mostly in the bc-plane. Furthermore, deuteration of the methyl groups significantly increases T_{CO}. The heavier mass slows the lattice dynamics and weakens the coupling between the anions and terminating endgroups (Figure 4); thus CO can be stabilized already at higher temperatures [23].

The second proposed translation of the anions [17] is towards the closest sulfur atom of a TMTTF molecule, indicated by black arrows in Figure 3c,d. This goes hand in hand with the enhanced π-hole density at the corresponding TMTTF molecule. NEXAFS measurements clearly show the strengthening of the S-F bonds [48].

Both shifts are supposed to increase the charge concentration on the corresponding TMTTF molecule. It is energetically favorable that the second neighboring TMTTF molecule contains less positive charge. Hence, by minimizing the electrostatic energy between the anion stacks and adjacent TMTTF stacks, a three-dimensional charge pattern is established that is crucial for stabilizing CO [23,43]. In fact, temperature-dependent transport measurements [4] clearly show the opening of a CO gap along all three crystallographic directions providing evidence for this three-dimensional charge pattern due to inter-stack coupling mediated by the anions.

Some further considerations have been published, which should be mentioned here. First, the anion shift accompanying CO does not introduce the tetramerization [23]. Second, although the anions do shift at the CO transitions, they are still considered to exhibit a random rotational motion depending on the temperature [17].

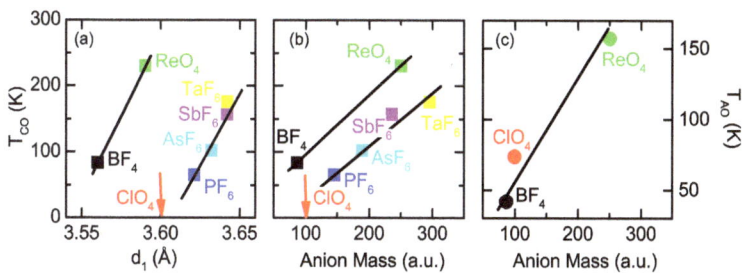

Figure 4. Dependence of the charge order temperature T_{CO} (colored squares) and anion order temperature T_{AO} (colored circles) for various Fabre salts $(TMTTF)_2X$ with different X as indicated. (**a**) T_{CO} plotted versus the distance between the TMTTF molecules d_1 along stacking direction determined at room temperature. Here, octahedral and tetrahedral salts fall on two different lines. Although the octahedral anions exhibit larger values for d_1, T_{CO} is lower compared to the tetrahedral salts. The octahedral anions are larger (Table 1) resulting in an increased value of d_1. The anion-TMTTF stack coupling increases due to the larger anions, and the match between anion and cavity symmetry; thus thermal fluctuations prevail upon the electronic correlations giving rise to CO down to lower temperatures. (**b**) Plot of T_{CO} versus the mass of the anions. A monotonic increase of T_{CO} with anion mass is observed, which is attributed to less effective motion of heavy anions due to thermal fluctuations and, consequently, a less disordered anion potential influencing electronic correlations within the TMTTF stacks. The octahedral and tetrahedral salts fall on two distinct lines due to different coupling between the anions and the TMTTF stacks. (**c**) For the anion ordering temperature observed in the tetrahedal $(TMTTF)_2X$ salts, an increase of T_{AO} with the anion mass is observed. At a specific temperature, heavy anions are less affected by thermal fluctuations. Consequently, the anion orientation is locked permanently for heavier anions at higher temperatures. As a result, anion order occurs at higher temperatures for heavy anions.

Having established the anion shift, we now have to pose the question whether charge order can still be understood as a phenomenon inherent to the one-dimensional TMTTF stacks as described by the extended Hubbard model, or whether a more advanced theoretical treatment is required including the contribution of the lattice in general and the anions in particular. In other words, we have to find out whether the Coulomb attraction between the negative charge located on the anion-ligand with shortest S-F contact is crucial for charge order, or whether it only stabilizes the arrangement. It has been proposed, that the shortest anion-ligand to sulfur distance d_{AL-S} is the parameter determining CO [4,18], in contrast to a purely electronic picture, which implies that the distances between the TMTTF molecules, d_1 and d_2, are decisive. Recent spectroscopic and structural investigations could shed light on this issue [18,23]. In the following we give a short overview on the different approaches and on the studies focusing on the structural aspects of the CO transition.

Table 1. Overview on the transition temperatures T_{CO} and T_{AO}, as well as on the room-temperature structural data of various $(TMTTF)_2 X$ salts with octahedral and tetrahedral anions X. Here d_{AL-S} is the shortest distance between the anion-ligand (F or O) and the sulfur atom. d_1 and d_2 denote the distances between planes defined by the TMTTF molecules within one stack [9,11–13,18]. Hence, $(d_1 - d_2)/(d_1 + d_2)$ denotes the structural dimerization along the stacks determined by the distance between the TMTTF molecular planes; it does not take into account any displacement of the molecules along the b or c direction. In contrast to that, δ_{struc} and δ_{elec} are determined by ab-initio DFT calculations based on X-ray structural data [49,50]. δ_{struc} corresponds to a structural dimerization as well, which is defined by the distance between the centers of mass in each TMTTF molecule. $\delta_{elec} = 2|t_1 - t_2|/(t_1 + t_2)$ explicitly takes into account the overlap of the transfer integrals t_1 and t_2. The unit cell volume is listed as $V_{unitcell}$ [9,11–13]. The parameter R_0 denotes the thermochemical radius of the monovalent anion [51]; although calculated for a salt with 1:1 stoichiometry, we consider this as a valid estimate for the anion size. The anion volume is listed as V_{anions} following [52]. In the last row the anion mass m_A is given in atomic units.

Anion X	PF_6	AsF_6	SbF_6	TaF_6	BF_4	ClO_4	ReO_4
T_{CO} (K)	67	102	157	175	84	-	230
T_{AO} (K)	-	-	-	-	41.5	73.4	157
d_{AL-S} (Å)	3.30	3.27	3.21	3.215	3.28	3.45	3.05
d_1 (Å)	3.621	3.632	3.642	3.642	3.56	3.60	3.59
d_2 (Å)	3.527	3.524	3.526	3.534	3.54	3.51	3.57
$\dfrac{d_1 - d_2}{d_1 + d_2}$ (10^{-2})	1.32	1.51	1.62	1.51	0.28	1.27	0.28
δ_{struc}	0.040	0.041	0.041	-	0.028	0.04	-
δ_{elec}	0.230	0.110	0.298	-	0.336	0.616	-
$V_{unitcell}$ (Å^3)	676.6	697.7	702.9	706.52	648.5	654.8	679.5
R_0 (pm)	242	243	252	250	205	225	227
V_{anions} (Å^3)	70.6	-	81.8	-	51.6	54.7	64.8
m_A (au)	144.96	188.91	235.75	294.94	86.80	99.45	250.21

In Table 1 we summarize electronic and structural parameters determined from literature data. For the octahedral anions, T_{CO} rises with anion size (and mass) $d(PF_6) < d(AsF_6) < d(SbF_6), < d(TaF_6)$; the same is found for the salts with tetrahedral anions. Moreover, a larger anion diameter correlates with a shorter sulfur-anion-ligand distance d_{AL-S} and larger intra-stack distances between the TMTTF molecules d_1 and d_2. Regarding the structural dimerization $\dfrac{d_1 - d_2}{d_1 + d_2}$ and T_{CO}, no clear trend can be seen [18,49,50]. So far, T_{CO} scales with both, d_{AL-S} and d_1, impeding a simple answer of the question posed above.

In Figure 5a the charge disproportionation 2δ is plotted as a function of the CO temperature T_{CO}. Interestingly, both the salts with tetrahedral and with octahedral anions fall onto the very same line, providing strong indications that CO is independent on anion symmetry and emerges from the electronic interaction and one-dimensional physics inherent to the TMTTF stacks [23]. This conclusion complies with the extended Hubbard model and theories based on it; where structural peculiarities are disregarded [6]. Along these lines the competition between the nearest-neighbor Coulomb repulsion V and the bandwidth $W = 4t$ is crucial for CO, wherein the transfer integral t sensitively depends on the distance between TMTTF molecules.

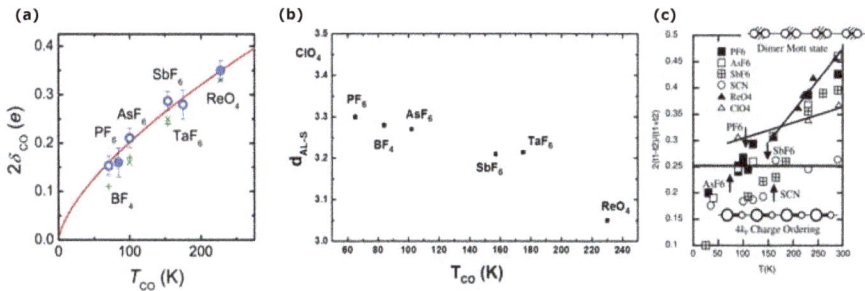

Figure 5. (a) Charge disproportionation 2δ of the $(\text{TMTTF})_2 X$ salts with different anions X plotted versus the CO transition temperature T_{CO}. The open circles [38,53] and closed circles [23] were determined by optical spectroscopy on the octahedral anions PF_6, AsF_6, SbF_6, TaF_6, and tetrahedral anions BF_4 and ReO_4, respectively; the green crosses and green plus signs were measured by NMR [37,54,55] and Raman spectroscopy [38,56], respectively. The red line corresponds to $2\delta \propto T_{CO}^{2/3}$. Most important, all salts fall onto this line independent on the anion symmetry. (b) The shortest distance d_{AL-S} between the sulfur atom of the organic molecule and the anion-ligand is evaluated at room temperature and plotted in dependence on the charge-order transition temperature T_{CO}. All salts exhibiting CO fall on a line, indicating a linear decrease of T_{CO} as the spacing d_{AL-S} increases. (c) Temperature dependence of the bond dimerization $\frac{2|t_1 - t_2|}{(t_1 + t_2)}$ taken from Reference [15], wherein the transfer integrals t_2 and t_1 were obtained by applying the extended Hückel model on temperature dependent X-ray data.

To countercheck, in Figure 5b the distance d_{AL-S} is plotted versus T_{CO}. For all salts we find a linear relation of falling transition temperature and increasing distance, independent on the symmetry of the anions. Note, the values given in Table 1 and plotted in Figure 5b are determined at ambient conditions; they change upon cooling and the application of pressure.

Figure 5c presents the temperature dependence of the electronic bond dimerization $\delta_{elec} = \frac{2|t_1 - t_2|}{(t_1 + t_2)}$, which is proportional to the electronic dimerization δ_{elec}, considering that the transfer integrals t_2 and t_1 are calculated by the extended Hückel model on temperature dependent X-ray data [15]. Most importantly, the dimerization decreases with lowering the temperature, fully consistent with the findings of Reference [50]. It is concluded that CO sets in below a critical bond dimerization of 0.25. This picture explains the absence of CO in $(\text{TMTTF})_2 \text{ClO}_4$, however, $(\text{TMTTF})_2 \text{BF}_4$ is not included. There also remains a discrepancy between the observed value of about 0.37 for $(\text{TMTTF})_2 \text{ReO}_4$ with $T_{CO} = 230$ K.

In a more advanced approach, the influence of structural effects on charge ordering in the $(\text{TMTTF})_2 X$ salts was investigated by ab-initio DFT calculations based on pressure- and temperature-dependent X-ray data [49,50]. No systematic relation was observed as far as T_{CO} and the structural dimerization is concerned, i.e., the distance between the TMTTF molecules. For the electronic dimerization δ_{elec}, however, they find an increased value for $(\text{TMTTF})_2 \text{ClO}_4$ compared to the other salts, that explains the absence of CO in this particular salt. Besides the distance of the molecular planes, δ_{elec} also takes into account the differences in the orientation and size of the transfer integrals. When the electronic dimerization increases, CO is suppressed and can even be absent as in the case of $(\text{TMTTF})_2 \text{ClO}_4$.

We can conclude this short overview by emphasizing the importance of ab-initio DFT calculations based on detailed temperature-dependent X-ray data for all $(TMTTF)_2X$ salts. This way we can unambiguously determine which role lattice effects play in general and the anions in particular on the charge-order transition.

2.3. Anion Order

In addition to charge order, the $(TMTTF)_2X$ salts with non-centrosymmetric anions $X =$ SCN, SFO_3, NO_3, BF_4, ClO_4, and ReO_4 undergo an ordering of the anions at T_{AO}; here we focus on the tetrahedral counter ions. At room temperature, the octahedral and tetrahedral anions exhibit random translational and rotational movement caused by thermal fluctuations [22]. As the temperature is lowered, the tetrahedra do not fit into the cavities formed by the methyl end-groups due to their non-matching symmetry resulting in an stochastic anion orientation [17]. This orientational disorder is resolved at T_{AO} by AO, a structural phase transition orienting the anions within the cavities.

For the $(TMTTF)_2X$ salts with $X = BF_4$, ClO_4 and ReO_4, the ordering of the anions is a first-order transition breaking the room temperature symmetry and doubling the lattice periodicity [4,18]. The resulting superstructure exhibits a reduced wave vector of $q_{AO} = (\frac{1}{2}, \frac{1}{2}, \frac{1}{2})$ [16,17,57]. AO is detected by several methods: in resistivity measurements a kink-like feature is observed with a corresponding hysteresis [4], a pronounced peak or shoulder is seen in dielectric spectroscopy [32,36,58], and the optical spectra are distinctly modified [23]. In the case of $(TMTTF)_2ClO_4$ charge disproportionation 2δ occurs right at the anion order [23]. As can be seen from Table 1, T_{AO} rises with increasing anion volume and mass.

It is instructive to look at the sibling compound $(TMTSF)_2X$, where the AO takes place at $T_{AO} = 24$ K for $X = ClO_4$ and 180 K for ReO_4 [57,59–61]. In general the selenium analogues possess a larger bandwidth, compared to the $(TMTTF)_2X$ salts and therefore the anion potential has only smaller influence. While the former compound orders in a $q_{AO} = (0, \frac{1}{2}, 0)$ fashion, the latter one also exhibits a first-order phase transition with a wave vector $q_{AO} = (\frac{1}{2}, \frac{1}{2}, \frac{1}{2})$; interestingly a high-pressure phase with $(0, \frac{1}{2}, \frac{1}{2})$ coexists for $p > 10$ kbar [62]. For both compounds, the solid solution of TMTTF and TMTSF reveals several instabilities, broadens and shifts the AO transition [61,63–65]. A large number of studies have been devoted to the occurrence of superconductivity in $(TMTSF)_2ClO_4$ as the cooling rate is varied in order to retain anion disorder to low temperatures [66–69]. To the best of our knowledge, similar studies have not been performed on the Fabre salts.

Figure 6 illustrates our present understanding of the structural modifications taking place at the AO transition [17,18,23]. Due to their non-centrosymmetric shape, the tetrahedral anions exhibit only a single anion-ligand to sulfur link along the $(-b + c)$-direction, indicated by the black dashed line. This defines the direction along which the tetrahedra preferably reorient: one corner points along the anion-ligand to sulfur link towards the nearest S atom of the closest neighboring TMTTF molecule. Simultaneously, the other three anion ligands lie in a plane matching the symmetry of the methyl cavity.

A great many experiments provided compelling evidence that the anions are not statistically oriented along one of the two possible directions towards the nearest S atoms but exhibit long-ranged order, resulting in a 0110 charge pattern and tetramerization of the TMTTF stacks as sketched in Figure 6. Structural refinements on $(TMTTF)_2ReO_4$ revealed the staggered orientation of the anions along the [111]-direction and the tetramerization with intra-stack distances $d_1 \approx 3.45$ Å, $d_2' \approx 3.6$ Å and $d_2'' \approx 3.48$ Å [70]. The molecular tetramerization was also observed by angular-dependent X-ray near-edge structure (XANES) measurements [27]; in addition, they reveal a slight shift and deformation of the ReO_4 anions. For all salts with tetrahedral anions $X = BF_4$, ClO_4 and ReO_4, electron spin resonance (ESR) spectroscopy sees a drop of the spin susceptibility right at T_{AO} that is attributed to the formation of a singlet-triplet gap upon tetramerization [71–73]. Most recently, infrared spectroscopy could confirm the 0110 charge pattern by looking at the molecular vibrations [23]; more comprehensive investigations on this issue are under way [74].

Albeit a precise and complete microscopic picture of the AO transition could not be reached yet, there is wide agreement that Coulomb interactions between the anions and the TMTTF stacks are the driving force [17]. The orientation and slight deformation of the anions gives rise to a periodic electrostatic potential influencing the charge distribution on the TMTTF molecules. This anion potential is most pronounced along the sulfur to anion-ligand link and thus the staggered orientation of the anions imposes the 0110 charge pattern onto the TMTTF stacks. As seen from Table 1, there is no monotonic dependence of T_{AO} on d_{AL-S}; we should mention, however, that d_{AL-S} is determined at room temperature and one should consider the distance right at the respective AO transition. It is not fully clear to what extent the 0110 charge pattern actually minimizes the Coulomb repulsion within the TMTTF stacks. The contribution of the anion potential cannot be determined precisely without detailed structural data in dependence on temperature covering the AO transition. Finally we would like to draw the attention on the effect of isotope substitution, since T_{AO} seems not to change appreciably upon deuteration [36].

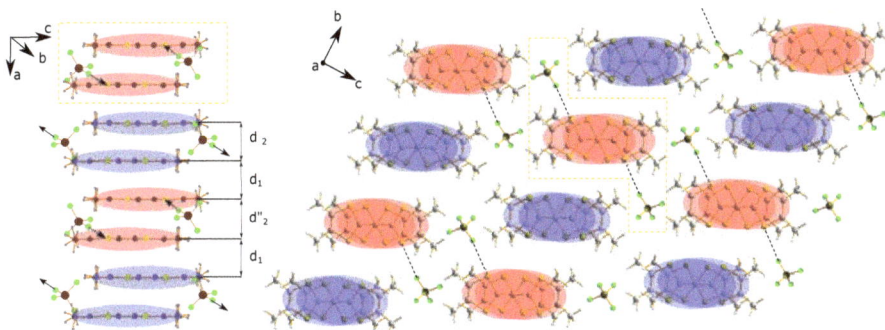

Figure 6. Scheme of the anion arrangement for the tetrahedral (TMTTF)$_2X$ salts as $T < T_{AO}$. Upon AO, the anions re-orient in the methyl cavities along the anion-ligand sulfur link in $(-b + c)$-direction, represented by the black dashed line, forming long-range order. Coulomb attraction between the anion-ligand and the charge on the TMTTF molecule results in a 0110 charge pattern; the red and blue ellipsoids mark charge-rich and charge-poor TMTTF molecules, respectively. The polygon formed by the dashed orange line indicates which anion stacks are shown on the left side, illustrating the charge pattern along the stacking direction. In addition to a reorientation of the anions, a shift along the anion-ligand sulfur link occurs as well; this is not included here.

2.4. Deuteration

The replacement of hydrogen by deuterium in the methyl end-groups of the TMTTF molecule has only minor influence on the anion order but increases T_{CO} [75–77]; hence deuterated (TMTTF)$_2X$ salts provide valuable information on the ordering mechanisms. It was pointed out recently [23] that with increasing anion size or mass, the effect of deuteration on the CO transition diminishes. Defining the temperature shift $\Delta T_{CO} = T_{CO,D} - T_{CO}$, where $T_{CO,D}$ and T_{CO} are the transition temperatures for the deuterated and hydrogenated crystals, respectively, the change in transition temperature is strongest for small anions, as illustrated in Figure 7a. This important relation provides compelling evidence that charge order – considered as a purely electrostatic effect – is crucially affected by dynamic anion fluctuations. A simple relation is found for all TMTTF salts studied: $T^2_{CO,D} = T^2_{CO} + D^2$.

It is important to note the structural aspect that dimerization decreases upon deuteration. Starting with equally spaced TMTTF molecules in an isolated stack, dimerization is imposed by the anions. The coupling of the thermally fluctuating anions to the TMTTF stacks takes place via the terminal methyl groups, i.e., it is mediated by hydrogen bonds. At elevated temperatures, the methyl groups and the anions both require more space due to thermal fluctuations; their stronger interaction enhances the anion-TMTTF coupling. By deuterium substitution the methyl-group motion is slowed down, reducing the coupling to the counterions. As a consequence, the dimerization is smaller for d_{12}-TMTTF salts compared to the hydrogenated analogues. This also explains the concomitant enhancement of T_{CO}.

Anion order takes place at lower temperatures when the anions and the methyl groups fluctuate less and their dynamic interaction is of minor relevance. Instead, it is more important to consider the static Coulomb interaction between the monovalent anions and the charge spread on the TMTTF molecules, mainly located at the S atom and the C=C double bonds. Since the electrostatic interaction is not affected by deuteration, the influence of the anions does not change due to this isotope substitution.

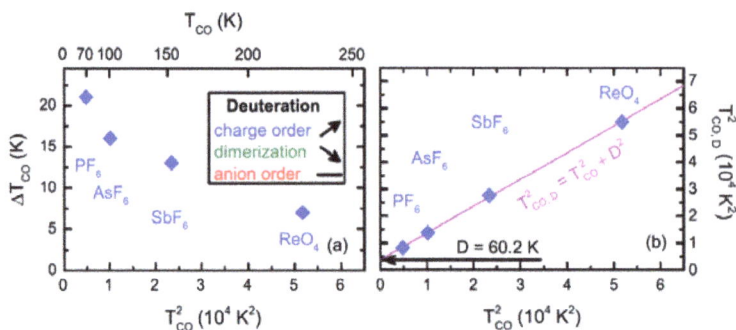

Figure 7. Effect of deuteration on charge order observed in (TMTTF)$_2$X. (**a**) For several anions X = PF$_6$, AsF$_6$, Sb$_6$ and ReO$_4$, the change in the transition temperature $\Delta T_{CO} = T_{CO,D} - T_{CO}$ is plotted versus the square of the CO transition temperature T_{CO}^2 of pristine crystals. The increase is most pronounced for small and light anions. The inset indicates that deuteration rises the CO temperature, diminishes dimerization, but has no effect on AO. (**b**) Plotting $T_{CO,D}$ over T_{CO} clearly reveals the dependence $T_{CO,D}^2 = T_{CO}^2 + D^2$ with $D = 60.2$ K. The data are taken from Reference [75,77]; the plot follows Reference [23].

3. Experimental Details

Single crystals of TMTTF$_2$X salts with X = BF$_4$, ClO$_4$ and ReO$_4$ were grown by the standard electrochemical methods using an H-type glass cell at ambient temperature and inert atmosphere. Platinum plates with an area of approximately 3 cm^2 served as electrodes and a sand barrier was introduced to reduce diffusion. By applying a constant voltage of 1.5 V, a current between 9.2 and 13.4 μA was drawn through the solution. The growth of the needle-shaped single crystals of typical dimensions $(2 \times 0.5 \times 0.1)$ mm^3 took several months. Since the crystal shape does not correspond to the triclinic symmetry of the crystal structure, b' denotes the projection of the b-axis perpendicular to the a axis, and c^* is chosen orthogonal to the ab-plane.

In order to perform the electrical transport measurements, small gold contacts were evaporated onto the natural crystal surface and thin gold wires were attached to them by carbon paste. The dc experiments applied the four-point method. Pressure-dependent measurements were carried out by means of a clamp-type pressure cell using Daphne 7373 oil as pressure medium. Daphne 7373 is the favored pressure medium for this kind of experiments because it conveys hydrostatic pressure and (TMTTF)$_2$X crystals are inert to it. For all applied pressures the pressure medium stays fluid at room temperature, relevant shearing forces seem to appear only when cooling through the freezing temperature at pressures between 0–3 kbar. Temperature-dependent transport measurements were carried out in a standard helium bath cryostat. The inherent pressure loss upon cooling was recorded continuously in-situ by an InSb semiconductor pressure gauge, as illustrated in Figure 8. The dielectric measurements were performed with two contacts in a pseudo four-point configuration by means of an Agilent 4284 Impedance Analyzer (20 Hz–1 MHz). For both measurement techniques, the applied voltage was in the mV range to avoid heating and ensure measurements in the Ohmic regime.

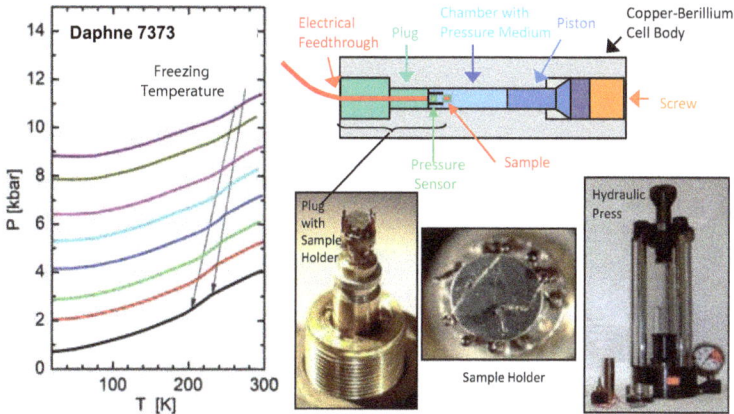

Figure 8. Left panel: Pressure loss of Daphne 7373 oil during cooling; the arrows mark the temperature range of freezing. The data were recorded in a Copper-Berrilium cell by four-point resistance measurement of an InSb pressure sensor. Right panel: Sketch of the actual pressure cell used and photos of the sample holder, and the hydraulic press.

4. Electrical Transport under Pressure

The physical properties of the $(TMTTF)_2X$ family have been widely studied for several decades [78–81], however most of the work dealt with centrosymmetric anions. Here we advocate that the structural phase transition should not be considered an additional complication but an opportunity to gain more insight on the effect of the lattice on the electronic properties; there is no question, however, that further investigations are needed. In the following we will present the first complete set of systematic dc measurements under hydrostatic pressure of the three most common $(TMTTF)_2X$ with tetrahedral anions ($X = BF_4$, ClO_4, ReO_4). After detailed discussions of $\rho(T, p)$ for each particular compounds, we will compare the rather diverse characteristics.

4.1. (TMTTF)₂ClO₄

From Table 1 we see that $X = ClO_4$ is placed between $X = BF_4$ and ReO_4 with respect to the anion size. From the charge transfer salts with octahedral anions we know that T_{CO} increases with anion radius [4,18]. The same is expected for the tetrahedral anions. $(TMTTF)_2ClO_4$ is in so far particular, as it does not exhibit the charge order typical for all other Fabre salts. As far as the anion transition is concerned, it follows strictly the expected relation for the tetrahedral anions: $T_{AO}(BF_4) < T_{AO}(ClO_4) < T_{AO}(ReO_4)$.

The pressure dependence of $\rho(T)$ is affected by the absence of CO showing extraordinary features. In Figure 9 the electrical resistivity $\rho_a(T, p)$ and $\rho_c(T, p)$ are displayed for pressures up to 11.6 kbar in panels a,b and d respectively. At ambient pressure, $\rho_a(T)$ develops the well-known activated temperature dependence for $T < T_\rho \approx 260$ K (localization temperature) and shows a distinct jump to lower values at $T_{AO} \approx 73$ K [4]. It is interesting to note a change in slope of $\rho_a(T)$ that appears around 40 K. Applying hydrostatic pressure, the local minimum at T_ρ disappears as well as the jump at T_{AO}. Whereas the former may be assessed as an artifact caused by the temperature-dependent pressure loss due to the pressure media in the cell (Daphne oil 7373, see Figure 8), the latter is inherent to (TMTTF)$_2$ClO$_4$. For increasing pressure, the AO transition smears, broadens and moves to lower temperatures, developing a distinct valley of low resistivity. Concurrently the onset of the transition T_{AO-max} follows the trend of T_{AO}, accompanied by the development of steep increase of $\rho(T, p)$ at T_M, which is related to the lower bound of the minimum.

At $p \approx 10$ kbar the trend stops; T_{AO-max} and T_M do not significantly shift in temperature any more when p increases further. The mechanism triggering the valley seems to come to a halt. As can be seen in Figure 9b, the valley is gradually filled. At the maximum pressure applied, $p = 11.6$ kbar, the minimum has somehow disappeared completely. Instead for high pressures a second peak develops at about 40 K, which does not shift in temperature for $p > 11$ kbar; this feature progressively dominates the temperature dependance of $\rho(T)$. This behavior is observed in both directions, for $\rho_a(T, p)$ and $\rho_c(T, p)$, as seen by comparison of panels b and d. In Figure 10 we take a closer look at the pressure range from 10.0 to 11.6 kbar. The appearance of the second peak is accompanied by a hysteretic behavior in the temperature-dependent resistivity on both sides of the valley. Multiple temperature cycles reveal fluctuations in $\rho_a(T)$ not observed in other cases.

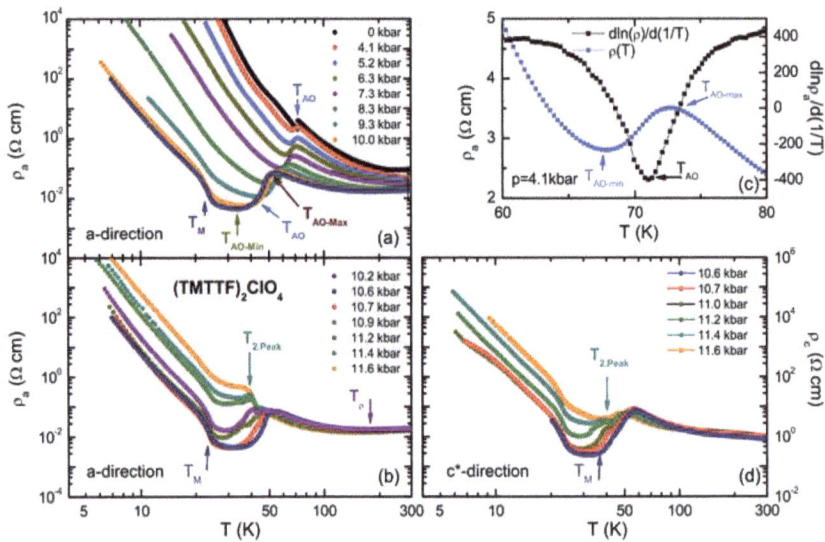

Figure 9. Pressure evolution of the temperature-dependent electronic transport of (TMTTF)$_2$ClO$_4$. (**a**) dc resistivity as a function of temperature for ambient conditions and for hydrostatic pressures up to 10.0 kbar measured along the stacking direction a. (**b**) For higher pressure values a more complex behavior of $\rho_a(T, p)$ is observed. (**c**) For the example taken at $p = 4.1$ kbar, we illustrate how the different parameters of the anion ordering transition were extracted from $\rho(T, p)$ and its derivative with respect to inverse temperature $1/T$. (**d**) Temperature dependence of the perpendicular resistivity $\rho_c(T, p)$ for the high-pressure range.

In the phase diagram displayed in Figure 11a, we summarize the different phases, transitions, and appearing features as extracted from our $\rho(T, p)$ data. At elevated temperatures (TMTTF)$_2$ClO$_4$ behaves like a low-dimensional metal with $d\rho_a/dT > 0$ along the stacking direction and $d\rho_c/dT < 0$ perpendicular to the stacks. Between 250 and 150 K charge localization takes place at T_ρ. While the metal-like behavior in $\rho_a(T, p)$ for $T > T_\rho$ is obvious at $p = 0$, it gets masked for finite but intermediate pressures; interestingly, the localization temperature is recovered at high p values. According to Figure 8 the temperature-dependent pressure loss is reduced as the applied pressure increases; hence this effect influences $\rho(T)$ less. For the $p = 9.3$ kbar curve, we can clearly identify a range of metallic behavior by $d\rho_a(T,p)/dT > 0$. For increasing pressures, the minimum in $\rho(T)$ at T_ρ is reduced. T_ρ extracted in this manner specifies the upper limit of the inherent localization temperature of $(TMTTF)_2ClO_4$. Since in the perpendicular direction $d\rho_c(T,p)/dT < 0$ for the whole investigated pressure and temperature range, our data identify the high-temperature phase of $(TMTTF)_2ClO_4$ as a low-dimensional metal. The subsequent phase is an insulating one; nothing hints towards any inherent changes in the material down to the anion ordering. No indications towards charge order are detected.

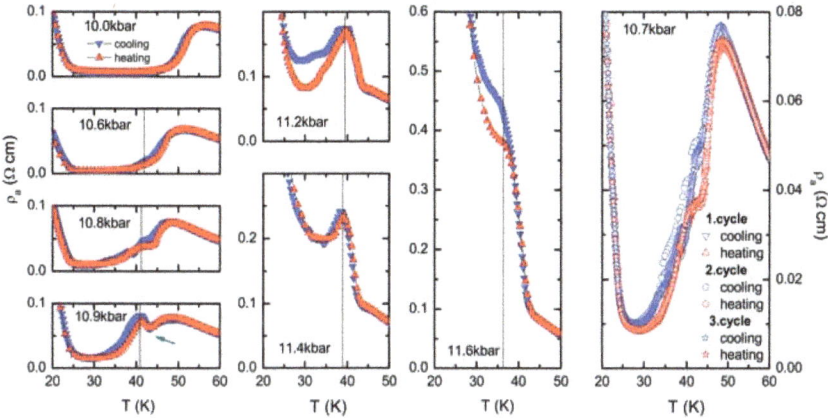

Figure 10. Magnification of the high-pressure resistivity along the *a*-axis of $(TMTTF)_2ClO_4$ in the temperature range of T_{AO}. For pressure between $10.2 - 11.7$ kbar, a second peak in $\rho_a(T)$ appears. Simultaneously at the anion ordering $\rho_a(T)$ exhibits a distinct hysteresis and fluctuations.

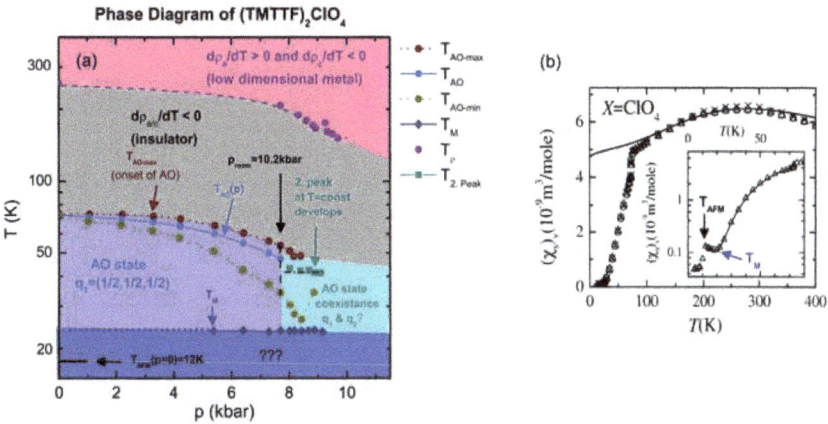

Figure 11. (a) The phase diagram of $(TMTTF)_2ClO_4$ as obtained from resistivity measurements indicates various phase transitions. (b) Temperature-dependent susceptibility of $(TMTTF)_2ClO_4$ as obtained from electron spin resonance (ESR) measurements (taken from Dumm et al. [82]). Two distinct drops in the ESR intensity can be identified, one at the AO transition and the other at the antiferrmagnetic order $T_{AFM} = 12$ K. The onset of the plateau appears around $T \approx 20$ K (indicated by the blue arrow), which coincides with the steep increase of the resistivity at T_M.

The onset of the AO transition, T_{AO-max}, is defined by the local maximum in $\rho(T)$ followed by a sharp drop; the AO temperature T_{AO} is given by the steepest decline of $\rho(T)$ between T_{AO-max} and the local minimum at T_{AO-min}. This is illustrated in Figure 9c in full detail, where also the derivative

$d \ln\{\rho_a\}/d\left(\frac{1}{T}\right)$ is plotted in the Arrhenius-like manner. The AO transition becomes obscured, when the second peak starts to grow. This additional feature is first visible as pressure exceeds 10 kbar. It always seems to be located at $T \approx 40$ K, with no significant shift for varying pressure. It is interesting to compare the structural properties with the sister compound $(TMTSF)_2ClO_4$, which can be turned from a $T_c = 1$ K superconductor to an antiferromagnetic SDW insulator by suppression of the $q_2 = (0, \frac{1}{2}, 0)$ anion order at $T_{AO} = 24$ K by rapid cooling [61]. The pronounced hysteresis and fluctuations might point towards the development of an AO with another superstructure at $T = 40$ K. Recalling the observation of different superstructures with pressure and disorder in $(TMTSF)_2ReO_4$ [62], we could imagine a coexistence of the conventional AO superstructure $q_1 = (\frac{1}{2}, \frac{1}{2}, \frac{1}{2})$ and a new, but weaker one with a different q vector also for $(TMTTF)_2ClO_4$ [13].

It might be worth to have a closer look on the temperature T_M defined by the lower bound of the resistivity valley, since this transition does not shift in temperature for the entire pressure range. The strong increase in $\rho(T)$ at T_M cannot be related to the transition into the antiferromagnetic state, since for $p = 0$ the corresponding transition occurs at $T_{AFM} = 12$ K [83] and no feature is visible in the resistivity data at that temperature. It is interesting to note, however, that the increase of $\rho(T)$ at $T = T_M$ coincides with the weak upturn of the magnetic susceptibility after the local minimum in the ESR data (reproduced in Figure 11b) [82]. Although T_M develops only for relatively high pressures, the ESR data are recorded at ambient pressure. Maybe this feature is present even at ambient conditions but masked by the high resistivity due to the strong anion order. We call for further investigations of these relations.

4.2. $(TMTTF)_2BF_4$

In Figure 12 we present our dc resistivity results $\rho_a(T, p)$ and $\rho_c(T, p)$ for $(TMTTF)_2BF_4$, the compound with the smallest tetrahedal anion. At first sight, the resistivity of the two crystal directions seems not to differ much, except of the absolute value. Dominated by an activated temperature dependence, a pronounced kink at the CO transition is present for all investigated hydrostatic pressure values. It is followed by an increase in slope for $T < T_{CO}$. At the anion ordering both $\rho_a(T)$ and $\rho_c(T)$ exhibit a step-like feature, but then continue with more or less the same derivative. As pressure rises, this step-down is washed out and has eventually disappeared completely. The activation energy is significantly reduced for rising pressure. A closer look reveals distinct differences between the transport in the different crystallographic directions: ρ_a and ρ_c. Most obvious, the temperature-dependent dynamics along the a-axis appears about two orders of magnitude higher for all pressures compared to the transport along the c^*-direction. Pressure has a much more severe effect on the perpendicular transport ρ_c than parallel to the stacks, ρ_a. For low temperatures, the high-pressure resistivity becomes almost isotropic.

The analysis of the charge and anion ordering transitions is explained in Figure 12. In panels c and d the derivatives $d \ln\{\rho_a\}/d\left(\frac{1}{T}\right)$ are plotted as a function of temperature in order to define the actual transition temperatures. Here we can also identify differences in the slopes, since this temperature derivative equals the gap size of an activated behavior. The kink in $\rho(T)$ at T_{CO} appears as a clear peak in the derivative. For increasing pressure the CO peak sharpens, it gets more pronounced and shifts to lower T.

For $T \leq T_{CO}$ and high pressures, we see a steep upturn in resistivity that occurs for $\rho_a(T, p)$ as well as for $\rho_c(T, p)$. It is followed by a clear change of the resistivity slope at T_{AO}. It is interesting to look at the ratio of the resistivity values at the CO transition and those at the AO transition: $\Delta\rho = \rho(T_{AO})/\rho(T_{CO})$. Along the stacks, the change is about five orders of magnitude at all pressures; for the c^*-direction, $\Delta\rho$ shrinks from 10,000 at ambient pressure to only about 100 at the highest investigated pressure of $p = 11.5$ kbar. Hydrostatic pressure has a stronger impact on $\rho_c(T, p)$ than on $\rho_a(T, p)$. This picture is supported by the more pronounced loss in activation energy with pressure at $T \leq T_{AO}$ for the transverse transport $\rho_c(T, p)$ compared to the longitudinal one $\rho_a(T, p)$.

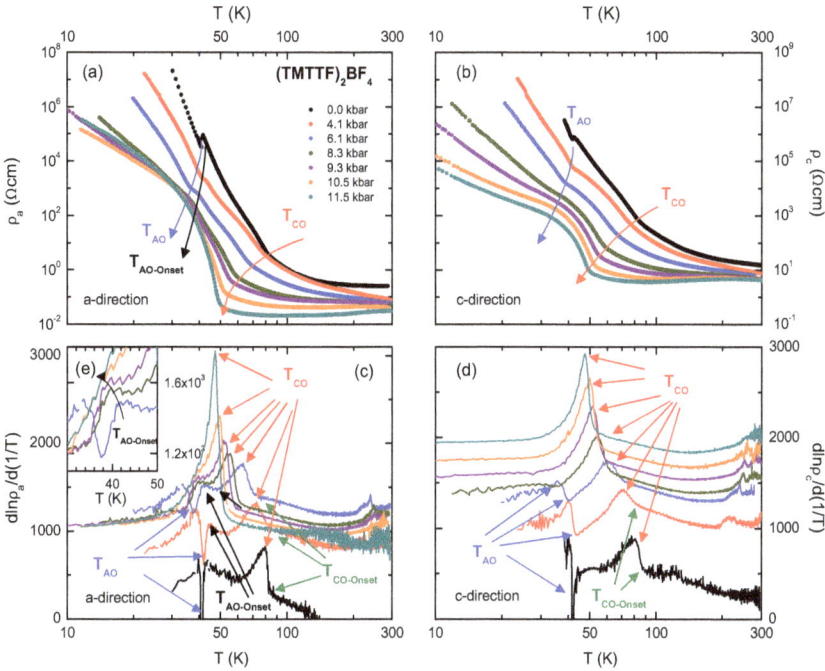

Figure 12. Temperature dependence of the dc resistivity of (TMTTF)$_2$BF$_4$ measured for the directions (**a**) parallel and (**b**) perpendicular to the stacks when increasing hydrostatic pressure is applied as indicated. Besides the minimum in resistivity at T_ρ, the charge-order transition at T_{CO} and the anion ordering at T_{AO} is clearly observed. The analysis of the data is illustrated in panels (**c**) and (**d**) where the temperature derivative d ln$\{\rho_a\}/$d $(1/T)$ is plotted for both orientations. The extrema indicate the ordering temperatures T_{AO} and T_{CO}. In addition we define an onset temperature for the charge and anion orders. The inset (**e**) enlarges the range around the AO temperature.

At T_{AO}, $\rho(T, p)$ undergoes another remarkable change in its behavior that strongly depends on pressure. Under ambient conditions, AO results in a discrete jump in $\rho(T)$ to lower values; since the slope is preserved, the gap size in an activated behavior remains unchanged at the transition. This observation was explained [4] by a reduction of scattering as the anions are locked in the methyl cavities. This jump becomes smooth when pressure is applied; a plateau starts to develop gradually. Increasing pressure further, $\rho(T)$ becomes rather broad for both directions, and eventually the AO transition cannot be identified as a discrete feature any more; what remains is an upturn in resistivity in a wide temperature range. Looking at the derivative in Figure 12c,d, T_{AO} is identified as a sharp peak in the low-pressure range. But for $p \geq 8.3$ kbar this characteristic vanishes. It is interesting to note that the onset of the AO transition at $T_{AO-onset}$ related to the upturn in d ln$\{\rho_a\}/$d $\left(\frac{1}{T}\right)$, seen just before T_{AO} (Figure 12c and inset e), can be distinguished up to the highest pressure. For $p = 11.5$ kbar the AO-onset peak is dominated by the strong CO peak coming very close in temperature. Surprisingly, no onset feature of AO is present in ρ_c. Nevertheless, these observations in ρ_a hint that the AO transition might survive even the high hydrostatic pressure. No doubt T_{AO} shifts down in temperature much slower with pressure than T_{CO}.

In Figure 13a,b we focus on the high-temperature regime in order to inspect the metallic behavior and charge localization. For high pressure, $p \geq 10.5$ kbar, both $\rho_a(T)$ as well as for $\rho_c(T)$ exhibit a metal-like temperature dependence. The derivatives plotted in Figure 13c,d evidence a crossing of the zero line at around $T = 100$ K. We should note that in the clamped cell pressure losses might occur when cooling down from high temperatures, masking the inherent metallic behavior. Thus the second crossing point should be considered with some reservations, when drawing the phase diagram in Figure 13e. We associate the change in slope $d\rho_a(T)/dT$ with the charge localization temperature $T_{loc} \approx 240$ K. With increasing pressure T_{loc} shifts to lower temperatures. For the perpendicular transport, $\rho_c(T)$ exhibits an activated behavior under ambient conditions. The boundary between the metallic regions in Figure 13e therefore indicates a lower limit.

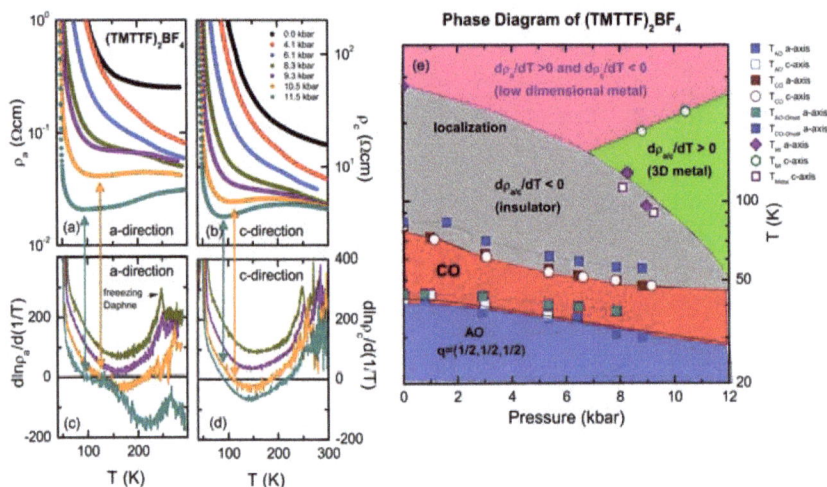

Figure 13. The panels on the left display the temperature dependence of the resistivity of $(TMTTF)_2BF_4$ measured for different hydrostatic pressure applied with the main focus on the metal-insulator transition and charge localization. (**a**) The on-chain resistivity $\rho_a(T)$ exhibits the development of a minimum as pressure increases. (**b**) Resistivity $\rho_c(T)$ perpendicular to the stacks for different pressure values. (**c**,**d**) Pressure evolution of the resistivity derivative $d\ln(\rho)/d(1/T)$ as a function of temperature for both orientations. The metal-insulator transition is indicated by the arrows. Also seen is the freezing of the pressure transmitting oil Daphne 7373. (**e**) In the phase diagram of $(TMTTF)_2BF_4$ different electronic states are depicted as derived from dc transport experiments. Upon cooling the low-dimensional metal exhibits a charge localization that moves from about room temperature down to below 100 K as pressure increases. For pressure above 8 kbar a three-dimensional metallic phase develops. At lower temperatures charge order and anion order appear subsequently. Both phase boundaries move towards lower temperature with increasing pressure.

4.3. $(TMTTF)_2ReO_4$

As listed in Table 1, $(TMTTF)_2ReO_4$ is the compound with the highest CO and AO transition temperatures in the whole family of Fabre salts. At ambient pressure $(TMTTF)_2ReO_4$ develops CO at $T_{CO} \approx 230$ K and the AO transition at $T_{AO} \approx 160$ K. Figure 14a displays the results of our pressure and temperature-dependent dc resistivity measurements along the perpendicular direction. The corresponding temperature derivative $d\ln\{\rho_c\}/d\left(\frac{1}{T}\right)$ is a measure of the energy gap for activated charge transport; its temperature dependence is presented in Figure 14b. The CO temperature T_{CO} can be identified as a kink in $\rho_c(T, p)$. Even more pronounced is the anion order at T_{AO}.

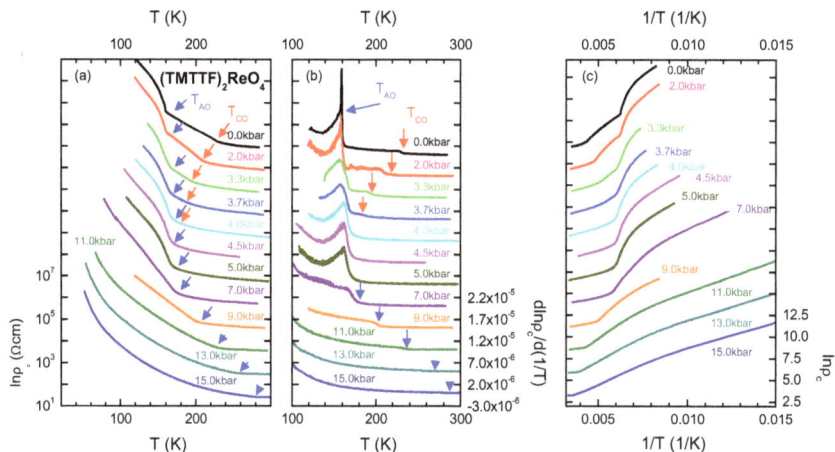

Figure 14. (**a**) Temperature-dependent resistivity of (TMTTF)$_2$ReO$_4$ measured along the c^*-direction for different values of external pressure. The red and black arrows indicate the charge order and anion order transition temperatures. (**b**) The temperature derivative $d\ln(\rho)/d(1/T)$ shows the charge-order transition as a step in the temperature dependence; the anion ordering T_{AO} is seen as a pronounced peak. Both exhibit a distinct pressure dependence. (**c**) Arrhenius plot of the resistivity $\rho_c(T)$ as the pressure increases. All curves are shifted with respect to each other for clarity reasons.

For $T < T_{AO}$ the resistivity increase can be described by an activated behavior with an energy gap that increases in a BCS-like fashion; more details on the dc transport at ambient pressure can be found by Köhler et al. [4]. When hydrostatic pressure is applied, T_{CO} shifts to lower values and resistivity changes at the transition get weaker. T_{AO} on the other hand does not move in temperature for $p < p_{crit} = 5$ kbar. Also the qualitative shape of $\rho_c(T)$ and its slope for $T \leq T_{AO}$ do not change significantly in this pressure range. In the derivative of the resistivity, the AO transition appears as a sharp peak, indicating a first-order phase transition for all values $p \leq p_{crit} = 5.0$ kbar. When the CO transition coincides with the anion ordering, a sudden change in the pressure dependence of T_{AO} is observed. Above a critical pressure $p_{crit} = 5.0$ kbar the AO related kink moves to higher temperatures when pressure increases further. In addition, the shape of $\rho_c(T)$ changes from a BCS-like to a rather straight and flat evolution of $\rho_c(T)$ for temperatures $T \leq T_{AO}$. For higher pressures, $p > p_{crit}$ the shape of the transition resembles a second-order phase transition. The strong reduction of the energy gap Δ at low temperatures for increasing pressures $p > p_{crit}$ can be seen best in the Arrhenius plot of Figure 14c.

In Figure 15c–e we show the temperature dependence of the charge and anion ordering gaps, Δ_{CO} and Δ_{AO}, respectively, for selected pressures. With increasing pressure $p < p_{crit}$, the energy gap Δ_{CO} continuously converges to zero at the critical pressure of 5.0 kbar is approached. On the contrary, Δ_{AO} stays fixed. For higher pressure $p > p_{crit}$, Δ_{AO} gets continuously suppressed as well. For comparison the dc measurements of Coulon et al. [41] along the chain direction, $\rho_a(T, p)$ are presented in Figure 15b. They exhibit qualitatively the same behavior as our dc results in the perpendicular direction.

By extracting the temperatures for the different phase transitions from our resistivity data we can compose a phase diagram of (TMTTF)$_2$ReO$_4$ presented in Figure 15a. The CO state is limited to low pressure. Pressure-dependent X-ray investigations may shed light on the origin of the sudden change at the critical pressure of $p_{crit} \approx 5.0$ kbar. By comparison with the sister compound (TMTSF)$_2$ReO$_4$, we can speculate about a change in anion superstructure. As shown in Figure 15f, Moret et al. [62] found a change from $q_2 = \left(\frac{1}{2}, \frac{1}{2}, \frac{1}{2}\right)$ to $q_3 = \left(0, \frac{1}{2}, \frac{1}{2}\right)$ superstructure vector as pressure exceeds a certain critical pressure.

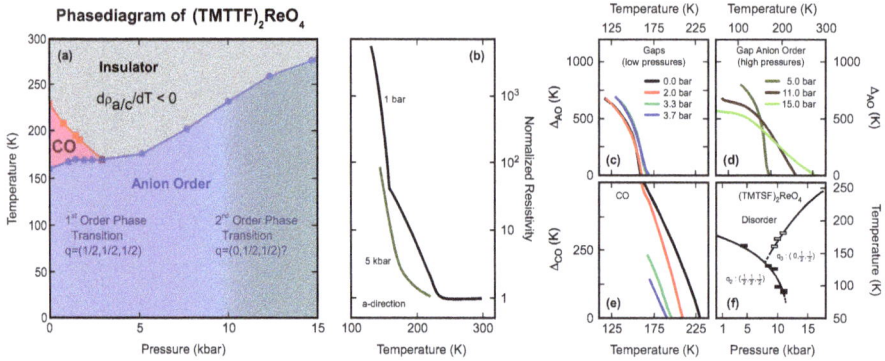

Figure 15. (**a**) Phase diagram of (TMTTF)₂ReO₄ as obtained from our pressure- and temperature-dependent transport measurements. Even for 15 kbar hydrostatic pressure, no metallic behavior is observed up to room temperature. The charge-ordered phase is limited to low pressure values. The transition into the anion-ordered state changes from first to second order. (**b**) Normalized dc resistivity versus temperature measured at 1 bar and 5 kbar along the chain direction of (TMTTF)₂ReO₄; reproduced from [41]. (**c–e**) Temperature change of the activation energy obtained below the AO transition and CO transition of (TMTTF)₂ReO₄ for different pressure. (**f**) Temperature-dependent anion-ordering phase diagram of (TMTSF)₂ReO₄ as obtained from X-ray scattering experiments of Moret et al. [62]. The q_2 and q_3 phase boundaries overlap at $p = 9.5$, 10, and 11 kbar and they have opposite slopes.

4.4. Discussion

For a better comparison, in Figure 16 we present next to each other the temperature-dependent resistivity of the three Fabre salts (TMTTF)₂X with tetrahedral anions, X = BF₄, ClO₄, and ReO₄, obtained at different pressure. In the case of centrosymmetric anions a simple relation between anions size and charge ordering could be derived [18]: the larger the anions, the larger the stack separation and the higher T_{CO}. This concept fails for the tetrahedral anions. Obviously the original idea of anions as pure spacers between the one-dimensional TMTTF stacks has to be discarded; instead we have to account for the three-dimensional lattice structure and in particular the interaction of the organic stacks with the anions.

Based on the size of the anions [52], $V_{Anion}^{BF_4} < V_{Anion}^{ClO_4} < V_{Anion}^{ReO_4}$, the CO temperature of (TMTTF)₂ClO₄ should fall right between $T_{CO}^{BF_4}$ and $T_{CO}^{ReO_4}$. Surprisingly, no CO transition is detected in (TMTTF)₂ClO₄. Provided it is not just an exception to the general rule, (TMTTF)₂ClO₄ may be the key compound for understanding the Fabre salts. Comparing our resistivity data for the three different compounds in Figure 16 reveals that the absence of CO has drastic consequences on the $\rho(T)$ for $T < T_{AO}$. At ambient pressure no qualitative difference is observed of X = BF₄ and ClO₄; at T_{AO} the resistivity exhibits a step to lower values followed by an activated temperature behavior. This common picture changes when pressure is applied. For (TMTTF)₂BF₄ we still find $\rho(T)$ to be thermally activated down to low temperatures. In (TMTTF)₂ClO₄, however, a large minimum in $\rho(T)$ develops for a temperature range of about 30 K. When the pressure has reached $p = 9.3$ kbar, T_{AO} has shifted down by 25 K for (TMTTF)₂ClO₄, while only 6 K in the case of (TMTTF)₂BF₄. The compound (TMTTF)₂ReO₄ does not show any appreciable temperature shift of T_{AO} for pressure up to $p = 5$ kbar; only after $T_{AO} \approx T_{CO}$ the AO temperature suddenly starts to move up quite fast with pressure, cf. Figure 15. At his point it is not clear whether this shift is due to the combination of charge and anion order or whether a new superstructure occurs, as in the case of (TMTSF)₂ReO₄ [62]. Ongoing spectroscopic investigations will clarify this point. In (TMTTF)₂BF₄ our data indicate the survival of both CO an AO under pressure. It would be of interest to follow the ordering temperatures to even higher pressure; we speculate that T_{AO} might move up in temperature as soon as it merges with T_{CO};

similar to (TMTTF)$_2$ReO$_4$. This would underline the competing influence of CO on the AO transition, as suggested by our transport data.

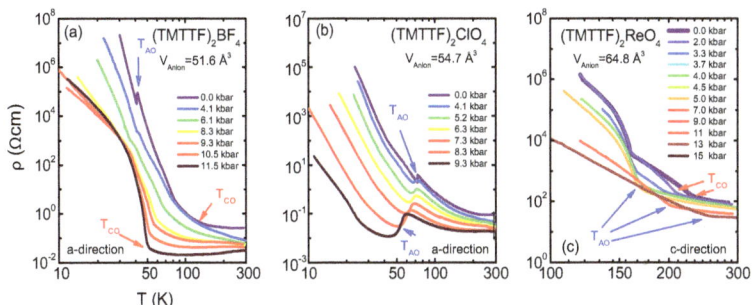

Figure 16. The temperature dependent resistivity for the (TMTTF)$_2$X salts with tetrahedral anions. The three compounds are sorted for increasing anion size from left to right. (**a**) (TMTTF)$_2$BF$_4$ is the compound with the smallest anion: $V_{Anion} = 51.6$ Å^3. It shows both transitions in the dc resistivity, charge and anion ordering. (**b**) In (TMTTF)$_2$ClO$_4$ with $V_{Anion} = 54.7$ Å^3 only anion ordering was detected but no separate charge order. (**c**) (TMTTF)$_2$ReO$_4$ has by far the largest tetrahedral anion with $V_{Anion} = 64.8$ Å^3; it also has the highest charge and anion ordering temperatures.

In order to gain further insight into the structural aspects on charge order, Figure 17 displays the distance between the anions and the TMTTF molecule. In the first frame T_{CO} is plotted as a function of the shortest distance between the anion-ligand (F or O) and the sulfur atom on the organic molecule for Fabre salts with tetrahedral as well as octahedral anions. In panel b we consider the distance of the anion center, defined by the B, Cl, Re, P, As or Sb-atoms, to the sulfur atom of the organic molecule. Finally, in Figure 17c we plot the distance of the anion-ligand to the nearest carbon atom. In those cases, T_{CO} behaves differently for (TMTTF)$_2$X with octahedral and tetrahedral X. It seems that the distance between the central atom of the anion and the nearest carbon atom reflects the cavity size or the anion size better than any coupling mechanism. The (theoretical) anion volume, listed in Table 1, follows the relation [52]: $V_{Anion}^{BF_4} < V_{Anion}^{ClO_4} < V_{Anion}^{ReO_4} < V_{Anion}^{PF_6} < V_{Anion}^{AsF_6} < V_{Anion}^{SbF_6}$. The correlations plotted in Figure 17 might help to understand some of the effects, but further temperature- and pressure-dependent investigations are needed to completely clarify the situation.

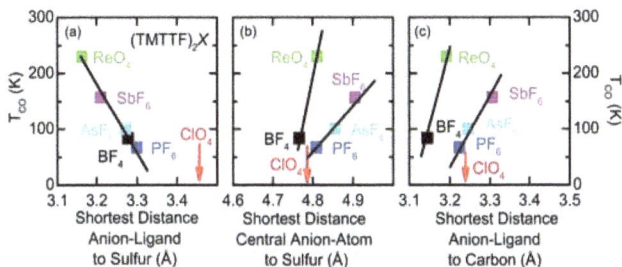

Figure 17. Dependence of the charge order temperature T_{CO} for various Fabre salts (TMTTF)$_2$X with different X as indicated. (**a**) Effect on the shortest distance between anion-ligand and TMTTF sulfur atom. (**b**) Change of T_{CO} with the distance between the central atom of the anion and the sulfur atom. (**c**) Influence on the shortest distance from the anion-ligand to the nearest carbon atom. The value for (TMTTF)$_2$ClO$_4$, which shows no CO, is indicated by a red arrow. All distances were determined at ambient conditions.

5. Dielectric Spectroscopy

Dielectric spectroscopy is a sensitive tool to detect changes in the charge degrees of freedom that have revealed evidence for CO in the (TMTTF)$_2$X salts [84,85]. Numerous studies performed over the years [30–36,86] identified a Curie-like peak of the permittivity $\epsilon_a(T)$ at T_{CO} reaching huge values up to 10^6 for $E \parallel a$. While the peak position is frequency-independent up to 1 MHz, its amplitude decreases with rising frequency. The Curie-behavior has drawn a lot of attention [84,85,87–90] because it is characteristic for ferroelectricity. Its origin is considered to be mainly electronic, and thus referred to as "electronic ferroelectricity" [33,91–93]. Nevertheless, the observation of a hysteresis in the electric polarization $P(E)$ is impeded by the high conductivity of the (TMTTF)$_2$X salts compared with canonical ferroelectrics.

Most of the studies dealt with Fabre salts of centrosymmetric anions, and there is only a single report on (TMTTF)$_2$ReO$_4$ and (TMTTF)$_2$BF$_4$ [36]. In the first compound, the AO transition is evidenced by a sharp drop of permittivity at T_{AO}; it is explained by the formation of TMTTF tetramers with a 0110 charge pattern, which are less polarizable compared to dimers [23]. Interestingly, for (TMTTF)$_2$BF$_4$ a small peak is revealed at T_{AO} with a frequency-dependent amplitude followed by a monotonous decrease of $\epsilon'_a(T)$ upon further cooling.

5.1. Results

We have conducted frequency-dependent measurements of the real part of the dielectric constant $\hat{\epsilon}_{b'} = \epsilon'_{b'} - i\epsilon''_{b'}$ with the electric field applied perpendicular to the chain direction, $E \parallel b'$, at different temperatures and ambient pressure. The findings for (TMTTF)$_2$X with X = BF$_4$, ClO$_4$ and ReO$_4$ taken at $f = 1, 10, 100$ kHz, and 1 MHz are plotted in Figure 18. For (TMTTF)$_2$BF$_4$ and (TMTTF)$_2$ReO$_4$ the charge and anion-order transitions are clearly observed with exactly the same signatures as reported for measurements along the stacking axis [36].

In the case of (TMTTF)$_2$BF$_4$, this corresponds to a peak at $T_{CO} \approx 80$ K, which is reduced in amplitude as frequency increases. For $f = 100$ kHz the peak height of 3000 is lower by factor of 200 compared to a value of $7 \cdot 10^5$ observed for $E \parallel a$ [36]. When the temperature is lowered, $\epsilon'_{b'}(T)$ drops until the AO transition, which is observed here by another peak around 36 K, in accord with the findings for $E \parallel a$. In contrast to the feature T_{CO}, the AO peak is strongly suppressed with increasing frequency. While for $f = 10$ kHz, a value of $\epsilon'_{b'} = 20$ is observed, the peak cannot be identified above 1 MHz.

For the ReO$_4$ salt, the CO peak appears at 228 K and exhibits a maximum of $4 \cdot 10^4$ when measured at $f = 100$ kHz perpendicular to the chains; this is one order of magnitude lower than the peak found for $E \parallel a$ [36]. With decreasing temperature $\epsilon'_{b'}(T)$ continuously decreases by one order of magnitude at $T_{AO} = 157$ K, where a step is observed with a frequency-dependent height. Upon further cooling, $\epsilon'_{b'}(T)$ approaches a constant value around 2000, which is slightly dependent on frequency.

(TMTTF)$_2$ClO$_4$ exhibits a strong and monotonous decrease in $\epsilon'_{b'}(T)$ from room temperature down to about 100 K. At T_{AO} an abrupt step with a frequency-independent height up to 50 marks the anion order. At lower temperatures $\epsilon'_{b'}(T)$ changes in shape forming a shoulder-like drop of one order of magnitude. Interestingly, for higher frequencies this drop occurs at higher temperatures reminiscent of a dielectric relaxation.

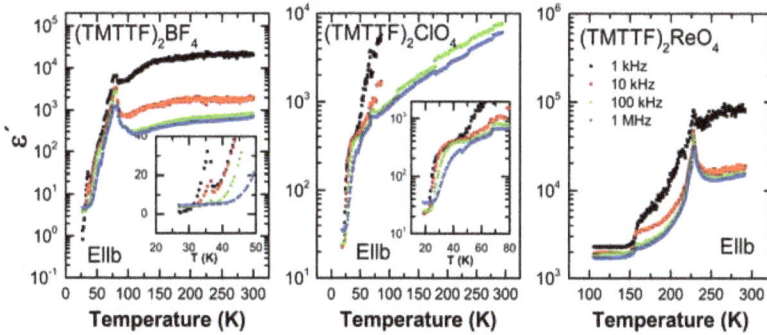

Figure 18. Temperature dependence of the real part of the dielectric permittivity $\epsilon'(T)$ measured with $E \parallel b'$ for $(TMTTF)_2 X$ with $X = BF_4$, ClO_4 and ReO_4 at different frequencies as indicated. The insets emphasize the changes close T_{AO} and below.

5.2. Analysis

In general, a ferroelectric phase transition is accompanied by a soft mode strongly slowing down near the critical temperature. It is usually observed in optical, infrared or microwave spectroscopy for a displacive phase transition; or by a relaxation mode at lower frequencies, as seen in disordered systems [94,95]. Hence, investigating the relaxation time can provide valuable insight into the dynamics related to the transitions. The Cole-Cole or generalized Debye model is a well established and useful description to analyze the frequency-dependent complex permittivity

$$\hat{\epsilon}(\omega) - \epsilon_{\inf} = \frac{\Delta\epsilon}{1 + (i\omega\tau_0)^{1-\alpha}} \quad , \tag{1}$$

where τ_0 is the relaxation time, $\omega = 2\pi f$ the angular frequency of the applied electric ac-field, $1 - \alpha$ a parameter describing symmetric broadening of the loss peak and $\Delta\epsilon = \epsilon_{static} - \epsilon_{\inf}$ the dielectric strength with ϵ_{static} and $\epsilon_{\inf}$ the values for low and high frequencies, respectively. The deviation from the Debye model ($\alpha = 0$) is attributed to the distribution of τ_0 arising from disorder [84]. If the materials are not completely insulating, but exhibit a considerable electronic background, such as the $(TMTTF)_2 X$ salts, the dc-conductivity has to be taken into account when the dielectric loss is determined:

$$\epsilon''(\omega) = \frac{\sigma'(\omega) - \sigma_{DC}}{\omega\epsilon_0} \quad . \tag{2}$$

As an example for $(TMTTF)_2 X$ salts with tetrahedral anions investigated here, Figure 19 shows the dielectric spectra of $(TMTTF)_2 ClO_4$ in the range of the relaxational mode for different temperatures close to the AO transition. The real part of the dielectric constant $\epsilon'_{b'}(\omega)$ is plotted in the upper panel: with increasing frequency it exhibits a step that shift towards higher frequencies as the temperature increases. The point of inflection marks the relaxation rate $1/\tau_0$ and corresponds to the maximum of the broad peak in the imaginary part $\epsilon''_{b'}(\omega)$. The solid lines represent fits according to Equation (1) and agree well with the expected behavior.

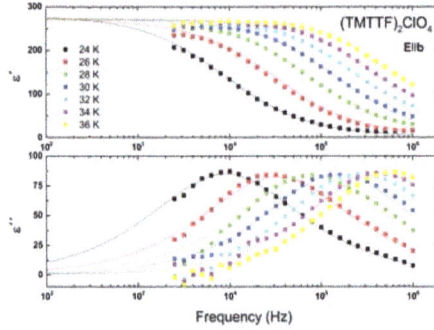

Figure 19. Frequency dependence of the permittivity $\hat{\varepsilon}(\omega) = \varepsilon'(\omega) - i\varepsilon''(\omega)$ of $(TMTTF)_2ClO_4$ measured with $E \parallel b'$ at various temperatures close to the AO transition. The real part $\varepsilon'(\omega)$ exhibits a roll-off that shifts to higher frequencies with increasing temperature. This behavior corresponds to a peak in $\varepsilon''(\omega)$ as shown in the lower panel. The solid lines represent fits according to the generalized Debye model (1).

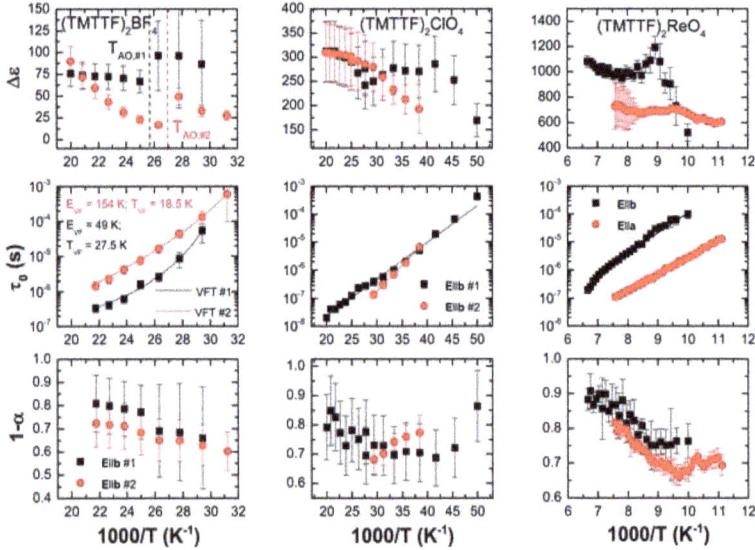

Figure 20. Temperature dependence of the parameters used to fit the dielectric spectra of $(TMTTF)_2BF_4$, $(TMTTF)_2ClO_4$ and $(TMTTF)_2ReO_4$ by the Cole-Cole model (1). The upper row shows the Arrhenius plots of the dielectric strength $\Delta\varepsilon$, followed by the relaxation time τ_0 and the broadening parameter $(1 - \alpha)$ in the lower row. For X = BF$_4$ and ClO$_4$ the data on two crystals each (indicated by black and red symbols) are taken for $E \parallel b'$. For X = ReO$_4$ we show data measured along the a and b'-directions. Most important, in $(TMTTF)_2ClO_4$ and $(TMTTF)_2ReO_4$ the relaxation time τ_0 exhibits an activated behavior; the obtained activation energies resembles the gap previously measured in dc transport [4]. For $(TMTTF)_2BF_4$, $\tau_0(T)$ increases with decreasing T in a fashion best described by the empirical Vogel-Fulcher-Tammann relation.

The parameters $\Delta\varepsilon$, τ_0 and $1 - \alpha$ extracted from the Cole-Cole fit of the dielectric spectra are plotted in Figure 20 as a function of inverse temperature. The data have been recorded along the b' direction of the $(TMTTF)_2BF_4$, $(TMTTF)_2ClO_4$ and $(TMTTF)_2ReO_4$ salts. We confined ourselves to this temperature region because our experimental frequency window is restricted to 1 MHz. For $(TMTTF)_2BF_4$ the relaxation mode moves through this frequency window for 50 K > T > 30 K, which includes the AO transition. Here $\Delta\epsilon$ exhibits a small peak around $T = 38$ K and 36 K for the two different samples, labeled 1 (black) and 2 (red symbols). These steps correspond to the features observed in $\hat{e}(T)$ (Figure 18) and are ascribed to the AO transition. The broadening parameter $(1 - \alpha)$ decreases upon cooling indicating an increased cooperativity in the relaxation. Interestingly, $\tau_0(T)$ rises with lowering temperature in a fashion best described by the empirical Vogel-Fulcher-Tammann relation

$$\tau_0 = \tau_{VF} \exp \left\{ \frac{E_{VF}}{T - T_{VF}} \right\} \quad , \tag{3}$$

which is known as a good parametrization for the slowing down of molecular motion in disordered systems and the glass-like freezing of dipolar order in relaxor ferroelectrics. The corresponding energy E_{VF} can be interpreted as a temperature-dependent activation energy for reorientational motion. T_{VF} denotes the temperature where τ_0 diverges, and τ_{VF} the time scale for the ac response in the high-temperature limit. The glass temperature T_G is defined as the temperature where $\tau_0 = 100$ s. The obtained fit parameters are listed in Table 2. There is a considerable difference in the values for E_{VF} and τ_{VF}, whereas there is a minor deviation for T_G which we attribute to different cooling rates during the measurements.

Table 2. Parameters obtained by fitting the temperature-dependent dielectric relaxation time $\tau_0(T)$ of $(TMTTF)_2BF_4$ with the Vogel-Fulcher-Tammann relation. E_{VF} is the activation energy, T_{VF} denotes the Vogel-Fulcher temperature, and τ_{VF} the corresponding time scale. The glass temperature is given by T_G.

$(TMTTF)_2BF_4$	Sample 1	Sample 2
E_{VF} (K)	49 ± 7	153.5 ± 11.5
T_{VF} (K)	27.5 ± 0.5	18.5 ± 0.5
τ_{VF} (s)	$(2.5 \pm 1.2) \cdot 10^{-8}$	$(6.1 \pm 2.5) \cdot 10^{-9}$
T_G (K)	30	25

For $(TMTTF)_2ClO_4$ the mode can be observed in our frequency window between $T = 20$ K and 50 K, which is below T_{AO}, but in the range of the dispersive shoulder-like drop in $\epsilon'_{b'}(T)$ shown in Figure 18. For the second crystal, we observe a monotonous decrease of $\Delta\epsilon$ with falling temperature, in agreement with the behavior of $\epsilon'_{b'}(T)$. There is an additional small bump around $T = 23$ K for sample 1. In both specimens an activated behavior in τ_0 is observed with activation energies of 300 K and 420 K, respectively. The fit parameters are listed in Table 3. We can compare our findings with the energy gap of (440 ± 60) K derived from dc-transport measurements reported in Reference [4].

Table 3. Parameters of the activated behavior of τ_0 for $(TMTTF)_2ClO_4$ derived from dielectric measurements of two different crystals.

$(TMTTF)_2ClO_4$	Sample 1	Sample 2
E_{act} (K)	300 ± 10	420 ± 8
τ_{act} (s)	$(5.7 \pm 1.9) \cdot 10^{-11}$	$(6.0 \pm 1.7) \cdot 10^{-13}$

The relaxation mode in $(TMTTF)_2ReO_4$ slows down more quickly; it enters our experimental frequency window for 90 K < T < 140 K, right below the AO transition. Surprisingly, there is a small peak in $\Delta\epsilon$ at $T = 112$ and 105 K for $E \parallel b'$ and $E \parallel a$, respectively, which superimposes a gradual increase with temperature. No corresponding counterpart is observed in $\epsilon'_{b'}(T)$. For both directions, τ_0 follows an activated behavior; the fit parameters are listed in Table 4 for the two polarization directions.

From dc transport a temperature-dependent energy gap was derived [4] that saturates at (1560 ± 80) K in the limit of $T \rightarrow 0$ for all crystal directions.

Table 4. Parameters of the activated behavior observed in τ_0 for two crystal directions of $(TMTTF)_2ReO_4$.

$(TMTTF)_2ReO_4$	$E \parallel a$	$E \parallel b'$
E_{act} (K)	1372 ± 4	1651 ± 27
τ_{act} (s)	$(3.1 \pm 0.1) \cdot 10^{-12}$	$(7.3 \pm 1.6) \cdot 10^{-12}$

Since for $(TMTTF)_2ClO_4$ and $(TMTTF)_2ReO_4$ the activation energies of the relaxation time τ_0 agree well with the values obtained by dc-transport, we conclude that the relaxation is determined by the free charge carries responsible for the dc conduction, which freeze out at low temperatures.

5.3. Discussion

For the $(TMTTF)_2X$ salts, the peak in ϵ' at T_{CO} obeys Curie's law as established for canonical order-disorder ferroelectrics [84]; it is attributed to a combination of charge disproportionation and ionic displacements [32,33].

As far as the anisotropy is concerned, the ratio of the peak amplitudes $\epsilon'_a / \epsilon'_{b'}$ probed at $f = 100$ kHz is about 200 and 1000 for $(TMTTF)_2BF_4$ and $(TMTTF)_2ReO_4$, respectively, higher than the anisotropy of the dc resistivity at T_{CO} [4]. It is interesting to compare our findings with the report of de Souza et al. [96] on the Fabre sals with octahedral anions. As reproduced in Figure 21, they found a strong peak in ϵ'_c at T_{CO} similar to previous reports on $\epsilon'_a(T)$ [30–32,34–36] and $\epsilon'_{b'}$ presented here (Figure 18). The peak value is sample-dependent and in general by several orders of magnitude lower compared the value observed along the stacks. While in the a-direction the permittivity is affected by critical fluctuations that become important when approaching the ordering transition in one dimension, the response in the perpendicular direction probes the involvement of ionic displacements and the stabilization of the three-dimensional charge pattern upon CO.

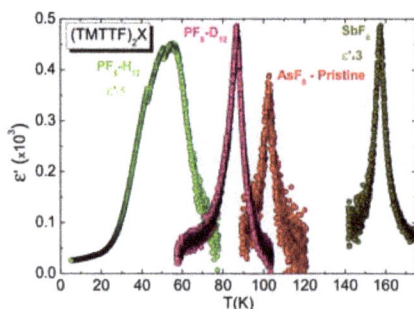

Figure 21. Temperature dependence of the dielectric constant of several $(TMTTF)_2X$ salts with octahedral anions: H_{12}-$(TMTTF)_2PF_6$ and D_{12}-$(TMTTF)_2PF_6$, $(TMTTF)_2AsF_6$ and $(TMTTF)_2SbF_6$. The experiments have been performed along the c^*-axis perpendicular to the chain direction [96]. The maximum in $\epsilon'_c(T)$ is located at T_{CO}, similar to measurements along the other orientations $\epsilon'_a(T)$ [30–32,34–36] and $\epsilon'_{b'}(T)$. The peak value is lower by several orders of magnitude compared to ϵ'_a, similar to the anisotropy determined in dc transport [4].

The rather high values of $\epsilon'(T)$ at room temperature, found in Figure 18, are attributed to the slight dimerization of the TMTTF molecules and the resulting enhanced polarizability, which scales with $\epsilon' \approx (\omega_p/\Delta)^2$, wherein ω_p is the plasma frequency and Δ the transport gap [97]. The temperature dependence of $\Delta(T)$ for $E \parallel b'$ is found [4] constant for $(TMTTF)_2ReO_4$ from $T = 300$ K to 250 K, whereas it is decreasing for (TMTTF)$_2$ClO$_4$ and (TMTTF)$_2$BF$_4$; the latter is explained by the decrease in dimerization. Accordingly, the reduction of $\epsilon'_{b'}(T)$ upon cooling observed here is induced by the decrease of $\omega_p(T)$ due to the freezing out of mobile charge carriers. This can be confirmed by detailed optical investigation of all (TMTTF)$_2$X salts.

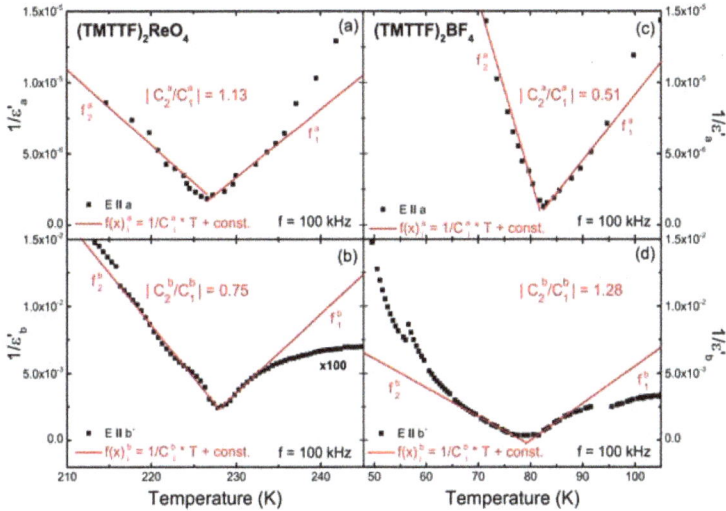

Figure 22. Temperature dependence of (a,c) $1/\epsilon'_a(T)$ (data from [36]) and (b,d) $1/\epsilon'_{b'}(T)$ for (a,c) (TMTTF)$_2$BF$_4$ and (b,d) (TMTTF)$_2$ReO$_4$. There is good agreement with the Curie-Weiss behavior (red solid lines) for $T \leq T_{CO}$, except for (TMTTF)$_2$BF$_4$ along b'-direction, where the peak in $\epsilon'_{b'}(T)$ is somewhat rounded. For canonical second-order ferroelectric transitions, mean-field theory predicts that the slopes on both side differ by a factor of 2 [94]. The deviation from this expected behavior is usually attributed to structural disorder [36,86].

The Curie-Weiss behavior $\epsilon'(T) = \frac{C}{T-T_{CO}}$ is nicely illustrated in Figure 22 where $1/\epsilon'_a(T)$ and $1/\epsilon'_{b'}(T)$ are plotted for (TMTTF)$_2$BF$_4$ and (TMTTF)$_2$ReO$_4$. In particular for $T \leq T_{CO}$ the agreement is rather good, except for (TMTTF)$_2$BF$_4$ along the b'-direction (cf. Figure 18). Above T_{CO} we observe rather large deviations from the expected behavior for both salts.

At this point it is important to mention that the amplitude and broadening of the CO peak in $\epsilon'_a(T)$ is known to be sample-dependent [36,86] and hence, in case a deviation from the Curie-behavior is observed below T_{CO}, it is ascribed to structural disorder. The latter also explains, why the slope ratio $|C_2^{a,b}/C_1^{a,b}|$ deviates in some reports [86,96] from the expected value 0.5, which is predicted by mean field theory for a canonical second-order phase transition [94].

The ratio of the slope obtained along the b'-direction does not necessarily infer a minor sample quality. The CO peak is weaker along the b'-direction, implying that at high temperatures the contribution of dimerization to the background could be appreciable. Similar deviations from this mean-field dependence are observed for (TMTTF)$_2$AsF$_6$ and (TMTTF)$_2$PF$_6$ when looking along the c^*-direction [96]. Additional ionic contributions are more pronounced in the perpendicular polarizations and should be investigated in more detail in future.

For (TMTTF)$_2$ReO$_4$ the abrupt drop of the permittivity at T_{AO} evidences the formation of TMTTF tetramers with a 0110 charge pattern; they are less polarizable compared to dimers. Interestingly, in (TMTTF)$_2$BF$_4$ a small peak is observed at T_{AO}, followed by the drop due to tetramerization. The absence of a comparable peak in the other two salts may be due to screening by free charge carriers. In (TMTTF)$_2$BF$_4$ the resistivity at T_{AO} is three orders of magnitudes higher compared to (TMTTF)$_2$ReO$_4$, for instance [4]. In the case of (TMTTF)$_2$ClO$_4$ there is no CO for $T > T_{AO}$. Below T_{AO}, around $T = 50$ K, there is a second drop which shifts to higher temperature for increasing frequency, reminiscent of a dielectric relaxation. Note that in this temperature range, ESR measurements reveal a drop of the spin susceptibility [73], which is attributed to the formation of a spin singlet-triplet gap upon tetramerization, and optical investigations [23] prove the growth of charge disproportionation with a 0110 charge pattern. Further investigations have to fully clarify the connection to the observed dielectric relaxation. The observation of a corresponding feature in $(TMTTF)_2BF_4$ and $(TMTTF)_2ReO_4$ might be obscured by a strong background due to the CO peak.

In $(TMTTF)_2AsF_6$ the mean relaxation time $\tau_0(T)$ contains a peak at T_{CO} due to the softening of the oscillating mode concomitant to the ferroelectric transition and an activated behavior with an activation energy similar to the gap obtained by dc transport [35,86]. Brazovskii and coworkers [89] report an additional relaxation mode in $(TMTTF)_2AsF_6$ for $T < T_{CO}$, which is ascribed to slow oscillations of pinned ferroelectric domains. Recent Raman spectroscopy evidences the presence of neutral $TMTTF^0$ and ionized $TMTTF^{-1}$ molecules, which are assigned to charged domain walls [56]. It is worthwhile to mention that in the two-dimensional charge-transfer salts κ-$(BEDT-TTF)_2X$, with $X = Cu[N(CN)_2]Cl$ [98], $Cu_2(CN)_3$ [99,100] and $Ag_2(CN)_3$ [101], as well as in α-$(BEDT-TTF)_2I_3$ [102,103], charge domain walls are also considered to give rise to an anomalous dielectric response in the audio- and radio-frequency range. For $(TMTTF)_2ClO_4$ and $(TMTTF)_2ReO_4$ the activation energies of the relaxation time $\tau_0(T)$ agree well with dc-transport measurements, in accord with the observations for $(TMTTF)_2PF_6$ [31] and $(TMTTF)_2AsF_6$ [35,86]. This indicates that the relaxation is determined by the free charge carries responsible for the dc-conduction, which freeze out at low temperatures. We assume that crystal defects and charged impurities locally change the charge pattern due to anion order, giving rise to domains walls. If charge carriers are thermally activated above the transport gap, which is composed of the contributions due to dimerization, charge and anion order [4], they move along the TMTTF chains adding to the dc-conductivity. This changes the charge pattern and results in an effective shift of the domain walls, explaining why the activation energy of the relaxation time corresponds to the dc-transport gap.

For $(TMTTF)_2BF_4$ with $T_{AO} = 41.5$ K, the relaxation mode is observed for $50 K > T > 30$ K and the relaxation time $\tau_0(T)$ follows the empirical Vogel-Fulcher-Tammann relation. In this temperature range, the resistivity is rather high and transport is ascribed to hopping conduction [4]. The drastic decreased number of free charge carriers results in glass-like freezing of domain walls, as evidenced by the observation of the Vogel-Fulcher behavior. Surprisingly, the results obtained here for $T < T_{AO}$ on the salts with tetrahedral anions agree well with the ones obtained for $T < T_{CO}$ on the centrosymmetric anions, indicating that the underlying order determining the charge pattern is of minor role for the relaxation dynamics.

6. Conclusions

Our pressure-dependent DC-transport measurements yield valuable information on the influence of the lattice on the ordering phenomena in the quasi one-dimensional charge-transfer salts $(TMTTF)_2X$ with tetrahedral anions $X = BF_4$, ClO_4 and ReO_4. We observe a decrease of the charge ordering temperature T_{CO} upon pressure for $(TMTTF)_2BF_4$ and $(TMTTF)_2ReO_4$; similar to the observation reported for salts with octahedral anions [104]. In contrast to CO, the pressure dependence of AO differs for each salt such that the construction of a generic phase diagram of $(TMTTF)_2X$ salts with tetrahedral anions is not possible. In $(TMTTF)_2BF_4$, CO is visible up to the highest pressure applied with $T_{CO}(p = 8.8kbar) = 47.5$ K whereas the signature indicating AO dissolves around 34.5 K at a pressure of 5.3 kbar. We call for structural studies that will yield information on the actual arrangement of the anions. In $(TMTTF)_2ReO_4$, AO is enhanced upon applying pressure with an increase of T_{AO}, at around 3.0 kbar $T_{AO} = T_{CO}$ is reached and for pressures above, CO seems to be fully suppressed and the AO shifted up to 277 K at $p = 14.7$ kbar. Interestingly in $(TMTTF)_2ClO_4$, the jump-like feature indicating AO shifts to lower temperatures and significantly broadens. For $p > 10.2$ kbar, we observe a second peak evolving right at T_{AO} whose origin is still unclear. The fact that no CO is detected in

(TMTTF)$_2$ClO$_4$ is probably the most puzzling observation that calls for explanation. We attribute the absence of CO in (TMTTF)$_2$ClO$_4$ to the polarizability of the anion. ClO$_4$ is less polarizable because Cl has an electronegativity comparable to O. Consequently, the bonds within the anion are more covalent making the anion more stiff and less deformable. As a result, the ClO$_4$ adapts less to the methyl cavity and deforms them more strongly. The deformation leads to a change of the π-orbital direction, which – considering one particular anion stack – applies to every second TMTTF molecule of a neighboring stack. This modulates the direction of the transfer integrals for every second TMTTF molecule and results in an increased value of the electronic dimerization δ_{elec} (cf. Table 1), preventing the emergence of CO in (TMTTF)$_2$ClO$_4$.

We furthermore performed measurements of the permittivity along b' direction, perpendicular to the stacking, and observe qualitatively the same signatures in $\epsilon'_{b'}$ as in literature for ϵ'_a. The ratio of the peak amplitudes $\epsilon'_a/\epsilon'_{b'}$ at 100 kHz is 200 and 1000 for (TMTTF)$_2$BF$_4$ and (TMTTF)$_2$ReO$_4$, respectively, are higher then the anisotropy at T_{CO} determined by dc resistivity measurements [4]. This reflects the one-dimensional nature of the critical fluctuations when approaching T_{CO}, whereas the reduced value along the b'-direction is due to the displacement of anions upon formation of the three-dimensional charge pattern. For all salts investigated, an abrupt drop of the permittivity at T_{AO} is detected, which is attributed to the formation of TMTTF tetramers with 0110 charge pattern. By analyzing the frequency dependence of ϵ_b, we observe a dielectric relaxation below T_{AO}. In (TMTTF)$_2$ClO$_4$ and (TMTTF)$_2$ReO$_4$, we observe an activated behavior of the relaxation time with activation energies resembling the gap measured in transport, indicating that the relaxation dynamics are determined by free charge carriers. This agree quite well with observations in (TMTTF)$_2$PF$_6$ [31] and (TMTTF)$_2$AsF$_6$ [35,86] for $T < T_{CO}$, for which the relaxation time is activated and corresponding activation energies match to the dc-transport gap as well. This indicates that the relaxation is determined by the free charge carries responsible for the dc conduction which freeze out at low temperatures and that the underlying order giving rise to the transport gap plays a minor role.

Acknowledgments: We appreciate financial support by the Deutsche Forschungsgemeinschaft (DFG) and Deutsche Akademischer Austauschdienst (DAAD).

Author Contributions: E.R. and K.S. conducted the pressure-dependent transport experiments, T.B., M.G., E.R. and T.I. performed the dielectric measurements, R.R., E.R. and A.P. contributed to the data analysis, R.R., E.R. and M.D. wrote the paper; all authors contributed to the discussion.

Conflicts of Interest: Declare conflicts of interest or state The authors declare no conflict of interest.

Abbreviations

The following abbreviations are used in this manuscript:

TMTTF	tetramethyltetrathiafulvalene
TMTSF	tetramethyltetraselenafulvalene
CO	charge ordering
AO	anion ordering
NEXAFS	X-ray absorption near edge structure
NMR	nuclear magnetic resonance
ESR	electron spin resonance
DFT	density functional theory

References

1. Giamarchi, T. *Quantum Physics in One Dimension*; Clarendon Press: Oxford, UK, 2004.
2. Dressel, M. Spin-charge separation in quasi one-dimensional organic conductors. *Naturwissenschaften* **2003**, *90*, 337–344.
3. Dressel, M. Ordering phenomena in quasi-one-dimensional organic conductors. *Naturwissenschaften* **2007**, *94*, 527–541.

4.   Köhler, B.; Rose, E.; Dumm, M.; Untereiner, G.; Dressel, M. Comprehensive transport study of anisotropy and ordering phenomena in quasi-one-dimensional $(TMTTF)_2X$ salts $(X=PF_6,AsF_6,SbF_6,BF_4,ClO_4,ReO_4)$. *Phys. Rev. B* **2011**, *84*, 035124.

5.   Dressel, M.; Hesse, P.; Kirchner, S.; Untereiner, G.; Dumm, M.; Hemberger, J.; Loidl, A.; Montgomery, L. Charge and spin dynamics of TMTSF and TMTTF salts. *Synth. Met.* **2001**, *120*, 719–720.

6.   Seo, H.; Hotta, C.; Fukuyama, H. Toward systematic understanding of diversity of electronic properties in low-dimensional molecular solids. *Chem. Rev.* **2004**, *104*, 5005–5036.

7.   Jérome, D. The physics of organic conductors. *Science* **1991**, *252*, 1509–1514.

8.   Granier, T.; Gallois, B.; Ducasse, L.; Fritsch, A.; Filhol, A. 4 K crystallographic and electronic structures of $(TMTTF)_2X$ salts $(X^-: PF_6^-, AsF_6^-)$. *Synth. Met.* **1988**, *24*, 343–356.

9.   Iwase, F.; Sugiura, K.; Furukawa, K.; Nakamura, T. Electronic properties of a TMTTF-Family Salt, $(TMTTF)_2$ TaF_6: New member located on the *modified* generalized phase-diagram. *J. Phys. Soc. Jpn.* **2009**, *78*, 104717.

10.   Liautard, P.B.; Peytavin, S.; Brun, G.M.M. Etude structurale du nitrate de tétraméthyltétrathiafulvaléne $(TMTTF)_2NO_3$. *Acta Cryst. B* **1982**, *38*, 2746–2749.

11.   Kobayashi, H.; Kobayashi, A.; Sasaki, Y.; Saito, G.; Inokuchi, H. The crystal structure of $(TMTTF)_2ReO_4$. *Bull. Chem. Soc. Jpn.* **1984**, 2025–2026.

12.   Galigné, J.L.; Liautard, B.; Peytavin, S.; Brun, G.; Maurin, M.; Fabre, J.M.; Torreilles, E.; Giral, L. Structure cristalline du fluoroborate de tetramethyltetrathiafulvalene $(TMTTF)_2BF_4$ a 100 K et a temperature ambiante. *Acta Cryst. B* **1979**, *35*, 1129–1135.

13.   Liautard, P.B.; Peytavin, S.; Brun, G. Structure du di(tétraméthyltétrathiafulvaĺenium)* perchlorate $[(TMTTF)_2ClO_4]$, $2C_{10}H_{12}S_4^{0,5}.ClO_4^-$. *Acta Cryst. C* **1984**, *40*, 1023–1026.

14.   Liautard, P.B.; Peytavin, S.; Brun, G.; Maurin, M. Structural correlations in the series $(TMTTF)_2X$. *J. Phys.* **1982**, *43*, 1453–1459.

15.   Nogami, Y.; Ito, T.; Yamamoto, K.; Irie, N.; Horita, S.; Kambe, T.; Nagao, N.; Shima, K.; Ikeda, N.; Nakamura, T. X-ray structural study of charge and anion orderings of TMTTF salts. *J. Phys. IV* **2005**, *131*, 39.

16.   Pouget, J.P.; Ravy, S. Structural aspects of the Bechgaard salts and related Compounds. *J. Phys. I* **1996**, *6*, 1501–1525.

17.   Pouget, J.P. Structural aspects of the Bechgaard and Fabre salts: An update. *Crystals* **2012**, *2*, 466–520.

18.   Rose, E.; Dressel, M. Coupling between molecular chains and anions in $(TMTTF)_2X$ salts. *Physica B* **2012**, *407*, 1787–1792.

19.   Kistenmacher, T.J. Anion-donor coupling in $(TMTSF)_2X$ salts: Symmetry considerations . *Solid State Commun.* **1984**, *51*, 931–934.

20.   Liautard, P.B.; Peytavin, S.; Brun, G.; Maurin, M. Structural studies and physical properties in the organic conductors series $(TMTTF)_2X$ and $(TMTSF)_2X$. *J. Phys.* **1983**, *44*, C3-951–C3-956.

21.   Kistenmacher, T.J. Cavity size versus anion size in $(TMTSF)_2X$ salts: Possible implications for the uniqueness of $(TMTSF)_2ClO_4$. *Solid State Commun.* **1984**, *50*, 729–733.

22.   Thorup, N.; Rindorf, G.; Soling, H.; Bechgaard, K. The structure of di(2,3,6,7- tetramethyl-1,4,5,8-tetrathiafulvalenium) hexafluorophosphate, $(TMTSF)_2PF_6$, the first superconducting organic solid. *Acta Cryst. B* **1981**, *37*, 1236–1240.

23.   Pustogow, A.; Peterseim, T.; Kolatschek, S.; Engel, L.; Dressel, M. Electronic correlations versus lattice interactions: Interplay of charge and anion orders in $(TMTTF)_2X$. *Phys. Rev. B* **2016**, *94*, 195125.

24.   Coulon, C.; Delhaes, P.; Flandrois, S.; Lagnier, R.; Bonjour, E.; Fabre, J.M. A new survey of the physical properties of the $(TMTTF)_2X$ series. Role of the counterion ordering. *J. Phys.* **1982**, *43*, 1059–1067.

25.   Beno, M.A.; Blackman, G.S.; Leung, P.C.W.; Williams, J.M. Hydrogen bond formation and anion ordering in superconducting $(TMTSF)_2ClO_4$ and $(TMTSF)_2AsF_6$. *Solid State Commun.* **1983**, *48*, 99–103.

26.   Granier, T.; Gallois, B.; Fritsch, A.; Ducasse, L.; Coulon, C. 135 K crystallographic and electronic structure of $(TMTTF)_2SbF_6$. In *Lower-Dimensional Systems and Molecular Electronics*; Metzger, R.M., Day, P., Papavassiliou, G.C., Eds.; Springer: Boston, MA, USA, 1990; Volume 248, pp. 163–168.

27.   Subías, G.; Abbaz, T.; Fabre, J.M.; Fraxedas, J. Characterization of the anion-ordering transition in $(TMTTF)_2ReO_4$ by X-ray absorption and photoemission spectroscopies. *Phys. Rev. B* **2007**, *76*, 085103.

28.   Chow, D.S.; Zamborszky, F.; Alavi, B.; Tantillo, D.J.; Baur, A.; Merlic, C.A.; Brown, S.E. Charge ordering in the TMTTF family of molecular conductors. *Phys. Rev. Lett.* **2000**, *85*, 1698–1701.

29. Zamborszky, F.; Yu, W.; Raas, W.; Brown, S.E.; Alavi, B.; Merlic, C.A.; Baur, A. Competition and coexistence of bond and charge orders in $(TMTTF)_2AsF_6$. *Phys. Rev. B* **2002**, *66*, 081103.

30. Nad, F.; Monceau, P.; Carcel, C.; Fabre, J.M. Charge ordering phase transition in the quasi-one-dimensional conductor $(TMTTF)_2AsF_6$. *J. Phys. Condens. Matter* **2000**, *12*, L435–L440.

31. Nad, F.; Monceau, P.; Carcel, C.; Fabre, J.M. Dielectric response of the charge-induced correlated state in the quasi-one-dimensional conductor $(TMTTF)_2PF_6$. *Phy. Rev. B* **2000**, *62*, 1753.

32. Nad, F.; Monceau, P.; Carcel, C.; Fabre, J.M. Charge and anion ordering phase transitions in $(TMTTF)_2X$ salt conductors. *J. Phys. Condens. Matter* **2001**, *13*, L717—L722.

33. Monceau, P.; Nad, F.Y.; Brazovskii, S. Ferroelectric Mott-Hubbard phase of organic $(TMTTF)_2X$ conductors. *Phys. Rev. Lett.* **2001**, *86*, 4080–4083.

34. Nagasawa, M.; Nad, F.; Monceau, P.; Fabre, J.M. Modification of the charge ordering transition in the quasi-one-dimensional conductor $(TMTTF)_2SbF_6$ under pressure. *Solid State Commun.* **2005**, *136*, 262–267.

35. Nad, F.; Monceau, P.; Kaboub, L.; Fabre, J.M. Divergence of the relaxation time in the vicinity of the ferroelectric charge-ordered phase transition in $(TMTTF)_2AsF_6$. *EPL* **2006**, *73*, 567–573.

36. Nad, F.; Monceau, P. Dielectric response of the charge ordered state in quasi-one-dimensional organic conductors. *J. Phys. Soc. Jpn.* **2006**, *75*, 1–12.

37. Hirose, S.; Kawamoto, A.; Matsunaga, N.; Nomura, K.; Yamamoto, K.; Yakushi, K. Reexamination of ^{13}C-NMR in $(TMTTF)_2AsF_6$: Comparison with infrared spectroscopy. *Phys. Rev. B* **2010**, *81*, 205107.

38. Dressel, M.; Dumm, M.; Knoblauch, T.; Masino, M. Comprehensive optical investigations of charge order in organic chain compounds $(TMTTF)_2X$. *Crystals* **2012**, *2*, 528–578.

39. Kitou, S.; Fujii, T.; Kawamoto, T.; Katayama, N.; Maki, S.; Nishibori, E.; Sugimoto, K.; Takata, M.; Nakamura, T.; Sawa, H. Successive dimensional transition in $(TMTTF)_2PF_6$ revealed by synchrotron X-ray diffraction. *Phys. Rev. Lett.* **2017**, *119*, 065701.

40. Nogami, Y.; Nakamura, T. X-ray observation of $2k_F$ and $4k_F$ charge orderings in$(TMTTF)_2ReO_4$ and $(TMTTF)_2SCN$ associated with anion orderings. *J. Phys. IV* **2002**, *12*, 145–148.

41. Coulon, C.; Parkin, S.S.P.; Laversanne, R. Structureless transition and strong localization effects in bis-tetramethyltetrathiafulvalenium salts $[(TMTTF)_2X]$. *Phys. Rev. B* **1985**, *31*, 3583–3587.

42. Yu, W.; Zhang, F.; Zamborszky, F.; Alavi, B.; Baur, A.; Merlic, C.A.; Brown, S.E. Electron-lattice coupling and broken symmetries of the molecular salt $(TMTTF)_2SbF_6$. *Phys. Rev. B* **2004**, *70*, 121101.

43. De Souza, M.; Foury-Leylekian, P.; Moradpour, A.; Pouget, J.P.; Lang, M. Evidence for lattice effects at the charge-ordering transition in $(TMTTF)_2X$. *Phys. Rev. Lett.* **2008**, *101*, 19–22.

44. De Souza, M.; Pouget, J.P. Charge-ordering transition in $(TMTTF)_2X$ explored via dilatometry. *J. Phys. Condens. Matter* **2013**, *25*, 343201.

45. Foury-Leylekian, P.; Petit, S.; Andre, G.; Moradpour, A.; Pouget, J.P. Neutron scattering evidence for a lattice displacement at the charge ordering transition of $(TMTTF)_2PF_6$. *Physica B* **2010**, *405*, 95–97.

46. Dumm, M.; Abaker, M.; Dressel, M. Mid-infrared response of charge-ordered quasi-1D organic conductors $(TMTTF)_2X$. *J. Phys. IV (France)* **2005**, *131*, 55–58.

47. Dumm, M.; Abaker, M.; Dressel, M.; Montgomery, L.K. Charge Order in $(TMTTF)_2PF_6$ Investigated by Infrared Spectroscopy. *J. Low Temp. Phys.* **2006**, *142*, 613–616.

48. Medjanik, K.; Chernenkaya, A.; Nepijko, S.A.; Ohrwall, G.; Foury-Leylekian, P.; Alemany, P.; Canadell, E.; Schonhense, G.; Pouget, J.P. Donor-anion interactions at the charge localization and charge ordering transitions of $(TMTTF)_2AsF_6$ probed by NEXAFS. *Phys. Chem. Chem. Phys.* **2015**, *17*, 19202–19214.

49. Rose, E.; Loose, C.; Kortus, J.; Pashkin, A.; Kuntscher, C.A.; Ebbinghaus, S.G.; Hanfland, M.; Lissner, F.; Schleid, T.; Dressel, M. Pressure-dependent structural and electronic properties of quasi-one-dimensional $(TMTTF)$ 2 PF 6. *J. Phys. Condens. Matter* **2013**, *25*, 014006.

50. Jacko, A.C.; Feldner, H.; Rose, E.; Lissner, F.; Dressel, M.; Valentí, R.; Jeschke, H.O. Electronic properties of Fabre charge-transfer salts under various temperature and pressure conditions. *Phys. Rev. B* **2013**, *87*, 155139.

51. Roobottom, H.K.; Jenkins, H.D.B.; Passmore, J.; Glasser, L. Thermochemical radii of complex ions. *J. Chem. Ed.* **1999**, *76*, 1570–1573.

52. Kaabel, S.; Adamson, J.; Topic, F.; Kiesila, A.; Kalenius, E.; Oeren, M.; Reimund, M.; Prigorchenko, E.; Lookene, A.; Reich, H.J.; et al. Chiral hemicucurbit[8]uril as an anion receptor: Selectivity to size, shape and charge distribution. *Chem. Sci.* **2017**, *8*, 2184–2190.

53. Oka, Y.; Matsunaga, N.; Nomura, K.; Kawamoto, A.; Yamamoto, K.; Yakushi, K. Charge order in $(TMTTF)_2TaF_6$ by infrared spectroscopy. *J. Phys. Soc. Jpn.* **2015**, *84*, 114709.

54. Nakamura, T.; Furukawa, K.; Hara, T. ^{13}C NMR analyses of successive charge ordering in $(TMTTF)_2ReO_4$. *J. Phys. Soc. Jpn.* **2006**, *75*, 013707.

55. Matsunaga, N.; Hirose, S.; Shimohara, N.; Satoh, T.; Isome, T.; Yamomoto, M.; Liu, Y.; Kawamoto, A.; Nomura, K. Charge ordering and antiferromagnetism in $(TMTTF)_2SbF_6$. *Phys. Rev. B* **2013**, *87*, 144415.

56. Świetlik, R.; Barszcz, B.; Pustogow, A.; Dressel, M. Raman spectroscopy evidence of domain walls in the organic electronic ferroelectrics $(TMTTF)_2X$ ($X = SbF_6, AsF_6, PF_6$). *Phys. Rev. B* **2017**, *95*, 085205.

57. Pouget, J.P.; Moret, R.; Comes, R.; Bechgaard, K.; Fabre, J.M.; Giral, L. X-Ray diffuse-scattering study of some $(TMTSF)_2X$ and $(TMTTF)_2X$ salts. *Mol. Cryst. Liq. Cryst.* **1982**, *79*, 129–143.

58. Nad, F.Y.; Monceau, P.; Carcel, C.; Fabre, J.M. Charge odering in $(TMTTF)_2X$ salts. *Synth. Met.* **2003**, *133*, 265–267.

59. Pouget, J.P.; Moret, R.; Comes, R.; Bechgaard, K. X-Ray diffuse scattering study of superstructure formation in tetramethyltetraselenafulvalenium perrhenate $(TMTSF)_2ReO_4$ and nitrate $(TMTSF)_2NO_3$. *J. Phys. Lett.* **1981**, *42*, doi:10.1051/jphyslet:019810042024054300.

60. Moret, R.; Pouget, J.P.; Comès, R.; Bechgaard, K. X-Ray scattering evidence for anion ordering and structural distortions in the low-temperature phase of di(tetramethyltetraselanafulvalenium) perrhenate $[(TMTSF)_2ReO_4]$. *Phys. Rev. Lett.* **1982**, *49*, 1008–1012.

61. Pouget, J.P.; Shirane, G.; Bechgaard, K.; Fabre, J.M. X-ray evidence of a structural phase transition in di-tetramethyltetraselenafulvalenium perchlorate $[(TMTSF)_2ClO_4]$, pristine and slightly doped. *Phys. Rev. B* **1983**, *27*, 5203–5206.

62. Moret, R.; Ravy, S.; Pouget, J.P.; Comes, R.; Bechgaard, K. Anion-ordering phase diagram of di(tetramethyltetraselenafulvalenium) perrhenate, $[(TMTSF)_2ReO_4]$. *Phys. Rev. Lett.* **1986**, *57*, 1915–1918.

63. Coulon, C.; Delhaes, P.; Amiell, J.; Manceau, J.P.; Fabre, J.M.; Giral, L. Effect of doping $(TMTSF)_2ClO_4$ with TMTTF - I. Ambient pressure results : A competition between the different possible ground states. *J. Phys.* **1982**, *43*, 1721–1729.

64. Ilakovac, V.; Ravy, S.; Pouget, J.P.; Lenoir, C.; Boubekeur, K.; Batail, P.; Babic, S.D.; Biskup, N.; Korin-Hamzic, B.; Tomic, S.; et al. Enhanced charge localization in the organic alloys $[(TMTSF)_{1-x}(TMTTF)_x]_2ReO_4$. *Phys. Rev. B* **1994**, *50*, 7136–7139.

65. Tomic, S.; Auban-Senzier, P.; Jérome, D. Charge localization in $[(TMTTF)_{0.5}(TMTSF)_{0.5}]_2ReO_4$: A pressure study. *Synth. Met.* **1999**, *103*, 2197–2198.

66. Tomić, S.; Jérome, D.; Monod, P.; Bechgaard, K. EPR and electrical conductivity of the organic superconductor di-tetramethyltetraselenafulvalenium-perchlorate, $(TMTSF)_2ClO_4$ and a metastable magnetic state obtained by fast cooling. *J. Phys. Lett.* **1982**, *43*, 839–844.

67. Takahashi, T.; Jérome, D.; Bechgaard, K. Observation of a magnetic state in the organic superconductor $(TMTSF)_2ClO_4$: Influence of the cooling rate. *J. Phys. Lett.* **1982**, *43*, 565–573.

68. Ishiguro, T.; Murata, K.; Kajimura, K.; Kinoshita, N.; Tokumoto, H.; Tokumoto, M.; Ukachi, T.; Anzai, H.; Saito, G. Superconductivity and metal-nonmetal transitions in $(TMTSF)_2ClO_4$. *J. Phys. Coll.* **1983**, *44*, C3-831–C3-838.

69. Yonezawa, S.; Marrache-Kikuchi, C.A.; Bechgaard, K.; Jérome, D. Crossover from impurity-controlled to granular superconductivity in $(TMTSF)_2ClO_4$. *Phys. Rev. B* **2018**, *97*, 014521.

70. Parkin, S.S.P.; Mayerle, J.J.; Engler, E.M. Anion ordering in $(TMTTF)_2ReO_4$: A displacive transition. *J. Phys. Colloq.* **1983**, *44*, C3-1105.

71. Dumm, M.; Loidl, A.; Fravel, B.W.; Starkey, K.P.; Montgomery, L.K.; Dressel, M. Electron spin resonance studies on the organic linear-chain compounds $(TMTCF)_2X (C = S, Se; X = PF_6, AsF_6, ClO_4, Br)$. *Phys. Rev. B* **2000**, *61*, 511–521.

72. Salameh, B.; Yasin, S.; Dumm, M.; Untereiner, G.; Montgomery, L.; Dressel, M. Spin dynamics of the organic linear chain compounds $(TMTTF)_2X$ ($X = SbF_6$, AsF_6, BF_4, ReO_4, and SCN). *Phys. Rev. B* **2011**, *83*, 205126.

73. Coulon, C.; Foury-Leylekian, P.; Fabre, J.M.; Pouget, J.P. Electronic instabilities and irradiation effects in the $(TMTTF)_2X$ series. *Eur. Phys. J. B* **2015**, *88*, 85.

74. Rohwer, A.; Dumm, M. Vibrational studies on TMTTF salts with non-centrosymmetric anions. To be published.

75. Nad, F.; Monceau, P.; Nakamura, T.; Furukawa, K. The effect of deuteration on the transition into a charge ordered state of $(TMTTF)_2X$ salts. *J. Phys. Condens. Matter* **2005**, *17*, L399.

76. Furukawa, K.; Hara, T.; Nakamura, T. Deuteration effect and possible origin of the charge-ordering transition of (TMTTF)$_2$X. *J. Phys. Soc. Jpn.* **2005**, *74*, 3288–3294.

77. Pouget, J.P.; Foury-Leylekian, P.; Le Bolloc'h, D.; Hennion, B.; Ravy, S.; Coulon, C.; Cardoso, V.; Moradpour, A. Neutron-scattering evidence for a spin-Peierls ground state in (TMTTF)$_2$PF$_6$. *J. Low Temp. Phys.* **2006**, *142*, 147–152.

78. Jérome, D.; Schulz, H. Organic conductors and superconductors. *Adv. Phys.* **1982**, *31*, 299–490.

79. Ishiguro, T.; Yamaji, K.; Saito, G. *Organic Superconductors*, 2nd ed.; Springer: Berlin, Germany, 1998.

80. Kagoshima, S.; Nagasawa, H.; Sambongi, T. *One-Dimensional Conductors*; Springer: Berlin, Germany, 1988.

81. Lebed, A. (Ed.) *The Physics of Organic Superconductors and Conductors*; Springer: Berlin, Germany, 2008.

82. Dumm, M.; Dressel, M.; Loidl, A.; Frawel, B.; Starkey, K.; Montgomery, L. Magnetic studies of (TMTTF)$_2$X (X=PF$_6$, ClO$_4$, and Br). *Synth. Met.* **1999**, *103*, 2068–2069.

83. Dumm, M.; Dressel, M.; Loidl, A.; Fravel, B.; Montgomery, L. Spin dynamics of organic linear chain compounds. *Physica B* **1999**, *259–261*, 1005–1006.

84. Lunkenheimer, P.; Loidl, A. Dielectric spectroscopy on organic charge-transfer salts. *J. Phys. Condens. Matter* **2015**, *27*, 373001.

85. Tomić, S.; Dressel, M. Ferroelectricity in molecular solids: A review of electrodynamic properties. *Rep. Prog. Phys.* **2015**, *78*, 096501.

86. Starešinić, D.; Biljaković, K.; Lunkenheimer, P.; Loidl, A. Slowing down of the relaxational dynamics at the ferroelectric phase transition in one-dimensional (TMTTF)$_2$AsF$_6$. *Solid State Commun.* **2006**, *137*, 241–245.

87. Brazovskii, S. Theory of the ferroelectric Mott-Hubbard phase in organic conductors. *J. Phys. IV* **2002**, *12*, 149–152.

88. Brazovskii, S. Ferroelectricity and charge-ordering in quasi-1d organic conductors. In *The Physics of Organic Superconductors and Conductors*; Lebed, A., Ed.; Springer: Berlin, Germany, 2008.

89. Brazovskii, S.; Monceau, P.; Nad, F.Y. Critical dynamics and domain motion from permittivity of the electronic ferroelectric (TMTTF)$_2$AsF$_6$. *Physica B* **2015**, *460*, 79–82.

90. Giovannetti, G.; Nourafkan, R.; Kotliar, G.; Capone, M. Correlation-driven electronic multiferroicity in (TMTTF)$_2$X organic crystals. *Phys. Rev. B* **2015**, *91*, 125130.

91. Brazovskii, S.; Monceau, P.; Nad, F. The ferroelectric Mott-Hubbard phase in organic conductors. *Synth. Met.* **2003**, *137*, 1331–1333.

92. Brazovskii, S. The theory for the ferroelectric Mott–Hubbard phase in organic conductors. *Synth. Met.* **2003**, *133–134*, 301–303.

93. Brazovskii, S. Theory of the ferroelectric phase in organic conductors: From physics of solitons to optics. *J. Phys. IV* **2004**, *114*, 9–13.

94. Lines, M.E.; Glass, A.M. *Principles and Applications of Ferroelectrics and Related Materials*; Clarendon Press: Oxford, UK, 1977.

95. Blinc, R. The soft mode concept and the history of ferroelectricity. *Ferroelectrics* **1987**, *74*, 301–303.

96. De Souza, M.; Squillante, L.; Sônego, C.; Menegasso, P.; Foury-Leylekian, P.; Pouget, J.P. Probing the ionic dielectric constant contribution in the ferroelectric phase of the Fabre salts. *Phys. Rev. B* **2018**, *97*, 045122.

97. Dressel, M.; Grüner, G. *Electrodynamics of Solids*; Cambridge University Press: Cambridge, UK, 2002.

98. Pinterić, M.; Ivek, T.; Čulo, M.; Milat, O.; Basletić, M.; Korin-Hamzić, B.; Tafra, E.; Hamzić, A.; Dressel, M.; Tomić, S. What is the origin of anomalous dielectric response in 2D organic dimer Mott insulators κ-(BEDT-TTF)2Cu[N(CN)2]Cl and κ-(BEDT-TTF)$_2$Cu$_2$(CN)$_3$. *Physica B* **2015**, *460*, 202–207.

99. Pinterić, M.; Čulo, M.; Milat, O.; Basletić, M.; Korin-Hamzić, B.; Tafra, E.; Hamzić, A.; Ivek, T.; Peterseim, T.; Miyagawa, K.; et al. Anisotropic charge dynamics in the quantum spin-liquid candidate $\kappa-$(BEDT-TTF)$_2$Cu$_2$(CN)$_3$. *Phys. Rev. B* **2014**, *90*, 195139.

100. Dressel, M.; Lazić, P.; Pustogow, A.; Zhukova, E.; Gorshunov, B.; Schlueter, J.A.; Milat, O.; Gumhalter, B.; Tomić, S. Lattice vibrations of the charge-transfer salt κ-(BEDT-TTF$_2$Cu$_2$(CN)$_3$: Comprehensive explanation of the electrodynamic response in a spin-liquid compound. *Phys. Rev. B* **2016**, *93*, 081201.

101. Pinterić, M.; Lazić, P.; Pustogow, A.; Ivek, T.; Kuveždić, M.; Milat, O.; Gumhalter, B.; Basletić, M.; Čulo, M.; Korin-Hamzić, B.; Löhle, A.; et al. Anion effects on electronic structure and electrodynamic properties of the Mott insulator $\kappa - (BEDT - TTF)_2Ag_2(CN)_3$. *Phys. Rev. B* **2016**, *94*, 161105.

102. Ivek, T.; Korin-Hamzić, B.; Milat, O.; Tomić, S.; Clauss, C.; Drichko, N.; Schweitzer, D.; Dressel, M. Collective excitations in the charge-ordered phase of α-(BEDT-TTF)$_2$I$_3$. *Phys. Rev. Lett.* **2010**, *104*, 206406.

103. Ivek, T.; Korin-Hamzić, B.; Milat, O.; Tomić, S.; Clauss, C.; Drichko, N.; Schweitzer, D.; Dressel, M. Electrodynamic response of the charge ordering phase: Dielectric and optical studies of α-(BEDT-TTF)$_2$I$_3$. *Phys. Rev. B* **2011**, *83*, 165128.

104. Voloshenko, I.; Herter, M.; Beyer, R.; Pustogow, A.; Dressel, M. Pressure-dependent optical investigations of Fabre salts in the charge-ordered state. *J. Phys. Condens. Matter* **2017**, *29*, 115601.

Sample Availability: Samples of the (TMTTF)$_2$X compounds are available from the authors.

crystals

MDPI

Article

Systematics of the Third Row Transition Metal Melting: The HCP Metals Rhenium and Osmium

Leonid Burakovsky [1,*,†,‡], **Naftali Burakovsky** [2,‡], **Dean Preston** [2,‡] and **Sergei Simak** [3,‡]

1   Theoretical Division, Los Alamos National Laboratory, Los Alamos, NM 87545, USA
2   Computational Physics Division, Los Alamos National Laboratory, Los Alamos, NM 87545, USA;
    nburakov@lanl.gov (N.B.); dean@lanl.gov (D.P.)
3   Department of Physics, Chemistry and Biology, Linköping University, 58183 Linköping, Sweden;
    sergeis@ifm.liu.se
*   Correspondence: burakov@lanl.gov; Tel.: +1-505-667-5222
†   Current address: Theoretical Division, Los Alamos National Laboratory, Los Alamos, NM 87545, USA.
‡   These authors contributed equally to this work.

Received: 6 April 2018 ; Accepted: 25 May 2018; Published: 6 June 2018

Abstract: The melting curves of rhenium and osmium to megabar pressures are obtained from an extensive suite of ab initio quantum molecular dynamics (QMD) simulations using the Z method. In addition, for Re, we combine QMD simulations with total free energy calculations to obtain its phase diagram. Our results indicate that Re, which generally assumes a hexagonal close-packed (hcp) structure, melts from a face-centered cubic (fcc) structure in the pressure range 20–240 GPa. We conclude that the recent DAC data on Re to 50 GPa in fact encompass both the true melting curve and the low-slope hcp-fcc phase boundary above a triple point at (20 GPa, 4240 K). A linear fit to the Re diamond anvil cell (DAC) data then results in a slope that is 2.3 times smaller than that of the actual melting curve. The phase diagram of Re is topologically equivalent to that of Pt calculated by us earlier on. Regularities in the melting curves of Re, Os, and five other 3rd-row transition metals (Ta, W, Ir, Pt, Au) form the 3rd-row transition metal melting systematics. We demonstrate how this systematics can be used to estimate the currently unknown melting curve of the eighth 3rd-row transition metal Hf.

Keywords: quantum molecular dynamics; phase diagram; melting curve; transition metal

1. Introduction

Uncertainty in experimental stress conditions is an important source of error in measurements of phase equilibria and physical properties at deep Earth pressures. This error is evident in the range of reported pressures for key mantle phase transitions such as the perovskite–post-perovskite boundary in $MgSiO_3$. Accurate pressures are also necessary for measurements of the equation of state and compressibility of different materials. Differences of up to 15% have been observed among equation of state measurements for materials such as $MgSiO_3$ post-perovskite and iron largely due to pressure scale differences at Mbar pressure. Advances in high-pressure techniques require standards which are applicable at multi-Mbar pressures.

An ideal pressure standard is inert and compressible, with a simple crystal structure and strong X-ray diffraction pattern. Ideally, it should have no phase transitions over the experimental pressure range. In reality, it may have phase transitions, but its phase diagram should be relatively simple, with all the solid phases being described by simple and well-defined equations of state. Both gold and platinum have been in focus of high-pressure research for several decades as primary equation of state standards. Supposedly large pressure and temperature stability ranges of the face-centered cubic (fcc) phases of the third row transition metals Au and Pt and their large isothermal compressibility make them very attractive materials to be used as pressure markers above 1 Mbar. However, the newly discovered structural transformation in gold [1] severely limits the applicability of the Au standard. Likewise, the structures instability in Pt at high pressure (P) and temperature (T) [2] limits the applicability of the Pt standard to low T.

Another consideration in pressure calibrant choice is overlap of diffraction peaks of the standard with those of the sample. For Au and Pt (as well as, e.g., another potential pressure marker MgO) that have fcc structure, their diffraction peak positions are similar at Mbar conditions. In this respect, another pair of the third row transition metals, namely, Re and Os, that have hexagonal close-packed (hcp) structure, may offer potential to be a useful alternative to fcc standards. At present, the phase diagrams of either Re or Os are virtually unknown. We address the issue of their phase diagrams (specifically, their melting curves) in this paper. We calculate the melting curves of Re to 3.5 Mbar and of Os to 5 Mbar using ab initio quantum molecular dynamics (QMD) simulations.

In what is now following, we summarize the existing information on both Re and Os that can be found in the literature.

1.1. Rhenium

The equation of state (EOS) of solid Re has been extensively studied to P as high as 640 GPa [3], and there are isothermal compression data at Ts up to 3000 K [4]. Comparison of the $T = 0$ free energies of candidate crystal structures for Re shows that hexagonal close-packed (hcp) is the most stable structure up to at least a compression of two; face-centered cubic (fcc), the closest structure to hcp, is 4 to 12 mRy/atom (55–165 meV/atom) higher in energy [5]. The $T = 0$ fcc-hcp free energy difference is a quasi-linear function of P up to ~1000 GPa: $\Delta F_{\text{fcc−hcp}}(P) \approx 55 + 0.11\,P$ meV/atom. Unless a free energy difference becomes sufficiently large (typically $\gtrsim 0.1$ eV/atom, which corresponds to $P \sim 400$ GPa for Re), it can be overcome by the entropy term at finite T, hence we expect fcc-Re to become energetically competitive with hcp-Re with increasing T over some limited range of P.

In 2012, Yang, Kandikar, and Boehler (YKB) carried out DAC measurements of the melting temperature of Re to 50 GPa [6]. Their melting curve is quasi-linear with a slope of 17 K/GPa, which is 2.3 times smaller than the slope (40 K/GPa) of the melting curve measured by Vereshchagin and Fateeva (VF) in 1975 using electrical heating in a belt apparatus to 8 GPa [7]. Other, theoretical, values for this slope are 58 K/GPa [8] and 29.2 K/GPa [9,10]. If 17 K/GPa were the correct value of the initial slope of its melting curve, Re would be unique among the transition metals of the third period since the melting curves of the other metals in this group that have been reliably determined either experimentally or theoretically, or both, have initial slopes in the range 40–55 K/GPa: ~55 for Au [11–13], ~50 for Ir [14], ~45 for Ta [15] and W [16], and ~40 for Pt [2]. Below we show that it is the VF melting curve that is correct, and with which our ab initio QMD simulations are in excellent agreement, rather than the recent YKB DAC curve which in fact includes part of the Re s-s (specifically, hcp-fcc) phase boundary.

1.2. Osmium

Comparison of the $T = 0$ free energies of candidate crystal structures for Os shows that hexagonal close-packed (hcp) is the most stable structure up to at least a compression of two; face-centered cubic, the closest structure to hcp, is ~10 mRy/atom (~0.14 eV/atom) higher in energy [17]. The very recent experimental study [18] reveals that Os retains its hcp structure upon compression to ~800 GPa. Hence, we assume that in the pressure range considered in this work Os is a single-phase (hcp) material.

The equation of state (EOS) of solid osmium has been extensively studied [19]. The very recent experimental data go up to a pressure of 700 GPa [18], and there are isothermal compression data at temperatures up to 3000 K [20], but its melting curve, $T_m(\rho)$ or $T_m(P)$, has never been measured. A theoretical melting curve of Os to 800 GPa [17] has been constructed on the basis of first-principles calculations of the Grüneisen parameter, $\gamma(\rho)$, and the use of the Lindemann formula for the melting temperature as a function of density, i.e., $d \ln T_m(\rho)/d \ln \rho = 2[\gamma(\rho) - 1/3]$. With this theoretical melting curve converted into the *P-T* coordinates, using the corresponding EOS, melting on the Hugoniot is predicted to occur at ~450 GPa and ~9200 K [17]. Most recently, Kulyamina et al. [21] analyzed all of the isobaric-heating data on Os available in the literature and extracted the initial slope of the Os melting curve: $dT_m(P)/dP = 40.4$ K/GPa. We note that their determination is based on the Clausius-Clapeyron relation $dT_m/dP = \Delta V_m/\Delta S_m$, but since neither the volume change at melt, ΔV_m, nor the melting entropy, ΔS_m, are known from experiment, their values can only be estimated. Other, theoretical, values for this slope are 65 K/GPa [8] and 53.4 K/GPa [9,10].

2. Z Method Calculations

In the present work we determine the melting curves of rhenium to ~350 GPa and osmium to ~500 GPa using the Z method which we describe in what follows.

The Z method was developed to calculate melting curves using first-principles based software, specifically VASP (Vienna Ab initio Simulation Package); it was introduced and used for the first time in our paper on the ab initio melting curve of Mo [22]. The method has since been applied to the study of a large number of melting curves of different materials [23–25], and comparisons with experimental data on Pb [26], Ta [15], Fe [27], and Pt [28] at European Synchrotron Radiation Facility (ESRF) show good agreement. If a material has more than one thermodynamically stable crystal structure, the Z method yields the solid-liquid equilibrium boundaries of those structures. The phase having the highest solid-liquid equilibrium temperature over some pressure range is the most stable, thus the physical melting curve, including triple points, is the envelope of the solid-liquid equilibrium boundaries. We note that until recently, VASP could only be run for NVE or NVT ensembles, but with the release of the latest version, VASP5.3, we now have the option of running NPT, hence the so-called 2-phase simulations are now an alternative to the Z method.

Figure 1 shows a typical Z isochore, there comprised of the three green segments AC-CD-DE. It can be approximately mapped out by performing a sequence of QMD runs at progressively higher temperatures and pressures, typically 6–8 points, starting in the solid (segment AB), progressing to the superheated (SH) solid (segment BC), and finally to the liquid (segment DE). If the total energy in a QMD run in the superheated solid is such that the equilibrium temperature $T < T_C$, the final state is on segment AC, but if $T > T_C$ the system melts and the final state is a point (P_l, T_l) on segment DE above the melting curve (dark blue); a further increase in the initial system energy moves the final state up segment DE. Ideally, the QMD runs in the superheated solid would differ by only small temperature increments so that the upper vertex C would be precisely determined, and then a run starting from C would take the system to the point D on the melting curve, but generally this cannot be achieved in practice. The standard implementation of the Z method involves bounding the vertex D from below by the highest calculated state (P_s, T_s) on solid segment AB, and from above by the lowest state (P_l, T_l) on liquid segment DE. Then the melting point can be approximated as $(P_m, T_m) \approx ((P_s + P_l)/2, (T_s + T_l)/2)$. The true melting point must be close to (P_m, T_m) because the actual melting curve crosses the box formed by $P_m \pm (P_l - P_s)/2$ and $T_m \pm (T_l - T_s)/2$.

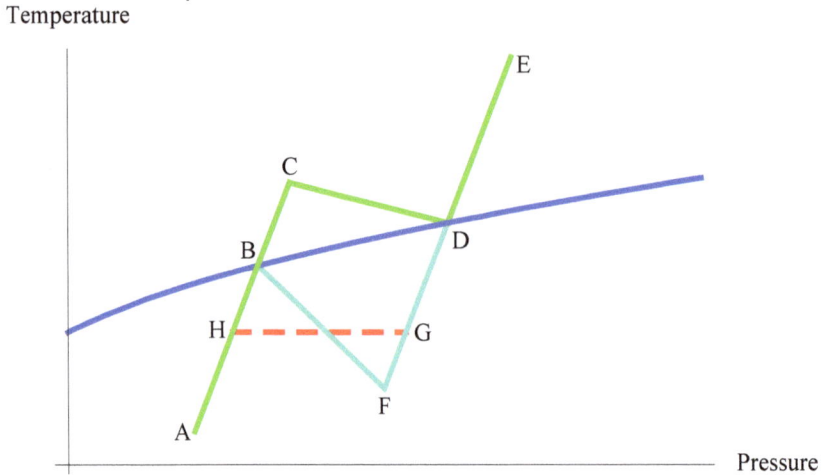

Figure 1. Typical isochore used in the Z methodology. Different segments of the isochore correspond to solid (AB), superheated solid (BC), liquid (DE), and supercooled liquid (DF) states. Melting corresponds to segment CD. Isochoric and isothermal solidification processes correspond to segments FB and GH, respectively, and are used in the so-called inverse Z method [2].

3. QMD Simulations of the EOS and the Melting Curve of Os

Our Z method calculations are carried out using the QMD code VASP, which is based on density functional theory (DFT). We use the generalized gradient approximation (GGA) with the Perdew-Burke-Ernzerhof (PBE) exchange-correlation functional. We model Os using the electron core-valence representation $[^{48}Cd\,4f^{14}]\,5p^6\,5d^6\,6s^2$, i.e., we assign the 14 outermost electrons of Os to the valence. The valence electrons are represented with a plane-wave basis set with a cutoff energy of 400 eV, while the core electrons are represented by projector augmented-wave (PAW) pseudopotentials.

3.1. T = 0 Isotherm

We first calculated the $T = 0$ isotherm of Os. This was done by optimizing the value of c/a, i.e., determining the c/a that minimizes the energy, at a fixed volume of a hexagonal supercell, and extracting the corresponding value of P. We used a $8 \times 8 \times 5$ (640-atom) supercell with a single Γ-point. With such a large supercell, energy convergence to $\lesssim 2.5$ meV/atom is achieved, which was verified by performing short runs with $2 \times 2 \times 2$, $3 \times 3 \times 3$, and $4 \times 4 \times 4$ k-point meshes and comparing their output with that of the 640-atom run with a single Γ-point. This introduces additional uncertainty of $\sim$30 K for the numerical values of the corresponding T_m which is small compared to the uncertainties of the Z method itself (75 K or 125 K, see below).

In the ab initio approach, the density of osmium at $(P, T) = (0, 0)$ is 22.17 g/cc, whereas the experimental value is 22.66 g/cc [29]. Specifically, VASP predicts the lattice constants of the hexagonal unit cell to be (upon optimizing its c/a ratio) $a = 2.7319$ Å and $c = 4.3134$ Å ($c/a = 1.5789$), which corresponds to 22.17 g/cc. The experimental values are [30] $a = 2.7315$ Å $c = 4.3148$ Å ($c/a = 1.5797$), and $\rho = 22.661$ g/cc. Alternatively, with VASP, the experimental $(P, T) = (0, 0)$ density of 22.66 g/cc corresponds to $(P, T) = (9.4$ GPa, 0). This 2.2% density mismatch, or 9.4 GPa pressure mismatch, is due to the specifc implementation of DFT, namely PBE, in our VASP simulations. In order to directly compare our QMD results to experiment, we will apply the 2.2% density correction (9.4 GPa pressure correction) to all VASP results in the ρ-T (P-T) plane. Specifically, we will multiply VASP densities by 1.022, or subtract 9.4 GPa from VASP pressures, whichever is relevant.

Our results on the $T = 0$ isotherm, as well as the value of c/a as a function of P, are shown in Figures 2 and 3, respectively. We note that each of the papers that discuss Os EOS data [18–20,31–34] uses the third-order Birch-Murnaghan (BM3) EOS

$$P(\rho) = \frac{3}{2} B_0 \left(\left(\frac{\rho}{\rho_0} \right)^{7/3} - \left(\frac{\rho}{\rho_0} \right)^{5/3} \right) \left[1 + \frac{3}{4} (B_0' - 4) \left(\left(\frac{\rho}{\rho_0} \right)^{2/3} - 1 \right) \right],$$

(1)

where B_0 and B_0' are the values of the bulk modulus and its pressure derivative at the reference point $\rho = \rho_0$. Since the $P = 0$ values of the density of Os at $T = 0$ and 300 K differ by ~0.3% (22.66 vs. 22.59 g/cc [29]), and $T = 300$ K introduces a negligibly small thermal pressure correction, the $T = 0$ and $T = 300$ K isotherms can be described by the same values of B_0 and B_0'. Consequently, we can compare room-temperature isotherm data to our zero-temperature isotherm as determined from QMD. A comparison is shown in Figure 2. It is seen that BM3 with $B_0' = 5$ fits both the experimental and QMD data up to ~200 GPa as well as the QMD data at higher P. Our best fit to the QMD data in the whole pressure range gives $B_0 = 415.0$ GPa and $B_0' = 4.87$.

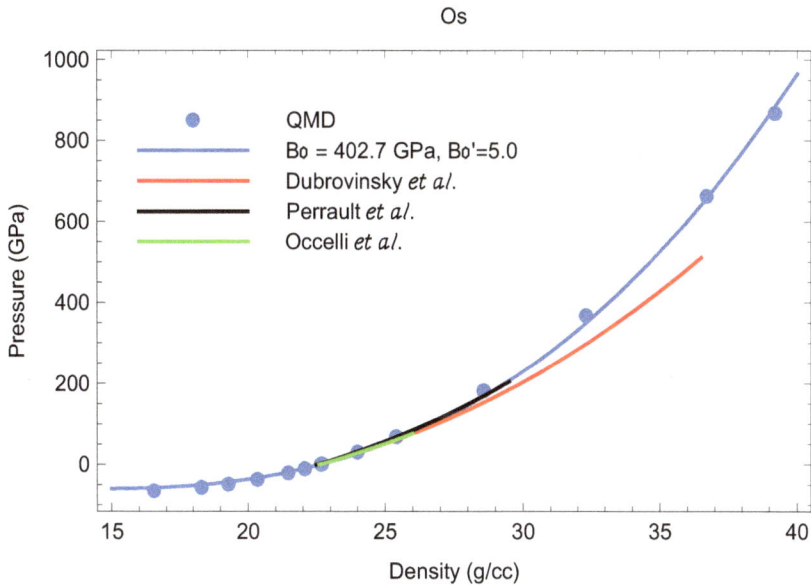

Figure 2. The $T = 0$ Os EOS from VASP compared to the experimental 300 K 3rd-order Birch-Murnagan (BM3) EOSs from different literature sources [18,31,34]. The pressure ranges of the experimental EOSs are those in which they were measured.

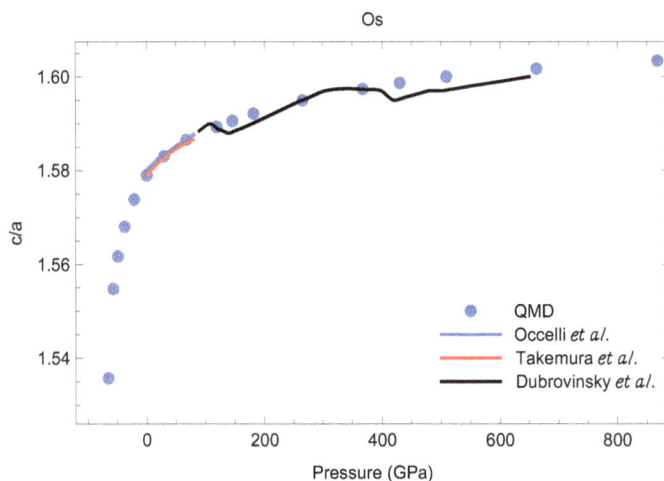

Figure 3. The *c*/*a* ratio as a function of *P*: QMD vs. the available experimental data from different literature sources [18,31,32].

3.2. c/a Ratio as a Function of P

Based on their experimental results on *c*/*a* as a function of *P*, Occelli et al. [31] suggested the existence of an isostructural phase transition in Os at about 25 GPa associated with an anomaly in the compressibility. This anomaly was in turn associated with a discontinuity in the first pressure derivative of the *c*/*a* ratio which may arise from the collapse of the "small hole-ellipsoid" [31] in the Fermi surface near the *L* point. Subsequent experimental studies did not confirm this *c*/*a* anomaly [19,32,35]. However, the very recent experiments by Dubrovinsky et al. [18] reveal two different anomalies, around 150 and 400 GPa, each of which represents a small, $\sim$0.2%, reduction in the *c*/*a* value (Figure 3). As discussed in [18], the first of these two anomalies may be the signature of a topological change in the Fermi surface for valence electrons, while the second might be related to an electronic transition associated with pressure-induced interactions between core electrons. No such *c*/*a* anomalies are seen in our ab initio study (Figure 3). As noted in [18], in the case of Os, direct comparison between theory and experiment is not legitimate, because the calculations are carried out at $T = 0$ whereas the experimental data are taken at room temperature. Moreover, for hcp metals, direct comparison between theory and experiment is generally nontrivial: in hcp metals the effect of the electronic transitions on the *c*/*a* ratio should become visible at finite *T* due to the anisotropy of the thermal expansion of the hcp lattice. In any event, for these two anomalies, reductions in the *c*/*a* value are so small that each of the corresponding pressures (related to the volume derivative of the total energy as a function of *c*/*a* at fixed volume) only changes by a few tens of GPa. Therefore, even if they are real, these anomalies do not influence the room-*T* isotherm of Os that does not take them into account; consequently, they do not influence comparison of our cold EOS to this room-*T* isotherm. Finally, we note that the most recent expeimental EOS of Os [34] does not confirm the occurence of either of the two anomalies of Ref. [18].

3.3. Melting Curve

In contrast to previous melting curve calculations based on the Z method, here the method was utilized as closely as possible to the original concept, but at the expense of an extensive suite of QMD simulations. We calculated eight melting points. At a given density we performed a sequence of very long runs, each up to 25,000 time steps or 25 ps, with initial temperatures separated by relatively small increments: 150 K for the first three points on the Os melting curve and 250 K for the remaining five points. We performed 10 such runs for each of the first three points, and 14 runs for each of the remaining five points for a total of 100 runs and ~2 million time steps. In the course of these extensive computer simulations, our strategy for detecting the melting point was as follows. The conversion of the initially ordered solid state into a disordered liquid during a QMD run was detected in one of three ways: (i) visual observation of atomic motion in the computational cell (vibrations around equilibrium sites in a solid vs. diffusion between the sites in a liquid); (ii) a drop in the value of the equilibrium T and the corresponding jump in the value of the equilibrium P; (iii) change in the radial distribution function (a long sequence of well-pronounced peaks in a solid vs. a few peaks in a liquid). If the system did not melt during the 25 ps of running time, we started the next run with an initial T higher by 150–250 K than the previous one, etc. The first run in which the system melts during the 25 ps of running time was assumed to correspond to the upper vertex C; during this run the complete melting process corresponding to the C→D transition in Figure 1 is usually observed. We refer to such a run as the melting run. With an even higher initial T, the system melts at an earlier time than in the melting run, and the duration of the melting process shortens; both the time when melting begins and the duration of the process decrease with increasing T, and for a sufficiently high initial T the system melts immediately. Examples of the QMD melting simulations using the Z method are shown in Figures 4–7.

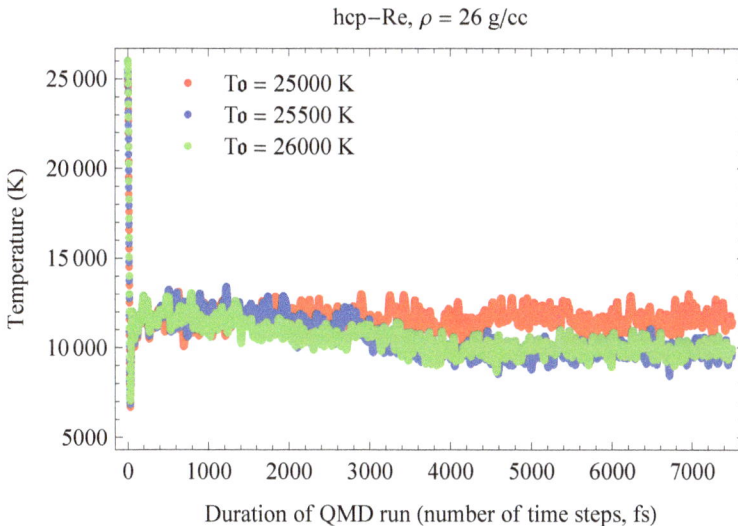

Figure 4. Time evolution of T for hcp-Re in three QMD runs with initial temperatures (T_0s) separated by 500 K. The middle run is the melting run, during which T decreases from ~12,000 K for the superheated state to ~10,000 K for the liquid at the corresponding melting point.

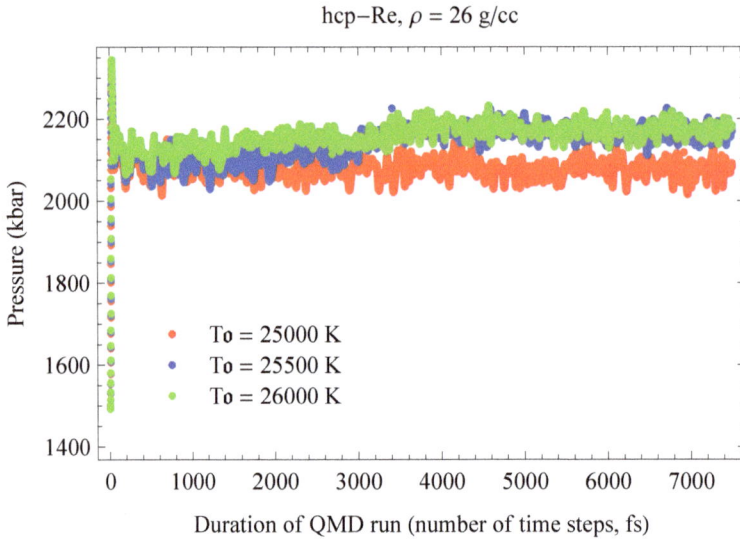

Figure 5. The same as in Figure 4 for the time evolution of P. During melting P increases from ~210 GPa for the superheated state to ~217 GPa for the liquid at the corresponding melting point.

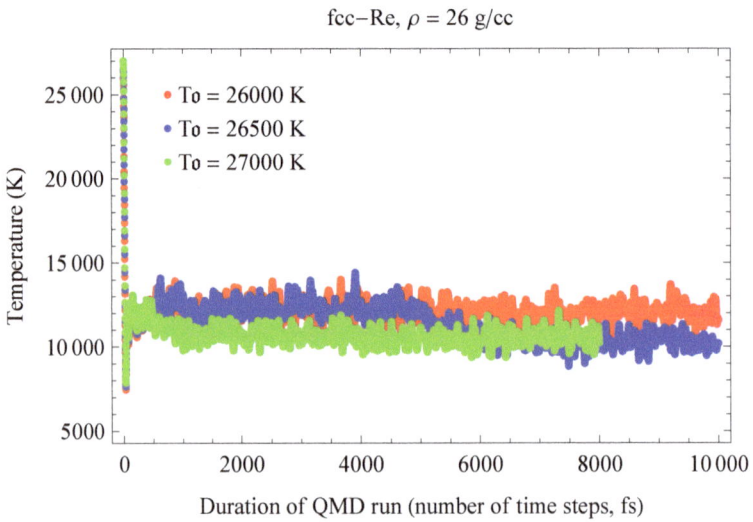

Figure 6. The same as in Figure 4 for fcc-Re.

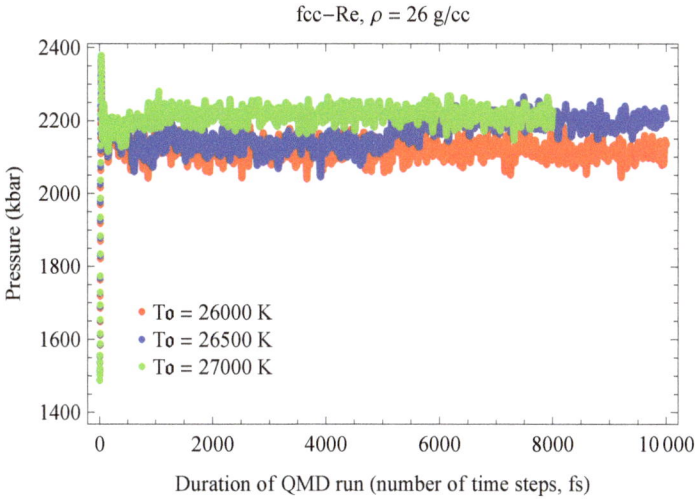

Figure 7. The same as in Figure 4 for fcc-Re.

An analytic expression for the melting curve of Os in the *P-T* plane can be constructed as the best fit of the Simon form, $T_m(P) = T_m(0)(1 + P/a)^b$, to all eight QMD points with $T_m(0) = 3370$ K. The result is (*T* in K, *P* in GPa)

$$T_m(P) = 3370 \left(1 + \frac{P}{36.1}\right)^{0.53}. \tag{2}$$

The melting curve (2) is shown in Figure 8 along with the eight QMD melting points and the theoretical melting curve of Ref. [17]. The (only) reason for a difference between our melting curve and the one from [17] is most likely the calculation of the Grüneisen $\gamma(\rho)$, which the latter is based on, not being accurate enough. For (2), $dT_m(P)/dP = 49.5$ K/GPa at $P = 0$, in good agreement with the values from [8–10,21].

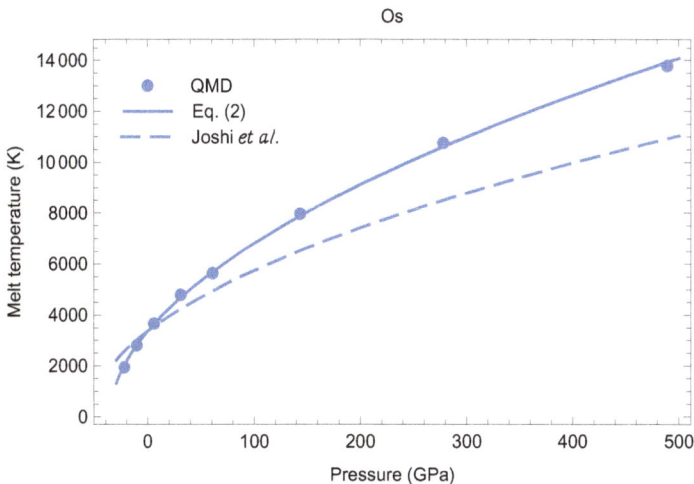

Figure 8. The melting curve of Os: VASP results (bullets) and the corresponding Simon equation, Equation (2), vs. the theoretical model of Joshi et al. [17].

4. Uncertainties in the Values of T_m and P_m

We now estimate the uncertainties in the melting temperature and pressure for the computational procedure outlined above.

First of all, changes in P are typically much smaller than that in T. For instance, as seen in Figures 4–7, which typify T and P changes during simulated melting, the pressure changes by less than 10% while $\Delta T \sim 20\text{–}25\%$. We estimate the errors in P to range from a few GPa at low pressures to $\sim$10 GPa at the highest pressure considered; such errors do not exceed the size of the points in Figure 8. As a reasonable approximation, we can ignore errors in P.

The error in the melting temperature is due to the uncertainty in the maximum temperature for which the Os remains a superheated solid, i.e., the temperature at vertex C in Figure 1. Assume that the melting occurs from a superheated solid state C′ which lies on the continuation of segment BC in Figure 1 beyond point C; the temperature at C is T_{SH} and that at C′ is $T_{SH} + \Delta T_{SH}$. Melting from C takes the system to point D at temperature T_m, while melting from C′ leaves the system in a liquid state D′ on segment DE at temperature $T_m + \Delta T_m$. Following [36], we now consider energy balance for the virtual transitions B→C and B→D$_{liq}$. The energy increase for B→C is $C_{VS}\left(T_{SH} - T_m^{(B)}\right)$, where C_{VS} is the solid heat capacity at constant volume and $T_m^{(B)}$ is the melting temperature at B. The transition B→D$_{liq}$ can be decomposed into B→D$_{sol}$ with energy change $C_{VS}\left(T_m - T_m^{(B)}\right)$, and then melting at D with an energy increase of $T_m \Delta S_m$, where ΔS_m is the entropy of melting; thus the total energy change for B→D$_{liq}$ is $C_{VS}\left(T_m - T_m^{(B)}\right) + T_m \Delta S_m$. Since points C and D$_{liq}$ have the same energy, we can equate the energy changes for B→C and B→D$_{liq}$ ($T_m^{(B)}$ drops out):

$$C_{VS}\left(T_{SH} - T_m\right) = T_m \Delta S_m. \tag{3}$$

Similarly, for B→C′ and B→D′ we have

$$C_{VS}\left(T_{SH} + \Delta T_{SH} - T_m\right) = T_m \Delta S_m + C_{VL} \Delta T_m. \tag{4}$$

Here C_{VL} is the liquid heat capacity at constant volume. It then follows from (1) and (2) that

$$\Delta T_m = \frac{C_{VS}}{C_{VL}} \Delta T_{SH}. \tag{5}$$

In the vicinity of the melting point the solid and liquid heat capacities are known to be approximately equal [37,38], hence

$$\Delta T_m \approx \Delta T_{SH}. \tag{6}$$

We have the simple result that the error in T_m is approximately equal to the difference in the temperatures for the first run during which melting occurs and the true melting run.

The temperature difference between two solid states on segment AC in Figure 1 is about half of the difference of the initial temperatures in the QMD runs. This is so because during a QMD run the initial thermal energy, which is the total system energy, is divided almost equally into potential and kinetic energies, and the latter is the thermal energy of the final state. Therefore, ΔT_{SH} cannot exceed one half of the difference of the initial temperatures for the first run during which the system melts and the last run during which the system remains superheated. For the present Os calculations, this implies a maximum error of $\sim$75 K for the first three points on the ab initio melting curve and $\sim$125 K for the remaining five points. All these errors are less than the size of the corresponding symbols in Figure 8.

5. QMD Simulations of the EOS and the Phase Diagram of Re

We now determine the phase diagram of Re to $\sim$350 GPa using the Z methodology, which is described in detail in [2,14]. In addition to the Z method described above it consists in the so-called inverse Z method, that is, the solidification of a liquid into a crystalline structure. The inverse Z method allows one to find solid-solid transition boundaries, by solidifying liquid into different solid structures on both sides of the boundary and thereby bracketing its location on the *P-T* plane.

Our Z methodology calculations were carried out using VASP. We again used the GGA approximation with the PBE exchange-correlation functional. We modeled Re using the electron core-valence representation $[^{48}Cd\,4f^{14}]\,5p^6\,5d^5\,6s^2$, i.e., we assigned the 13 outermost electrons of Re to the valence. The valence electrons were represented with a plane-wave basis set with a cutoff energy of 400 eV, while the core electrons were represented by PAW pseudopotentials.

5.1. T = 0 Isotherm

We first calculated the $T = 0$ isotherm of Re. Just like for Os, this was done by optimizing the value of c/a, i.e., determining the c/a that minimizes the energy at a fixed volume of a hexagonal supercell, and extracting the corresponding value of P. We used a $7 \times 7 \times 6$ (588-atom) supercell with a single Γ-point. With such a large supercell, energy convergence to $\lesssim$4.5 meV/atom is achieved, which was verified by performing short runs with $2 \times 2 \times 2$, $3 \times 3 \times 3$, and $4 \times 4 \times 4$ *k*-point meshes and comparing their output with that of the 588-atom run with a single Γ-point. This introduces additional uncertainty of ~50 K for the numerical values of T_m which is small compared to uncertainites of the Z method itself (250 K for Re, see below).

In the ab initio approach, the density of Re at $(P, T) = (0, 0)$ is 20.686 g/cc, whereas the experimental value is 21.12 g/cc [4]. Alternatively, with VASP, the experimental $(P, T) = (0, 0)$ density of 21.12 g/cc corresponds to $(P, T) = (8.2\ \text{GPa}, 0)$. Just like for Os, this 2.1% density mismatch, or 8.2 GPa pressure mismatch, is due to the PBE implementation of DFT in our VASP simulations. In order to directly compare our QMD results to experiment, we will apply the 2.1% density correction (8.2 GPa pressure correction) to all VASP results in the ρ-T (P-T) plane. Specifically, we will multiply VASP densities by 1.021, or subtract 8.2 GPa from VASP pressures, whichever is relevant.

Since the $P = 0$ values of the density of Re at $T = 0$ and 300 K differ by ~0.5% (21.12 g/cc vs. 21.02 g/cc [4]), and there is a negligibly small thermal pressure correction at $T = 300$ K, the $T = 0$ isotherm is a very good approximation to the $T = 300$ K isotherm. In Figure 9 we compare room-T isotherm data to our $T = 0$ isotherm as determined from QMD; the experimental data are from [3,4,39–41]. It is seen that all of the available data and the QMD points are in excellent agreement up to $\rho\sim$26 g/cc ($P\sim$150 GPa) beyond which the different sources of data begin to depart from one another. We note that our QMD results are very accurately represented by a fourth-order Birch-Murnaghan EOS (BM4)

$$P(\rho) = \tfrac{3}{2} B_0 \left(\left(\tfrac{\rho}{\rho_0}\right)^{7/3} - \left(\tfrac{\rho}{\rho_0}\right)^{5/3} \right) \left[1 + \tfrac{3}{4}(B_0' - 4) \left(\left(\tfrac{\rho}{\rho_0}\right)^{2/3} - 1 \right) \right.$$
$$\left. + \tfrac{3}{8} \left(\tfrac{143}{9} + (B_0' - 7)B_0' + B_0 B_0'' \right) \left(\left(\tfrac{\rho}{\rho_0}\right)^{2/3} - 1 \right)^2 \right], \tag{7}$$

where B_0, B_0' and B_0'' are the values of the bulk modulus and its first and second pressure derivatives at the reference point $\rho = \rho_0$. For practical purposes BM3 (1) can be used as well if B_0' is changed from 6 to 5 (see Figure 9).

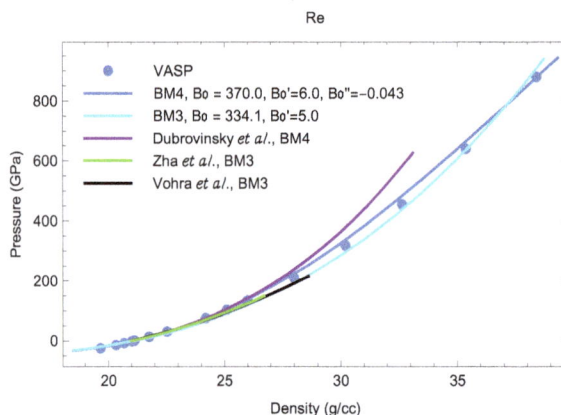

Figure 9. The $T = 0$ Re EOS from VASP compared to the experimental 300 K EOSs from different literature sources [3,4,39]. The pressure ranges of the experimental EOSs are those in which they were measured.

5.2. c/a Ratio as a Function of P

Our results for c/a as a function of P are shown in Figure 10. We find that our theoretical c/a lies in the range 1.615 ± 0.006 for $P \in [0, 900$ GPa] (so that its variation is less than 1%), consistent with experimental data [3,39]. Other theoretical data from [5] have c/a lie in the narrower range of 1.6175 ± 0.003 (Figure 10) but exhibit quite irregular behavior as a function of P. Our c/a exhibits a P-dependence consistent with that from [3]. Although our calculations demonstrate that hcp-Re remains the most thermodynamically stable solid phase up to at least 900 GPa, in agreement with [5] (a somewhat more detailed discussion will follow), the decreasing c/a may signal its dynamic instability, and the corresponding s-s phase transition, at higher P.

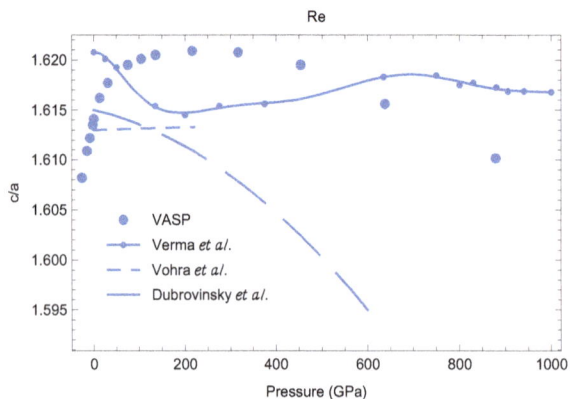

Figure 10. The c/a ratio as a function of P: VASP vs. the available experimental data from different literature sources [3,5,39].

5.3. Melting Curve

We then calculated the melting curves of both hcp- and fcc-Re. For fcc-Re, we also used a 588-atom supercell in a hexagonal setting: 3 atoms per unit cell with ABC stacking (the hcp counterpart has 2 atoms per unit cell with AB stacking); the supercell dimension was $7 \times 7 \times 4$. By using supercells with the same number of atoms we could make direct comparisons of the hcp and fcc free energies.

We simulated 11 melting points: 6 for hcp-Re and 5 for fcc-Re. At a given density we performed a sequence of long runs, each of 7500–10,500 time steps or 7.5–10.5 ps, with initial temperatures separated by increments of 500 K. We performed 10 QMD runs for each of the 11 melting points; our simulations covered a range of initial *T* of 4500 K in each case. We carried out a total of 110 runs which, with an average of ~9000 time steps per run, amounted to a total of ~1 million time steps for our melting simulations.

Figures 4 and 5 illustrate the *T*- and *P*-evolution of hcp-Re runs with initial temperatures of 25,500, 26,000 and 26,500 K; these runs correspond to the ab initio hcp-Re melting point at ~209 GPa shown in Figure 11 as an open blue circle. Similarly, Figures 9 and 10 illustrate the *T*- and *P*-evolution of the fcc-Re runs that correspond to its ~212 GPa ab initio melting point shown in Figure 11 as an open green circle.

The 11 melting points of Re are shown in Figure 11. We note that their *T* error bars are half of the *T* increment, i.e., ±250 K, basically the size of the points in Figure 11; the *P* error bars are very small which we ignore, just like for Os. The melting curve of hcp-Re, as the best fit to the corresponding 6 QMD points, is given by (*T_m* in K, *P* in GPa)

$$T_m^{hcp}(P) = 3460 \ (1 + P/58.5)^{0.69}, \tag{8}$$

for which the initial slope is 40.8 K/GPa, in excellent agreement with the VF melting curve obtained by electrical heating [7]. The melting curve of fcc-Re, as the best fit to the corresponding 5 QMD points, is

$$T_m^{fcc}(P) = 2466 \ (1 + P/7.6)^{0.42}. \tag{9}$$

The two melting curves cross each other at (20 GPa, 4240 K) and (240 GPa, 10,650 K), therefore Re melts from fcc over the pressure range 20–240 GPa. In order to determine the exact locaton of the hcp-fcc phase boundary, we carried out full free energy calculations on the two solid structures.

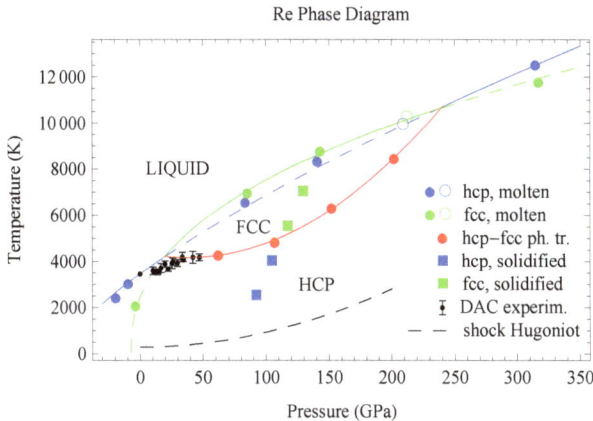

Figure 11. The phase diagram of Re based on the theoretical results of this work vs. the experimental DAC data of Ref. [6].

5.4. Full Free Energy Calculations on hcp-Re and fcc-Re

In Figures 12 and 13 we show the phonon spectra of hcp- and fcc-Re, respectively, at one fixed density, namely, 25.1 g/cc (*P* ~ 115 GPa), at five different temperatures calculated using the temperature-dependent effective potential (TDEP) method [42,43], which takes into account anharmonic lattice vibrations. It is seen that fcc-Re is dynamically stable (no imaginary phonon branches) at all temperatures, and the *T*-dependence of its phonon spectra is weak. Similar calculations show that fcc-Re is dynamically stable in the whole range of densities considered in this work.

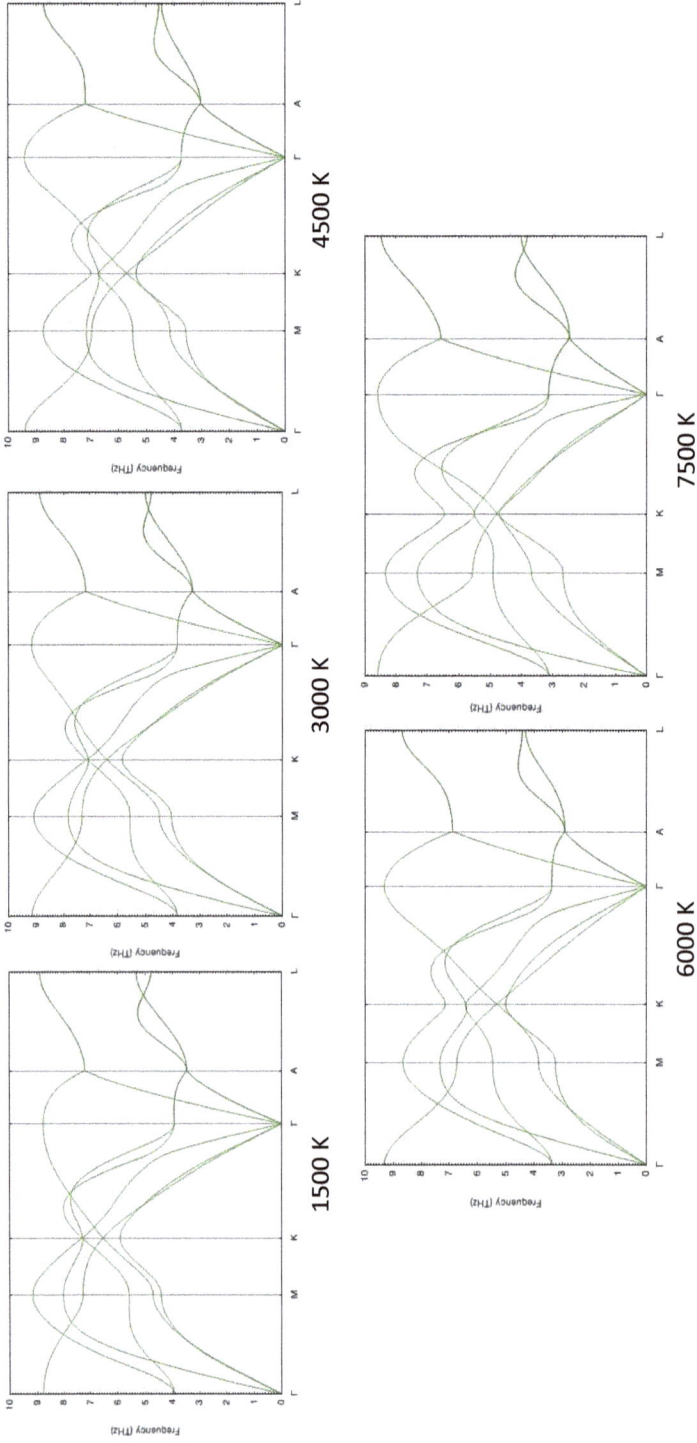

Figure 12. Phonon spectra of hcp-Re as functions of *T* at a fixed density of 25.1 g/cc.

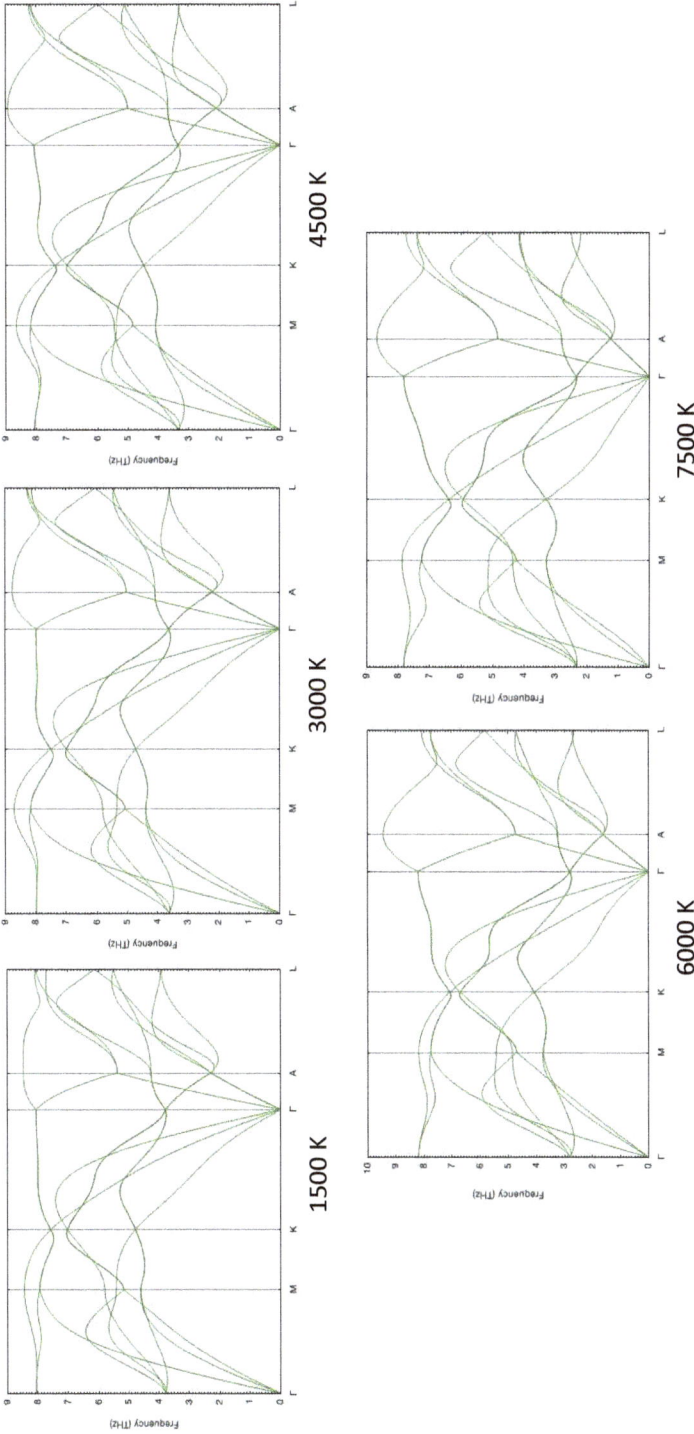

Figure 13. Phonon spectra of fcc-Re as functions of *T* at a fixed density of 25.1 g/cc.

We calculated the full free energies of both fcc-Re and hcp-Re using TDEP. The differences between the full free energies of fcc- and hcp-Re as functions of T at four pressures are shown in Figure 14. The $T = 0$ differences between the enthalpies of fcc- and hcp-Re are in excellent agreement with previous calculations (on both hcp-fcc and hcp-bcc free energy differences) [5]; they are the starting points of the four curves in Figure 14. There are four hcp-fcc transition points ($\Delta F = 0$). The best fit to these four transition points is $T(P) = 4415 - 11.9\,P + 0.158\,P^2$ (T_m in K, P in GPa); this fit is plotted in Figure 11 as a red curve. It crosses the hcp and fcc melting curves at the (P, T) points $(20, 4240)$ and $(240, 10{,}650)$, which are hcp-fcc-liquid triple points.

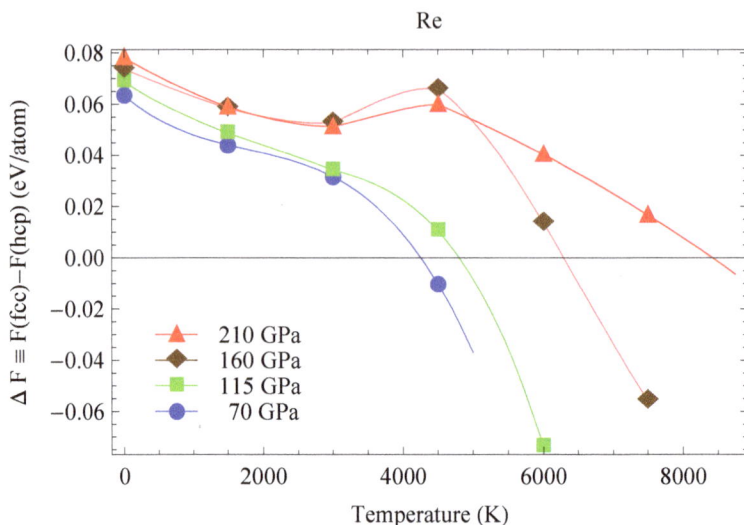

Figure 14. The hcp-Re–fcc-Re full free energy differences as functions of T at four pressures.

In order to validate the location of the hcp-fcc s-s phase boundary we also carried out a set of four independent inverse Z runs [2] to solidify liquid Re into hcp below the boundary and fcc above it. We used a computational cell of 686 atoms prepared by melting a $7 \times 7 \times 7$ solid body-centered cubic (bcc) supercell which would eliminate any bias towards solidification into either hcp or fcc. We carried out NVT simulations using the Nosé-Hoover thermostat with a timestep of 1 fs. Complete solidification typically required from 15 to 25 ps, or 15,000–25,000 timesteps. The inverse Z runs indicate that Re does solidify into hcp below the red curve in Figure 11, while above this curve it solidifies into fcc. The radial distribution functions (RDFs) of the final solid states are noisy; upon fast quenching of the four structures to low T, where RDFs are more discriminating, we clearly observed both hcp and fcc; see Figure 15.

Figure 11 also shows the Re principal Hugoniot [44], $T(P) = 293 + 0.08\,P^{1.955}$, which crosses the melting curve at $(525\,\text{GPa}, 16{,}920\,\text{K})$; this is our prediction for the Re melting point in a SW experiment.

Re, solidified from bcc−based liquid

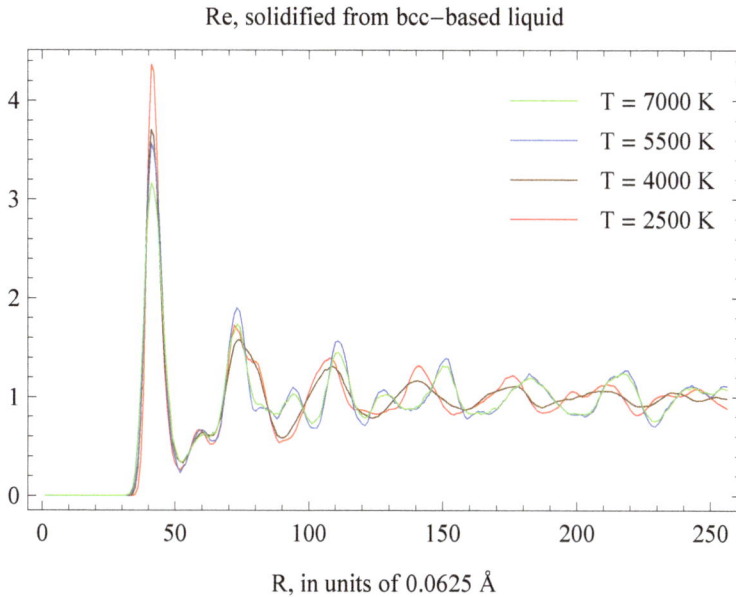

Figure 15. RDFs of the final states in the inverse Z simulations described in the text.

5.5. Topological Equivalence of the Phase Diagrams of Re and Pt

We note the topological equivalence of the phase diagrams of Re calculated in this work and that of Pt from our earlier study [2]. In both cases there is a second solid phase—fcc for Re and random hcp (rhcp) for Pt—along the melting curve over a limited range of pressures. The *P* intervals for the second phase are similar: 20–240 GPa for Re and 35–300 GPa for Pt. The $T = 0$ free energy differences between the high-*PT* and ambient structures grow with *P* in a similar way: $55 + 0.11\,P$ meV/atom for Re and $60 + 0.07\,P$ meV/atom for Pt (Supplementary Materials of Ref. [1]). At the corresponding transition points the $T = 0$ free energy differences are nearly equal: 57.2 meV/atom for Re vs. 62.5 meV/atom for Pt at the lower-*P* transition points, and 81.4 meV/atom for Re vs. 81.0 meV/atom for Pt at the upper-*P* points. Both DAC melting curves are also similar: in both cases the slope of the DAC melting curve is 2.3 times smaller than the actual one (in K/GPa): 17 [6] vs. 40 [7] for Re, and 19.3 [45] vs. 43.8 [2] for Pt. Our results suggest that the DAC melting curve in Figure 11 can be split into two different segments, just like the one for Pt [2]. The first, which lies between the origin and the triple point, is consistent with the ab initio melting curve after taking the DAC error bars into account, and the second segment is above the triple point on the hcp-fcc s-s phase boundary. Fitting a single linear form to these two segments results in a much lower Clapeyron slope for the DAC melting curve, which is apparently the case for both Re and Pt.

6. 3rd-Row Transition Metal Melting Systematics

Of all the 3rd-row transition metals, only the melting curve of Hf has never been measured or calculated yet. In addition to the melting curves of Re and Os calculated in this work, in Figure 16 we also plot the melting curves of Ta [15,23], W [16], Ir [14], Pt [2,25,28], and Au [11–13]. As seen in Figure 16, all the seven 3rd-row transition metal melting curves have low curvature and comparable slopes, roughly 45 K/GPa. These regularities form the 3rd-row transition metal melting systematics. The initial slopes of the seven melting curves can be grouped accoring to (i) crystal structure, (ii) relative location in the periodic table (PT), and (iii) topological equivalence of the phase diagrams (for Re and Pt). More specifically, these initial slopes are (in K/GPa): ∼55 for Au, ∼50 for Os and Ir (PT neighbors that have very similar mechanical and bulk properties, mass density in particular; although their ambient crystal structures are different, fcc iridium transforms into hexagonal structure at high-*PT* [14]), ~45 for W and Ta (PT neighbors, both bcc), and ~40 for Re and Pt (that have topologically equivalent phase diagrams albeit different ambient crystal structures). Even the numerical values of the parameters (*a*, *b*) in the Simon form of the melting curve, $T_m = T_m(0)(1 + P/a)^b$, can be grouped accordingly: (36.1, 0.53) for Os and (31.2, 0.59) for Ir, and (35.1, 0.47) for Ta and (41.8, 0.5) for W. The exception is the pair Re and Pt for which the corresponding sets are somewhat different: (58.5, 0.69) for Re and (44.3, 0.85) for Pt. We now demonstrate how this melting systematics combined with the *P* = 0 value of the melting temperature can be used to estimate the currently unknown melting curve of the 8th 3rd-row transition metal Hf.

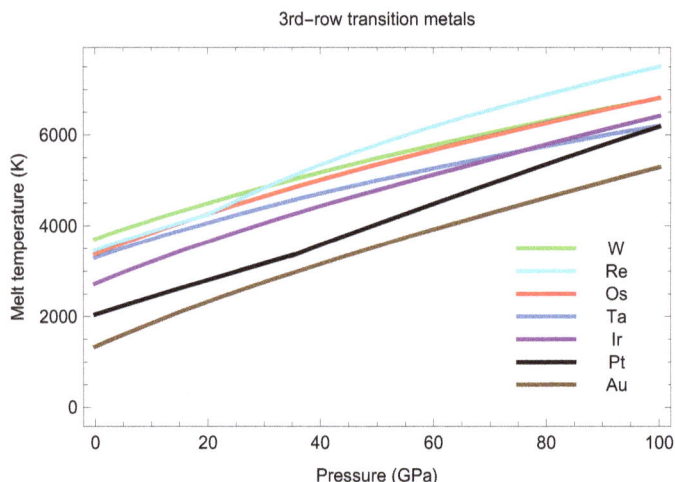

Figure 16. The melting systematics of the 3rd-row transition metals based on the available experimental and/or theoretical data.

Hf Melting Curve Estimate

At ambient *P* Hf is a hcp metal; on increasing *T* it undergoes a hcp-bcc structural transformation at ~2000 K and then melts (from bcc) at 2506 K [46]. To estimate the numerical values of the parameters (*a*, *b*) in the Simon form of its melting curve, $T_m = T_m(0)(1 + P/a)^b$, we refer to its 3rd-row bcc counterparts Ta and W for which the corresponding values are given above. Since the values of *b* are close to each other, and all three are PT neighbors with *Z* =72, 73, and 74, the value of *b* for Hf should be close to those for Ta and W. If there is any *Z* dependence of *b*, perhaps it may be approximated by a linear function over a short *Z* interval (72–74) to yield $b(72) \approx 0.44$. We therefore assume that for Hf $b = 0.47 \pm 0.03$, which seems to encompass all the possibilities. As regards *a*, a linear extrapolation of the corresponding Ta and W values would give 28.4 for Hf, but $a = 35.1 \pm 6.7$ (to cover all possible values from 28.4 to tungsten's 41.8) will produce a large uncertainty in the corresponding values of T_m. We assume that *a* for Hf is not larger than that for Ta (35.1) but its error is essentially smaller than $6.7(= (41.8 - 28.4)/2)$; speficially, we choose $a = 32.5 \pm 2.5$. Thus, our estimate for the melting curve of Hf is

$$T_m(P) = 2506 \left(1 + \frac{P}{32.5 \pm 2.5}\right)^{0.47 \pm 0.03}. \tag{10}$$

In order to test this estimate, we carried out the calculation of two melting points of Hf using the Z method implemented with VASP. We ran bcc-Hf cells of 432 atoms ($6 \times 6 \times 6$), having lattice constants of 3.26 Å and 2.65 Å, or densities of 17.11 g/cc and 31.85 g/cc, which correspond to ~50 and 500 GPa, respectively, with a single Γ point. We chose a timestep of 2 fs and an initial T increment of 375 K (so that the T_m error is roughly ± 200 K) in each case. Figures 17 and 18 show the time evolution of T and P in the sets of runs that include the melting run, in one of these cases, namely, $\rho = 17.11$ g/cc. Our simulated (P, T) melting points are $(T$ in K, P in GPa) (51, 3940) and (497, 9465). The corresponding values of T from (10) are 3905 ± 140 and 9303 ± 840, in excellent agreement with our VASP results. Thus, our estimate (10) is at least a very good approximation for the actual melting curve of Hf over a wide range of P, 0–500 GPa. A more consistent calculation of the melting curve of Hf, which requires simulating additional number of its melting points, goes far beyond the scope of this work; it will be undertaken in one of our future research projects.

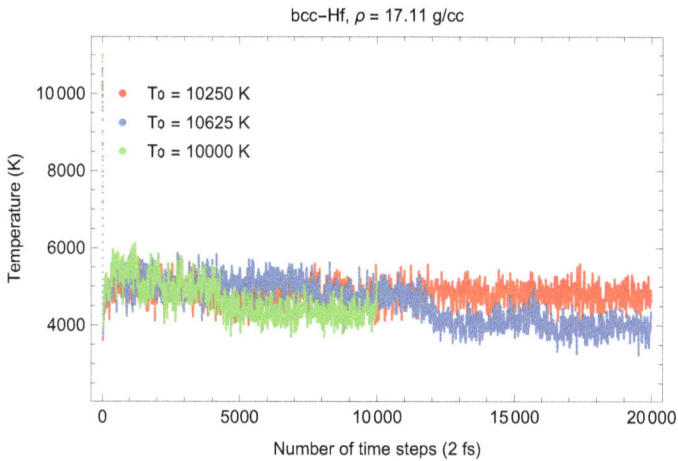

Figure 17. Time evolution of T for bcc-Hf with a density of 17.11 g/cc in three QMD runs with initial T_0s separated by 375 K.

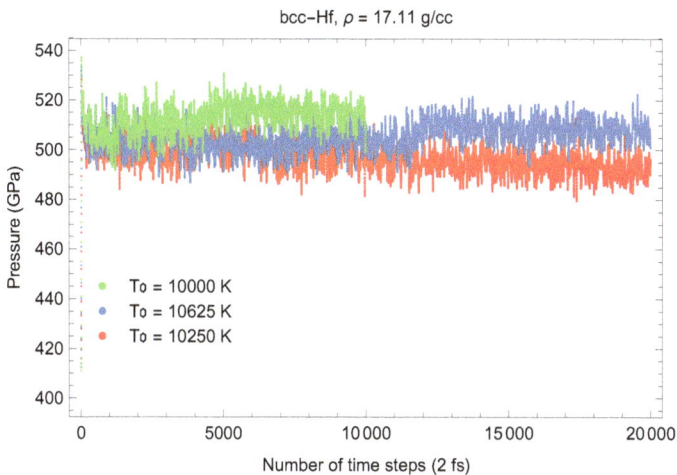

Figure 18. The same as in Figure 17 for the time evolution of P.

7. Concluding Remarks

We have calculated the melting curve of Os using the Z method based on first-principles QMD implemented with VASP. We have also calculated the phase diagram of Re to 350 GPa including the melting temperatures of both hcp-Re and fcc-Re. We have run a total of about 2 million time steps in our QMD simulations on Os, and over 1 million time steps in those on Re; however, the high accuracy of the results and their importance to the field of phase diagram studies justifies the computational cost. Our calculated melting curve of hcp-Re is in excellent agreement with the low-pressure experimental data of VF [7]. Free energy calculations using TDEP and inverse Z simulations yield the same hcp-fcc phase boundary. We have shown that the recent DAC data of YKB [6] in fact map out the hcp-fcc s-s phase boundary. The phase diagram of Re is topologically equivalent to that of Pt. The two DAC melting curves are also similar: in both cases the slope of the DAC melting curve is 2.3 times smaller than the correct one because of fitting a single linear form to its two segments; the first is along the melting curve between the origin and the triple point, and the second is above the triple point along the s-s phase boundary. Our findings suggest that the DAC melting curve may be erroneous in some other cases. In fact, the older low-slope DAC melting curve may map out either the corresponding s-s phase boundary, similar to the cases of Re and Pt, or a solid texture transition line, similar to the case of Mo for which such a texture transition was discovered in the most recent experimental study [47]. The resolution of this issue calls for additional study, both experimental and theoretical. Regularities in the melting curves of Re, Os, and five other 3rd-row transition metals form the 3rd-row transition metal melting systematics. We have demonstrated how this systematics can be used to estimate the currently unknown melting curve of the eighth 3rd-row transition metal Hf.

Author Contributions: L.B., N.B. and D.P. carried out QMD simulations using the Z methodology. S.S. performed full free energy calculations using TDEP method.

Funding: This research received no external funding.

Acknowledgments: The work was done under the auspices of the US DOE/NNSA. The QMD simulations were performed on the LANL clusters Pinto and Badger as parts of the Institutional Computing project w18_rhenium.

Conflicts of Interest: The authors declare no conflict of interest.

References

1. Dubrovinsky, L.; Dubrovinskaya, N.; Crichton, W.A.; Mikhaylushkin, A.S.; Simak, S.I.; Abrikosov, I.A.; de Almeida, J.S.; Ahuja, R.; Luo, W.; Johansson, B. Noblest of all metals is structurally unstable at high pressure. *Phys. Rev. Lett.* **2007**, *98*, 045503. [CrossRef] [PubMed]
2. Burakovsky, L.; Chen, S.P.; Preston, D.L.; Sheppard, D.G. Z methodology for phase diagram studies: Platinum and tantalum as examples. *J. Phys. Conf. Ser.* **2014**, *500*, 162001. [CrossRef]
3. Dubrovinsky, L.; Dubrovinskaia, N.; Prakapenka, V.B.; Abakumov, A.M. Implementation of micro-ball nanodiamond anvils for high-pressure studies above 6 Mbar. *Nat. Commun.* **2012**, *3*, 1163. [CrossRef] [PubMed]
4. Zha, C.-S.; Bassett, W.A.; Shim, S.-H. Rhenium, an in situ pressure calibrant for internally heated diamond anvil cells. *Rev. Sci. Instrum.* **2004**, *75*, 2409–2418. [CrossRef]
5. Verma, A.K.; Ravindran, P.; Rao, R.S.; Godwal, B.K.; Jeanloz, R. On the stability of rhenium up to 1 TPa pressure against transition to the bcc structure. *Bull. Mater. Sci.* **2003**, *26*, 183–187. [CrossRef]
6. Yang, L.; Karandikar, A.; Boehler, R. Flash heating in the diamond cell: Melting curve of rhenium. *Rev. Sci. Instrum.* **2012**, *83*, 063905. [CrossRef] [PubMed]
7. Vereshchagin, L.F.; Fateeva, N.S. Melting curve of rhenium up to 80 kbar. *JETP Lett.* **1975**, *22*, 106.
8. Regel', A.R.; Glazov, V.M. Periodicheskiy zakon i fizicheskie svoistva elektronnykh rasplavov. In *The Periodic Law and Physical Properties of Electronic Melts*; Nauka: Moscow, Russia, 1978.
9. Gorecki, T. Vacancies and melting curves of metals at high pressure. *Zeitschrift für Metallkunde* **1977**, *68*, 231–236.
10. Gorecki, T. Vacancies and a generalised melting curve of metals. *High Temp. High Press.* **1979**, *11*, 683–692.

11. Decker, D.L.; Vanfleet, H.B. Melting and high-temperature electrical resistance of gold under pressure. *Phys. Rev.* **1965**, *138*, A129. [CrossRef]

12. Tsuchiya, T. First-principles prediction of the P-V-T equation of state of gold and the 660-km discontinuity in Earth's mantle. *J. Geophys. Res.* **2003**, *108*, 2462. [CrossRef]

13. Pippinger, T.; Dubrovinsky, L.; Glazyrin, K.; Miletich, R.; Dubrovinskaya, N. Detection of melting by in-situ observation of spherical-drop formation in laser-heated diamond-anvil cells. *Física de la Tierra* **2011**, *23*, 2011.

14. Burakovsky, L.; Burakovsky, N.; Cawkwell, M.J.; Preston, D.L.; Errandonea, D.; Simak, S.I. Ab initio phase diagram of iridium. *Phys. Rev. B* **2016**, *94*, 094112. [CrossRef]

15. Dewaele, A.; Mezouar, M.; Guignot, N.; Loubeyre, P. High melting points of tantalum in a laser-heated diamond anvil cell. *Phys. Rev. Lett.* **2010**, *104*, 255701. [CrossRef] [PubMed]

16. Baty, S.R.; Burakovsky, L.; Preston, D.L. Ab initio melting curve of tungsten and W93 alloy. **2018**, in preparation.

17. Joshi, K.D.; Gupta, S.C.; Banerjee, S. Shock Hugoniot of osmium up to 800 GPa from first principles calculations. *J. Phys. Condens. Matter* **2009**, *21*, 415402. [CrossRef] [PubMed]

18. Dubrovinsky, L.; Dubrovinskaya, N.; Bykova, E.; Bykov, M.; Prakapenka, V.; Prescher, C.; Glazyrin, K.; Liermann, H.-P.; Hanfland, M.; Ekholm, M.; et al. The most incompressible metal osmium at static pressures above 750 gigapascals. *Nature* **2015**, *525*, 226. [CrossRef] [PubMed]

19. Godwal, B.K.; Yan, J.; Clark, S.M.; Jeanloz, R. High-pressure behavior of osmium: An analog for iron in Earth's core. *J. Appl. Phys.* **2012**, *111*, 112608. [CrossRef]

20. Armentrout, M.M.; Kavner, A. Incompressibility of osmium metal at ultrahigh pressures and temperatures. *J. Appl. Phys.* **2010**, *107*, 093528. [CrossRef]

21. Kulyamina, E.Y.; Zitserman, V.Y.; Fokin, L.R. Osmium: Melting curve and matching of high temperature data. *High Temp.* **2015**, *53*, 151–154. [CrossRef]

22. Belonoshko, A.B.; Burakovsky, L.; Chen, S.P.; Johansson, B.; Mikhaylushkin, A.S.; Preston, D.L.; Simak, S.I.; Swift, D.C. Molybdenum at high pressure and temperature: Melting from another solid phase. *Phys. Rev. Lett.* **2008**, *100*, 135701. [CrossRef] [PubMed]

23. Burakovsky, L.; Chen, S.P.; Preston, D.L.; Belonoshko, A.B.; Rosengren, A.; Mikhaylushkin, A.S.; Simak, S.I.; Moriarty, J.A. High-pressure–high-temperature polymorphism in Ta: Resolving an ongoing experimental controversy. *Phys. Rev. Lett.* **2010**, *104*, 255702. [CrossRef] [PubMed]

24. Belonoshko, A.B.; Rosengren, A.; Burakovsky, L.; Preston, D.L.; Johansson, B. Melting of Fe and $Fe_{0.9375}Si_{0.0625}$ at Earth's core pressures studied using ab initio molecular dynamics. *Phys. Rev. B* **2009**, *79*, 220102. [CrossRef]

25. Belonoshko, A.B.; Rosengren, A. High-pressure melting curve of platinum from ab initio Z method. *Phys. Rev. B* **2012**, *85*, 174104. [CrossRef]

26. Dewaele, A.; Mezouar, M.; Guignot, N.; Loubeyre, P. Melting of lead under high pressure studied using second-scale time-resolved x-ray diffraction. *Phys. Rev. B* **2007**, *76*, 144106. [CrossRef]

27. Anzellini, S.; Dewaele, A.; Mezouar, M.; Loubeyre, P.; Morard, G. Melting of iron at Earth's inner core boundary based on fast x-ray diffraction. *Science* **2013**, *340*, 464–466. [CrossRef] [PubMed]

28. Errandonea, D. High-pressure melting curves of the transition metals Cu, Ni, Pd, and Pt. *Phys. Rev. B* **2013**, *87*, 054108. [CrossRef]

29. Arblaster, J.W. Is osmium always the densest metal? A comparison of the densities of iridium and osmium. *Johns. Matthey Technol. Rev.* **2014**, *58*, 137–141. [CrossRef]

30. Arblaster, J.W. Crystallographic properties of osmium. *Platin. Metals Rev.* **2013**, *57*, 177–185. [CrossRef]

31. Occelli, F.; Farber, D.L.; Badro, J.; Aracne, C.M.; Teter, D.M.; Hanfland, M.; Canny, B.; Couzinet, B. Experimental evidence for a high-pressure isostructural phase transition in osmium. *Phys. Rev. Lett.* **2004**, *93*, 095502. [CrossRef] [PubMed]

32. Takemura, K.; Arai, M.; Kobayashi, K.; Sasaki, T. Bulk modulus of Os by experiments and first-principles calculations. In Proceedings of the 20th AIRAPT—43th EHPRG, Karlsruhe, Germany, 27 June 27–1 July 2005. Available online: http://bibliothek.fzk.de/zb/verlagspublikationen/AIRAPT_EHPRG2005/Posters/P139.pdf (accessed on 10 December 2017).

33. Kenichi, T. Bulk modulus of osmium: High-pressure powder x-ray diffraction experiments under quasihydrostatic conditions. *Phys. Rev. B* **2004**, *70*, 012101. [CrossRef]

34. Perreault, C.S.; Velisavljevic, N.; Vohra, Y.K. High-pressure structural parameters and equation of state of osmium to 207 GPa. *Cogent Phys.* **2017**, *4*, 1376899. [CrossRef]

35. Sahu, B.R.; Kleimann, L. Osmium is not harder than diamond. *Phys. Rev. B* **2005**, *72*, 113106. [CrossRef]

36. Belonoshko, A.B.; Skorodumova, N.V.; Rosengren, A.; Johansson, B. Melting and critical superheating. *Phys. Rev. B* **2006**, *73*, 012201. [CrossRef]

37. Grover, R. Liquid metal equation of state based on scaling. *J. Chem. Phys.* **1971**, *55*, 3435; [CrossRef]

38. Grover, R. Metallic high pressure equation of state derived from experimental data. In *High Pressure Science and Technology, Proceedings of the 6th AIRAPT Conference, Boulder, CO, USA, 25–29 July 1977*; Plenum Press: New York, NY, USA, 1979; Volume 1, p. 33. In this paper it is suggested that, for C_{VL} at a given density, $C_{VL} = \frac{3}{2} R \left[1 + (1 + 0.1\, T/T_m)^{-1} \right]$ is a good average representation of all the available experimental data and computer simulations on liquid metals, including those for the OCP. Hence, at $T = T_m$, $C_{VL} = \frac{63}{22} R$. Since $C_{VS} \cong 3R$, $C_{VL} \cong \frac{21}{22} C_{VS} \approx C_{VS}$.

39. Vohra, Y.K.; Duclos, S.J.; Ruoff, A.L. High-pressure x-ray diffraction studies on rhenium up to 216 GPa (2.16 Mbar). *Phys. Rev. B* **1987**, *36*, 9790. [CrossRef]

40. Anzellini, S.; Dewaele, A.; Occelli, F.; Loubeyre, P.; Mezouar, M. Equation of state of rhenium and application for ultra high pressure calibration. *J. Appl. Phys.* **2014**, *115*, 043511. [CrossRef]

41. Jeanloz, R.; Godwal, B.K.; Meade, C. Static strength and equation of state of rhenium at ultra-high pressures. *Nature* **1991**, *349*, 687. [CrossRef]

42. Hellman, O.; Abrikosov, I.A.; Simak, S.I. Lattice dynamics of anharmonic solids from first principles. *Phys. Rev. B* **2011**, *84*, 180301. [CrossRef]

43. Hellman, O.; Steneteg, P.; Abrikosov, I.A.; Simak, S.I. Temperature dependent effective potential method for accurate free energy calculations of solids. *Phys. Rev. B* **2013**, *87*, 104111. [CrossRef]

44. Kinslow, R. (Ed.) *High-Velocity Impact Phenomena*; Academic Press: New York, NY, USA; London, UK, 1970; p. 544.

45. Kavner, A.; Jeanloz, R. High-pressure melting curve of platinum. *J. Appl. Phys.* **1998**, *83*, 7553–7559. [CrossRef]

46. Ostanin, S.A.; Trubitsin, V.Y. Calculation of the P-T phase diagram of hafnium. *Comput. Mater. Sci.* **2000**, *17*, 174. [CrossRef]

47. Hrubiak, R.; Meng, Y.; Shen, G. Microstructures define melting of molybdenum at high pressures. *Nat. Commun.* **2017**, *8*, 14562. [CrossRef] [PubMed]

crystals

MDPI

Article

Pressure-Induced Transformation of Graphite and Diamond to Onions

Vladimir D. Blank [1,2,3,*], **Valentin D. Churkin** [1,3], **Boris A. Kulnitskiy** [1,3], **Igor A. Perezhogin** [1,3,4], **Alexey N. Kirichenko** [1], **Sergey V. Erohin** [1,3], **Pavel B. Sorokin** [1,2,5] and **Mikhail Yu. Popov** [1,2,3]

1   Technological Institute for Superhard and Novel Carbon Materials, Centralnaya Str. 7a, 108840 Troitsk, Moscow, Russia; churkin_valentin@rambler.ru (V.D.C.); boris@tisnum.ru (B.A.K.); iap1@mail.ru (I.A.P.); akir73@mail.ru (A.N.K.); sverohin@tisnum.ru (S.V.E.); PBSorokin@tisnum.ru (P.B.S.); mikhail.popov@tisnum.ru (M.Y.P.)
2   Department of Materials Science of Semiconductors and Dielectrics, National University of Science and Technology MISiS, Leninskiy prospekt 4, 119049 Moscow, Russia
3   Department of Molecular and Chemical Physics, Moscow Institute of Physics and Technology State University, Institutskiy per. 9, 141700 Dolgoprudny, Moscow Region, Russia
4   International Laser Center, M.V. Lomonosov Moscow State University, Leninskie Gory 1, 119991 Moscow, Russia
5   Inorganic Nanomaterials Laboratory, National University of Science and Technology MISiS, Leninskiy Prospekt 4, 119049 Moscow, Russia
*   Correspondence: vblank@tisnum.ru; Tel.: +7-499-272-2313

Received: 7 December 2017; Accepted: 29 January 2018; Published: 31 January 2018

Abstract: In this study, we present a number of experiments on the transformation of graphite, diamond, and multiwalled carbon nanotubes under high pressure conditions. The analysis of our results testifies to the instability of diamond in the 55–115 GPa pressure range, at which onion-like structures are formed. The formation of interlayer sp^3-bonds in carbon nanostructures with a decrease in their volume has been studied theoretically. It has been found that depending on the structure, the bonds between the layers can be preserved or broken during unloading.

Keywords: onions; carbon; high pressure

1. Introduction

On the basis of experimental data, we have recently proposed a new phase diagram of carbon with a region of diamond instability in the 55–115 GPa pressure range [1,2]. In this range, the data on shock compression of single-crystal graphite [3] indicate formation of a phase denser (by 2%) than diamond. At pressures less than 55 GPa and above 115 GPa, graphite transforms to diamond. No diamond formation in the intermediate pressure range is observed, while graphite transforms to onion-like structures [4]. Multiwall carbon nanotubes (MWCNT) also transform into onions with layers cross-linked by sp^3-bonds at around 65 GPa [5]. For comparison, the outer shells of MWCNT form some sp^3-hybridized regions at pressures below 50 GPa [6]. In this case, the inner layers of the nanotubes are retained. In shock wave experiments, diamond formation from MWCNT was observed at a pressure of $\leq$50 GPa and temperature of 1500 °C [7,8]. According to the data on shock compression of graphite [3], formation of a phase denser than diamond is observed at 55–100 GPa (at a temperature of about 3000 K). However, the structure of this phase is underinvestigated.

The existence of denser phases should lead to a loss of stability in diamond [9,10], which was also observed in [1–3]. The loss of stability of diamond can be initiated by a critical shear stress at room temperature. In [11], a phase transformation from diamond to the intermediate carbon phase was observed [12], the latter being composed of graphene plates cross-linked by sp^3-bonds.

The transition was stimulated by additional stresses applied to the compressed diamond anvils with torque by rotation of the anvil around the anvil's axis; the maximal shear stress approached 55 GPa during rotation under a hydrostatic part of the stress tensor around 40 GPa [11].

The aim of this work is to study the onion-like structures that are formed in the diamond instability region in the 55–115 GPa pressure range.

2. Materials and Methods

We used a shear diamond anvil cell (SDAC) for our high-pressure study. In the SDAC, controlled shear deformation was applied to the compressed sample by rotation of one of the anvils around the anvil's symmetry axis [13]. Pressure was measured from the stress-induced shifts of the Raman spectra from the diamond anvil tip [14]. The sample (graphite or MWCNT) was placed in a hole of a pre-pressed tungsten gasket without any pressure-transmitting media. We used the samples made of synthetic single-crystal graphite. The MWCNT were synthesized using a chemical vapor deposition procedure [15]. To study the diamond stability, we used a diamond with a mean crystal size of 25 nm produced at the Microdiamant AG (Lengwil, Switzerland). The diamond grain size must be much bigger than 2–5 nm when quantum confinement effect is not significant. As a result of quantum confinement effect, a bandgap of nanodiamond increases in the 2–5 nm size interval, along with discrete energy levels arising at the band edges. In a case of covalently bonded solids, the bandgap growth means an increase in the chemical bond energy which means an increase in elastic moduli. Indeed, bulk modulus of the 2–5 nm nanodiamond is around 560 GPa, while bulk modulus of the 25 nm diamond being identical to bulk diamond (443 GPa) [16].

On the other hand, the smaller the grain size, the larger the specific surface area that is the source of structural instability. The mixture of 25 nm nanodiamond and 25 wt % of NaCl as a pressure-transmitting media was placed in the gasket hole. The mixture has been preliminary treated in a planetary mill. A Fritsch planetary mill with ceramic silicon nitride (Si_3N_4) bowls and balls of 10 mm in diameter was used. Treatment in a planetary mill provides preparation of homogeneous nanocomposites without contamination by material of the balls [16]. The Raman spectra were recorded with a Renishaw inVia Raman microscope, excitation wavelength 532 nm (Renishaw plc, New Mills, Wotton-under-Edge, Gloucestershire, UK), and a TRIAX 552 (Jobin Yvon Inc., Edison, NJ, USA) spectrometer, equipped with a CCD Spec-10, 2KBUV Princeton Instruments 2048 × 512 detector and razor edge filters (excitation wavelength 257 nm). The transmission electron microscope (TEM) studies were done by a JEM 2010 high-resolution microscope (JEOL Ltd., Tokyo, Japan).

Adaptive Intermolecular Reactive Empirical Bond Order (AIREBO) potential was used [17] to theoretically study the atomic structure and mechanical properties of the proposed models. Simulation was carried out using the LAMMPS software package for molecular dynamics (Sandia National Laboratories, Albuquerque, NM, USA), which allows calculating structures containing up to 10^6 atoms. An undoubted advantage of this method is the ability to simulate huge systems with a sufficiently high calculation speed and acceptable quality.

3. Results

3.1. Onion Formation from Graphite and MWCNT

The Raman spectra of the initial graphite and MWCNT samples are characterized by a G-band at 1581 cm^{-1}. A disorder-induced D-band in fact is absent in both samples (Figure 1a). The graphite or MWCNT samples were loaded into a shear diamond anvil cell (SDAC), and their Raman spectra (Figure 1b) were studied in situ.

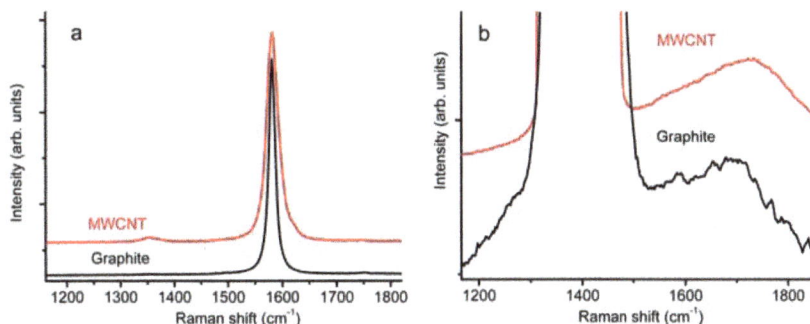

Figure 1. (**a**) Raman spectra of the initial samples of graphite and MWCNT; (**b**) Raman spectra of the samples of graphite and MWCNT under a 62 GPa pressure (note the Raman G band shifts to ~1700 cm^{-1}). A part of the spectra between 1310 cm^{-1} and ~1490 cm^{-1} is covered by the stressed diamond anvil.

Pressure dependences of the Raman G-band of graphite and MWCNT (Figure 2) are quite similar. Both dependences are linear below 25–30 GPa. With further increase in pressure, both dependences display the instability of sp^2-bonding [18] at a pressure above 30 GPa. Slowing in the change of Raman frequency indicates that the graphite and MWCNT structures become unstable at pressures above 30–35 GPa.

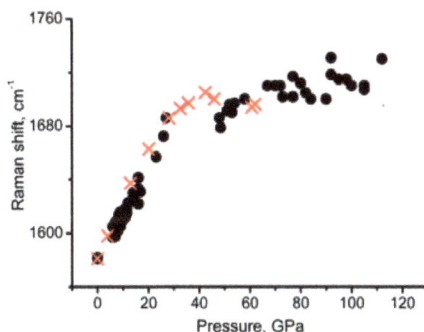

Figure 2. Pressure dependences of the Raman G-band of graphite (circles) and MWCNT (crosses).

Application of shear deformations in the beginning of the instability region (around 30–35 GPa) activates the phase transition. In the case of graphite we observe a phase transition to diamond [19]. If we use MWCNT, the outer shells of the nanotubes form some sp^3-hybridized regions [6]. In this situation, the inner layers of the nanotubes are retained.

According to the results of our simulation, the onion-like structures containing sp^3-bonds have a density higher than that of diamond in the 50–100 GPa pressure region. Consequently, their formation is preferable in comparison to diamond in this pressure range [9,10].

The results of the simulation are confirmed by the TEM studies. Figure 3 illustrates the onions obtained from graphite and MWCNT. The resulting structures have some common features: an onion with radial disorders (linear defects) responsible for the formation of sp^3-bonds.

The defects indicated by arrows in Figure 3 are formed inside the sp^2 network and are bound by sp^3-bonds. This was first reported in Refereces [20,21]. The relative content of these defects in the onion structure (carbon onion) is small compared to sp^2-hybridized carbon, so the presence of these defects with sp^3-bonds does not lead to appreciable changes in the EELS spectra obtained by us.

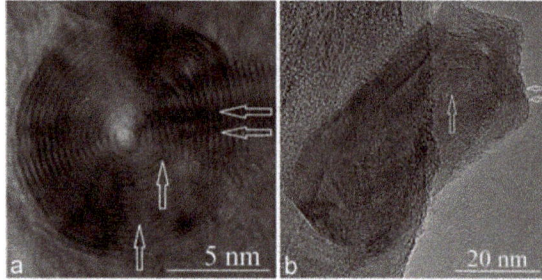

Figure 3. Onions formed from different precursors. (**a**) An onion obtained from graphite at 70 GPa. The arrows denote linear and point defects. There is a splitting of some graphene layers in multi-layered onions. According to [20,21] in the places of splits, the atoms of adjacent layers join and form sp³-bonds; (**b**) An onion obtained from MWCNT. The horizontal arrows indicate the radial disorders (linear defects) responsible for the formation of sp³-bonds, the vertical arrow points to the fragments of nanotubes inside the onion.

3.2. Onion Formation from Diamond

The presence of a 55–115 GPa diamond instability region in the phase diagram of carbon indicates the possibility of phase transformations of diamond into onion-like structures. However, in view of the large hysteresis of the phase transitions in carbon [9] and the high strength of diamond, activation of the phase transition by application of shear strains in a SDAC is a technically challenging problem [11]. The transition of diamond to onion-like structures can be facilitated if the size of the diamond particle is close to the ~20 nm size of the onions formed in this pressure range. The choice of diamond grain sizes is discussed in more detail in the Materials and Methods section.

The mixture of 25 nm nanodiamond and 25 wt % of NaCl as a pressure-transmitting media was placed in the gasket hole. The Raman band of the diamond loaded in a NaCl medium to 52 GPa is located at 1483 cm^{-1}, which is appropriate to the known data [22]. The Raman band of diamond shifts from 1333 cm^{-1} to 1483 cm^{-1} under a 52 GPa pressure. The stress tensor conditions in the diamond sample differ from the ones in the diamond anvils, so the Raman band of the diamond loaded between the anvils is separated from the Raman band of the diamond anvil [22]. An increase in pressure from 52 GPa to 57 GPa leads to the disappearance of the diamond Raman band (Figure 4) under the influence of laser radiation (the Raman spectra were registered when excited by a 1 mW laser beam with a 532 nm wavelength focused to a 1–2 μm spot). The G-band does not appear in the spectra (according to Figure 2, one could be expected around 1700 cm^{-1} at pressure ~60 GPa). Under ambient conditions, the same irradiation does not lead to degradation of diamond powder with a 25 nm grain size.

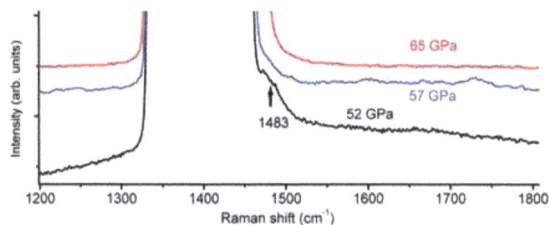

Figure 4. Raman spectra of the 25 nm diamond under a 52 GPa pressure, and the disappearance of the Raman band under 57 GPa. A part of the spectra between 1330 cm^{-1} and ≥1460 cm^{-1} is covered by the stressed diamond anvil.

For a more detailed study of the observed effect, four samples of the 25 nm diamond in a NaCl medium were loaded to pressures of 57, 60, 70, and 120 GPa, respectively. The samples were irradiated under pressure by a 15 mW laser beam with a 532 nm wavelength focused to a 1–2 μm spot in steps of 2 μm for 100 s in each position. The grain temperature of the 25 nm diamond in NaCl could reach 2000–3000 K.

The diamonds irradiated at a pressure of 120 GPa are preserved without any noticeable changes in the phase composition. The diamonds irradiated at 57, 60, and 70 GPa are transformed into onions. Figure 5 depicts two types of onions created from diamond under 57 and 60 GPa, and the onion-like structures with nuclei of 5 nm nanodiamonds in the center obtained under laser radiation at a 70 GPa pressure (Figure 6a). The latter structure could be also considered as 5 nm nanodiamonds covered by a few graphitized layers of carbon created from a 25 nm nanodiamond. Nevertheless, comparison of Figures 5 and 6 prompts consideration of the structures presented in Figure 6a as onion-like structures with nuclei of 5 nm nanodiamonds in the center. The images of the initial 25 nm diamonds are provided for comparison (Figure 6b). No graphitized layers of carbon can be seen on the surface of the initial diamonds.

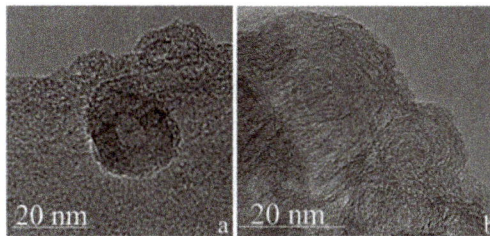

Figure 5. Onions, formed from diamond: (**a**) onion created from diamond under 57 GPa; (**b**) onion-like structure created from diamond under 60 GPa. This onion type corresponds to the simulation results.

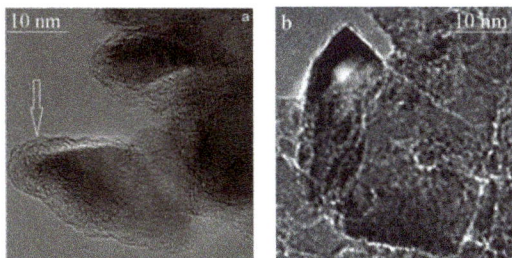

Figure 6. (**a**) onion-like structures with nanodiamond nuclei (of around 5 nm) created from 25 nm diamond under a 70 GPa pressure (the onion-like structure is marked with an arrow); (**b**) initial 25 nm diamonds.

4. Discussion

There are various traditional methods for producing carbon onions. For example, carbon onions are synthesized using the high-energy electron irradiation of carbon particles [23], annealing of nanodiamonds [24], an arc discharge between two graphite electrodes submerged in water [25], the carbon-ion implantation of silver [26] or copper [27] substrates, etc. In recent years, the onions have been obtained under high-pressure conditions. For example, carbon onions made by four different methods, three of them using high-pressure techniques, have been investigated in [28]. In the present work, we continue investigations of the onions obtained using high pressures.

We have studied two types of onion structures that demonstrate different behavior under the stress. We designed onions containing interlayer sp^3-bonds built on C_{20} and C_{36} fullerenes. We investigated stability of the four-layered structures $C_{36}@C_{144}@C_{324}@C_{576}$ and $C_{20}@C_{80}@C_{180}@C_{320}$.

Figure 7 shows the energy dependence on the volume of intermediate structures during the onion–nanodiamond transition. We observe that the behavior of the two nanostructures under investigation is fundamentally different. In particular, the onions containing the chemically active C_{20} fullerene in their center (orange curve) tend to retain the energy-favorable sp^3-hybridized interlayer bonds. Consequently, these onions remain in a compressed state even after unloading. This is due to the fact that such a sequence of fullerene shells is obtained with a ~2.5 Å distance between the layers, which is much less than the interlayer distance in graphite (3.4 Å). While in the structure containing some interlayer sp^3-bonds their lengths are ~1.6 Å, with an increase in the number of layers this value tends more and more toward the characteristic bond length in cubic diamond (1.54 Å). The appearance of a local minimum of the dependence near 0.92 V/V_{onion} in the figure is explained by the fact that the fourth layer of the transition structure at this point is already at a distance significantly exceeding the characteristic length of the carbon sp^3-bond, while the three-layer nucleus still contains some interlayer bonds.

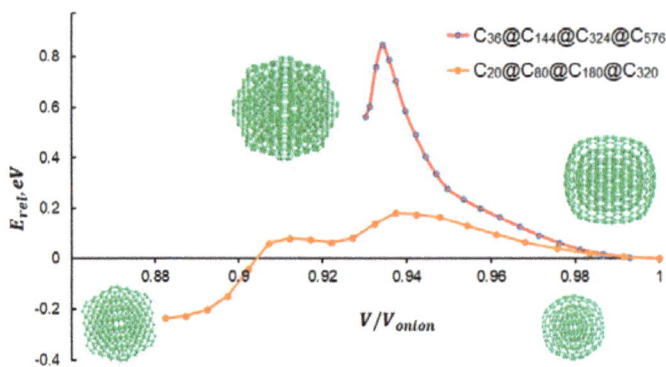

Figure 7. Behavior of the onion structures under pressure. The x-axis represents the volume of structures constructed with respect to the volume of the onion sp^2-hybridized structure. Along the y-axis, we plot the energy (per atom) given with respect to the energy of the onion sp^2-hybridized structure.

In the case of C_{36}-based structures, the dependence character changes to the opposite, that is, the structure containing interlayer sp^3-bonds becomes less energetically favorable than the corresponding onion in the absence of loading. The distance between the layers of fullerenes in this case is ~3 Å. Due to this, increasing number of layers in the structure is accompanied by quick accumulation of mechanical stresses by the interlayer bonds, which leads to structure destabilization and stratification into sp^2-hybridized onions when the load is removed.

5. Conclusions

It has been shown experimentally that diamond is unstable in the pressure range of 55–115 GPa. At these pressures, graphite, diamond, and multi-walled carbon nanotubes transform into onion-like structures. According to the results of our simulation, when the volume of the onion decreases, some sp^3-bonds can form between its layers, which can be retained when the load is removed from the material (depending on the structure of the onion).

Acknowledgments: This work was supported by the Ministry of Education and Science of the Russian Federation (project ID RFMEFI59317X0007; the agreement No. 14.593.21.0007); the work was done using the Shared-Use Equipment Center "Research of Nanostructured, Carbon and Superhard Materials" in FSBI TISNCM.

Author Contributions: Mikhail Yu. Popov, Valentin D. Churkin, and Alexey N. Kirichenko, prepared the samples and performed high-pressure and Raman studies. Boris A. Kulnitskiy, Igor A. Perezhogin, and Vladimir D. Blank carried out TEM studies. Pavel B. Sorokin, Sergey V. Erohin, and Vladimir D. Blank performed modeling. All the authors have taken part in discussions and in the interpretation of the results; and have read and approved the final manuscript.

Conflicts of Interest: The authors declare no conflict of interest.

References

1. Blank, V.D.; Churkin, V.D.; Kulnitskiy, B.A.; Perezhogin, I.A.; Kirichenko, A.N.; Denisov, V.N.; Erohin, S.V.; Sorokin, P.B.; Popov, M.Y. Phase diagram of carbon and the factors limiting the quantity and size of natural diamonds. *Nanotechnology* **2018**, in press. [CrossRef] [PubMed]

2. Popov, M.; Kulnitskiy, B.; Blank, V. Superhard materials, based on fullerenes and nanotubes. *Compr. Hard Mater.* **2014**, *3*, 515–538.

3. Gust, W.H. Phase transition and shock compression parameters to 120 GPa for three types of graphite and for amorphous carbon. *Phys. Rev. B* **1980**, *22*, 4744–4756. [CrossRef]

4. Blank, V.D.; Denisov, V.N.; Kirichenko, A.N.; Kulnitskiy, B.A.; Martushov, S.Y.; Mavrin, B.N.; Perezhogin, I.A. High pressure transformation of single-crystal graphite to form molecular carbon–onions. *Nanotechnology* **2007**, *18*, 345601. [CrossRef]

5. Pankov, A.M.; Bredikhina, A.S.; Kulnitskiy, B.A.; Perezhogin, I.A.; Skryleva, E.A.; Parkhomenko, Y.N.; Blank, V.D. Transformation of multiwall carbon nanotubes to onions with layers cross-linked by sp^3 bonds under high pressure and shear deformation. *AIP Adv.* **2017**, *7*, 085218. [CrossRef]

6. Pashkin, E.Y.; Pankov, A.M.; Kulnitskiy, B.A.; Perezhogin, I.A.; Karaeva, A.R.; Mordkovich, V.Z.; Popov, M.Y.; Sorokin, P.B.; Blank, V.D. The unexpected stability of multiwall nanotubes under high pressure and shear deformation. *Appl. Phys. Lett.* **2016**, *109*, 081904. [CrossRef]

7. Zhu, Y.Q.; Sekine, T.; Kobayashi, T.; Takazawa, E.; Terrones, M.; Terrones, H. Collapsing carbon nanotubes and diamond formation under shock waves. *Chem. Phys. Lett.* **1998**, *287*, 689–693. [CrossRef]

8. Zhu, Y.Q.; Sekine, T.; Brigatti, K.S.; Firth, S.; Tenne, R.; Rosentsveig, R.; Kroto, H.W.; Walton, D.R. Shock-Wave Resistance of WS2 Nanotubes. *J. Am. Chem. Soc.* **2003**, *125*, 1329–1333. [CrossRef] [PubMed]

9. Blank, V.D.; Estrin, E.I. *Phase Transitions in Solids under High Pressure*; CRC Press: Boca Raton, FL, USA, 2014.

10. Lobodyuk, V.A.; Estrin, E.I. *Martensitic Transformations*; Cambridge International Science Publishing: Cambridge, UK, 2014; p. 538.

11. Popov, M. Stress-induced phase transition in diamond. *High Press. Res.* **2010**, *30*, 670–678. [CrossRef]

12. Blank, V.D.; Aksenenkov, V.V.; Popov, M.Y.; Perfilov, S.A.; Kulnitskiy, B.A.; Tatyanin, Y.V.; MZhigalina, O.; Mavrin, B.N.; Denisov, V.N.; Ivlev, A.N.; et al. A new carbon structure formed at MeV neutron irradiation of diamond: Structural and spectroscopic investigations. *Diam. Relat. Mater.* **1999**, *8*, 1285–1290. [CrossRef]

13. Blank, V.; Popov, M.; Buga, S.; Davydov, V.; Denisov, V.N.; Ivlev, A.N.; Marvin, B.N.; Agafonov, V.; Ceolin, R.; Szwarc, H.; et al. Is C 60 fullerite harder than diamond? *Phys. Lett. A* **1994**, *188*, 281–286. [CrossRef]

14. Popov, M. Pressure measurements from Raman spectra of stressed diamond anvils. *J. Appl. Phys.* **2004**, *95*, 5509–5514. [CrossRef]

15. Karaeva, A.R.; Khaskov, M.A.; Mitberg, E.B.; Kulnitskiy, B.A.; Perezhogin, I.A.; Ivanov, L.A.; Denisov, V.N.; Kirichenko, A.N.; Mordkovich, V.Z. Longer Carbon Nanotubes by Controlled Catalytic Growth in the Presence of Water Vapor. *Fuller. Nanotub. Carbon Nanostruct.* **2012**, *20*, 411–418. [CrossRef]

16. Popov, M.; Churkin, V.; Kirichenko, A.; Denisov, V.; Ovsyannikov, D.; Kulnitskiy, B.; Perezhogin, I.; Aksenenkov, V.; Blank, V. Raman spectra and bulk modulus of nanodiamond in a size interval of 2–5 nm. *Nanoscale Res. Lett.* **2017**, *12*, 561. [CrossRef] [PubMed]

17. Stuart, J.S.; Tutein, B.A.; Harrison, J.A. A Reactive Potential for Hydrocarbons with Intermolecular Interactions. *J. Chem. Phys.* **2000**, *112*, 6472–6486. [CrossRef]

18. Weinstein, B.A.; Zallen, R. Pressure-Raman Effects in Covalent and Molecular Solids. In *Light Scattering in Solids*; Cardona, M., Guntherodt, G., Eds.; Springer: Berlin, Germany, 1984; Volume IV, p. 543.

19. Blank, V.D.; Kulnitskiy, B.A.; Perezhogin, I.A.; Tyukalova, E.V.; Denisov, V.N.; Kirichenko, A.N. Graphite-to-diamond (13C) direct transition in a diamond anvil high-pressure cell. *Int. J. Nanotechnol.* **2016**, *13*, 604–611. [CrossRef]

20. Balaban, A.T.; Klein, D.J.; Folden, C.A. Diamond-graphite hybrids. *Chem. Phys. Lett.* **1994**, *217*, 266–270. [CrossRef]

21. Hiura, H.; Ebbesen, T.W.; Fujita, J.; Tanigaki, K.; Takada, T. Role of sp^3 defect structures in graphite and carbon nanotubes. *Lett. Nat.* **1994**, *367*, 148–151. [CrossRef]

22. Hanfland, M.; Syassen, K.; Fahy, S.; Louie, S.G.; Cohen, M.L. Pressure dependence of the first-order Raman mode in diamond. *Phys. Rev. B* **1985**, *31*, 6896–6899. [CrossRef]

23. Ugarte, D. Curling and closure of graphitic networks under electron-beam irradiation. *Nature* **1992**, *359*, 707–709. [CrossRef] [PubMed]

24. Kuznetsov, V.L.; Chuvilin, A.L.; Butenko, Y.V.; Malkov, I.L.; Titov, V.M. Onion-like carbon from ultra-disperse diamond. *Chem. Phys. Lett.* **1994**, *222*, 343–348. [CrossRef]

25. Sano, N.; Wang, H.; Chhowalla, M.; Alexandrou, I.; Amaratunga, G.A.J. Nanotechnology: Synthesis of carbon 'onions' in water. *Nature* **2001**, *414*, 506–507. [CrossRef] [PubMed]

26. Cabioc'h, T.; Kharbach, A.; Roy, A.L.; Riviere, J.P. Fourier transform infra-red characterization of carbon onions produced by carbon-ion implantation. *Chem. Phys. Lett.* **1998**, *285*, 216–220. [CrossRef]

27. Abe, U.; Yamamoto, S.; Miyashita, A. In situ TEM observation of nucleation and growth of spherical graphitic clusters under ion implantation. *J. Electron. Microsc.* **2002**, *51*, S183–S187. [CrossRef]

28. Blank, V.D.; Kulnitskiy, B.A.; Perezhogin, I.A. Structural Peculiarities of Carbon Onions, Formed by Four Different Methods: Onions and Diamonds, Alternative Products of Graphite High-Pressure Treatment. *Scr. Mater.* **2009**, *60*, 407–410. [CrossRef]

![crystals logo] *crystals*

MDPI

Review

High-Pressure, High-Temperature Behavior of Silicon Carbide: A Review

Kierstin Daviau * and Kanani K. M. Lee (iD)

Department of Geology & Geophysics, Yale University, New Haven, CT 06511, USA; kanani.lee@yale.edu
* Correspondence: kierstin.daviau@yale.edu

Received: 26 April 2018; Accepted: 11 May 2018; Published: 16 May 2018

Abstract: The high-pressure behavior of silicon carbide (SiC), a hard, semi-conducting material commonly known for its many polytypic structures and refractory nature, has increasingly become the subject of current research. Through work done both experimentally and computationally, many interesting aspects of high-pressure SiC have been measured and explored. Considerable work has been done to measure the effect of pressure on the vibrational and material properties of SiC. Additionally, the transition from the low-pressure zinc-blende B3 structure to the high-pressure rocksalt B1 structure has been measured by several groups in both the diamond-anvil cell and shock communities and predicted in numerous computational studies. Finally, high-temperature studies have explored the thermal equation of state and thermal expansion of SiC, as well as the high-pressure and high-temperature melting behavior. From high-pressure phase transitions, phonon behavior, and melting characteristics, our increased knowledge of SiC is improving our understanding of its industrial uses, as well as opening up its application to other fields such as the Earth sciences.

Keywords: silicon carbide; high pressure; high temperature; review

---

1. Introduction

A hard and refractory semi-conductor, silicon carbide (SiC) is both a useful industrial material as well as an interesting component of naturally occurring systems. Known for its many polytypic structures appearing at ambient conditions, SiC is composed of stacked Si_4C (or SiC_4) tetrahedra and is found in cubic, hexagonal or rhombohedral forms depending on the stacking sequence [1]. The most well studied and/or naturally occurring structures are the zinc-blende (B3), also known as the 3C polytype, and the hexagonal wurtzite structured 6H polytype. The cubic structure is also known as beta (β) SiC while the hexagonal and rhombohedral structures are all classified under the umbrella term alpha (α) SiC [1]. Although not exhaustive, the range of typical SiC polytype structures is depicted in Figure 1.

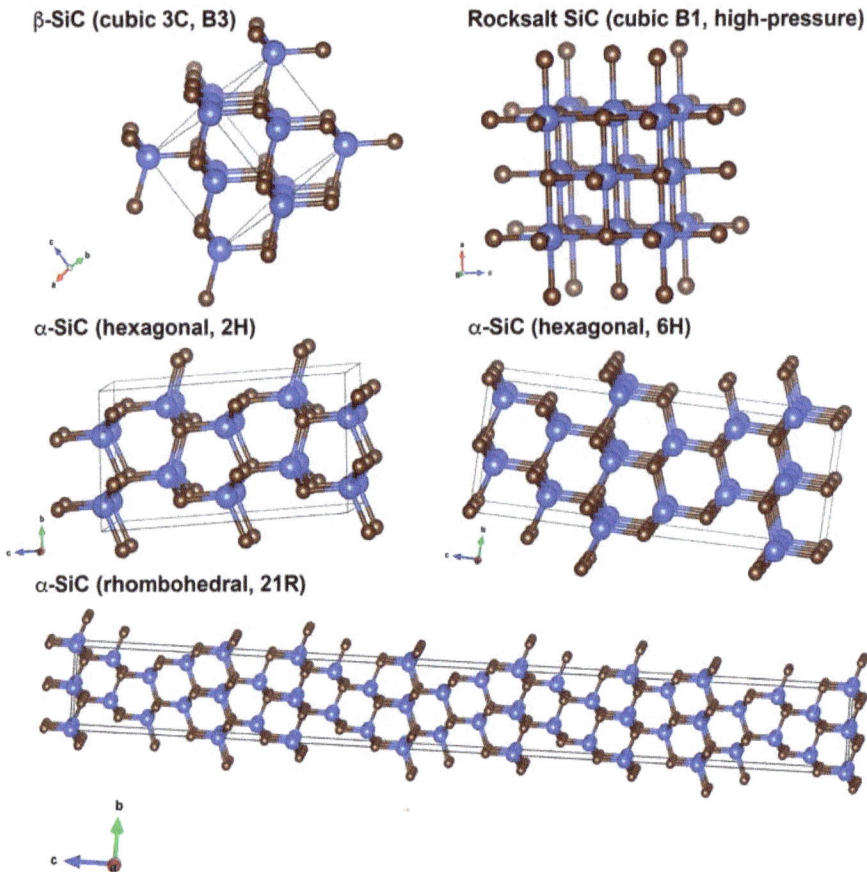

Figure 1. Structure diagram of several low pressure polytypes of SiC as well as the high-pressure rocksalt structure. The larger blue spheres represent the Si atoms while the smaller brown spheres represent the C atoms. Structures visualized using the program VESTA [2] with 3C from [3], 2H from [4], 6H from [5], and 21R from [6].

The effect of high pressure on the crystal structure, material properties and melting characteristics of SiC has been an active area of research for many years. In addition to better understanding the industrial uses of SiC, such work has also allowed a better understanding of SiC in a planetary context. Known under the mineral name moissanite [7], SiC has been found in small amounts in many geologic settings on Earth [8], as well as in meteorites and other astronomical bodies [9]. It was recently proposed that star systems more carbon-rich than the Solar System may harbor entire planets composed of significant quantities of SiC [10,11]. This review aims to summarize and discuss aspects of high-pressure work on SiC that aids in our understanding of both its industrial uses as well as its place in the natural world.

We begin with a discussion of the high-pressure structure of SiC, including the stability of different polytypes. Following, we present a discussion of the large body of work on the high-pressure phase transition in SiC, including observations and predictions of the transition conditions, predicted intermediate structures and transition mechanisms, as well as the kinetics across the transition. We then discuss proposed alternative stoichiometries at pressure. Work on the vibrational modes of SiC at pressure, as well as absorption measurements, and both measurements and computations of the equation of state (EOS) parameters and elastic constants are discussed in the following section. We end with a discussion of thermal expansivity and the thermal equation of state, as well as a discussion of high-pressure and high-temperature (high P-T) melting and decomposition. We find that thermodynamic equilibrium is often difficult to attain in high P-T experiments on SiC, meaning that time-dependent kinetic effects and hysteresis are often observed across phase transitions and melting reactions. The observed sluggish kinetics of SiC can explain many of the discrepant findings between studies.

2. High-Pressure Crystal Structure

Over 250 polytypes of SiC have been observed at ambient conditions [12]. Much work has been done to understand the formation of specific polytypes at ambient pressure and high temperature and to understand the transformation conditions and mechanisms between polytypes (i.e., [13]). As the structure of SiC may have an impact on material properties, identifying the stable structure at pressure is necessary to improve our understanding of SiC in high-pressure contexts, such as in dynamic applications or in planetary interiors. Several experiments have been performed exclusively on determining stable polytype structures at pressure [14–16]. It was found in [14] that the 3C cubic phase was stable at lower temperatures but became more stable with increased pressure, while the 6H phase was preferred at high temperatures (2300–2800 K), at least at pressures of 6.5 GPa and below. These findings were supported by observations in [17] at pressures below 3 GPa. The phase boundary marking the transition to 6H likely crosses the melting/decomposition line at high pressures [18,19], however, implying that 3C may be the more stable solid phase at higher-pressure conditions.

Shock studies on both α-SiC and β-SiC compare the proportion of polytypes present in samples after being shocked to P-T conditions of 5–25 GPa and 600–1500 K [16]. They find that in α-SiC, 6H begins to transform to 15R and a small amount of 3C after shock experiments as indicated by X-ray diffraction of recovered samples. β-SiC also transforms to rhombohedral structures (21R, 33R) during shock again based on X-ray diffraction after shock. This preference for rhombohedral polytypes at high P-T may be due to the effect of the shear stresses from the passing shockwave, however, rather than an indication that 15R, 21R or 33R are the equilibrium stable phases at pressure. The shockfront is associated with a reduction in particle size as well as potentially changing the stacking sequence of the Si-C layers, resulting in the formation of rhombohedral polytypes. Shock studies to higher pressures of over 100 GPa with in situ X-ray diffraction do not see a transition to rhombohedral structures but instead find a transition to the high-pressure rocksalt structure over 100 GPa [20]. These findings support previous shock work on 6H SiC [21] that infers a transition to the rocksalt structure from a density change in the sample.

Upon consideration of all the high P-T studies on 3C, 6H, and 15R, a consistent phase boundary marking transitions between polytypes is not agreed upon. This indicates that their relative enthalpy is perhaps small or that their transformation kinetics are slow. Impurities may also play a role in the expressed polytypes as has been observed in ambient pressure experiments [13]. In reported static experiments at high-pressures the starting polytype structure remains throughout pressure loading and unloading [22,23], at least up to the conditions of the transition to the rocksalt structure [20,24,25]. See Figure 2 for a compilation of experimental data on the stability of common polytypes at pressure. In addition to polytype transformations and the high-pressure transition to the rocksalt structure, other structural transitions have also been found in SiC, such as the transition to a high-density amorphous phase under large plastic shear and high pressure [26].

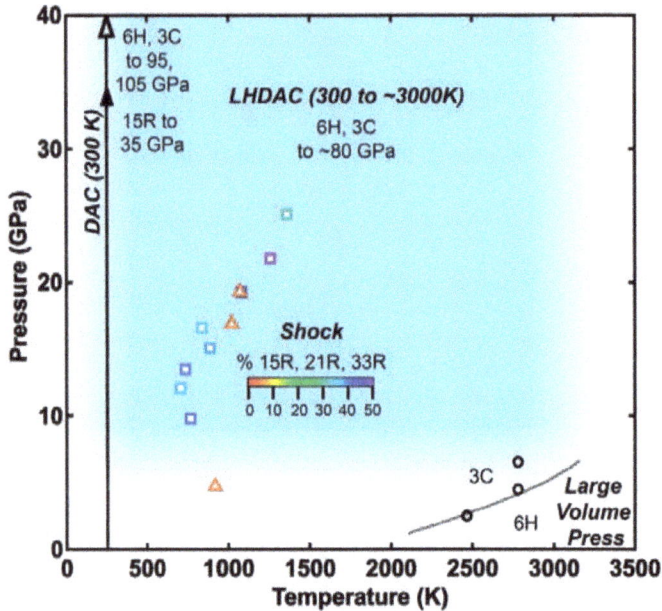

Figure 2. Compilation of experimental work on polytype transformations at pressure. DAC studies find that 15R remains under compression to at least 35 GPa at 300 K [22], 3C remains stable to 105 GPa at 300 K and 6H remains stable to 95 GPa at 300 K before transition to a dense, high-pressure rocksalt structure [25]. LHDAC experiments find that both 6H and 3C remain in experiments covering conditions of ~10–80 GPa and ~3000 K (shown in blue) before transition to the rocksalt structure [18,23,24,27,28]. Shock experiments up to 25 GPa and ~2000 K find an increase in the proportion of rhombohedral and cubic phases at high P-T and a decrease in hexagonal phases [16]. Rainbow colored shock data indicate the proportion of rhombohedral polytypes resulting from shock: 15R forming in the α-SiC sample (15% in starting sample), plotted with open squares; 21R and 33R forming in the β-SiC sample (0% in starting sample), plotted with open triangles. Multi-anvil, high pressure apparatus work finds that 3C transitions to 6H at high P-T over 2000 K [14]. The lack of clear boundaries between phases may indicate that either the kinetics of transition are sluggish or that the energy difference between polytypes is low at high P-T.

2.1. Transition to the Rocksalt Structure

The most well studied aspect of high pressure SiC is the transition from the cubic 3C (zinc-blende, B3) structure to the cubic rocksalt (B1) structure, with additional work considering the transition from the hexagonal 6H polytype to the rocksalt structure. Computational and experimental work has made significant progress in identifying the transition conditions, the intermediate structures and the mechanism of the transition in both cases. Such research has additionally illuminated the slow kinetics and hysteresis of the transition.

The transition from a four-fold coordinated zinc-blende structure to a six-fold coordinated rocksalt structure at high pressure occurs in many semi-conductors i.e., [29]. It was first predicted for the 3C-SiC system by [30] and first observed experimentally by [25] during room temperature compression of SiC in a static diamond-anvil cell (DAC). Since these early results, many computational studies have worked to better understand the transition conditions [31–38] and several experimental studies have further observed the transition using both laser-heated diamond-anvil cells (LHDAC) and shockwave experiments [20,21,24,27,28,39,40].

Calculations also predict the transition to the rocksalt structure from the 6H polytype as well [33,36,41]. Experimentally, this transition has been observed in shock studies at pressures of ~105 GPa [21]. Calculations find that the 2H and 4H polytypes also transform to the rocksalt structure at high pressures [33,42] although no experimental work has been done on the transition in these polytypes to date.

A hallmark feature in the transition to the rocksalt structure is the large unit cell volume drop, or density increase, across the transition. Both experimental and computational work indicates that the volume of SiC drops between 15% and 20% across the transition. Such a large density increase has implications for the role of SiC deep in planetary interiors as has been discussed by several studies [24,27,43]. A compilation of the transition parameters from computations and experiments is presented in Table 1, while a summary of the P-T transition conditions, with a focus on experimental work, can be found in Figure 3.

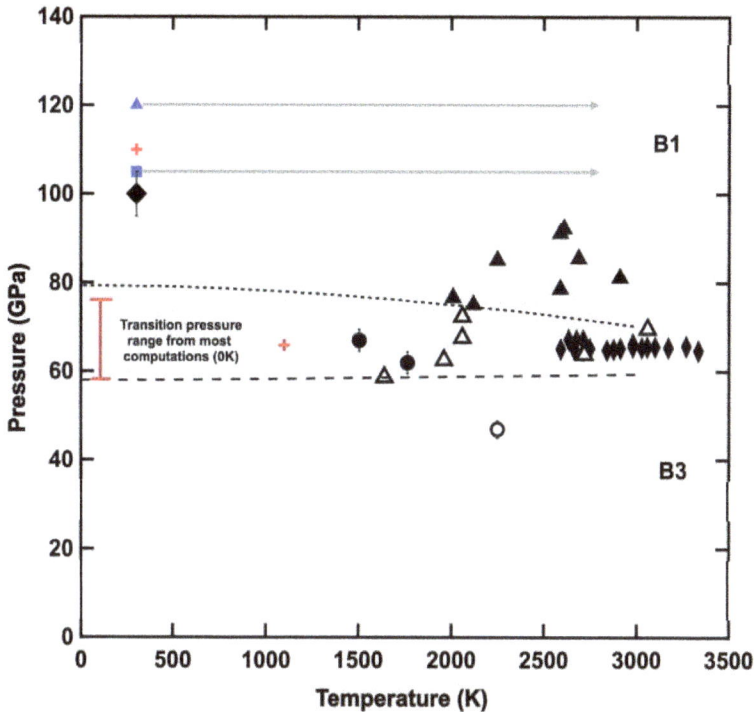

Figure 3. Experimental data from DAC (black), shock, (blue), and selected computational results (red) for the high-P-T transition conditions to the rocksalt (B1) structure. All data shown start with the cubic B3 polytype aside from the shock study [21], which starts with the 6H polytype. Bars extending from shock studies (blue triangle [20], blue square [21]) indicate the likely high sample temperature. Open symbols represent the low pressure B3 structure (open circle [24], open triangles [27]) while solid symbols indicate the appearance or stability of the B1 structure in (LH) DAC experiments (solid circles [24], solid triangles [27], solid thick diamond [25], solid thin diamonds [28]). Computations considering the hysteresis across the transition are plotted in red crosses [44], while the range of most transition temperatures found in computational studies is shown by the red bar. Long dashed line indicates Clapeyron slope proposed by [24] while short dashed line indicates Clapeyron slope proposed by [27].

Table 1. Parameters for the transition from zinc-blende (B3) to rocksalt (B1) SiC, compiled from both experiments and computations. V_t/V_0 represents the ratio of the volume of B3 SiC at the transition, to the zero pressure volume of B3. ΔV (%) shows the change in volume between B3 and B1 at the transition pressure.

Pressure (GPa)	V_t/V_0	ΔV (%)	Method	Reference
		Experiments		
74	0.813	17.3	LHDAC	[27]
65–70			LHDAC	[28]
62.4	0.811	16.5	LHDAC	[24]
66.6	0.809	17.4	LHDAC	
100	0.757	20.3	DAC	[25]
		Computations		
58	0.825	18.1	DFT (PBE)	[34]
67	0.811	18.2	DFT (PBEsol)	
75.4	0.799	18	DFT (LDA)	[35]
63	0.80	18	DFT (GGA)	[36]
140	0.78	21	MD	[45]
66	0.819	17.9	DFT (LDA)	[33]
66	0.81	18.5	Ab initio pseudopotential	[30]
65.9	0.823	18.3	DFT (GGA)	[44]

2.2. Intermediate Structures

In addition to identifying the transition conditions, work has been done to understand the mechanism of the transition to the high-pressure structure, particularly to identify the intermediate structures that come about during the transition from the four-fold coordinated structure to the six-fold coordinated structure. Experimental work has not identified an intermediate structure across the transition. This is perhaps not surprising as the intermediate structures are likely very transient, with computations indicating that the transition to the rocksalt structure occurs in 0.1 ps [46]. Despite the large amount of work however, the transitional structure still remains controversial.

In early first-principle calculations it was proposed that SiC passes through a rhombohedral *R3m* transition state with one formula unit per primitive unit cell [33,47]. Molecular dynamics (MD) simulations using a new interatomic potential model in [45] found that the MD cell changes from cubic to monoclinic (rather than rhombohedral) during the transition at 100 GPa, and is accompanied by the Si and C sublattices shifting relative to each other along the [100] direction in the zinc-blende structure. The following year it was proposed that the intermediate state actually has orthorhombic *Pmm2* symmetry with two formula units per cell [31]. These least-enthalpy calculations used a periodic linear-combination-of-atomic-orbitals scheme with a transition pressure of 92 GPa. It was found by [31] that this orthorhombic transition pathway has a much lower activation energy than the original *R3m* state. The *Pmm2* symmetry of the orthorhombic pathway was questioned in an active comment thread by [48,49] and it was instead proposed that the transition structure has an *Imm2* symmetry. The *Imm2* structure was later confirmed by the original author [50], although the results of the study were not affected by favoring the higher symmetry *Imm2* structure over the *Pmm2* structure. It was also suggested that the orthorhombic structure is body-centered rather than primitive [48], in contrast with the original study [51].

The orthorhombic transition pathway was seen to have the least enthalpy barrier by [52] along with seven other structures with low enthalpy barriers that could represent the transition state. Their computations were performed with density functional theory (DFT) using both the local density and generalized gradient approximations (LDA, GGA) in addition to spin polarization [53]. They propose a bilayer sliding mechanism of the [111] planes such that the bonding evolves from tetrahedral to octahedral without breaking any of the bonds. Such a mechanism is consistent with each of the lowest enthalpy barrier structures in their study.

This orthorhombic transition pathway was found to be very close to a generalized monoclinic pathway by [36], which was found to be a unified path for any tetrahedrally bonded SiC polytype to transition to the rocksalt structure. In a later work, it was suggested that this pathway would also hold true for the transition to a rocksalt structure in other similarly structured semiconductors regardless of the chemical components [29]. This transition state was proposed to be a unified path for all semi-conductors going from the zinc-blende to the rocksalt structure at high-pressure.

More recently, several studies have found that the transition pathway follows a tetragonal and an orthorhombic [32] path at very high transition pressures of 600 GPa. After further consideration it was found that the transition path follows a tetragonal and then a monoclinic intermediate state by [38] at 101 GPa and that the transition pressure could change the favored transition pathway, particularly in studies with over-applied pressure [46].

The disparate findings from the previous studies can be explained as follows. The various computational techniques likely contribute to the conflicting results. As discussed in [46], they could also be partly due to the differing transition pressures at which the zinc-blende to rocksalt transition occurs between studies. It is possible that different transition pathways stabilize at different pressure conditions. As will be discussed in the next section, there appears to be a strong kinetic barrier across the transition, meaning that a single transition pressure is difficult to agree on both experimentally and computationally. It is possible that several transition pathways could occur in nature or in experiment depending on the conditions and how close to or far from equilibrium the system is.

2.3. Transition Kinetics

One emerging result of the work on the transition to the rocksalt structure is the variability in the observed transition conditions both in computational, but particularly in the experimental studies (as can be seen in Table 1, Figure 3). The transition to the rocksalt structure has been observed at a range of pressures and temperatures from as low as 62 GPa at ~1750 K to as high as 100 GPa at room temperature [24,25,27]. Such a steep Clapeyron slope is not expected for the transition however, as the volume change is large (~17–18%) but the entropy change is likely small across this solid-solid transition [24,54]. Calculations performed in [24] based on the method proposed in [54] indicate that the Clapeyron slope is very nearly flat at equilibrium transition conditions. The steep experimental phase boundary as well as the offsets between different experimental results can be explained if there is a large kinetic barrier across the transition. This is particularly evident when comparing static diamond-anvil cell work with shock wave data. Even though shock wave experiments generate heat, the time scale of the experiment is so short that the transition to rocksalt occurs at higher pressures than in heated diamond-anvil cell studies (e.g., [21]).

Recent computational work considers the hysteresis across the B3 to B1 (zinc-blende to rocksalt) transition, finding that the equilibrium transition pressure is at 65.9 GPa but that a large enthalpy barrier is present [44]. This work uses a martensitic approach with an intermediate *Imm2* structure. The enthalpy barrier can be surmounted by over pressurizing SiC, as seen in room temperature static data [25] or by heating the sample as seen in laser-heated diamond-anvil cell studies [24,27]. This barrier also explains the hysteresis upon decompression in which rocksalt SiC has been seen to remain until ~35–40 GPa after which it transitions back to zinc-blende or B3 SiC [24,25,27]. The slow kinetics is perhaps not surprising as transitions in pure carbon are also quite slow, such as that of cold compressed graphite to M-carbon [55–57] or of metastable diamond to graphite. It does mean, however, that we must be aware of the experimental conditions at which high-pressure SiC is studied due to the difficulty in achieving equilibrium conditions.

2.4. Beyond Equimolar Compositions

At ambient pressure, 1:1 SiC is the only stable stoichiometry in the Si-C system with enrichments of Si or C remaining in their elemental form. Computations exploring the possibilities of other structures, such as Si_2C or SiC_2, find that these structures are all unstable relative to SiC, though may be formed metastably [58]. Other stoichiometries have also been explored computationally for monolayer 2D Si-C structures [59]. The Si-C system has been considered at pressure in both experimental and computational work. Laser-heated diamond-anvil cell experiments coupled with in situ synchrotron X-ray diffraction find that both Si- and C-rich compositions form SiC with the enrichment coming out in its elemental form up to pressures of 200 GPa and temperatures of 3500 K [28]. This indicates that alternative stoichiometries, such as Si_2C or SiC_2, are not stable at the experimental conditions considered [28]. It has been proposed however, that such stoichiometries become stable at much higher pressures than those accessible by current experiments [43]. It was found by [43] through a random structure search algorithm that SiC + Si forms *I4/mcm* structured Si_2C at pressures over 13 Mbar and that SiC + C forms *Cmmm* structured SiC_2 at pressures over 23 Mbar. Experimental confirmation of such findings is not easily achieved, though such high-pressure SiC may be applicable to the interior structure and composition of very large carbon-rich exoplanets [43].

3. High-Pressure Spectroscopy and Equations of State

3.1. High-Pressure Vibrational Spectroscopy

Raman spectroscopic measurements have improved our understanding of the lattice structure of SiC. At pressure, detailed Raman measurements have been performed on the 3C, 6H and 15R polytypes.

The first Raman measurements of SiC at pressure were performed nearly 50 years ago by [60], where the shift of the longitudinal optical (LO) and transverse optical (TO) phonon modes were measured on single crystal zinc-blende 3C-SiC. At ambient pressure, the LO mode was measured at 971 cm^{-1} and the TO mode was found at 795 cm^{-1}. Both modes were seen to shift with pressures up to ~1 GPa, with the LO mode shifting linearly to higher wave numbers and the TO mode shifting non-linearly to higher wave numbers [60]. Similar measurements on 3C were subsequently explored in many experimental studies, covering a pressure range up to 80 GPa [22,61–66]. Several observations have come from these Raman measurements. One is that the splitting between the LO and TO modes increases with pressure. This implies that the transverse effective charge is also increasing, indicating that it is quite sensitive to the electronic structure [64]. The pressure dependence of the LO and TO bands in 3C-SiC based on DAC studies is summarized in Figure 4a. The agreement between studies at lower pressures is exceptional. The two studies reaching higher pressures find a slightly different pressure dependence for the phonons [22,66], though the increased LO-TO splitting is still robust. Second order phonons are reported in [63], although the modes were not all identified. The modes labeled in Figure 4a indicate their interpretation [63].

Pressure dependence of the Raman phonons in 6H-SiC was measured by [67] up to ~10 GPa in a diamond-anvil cell. Five optical modes (LO, TO$_1$, TO$_2$, axial and planar modes) as well as four acoustic phonon modes were measured. While the optical modes and one acoustic mode all shifted linearly to higher wavenumbers with pressure, three of the measured acoustic modes showed no response to increased lattice compression. This is unexpected based on most tetrahedrally coordinated semi-conductors [68]. The mode-Grüneisen parameters were calculated for each mode with only one, the planar acoustic mode ($x = 1$), having a slightly negative γ, indicating a softening of that mode with pressure. This is again counter to findings of other similar semi-conductors [67]. Raman measurements of 6H-SiC were extended to 50 GPa [22] where fifteen fundamental bands were observed. It was again observed that the transverse acoustic (TA) modes were anomalously pressure-independent while the LO, TO and longitudinal acoustic (LA) modes have a strong positive pressure dependence. Raman measurements of 15R-SiC were also carried out to 35 GPa by [22] where eighteen bands were observed. The behavior of the LO, TO, LA, and TA modes are similar to the 6H polytypes. Work on 6H-SiC to even higher pressures of 95 GPa measured by [69] observes that the splitting between the LO and TO modes increases rapidly below 60 GPa but then flattens out at higher pressures. Based on their measurements, the transverse effective charge is seen to decrease at high pressures indicating an increasing covalent bonding. It is also seen that both the LO and TO mode show an anomalous decrease potentially due to the high-pressure phase transition. High-pressure IR measurements on 6H SiC were also carried out to 53 GPa where the LO and TO modes were measured and compared to the Raman data [70]. The shift in the LO and TO modes agree well with the Raman measurements and are shown as symbols on the plot in Figure 4b.

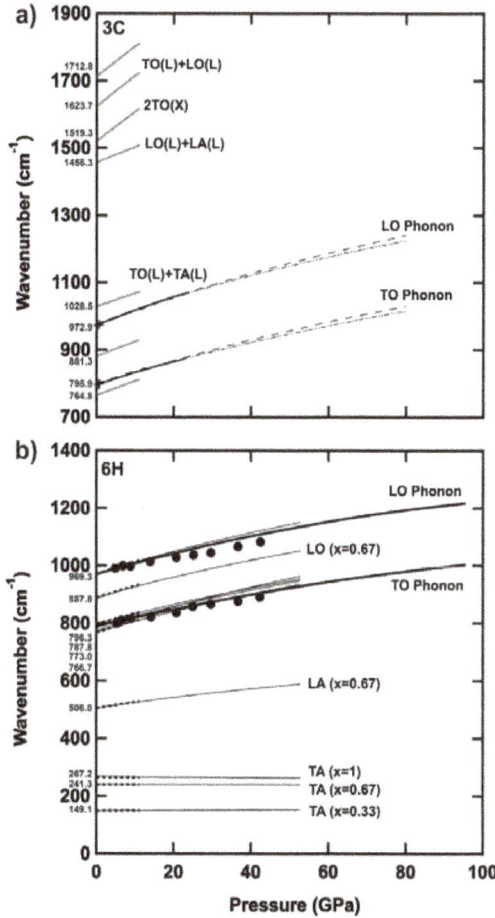

Figure 4. Summary of DAC studies on the first-order LO and TO phonon pressure dependence in SiC. (**a**) Summary of experimental Raman data of the first and second order phonons in 3C-SiC. The first-order LO and TO phonons are labeled and are consistent between studies. Crosses are from [60], thick black lines from [64], thin dot-dot-dash lines from [22], thick dashed lines from [62], and thin dashed lines from [66]. The second-order phonons as measured by [63] are plotted as dotted grey lines to 11 GPa, with zero pressure positions in order of increasing wavenumber listed: 764.8, 795.9, 881.3, 972.9, 1028.5, 1456.3, 1519.3, 1623.7, 1712.8 cm^{-1}. The assignment of second-order phonon modes from [63] are labeled. All measured phonons for 3C-SiC shift to higher wavenumbers with increased pressure with a slight increase in the LO-TO splitting; (**b**) Summary of experimental Raman data on the phonons of 6H-SiC. Phonons in 6H that have been measured, with zero pressure positions in order of increasing wavenumber listed: 149.1, 241.3, 267.2, 506, 766.7, 773, 787.8, 796.3, 887.8, 969.3 cm^{-1}. Thick black lines are from [69], thin black lines from [22], and dashed lines from [67]. IR measurements are plotted in solid black circles [70].

The pressure dependence of Raman linewidths in 3C-SiC have been explored in several studies. It was found by [63] that the linewidths of the first-order optical phonons increased dramatically at ~ 10 GPa. This increase was inferred to indicate a corresponding increase in the decay rates of the phonons with pressure. Raman linewidths was subsequently explored in several other works. It was observed experimentally that the linewidths remained constant with pressures up to 20 GPa [71] and up to ~15 GPa [65]. This result was consistent with ab initio computations [65] and it was proposed that the broadening observed in [63] was due to non-isotropic stress from the freezing of their alcohol pressure medium rather than from an increased phonon decay rate. Based on the results in [65], in which helium was used as a pressure medium for hydrostatic conditions, pressure appears to have little effect on the Raman linewidths of SiC, at least up to pressures of ~15–20 GPa.

High-pressure Raman measurements have not been carried out on the other polytypes of alpha SiC, although measurements at ambient conditions have been completed on 4H, and 21R in addition to 3C and 6H, and 15R structures and show a common set of strong phonon modes between polytypes [72,73].

3.2. High-Pressure Absorption

Absorption measurements of SiC particles, particularly in the infrared (IR) wavelengths, are essential for understanding the spectrum of carbon-stars where SiC is ubiquitous [74,75]. Detailed IR measurements at ambient pressure are necessary in order to interpret the stellar spectra of such stars. These IR absorption measurements indicate that there is an absorption feature in SiC at around 11.3 μm [76] which is also present in carbon stars [74,75]. At pressure, the absorption properties of SiC have been measured in reference to the use of moissanite anvils in high-pressure experiments [70]. Measurements on moissanite single crystals have been carried out up to 53 GPa, while absorption measurements have been performed up to 43 GPa on powdered SiC. It was observed that pressure did not significantly change the transmission properties of the moissanite anvils. Additionally, transmission through the anvils was up to an order of magnitude higher than through type II diamonds across a wavenumber range of 1900–2600 cm^{-1}, aside from the absorption feature in moissanite at 2300 cm^{-1} [70]. This range corresponds to the strong second order phonon absorption in diamond, and suggests that moissanite anvils may provide a solution to measuring absorption spectra across these wavenumbers in materials at high pressure [70]. The concentration of impurities in SiC may change the absorption over this region, however [70] did not observe differences in absorption between different moissanite anvils, in contrast to diamonds which can vary drastically depending on type [77]. These observations open up applications of moissanite anvils for high-pressure absorption measurements on materials.

Absorption measurements on additional polytypes of pure SiC at pressure are limited. Measurements on doped n-type 3C-SiC up to 14 GPa in a diamond-anvil cell were performed by [78] in which the pressure dependence of the luminescence spectra was investigated. Four luminescence lines were measured, all of which moved to higher wavenumbers with pressure [78]. To our knowledge, no studies have been carried out to higher pressure on pure 3C-SiC nor have they been made on the other polytypes.

The pressure dependence on the band gap of SiC has also been considered in several studies. The first experimental work measured the movement of the absorption edge of 3C-SiC [78] and found that the band gap had a very small pressure derivative of −1.9 MeV/GPa. The experimental data was later reanalyzed and found to be more consistent with a value of −3.4 MeV/GPa, consistent with computations finding −3.3 MeV/GPa [79]. High-pressure and low-temperature experiments were conducted on nitrogen-doped 6H- and 4H-SiC where positive pressure dependence was observed. The pressure dependence of the indirect gap in these cases was found to be 2.0 MeV/GPa for N doped 6H-SiC at 29 K and pressures up to 5 GPa [80], and 2.7 MeV/GPa for N doped 4H-SiC at 7 K and pressures up to 5 GPa [81].

3.3. High-Pressure Elasticity and Equation of State

SiC is known to be a hard and strong material [82]. The elastic properties and equation of state of SiC have been explored comprehensively at pressure both experimentally and computationally. The equation of state (EOS) has been found experimentally through X-ray diffraction, ultrasonic measurements, or Brillouin scattering by [25,66,83–85] and computationally through several techniques by [86–89]. SiC has been found to have a large room-pressure bulk modulus K_0 greater than 200 GPa with pressure derivative K_0' around 4. Table 2 compiles the experimental EOS data for each polytype of SiC that has been studied in the literature.

In addition to the EOS parameters, work has been performed on the pressure dependence of the elastic constants of SiC. Brillouin and Raman spectroscopy on the 3C polytypes done by [66] finds that the C_{11} and C_{12} constants increase by over 50% across the pressure range of 0 to 65 GPa (399 GPa increase to 672 GPa and 133 GPa increase to 339 GPa respectively). The C_{44} constant also increases, although much less dramatically and seeming to follow a second-order polynomial fit (251 increase to 316 GPa over the same pressure range). The pressure dependence of several of the elastic constants of the 6H polytype were predicted by first principles DFT calculations using LDA potentials [87]. A similar pressure dependence trend holds for the C_{11} and C_{44} constants, although the predicted value of C_{11} is a little higher than that measured for 3C while the predicted values of C_{12} and C_{44} are a little lower.

Table 2. Experimental equations of state for several polytypes of SiC.

Polytype	Max Pressure (GPa)	EOS	K_0	K_0'	Method	Ref.
3C	45	scale proposed by [90]	227 ± 3	4.1 ± 0.1	Raman	[61]
3C	25	Murnaghan (M) EOS	248 ± 9	4.0 ± 0.3	XRD	[85]
3C	75	primary scale	218 ± 1	3.75 ± 4	XRD, Brillouin, Raman	[66]
3C	8.1	Birch–Murnaghan (BM EOS)	237 ± 2	4 (fixed)	XRD	[91]
3C, 6H	95	BM EOS	260.9 ± 9	2.9 ± 0.3	XRD	[25]
6H	68.4	BM EOS	230.2 ± 4.0	4 (fixed)	XRD	[84]
6H	13.6	BM EOS	216.5 ± 1.1	4.19 ± 0.09	Ultrasonic	[83]
6H	27	BM EOS	218.4 ± 4.9	4.19 (fixed)	XRD	[83]
15R	35	scale proposed by [90]	224 ± 3	4.3 ± 0.3	XRD, Raman	[22,92]

3.4. Thermal Expansion and Equation of State

The combined effect of pressure and temperature on SiC has been the subject of several recent studies with a focus on the thermal expansion of both the 3C- and 6H-SiC polytypes. At ambient pressure the thermal expansion of SiC, particularly of 3C-SiC, has been extensively studied due to its importance in material applications. The thermal expansion has consistently been reported to be between 4×10^{-6} 1/K and 6×10^{-6} 1/K based on both X-ray diffraction and dilatometer measurements [93–99]. X-ray diffraction measurements taken over a temperature range of 300–1300 degrees K find that a second order polynomial better fits the thermal expansion giving a value of ~3.2×10^{-6} 1/K at ~300 K and a larger value of 5.1×10^{-6} 1/K at ~1300 K [100].

Determination of the thermal EOS of SiC and the thermal expansion at pressure has recently been carried out. Both DFT calculations and large volume press experiments coupled with in situ X-ray diffraction at conditions up to 8.1 GPa and 1100 K were performed by [91] to determine thermal EOS parameters. Fitting their diffraction data to a modified Birch–Murnaghan EOS gives a value of $\alpha = 5.77 \times 10^{-6} + 1.36 \times 10^{-8}$ T. Their DFT results give a similar thermal expansion, with a value of $\alpha = 5.91 \times 10^{-6} + 1.08 \times 10^{-8}$ T using LDA and $\alpha = 6.99 \times 10^{-6} + 1.11 \times 10^{-8}$ T using GGA. The values from each method seem to agree well, both with each other as well as with previous ambient pressure data. They find that the thermal expansion decreases with pressure and find the pressure derivative through two different methods. The modified Birch–Murnaghan EOS gives a pressure derivative of -6.53×10^{-7} GPa^{-1} K^{-1}, while a thermal pressure approach gives a slightly lower value of -7.23×10^{-7} GPa^{-1} K^{-1}.

Only one study to date has explored the thermal expansion of SiC at even higher pressures. A recent study [23] used the LHDAC coupled with in situ X-ray diffraction to measure the thermal expansion of both the 3C- and 6H-SiC polytypes [23]. Their measurements spanned a range of conditions up to 80 GPa and 1900 K for 3C-SiC, and up to 65 GPa and 1920 K for 6H-SiC. They determined pressure by a gold standard loaded in their sample chamber and considered three different gold EOS's [101–103] when calculating the thermal expansion from their measurements. Several interesting findings are presented, including a higher thermal expansion for SiC than previous measurements, although this is likely due to the high temperatures of their study. The thermal expansion found in [23] is on the order of 1×10^{-5} 1/K at 2500 K, nearly an order of magnitude higher than previous studies at room temperature. They also find that the thermal expansion changes very little with pressure. Based on the gold scale in [102] they find that the thermal expansion of SiC is nearly constant over the entire pressure range considered. The gold scale by [101] gives a decrease in the thermal expansion with pressure, though the change is still less than a factor of two. Further studies on the effect of pressure on the thermal expansion of SiC are needed to better understand these observations.

4. Melting Behavior and Decomposition

SiC is known to be a very refractory material with a high ambient pressure melting point [19]. Rather than melting congruently (i.e., solid SiC melting to liquid SiC), SiC has been observed to melt incongruently at ambient pressure with the Si fraction coming out as a liquid and the C fraction remaining as a solid [19]. The ambient pressure decomposition of SiC into solid C plus liquid Si begins at ~2840 K in experiments [19] but is predicted to occur at higher temperatures of 3100 K in computations [104,105]. Prior to recent diamond-anvil cell work up to ~80 GPa [18], explorations of high-pressure melting and decomposition have gone up to ~10 GPa while heating to temperatures as high as 3500 K [17,106–110]. Although confusion has arisen as to the nature of SiC melting at lower pressures, higher pressure studies indicate that 3C-SiC continues to decompose at high temperature, at least up to the transition to the rocksalt structure at ~60 GPa.

Many earlier studies have observed decomposition of SiC to Si + C at low pressure. SiC was seen to decompose up to 8 GPa based on quench texture and composition [110], although the temperature of the decomposition was not directly measured. Similar results were found in [17,106] in a high-pressure high-temperature cell at 3 GPa without a direct temperature measurement, though the power-temperature relation indicated that the sample was above 2800 K. Incongruent melting was also observed in [109] up to ~10 GPa. Decomposition was identified through a change in the resistivity of the sample as well as through Si and C diffraction signals upon quench. Temperature of decomposition was determined based on the inserted energy and it was found that SiC decomposed following a positive phase boundary. Based on the increase in the solubility of C in liquid Si with increasing pressure, however, [109] predicts that decomposition does not continue past about ~10 GPa, after which SiC melts congruently.

In contrast to these works, congruent melting was inferred by several studies over the same pressure-temperature conditions. Sokolov et al., 2012 [108] performed experiments in a high-temperature, high-pressure apparatus at 5 and 7.7 GPa, in which they identified melting of SiC by a change in the microstructure of recovered samples or by a jump in the electrical resistivity of the sample. They additionally performed X-ray diffraction measurements but did not see evidence of decomposition. Based on their experiments SiC melts congruently following a negative phase boundary. Congruent melting at pressure was also previously seen in [107].

Recent diamond-anvil cell work [18] finds that 3C-SiC continues to decompose at high pressures and high temperatures, following a phase boundary with a negative slope. The high-pressure decomposition temperatures measured are considerably lower than the decomposition temperature at ambient, with the measurements indicating that SiC begins to decompose at ~2000 K at 60 GPa as compared to ~2800 K at ambient pressure. Once 3C-SiC had transitioned to the high-pressure rocksalt structure, decomposition was no longer observed, despite heating to temperatures in excess of ~3200 K. Several methods were used to identify in situ decomposition in samples as temperature increased, including the appearance of diamond peaks in X-ray diffraction as well as changes in the optical character of the sample. Additionally, recovered samples were cross-sectioned and analyzed for composition, confirming the presence of SiC decomposition products. Raman measurements across decomposed regions indicated the presence of carbon, whereas measurements across un-decomposed regions did not. Kinetics appears to be a strong influence on the decomposition reaction as well, since complete decomposition was not observed on the timescale of the experiments (on the order of minutes) and the reaction was not observed to be reversible. Once the products of decomposition were observed, they remained in the sample both upon temperature/pressure quench and upon heating to temperatures below the observed decomposition onset boundary.

The temperature of decomposition and the nature of the decomposition phase boundary appear to be strongly influenced by the pressure-induced phase transitions to higher density structures in SiC, silicon and carbon, as is discussed further in [18]. However, additional work is necessary to understand the melting characteristics of the rocksalt structure at pressures above 60 GPa. Figure 5 summarizes the high P-T phase diagram of SiC to date including the data on melting and/or decomposition from each previous study. Above the transition to the B1 structure, it is still unclear whether or not SiC decomposes at high temperatures or melts congruently, as indicated by the arrow in the top right corner of the plot. As these measurements have not been performed, the temperatures required for melting may be much higher than those represented on the current phase diagram.

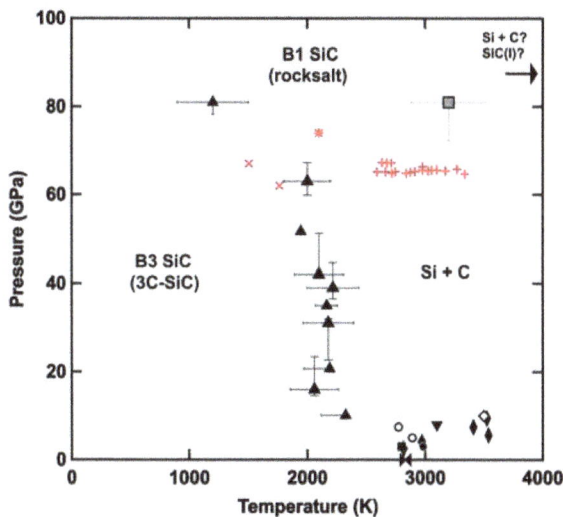

Figure 5. High-pressure and -temperature phase diagram of SiC melting and decomposition. Solid black symbols indicate studies finding incongruent melting (decomposition) (bowtie [19], circle [106], diamonds [109], upside down triangle [110], square [17], triangles (multi-wavelength imaging radiometry and X-ray diffraction) [18]) while open symbols indicate studies observing congruent melting to SiC liquid (open circles [108], diamond [107]). The solid grey square indicates P-T conditions where no melting of any kind was observed [18]. Red symbols indicate the experimentally observed conditions of the B3 to B1 transition in the LHDAC (red x's [24], red asterisk [27], red plus signs [28]).

5. Conclusions

High-pressure work on SiC has opened up many new questions as well as answering those discussed here. The issue of kinetics continues to reappear in high-pressure studies of SiC, whether it is on the transition between polytypes or in the melting behavior. Metastable states and the time scales needed to achieve equilibrium at high P-T conditions are topics that are not yet well understood but which may have important implications for industrial and naturally occurring SiC. More high-P-T studies above 10 GPa are certainly needed to confirm and expand upon thermal expansion and thermal equation of state measurements, as well as to explore decomposition/melting in other polytypes or in the B1 rocksalt structure.

Acknowledgments: We acknowledge support from the Carnegie DOE Alliance Center (CDAC) as well as from the NASA Connecticut Space Grant Consortium (CTSGC) Graduate Research Fellowship grant number P-1127.

Conflicts of Interest: The authors declare no conflicts of interest.

References

1. Shaffer, P. A review of the structure of silicon carbide. *Acta Crystallogr. Sect. B Struct. Crystallogr. Cryst. Chem.* **1969**, *25*, 477–488. [CrossRef]
2. Momma, K.; Izumi, F. VESTA 3 for three-dimensional visualization of crystal, volumetric and morphology data. *J. Appl. Crystallogr.* **2011**, *44*, 1272–1276. [CrossRef]
3. Braekken, H. Zur Kristallstruktur des kubischen Karborunds. *Zeitschrift für Kristallographie* **1930**, *75*, 572–573.
4. Wyckoff, R. Interscience Publishers, New York, New York rocksalt structure. *Cryst. Struct.* **1963**, *1*, 85–237.
5. Capitani, G.C.; Di Pierro, S.; Tempesta, G. The 6H-SiC structure model: Further refinement from SCXRD data from a terrestrial moissanite. *Am. Mineral.* **2007**, *92*, 403–407. [CrossRef]
6. Ramsdell, L.S. The crystal structure of α-SiC, type 4. *Am. Mineral.* **1944**, *29*, 431–442.
7. Moissan, H. Nouvelles recherches sur la météorité de Cañon Diablo. *C. R.* **1904**, *139*, 773–786.
8. Lyakhovich, V. Origin of accessory moissanite. *Int. Geol. Rev.* **1980**, *22*, 961–970. [CrossRef]
9. Amari, S.; Lewis, R.S.; Anders, E. Interstellar grains in meteorites: I. Isolation of SiC, graphite and diamond; size distributions of SiC and graphite. *Geochim. Cosmochim. Acta* **1994**, *58*, 459–470. [CrossRef]
10. Kuchner, M.J.; Seager, S. Extrasolar carbon planets. *arXiv* **2005**, arXiv:astro-ph/0504214.
11. Madhusudhan, N.; Lee, K.K.M.; Mousis, O. A possible carbon-rich interior in Super-Earth 55 Cancri e. *Astrophys. J. Lett.* **2012**, *759*, L40. [CrossRef]
12. Fisher, G.; Barnes, P. Towards a unified view of polytypism in silicon carbide. *Philos. Mag. B* **1990**, *61*, 217–236. [CrossRef]
13. Jepps, N.; Page, T. Polytypic transformations in silicon carbide. *Prog. Cryst. Growth Charact.* **1983**, *7*, 259–307. [CrossRef]
14. Whitney, E.; Shaffer, P. Investigation of the Phase Transformation between α-and β-Silicon Carbide at High Pressures. *High Temp. High Press.* **1969**, *1*, 107–110.
15. Sugiyama, S.; Togaya, M. Phase Relationship between 3C-and 6H-Silicon Carbide at High Pressure and High Temperature. *J. Am. Ceram. Soc.* **2001**, *84*, 3013–3016. [CrossRef]
16. Zhu, Y.; Sekine, T.; Kobayashi, T.; Takazawa, E. Shock-induced phase transitions among SiC polytypes. *J. Mater. Sci.* **1998**, *33*, 5883–5890. [CrossRef]
17. Bhaumik, S. Synthesis and sintering of monolithic and composite ceramics under high pressures and high temperatures. *Metals Mater. Process.* **2000**, *12*, 215–232.
18. Daviau, K.; Lee, K.K.M. Decomposition of silicon carbide at high pressures and temperatures. *Phys. Rev. B* **2017**, *96*, 174102. [CrossRef]
19. Dolloff, R. *Research Study to Determine the Phase Equilibrium Relations of Selected Metal Carbides at High Temperatures*; Period Covered January 1959–March 1960; WADD-TR-60-143; Union Carbide Corp. Parma Research Lab.: Parma, OH, USA, 1960.

20. Tracy, S.J.; Smith, R.F.; Wicks, J.K.; Fratanduono, D.E.; Gleason, A.E.; Bolme, C.; Speziale, S.; Appel, K.; Prakapenka, V.B.; Fernandez Panella, A.; et al. High-pressure phase transition in silicon carbide under shock loading using ultrafast X-ray diffraction. In Proceedings of the AGU Fall Meeting, New Orleans, LA, USA, 11–15 December 2017.

21. Sekine, T.; Kobayashi, T. Shock compression of 6H polytype SiC to 160 GPa. *Phys. Rev. B* **1997**, *55*, 8034. [CrossRef]

22. Aleksandrov, I.; Goncharov, A.; Yakovenko, E.; Stishov, S. High pressure study of diamond, graphite and related materials. *High Press. Res. Appl. Earth Planet. Sci.* **1992**, 409–416.

23. Nisr, C.; Meng, Y.; MacDowell, A.; Yan, J.; Prakapenka, V.; Shim, S.H. Thermal expansion of SiC at high pressure-temperature and implications for thermal convection in the deep interiors of carbide exoplanets. *J. Geophys. Res. Planets* **2017**, *122*, 124–133. [CrossRef]

24. Daviau, K.; Lee, K.K.M. Zinc-blende to rocksalt transition in SiC in a laser-heated diamond-anvil cell. *Phys. Rev. B* **2017**, *95*, 134108. [CrossRef]

25. Yoshida, M.; Onodera, A.; Ueno, M.; Takemura, K.; Shimomura, O. Pressure-induced phase transition in SiC. *Phys. Rev. B* **1993**, *48*, 10587. [CrossRef]

26. Levitas, V.I.; Ma, Y.; Selvi, E.; Wu, J.; Patten, J.A. High-density amorphous phase of silicon carbide obtained under large plastic shear and high pressure. *Phys. Rev. B* **2012**, *85*, 054114. [CrossRef]

27. Kidokoro, Y.; Umemoto, K.; Hirose, K.; Ohishi, Y. Phase transition in SiC from zinc-blende to rock-salt structure and implications for carbon-rich extrasolar planets. *Am. Mineral.* **2017**, *102*, 2230–2234. [CrossRef]

28. Miozzi Ferrini, F.; Morard, G.; Antonangeli, D.; Clark, A.N.; Edmund, E.; Fiquet, G.; Mezouar, M. On the Interior of Carbon-Rich Exoplanets: New Insight from SiC System at Ultra High Pressure. In Proceedings of the AGU Fall Meeting, New Orleans, LA, USA, 11–15 December 2017.

29. Miao, M.; Lambrecht, W.R. Universal transition state for high-pressure zinc blende to rocksalt phase transitions. *Phys. Rev. Lett.* **2005**, *94*, 225501. [CrossRef] [PubMed]

30. Chang, K.J.; Cohen, M.L. Ab initio pseudopotential study of structural and high-pressure properties of SiC. *Phys. Rev. B* **1987**, *35*, 8196. [CrossRef]

31. Catti, M. Orthorhombic intermediate state in the zinc blende to rocksalt transformation path of SiC at high pressure. *Phys. Rev. Lett.* **2001**, *87*, 035504. [CrossRef] [PubMed]

32. Durandurdu, M. Pressure-induced phase transition of SiC. *J. Phys. Condens. Matter* **2004**, *16*, 4411. [CrossRef]

33. Karch, K.; Bechstedt, F.; Pavone, P.; Strauch, D. Pressure-dependent properties of SiC polytypes. *Phys. Rev. B* **1996**, *53*, 13400. [CrossRef]

34. Lee, W.; Yao, X. First principle investigation of phase transition and thermodynamic properties of SiC. *Comput. Mater. Sci.* **2015**, *106*, 76–82. [CrossRef]

35. Lu, Y.-P.; He, D.-W.; Zhu, J.; Yang, X.-D. First-principles study of pressure-induced phase transition in silicon carbide. *Phys. B Condens. Matter* **2008**, *403*, 3543–3546. [CrossRef]

36. Miao, M.S.; Lambrecht, W.R.L. Unified path for high-pressure transitions of SiC polytypes to the rocksalt structure. *Phys. Rev. B* **2003**, *68*, 092103. [CrossRef]

37. Wang, C.-Z.; Yu, R.; Krakauer, H. Pressure dependence of Born effective charges, dielectric constant, and lattice dynamics in SiC. *Phys. Rev. B* **1996**, *53*, 5430. [CrossRef]

38. Xiao, H.; Gao, F.; Zu, X.T.; Weber, W.J. Ab initio molecular dynamics simulation of a pressure induced zinc blende to rocksalt phase transition in SiC. *J. Phys. Condens. Matter* **2009**, *21*, 245801. [CrossRef] [PubMed]

39. Gust, W.; Holt, A.; Royce, E. Dynamic yield, compressional, and elastic parameters for several lightweight intermetallic compounds. *J. Appl. Phys.* **1973**, *44*, 550–560. [CrossRef]

40. Vogler, T.; Reinhart, W.; Chhabildas, L.; Dandekar, D. Hugoniot and strength behavior of silicon carbide. *J. Appl. Phys.* **2006**, *99*, 023512. [CrossRef]

41. Eker, S.; Durandurdu, M. Phase transformation of 6H-SiC at high pressure: An ab initio constant-pressure study. *EPL (Europhys. Lett.)* **2008**, *84*, 26003. [CrossRef]

42. Eker, S.; Durandurdu, M. Pressure-induced phase transformation of 4H-SiC: An ab initio constant-pressure study. *EPL (Europhys. Lett.)* **2009**, *87*, 36001. [CrossRef]

43. Wilson, H.F.; Militzer, B. Interior phase transformations and mass-radius relationships of silicon-carbon planets. *Astrophys. J.* **2014**, *793*, 34. [CrossRef]

44. Salvadó, M.A.; Franco, R.; Pertierra, P.; Ouahrani, T.; Recio, J. Hysteresis and bonding reconstruction in the pressure-induced B3–B1 phase transition of 3C-SiC. *Phys. Chem. Chem. Phys.* **2017**, *19*, 22887–22894. [CrossRef] [PubMed]

45. Shimojo, F.; Ebbsjö, I.; Kalia, R.K.; Nakano, A.; Rino, J.P.; Vashishta, P. Molecular dynamics simulation of structural transformation in silicon carbide under pressure. *Phys. Rev. Lett.* **2000**, *84*, 3338. [CrossRef] [PubMed]

46. Xiao, H.; Gao, F.; Wang, L.M.; Zu, X.T.; Zhang, Y.; Weber, W.J. Structural phase transitions in high-pressure wurtzite to rocksalt phase in GaN and SiC. *Appl. Phys. Lett.* **2008**, *92*, 241909. [CrossRef]

47. Blanco, M.A.; Recio, J.; Costales, A.; Pandey, R. Transition path for the B3⇌B1 phase transformation in semiconductors. *Phys. Rev. B* **2000**, *62*, R10599. [CrossRef]

48. Perez-Mato, J.; Aroyo, M.; Capillas, C.; Blaha, P.; Schwarz, K. Comment on "Orthorhombic Intermediate State in the Zinc Blende to Rocksalt Transformation Path of SiC at High Pressure". *Phys. Rev. Lett.* **2003**, *90*, 049603. [CrossRef] [PubMed]

49. Miao, M.; Prikhodko, M.; Lambrecht, W.R. Comment on "orthorhombic intermediate state in the zinc blende to rocksalt transformation path of SiC at high pressure". *Phys. Rev. Lett.* **2002**, *88*, 189601. [CrossRef] [PubMed]

50. Catti, M. Cattie Replies. *Phys. Rev. Lett.* **2003**, *90*, 049604. [CrossRef]

51. Catti, M. Catti Replies. *Phys. Rev. Lett.* **2002**, *88*, 189602. [CrossRef]

52. Hatch, D.M.; Stokes, H.T.; Dong, J.; Gunter, J.; Wang, H.; Lewis, J.P. Bilayer sliding mechanism for the zinc-blende to rocksalt transition in SiC. *Phys. Rev. B* **2005**, *71*, 184109. [CrossRef]

53. Lewis, J.P.; Glaesemann, K.R.; Voth, G.A.; Fritsch, J.; Demkov, A.A.; Ortega, J.; Sankey, O.F. Further developments in the local-orbital density-functional-theory tight-binding method. *Phys. Rev. B* **2001**, *64*, 195103. [CrossRef]

54. Li, X.; Jeanloz, R. Measurement of the B1-B2 transition pressure in NaCl at high temperatures. *Phys. Rev. B* **1987**, *36*, 474–479. [CrossRef]

55. Wang, Y.; Lee, K. From soft to superhard: Fifty years of experiments on cold-compressed graphite. *J. Superhard Mater.* **2012**, *34*, 360–370. [CrossRef]

56. Wang, Y.; Panzik, J.E.; Kiefer, B.; Lee, K.K. Crystal structure of graphite under room-temperature compression and decompression. *Sci. Rep.* **2012**, *2*, 520. [CrossRef] [PubMed]

57. Montgomery, J.M.; Kiefer, B.; Lee, K.K. Determining the high-pressure phase transition in highly-ordered pyrolitic graphite with time-dependent electrical resistance measurements. *J. Appl. Phys.* **2011**, *110*, 043725. [CrossRef]

58. Gao, G.; Ashcroft, N.W.; Hoffmann, R. The unusual and the expected in the Si/C phase diagram. *J. Am. Chem. Soc.* **2013**, *135*, 11651–11656. [CrossRef] [PubMed]

59. Li, P.; Zhou, R.; Zeng, X.C. The search for the most stable structures of silicon–carbon monolayer compounds. *Nanoscale* **2014**, *6*, 11685–11691. [CrossRef] [PubMed]

60. Mitra, S.; Brafman, O.; Daniels, W.; Crawford, R. Pressure-induced phonon frequency shifts measured by Raman scattering. *Phys. Rev.* **1969**, *186*, 942. [CrossRef]

61. Aleksandrov, I.; Goncharov, A.; Jakovenko, E.; Sttshov, S. High pressure study of cubic BN and SIC (Raman scattering and EDS). *High Press. Sci. Technol.* **1990**, *5*, 938–940. [CrossRef]

62. Debernardi, A.; Ulrich, C.; Syassen, K.; Cardona, M. Raman linewidths of optical phonons in 3C-SiC under pressure: First-principles calculations and experimental results. *Phys. Rev. B* **1999**, *59*, 6774. [CrossRef]

63. Olego, D.; Cardona, M. Pressure dependence of Raman phonons of Ge and 3C-SiC. *Phys. Rev. B* **1982**, *25*, 1151. [CrossRef]

64. Olego, D.; Cardona, M.; Vogl, P. Pressure dependence of the optical phonons and transverse effective charge in 3C-SiC. *Phys. Rev. B* **1982**, *25*, 3878. [CrossRef]

65. Ulrich, C.; Debernardi, A.; Anastassakis, E.; Syassen, K.; Cardona, M. Raman linewidths of phonons in Si, Ge, and SiC under pressure. *Phys. Status Solidi B* **1999**, *211*, 293–300. [CrossRef]

66. Zhuravlev, K.K.; Goncharov, A.F.; Tkachev, S.N.; Dera, P.; Prakapenka, V.B. Vibrational, elastic, and structural properties of cubic silicon carbide under pressure up to 75 GPa: Implication for a primary pressure scale. *J. Appl. Phys.* **2013**, *113*, 113503–113512. [CrossRef]

67. Salvador, G.; Sherman, W. Pressure dependence of the Raman phonon spectrum in 6h-silicon carbide. *J. Mol. Struct.* **1991**, *247*, 373–384. [CrossRef]

68. Weinstein, B.A.; Zallen, R. Pressure-Raman effects in covalent and molecular solids. In *Light Scattering in Solids IV*; Springer: Berlin/Heidelberg, Germany, 1984; pp. 463–527.

69. Liu, J.; Vohra, Y.K. Raman modes of 6 h polytype of silicon carbide to ultrahigh pressures: A comparison with silicon and diamond. *Phys. Rev. Lett.* **1994**, *72*, 4105. [CrossRef] [PubMed]

70. Liu, Z.; Xu, J.; Scott, H.P.; Williams, Q.; Mao, H.-K.; Hemley, R.J. Moissanite (SiC) as windows and anvils for high-pressure infrared spectroscopy. *Rev. Sci. Instrum.* **2004**, *75*, 5026–5029. [CrossRef]

71. Kobayashi, M.; Akimoto, R.; Endo, S.; Yamanaka, M.; Shinohara, M.; Ikoma, K. Amorphous and Crystalline Silicon Carbide III. In *Springer Proceedings in Physics*; Springer: Berlin, Germany, 1992; Volume 56, p. 263.

72. Feldman, D.; Parker, J.H., Jr.; Choyke, W.; Patrick, L. Phonon Dispersion Curves by Raman Scattering in SiC, Polytypes 3C, 4H, 6H, 15R, and 21R. *Phys. Rev.* **1968**, *173*, 787. [CrossRef]

73. Feldman, D.; Parker, J.H., Jr.; Choyke, W.; Patrick, L. Raman Scattering in 6H SiC. *Phys. Rev.* **1968**, *170*, 698. [CrossRef]

74. Hackwell, J. Long wavelength spectrometry and photometry of M, S and C-stars. *Astron. Astrophys.* **1972**, *21*, 239–248.

75. Treffers, R.; Cohen, M. High-resolution spectra of cool stars in the 10-and 20-micron regions. *Astrophys. J.* **1974**, *188*, 545–552. [CrossRef]

76. Mutschke, H.; Andersen, A.; Clément, D.; Henning, T.; Peiter, G. Infrared properties of SiC particles. *Astron. Astrophys.* **1999**, *345*, 187–202.

77. Walker, J. Optical absorption and luminescence in diamond. *Rep. Prog. Phys.* **1979**, *42*, 1605. [CrossRef]

78. Kobayashi, M.; Yamanaka, M.; Shinohara, M. High-Pressure Studies of Absorption and Luminescence Spectra in 3C-SiC. *J. Phys. Soc. Jpn.* **1989**, *58*, 2673–2676. [CrossRef]

79. Cheong, B.; Chang, K.; Cohen, M.L. Pressure dependences of band gaps and optical-phonon frequency in cubic SiC. *Phys. Rev. B* **1991**, *44*, 1053. [CrossRef]

80. Engelbrecht, F.; Zeman, J.; Wellenhofer, G.; Peppermüller, C.; Helbig, R.; Martinez, G.; Rössler, U. Pressure Dependence of the Electronic Band Gap in 6H-Sic. *Phys. Status Solidi B* **1996**, *198*, 81–86. [CrossRef]

81. Zeman, J.; Engelbrecht, F.; Wellenhofer, G.; Peppermüller, C.; Helbig, R.; Martinez, G.; Rössler, U. Pressure Dependence of the Band Gap of 4H-SiC. *Phys. Status Solidi B* **1999**, *211*, 69–72. [CrossRef]

82. Harris, G.L. *Properties of Silicon Carbide*; IET: London, UK, 1995.

83. Amulele, G.M.; Manghnani, M.H.; Li, B.; Errandonea, D.J.H.; Somayazulu, M.; Meng, Y. High pressure ultrasonic and X-ray studies on monolithic SiC composite. *J. Appl. Phys.* **2004**, *95*, 1806–1810. [CrossRef]

84. Bassett, W.; Weathers, M.; Wu, T.C.; Holmquist, T. Compressibility of SiC up to 68.4 GPa. *J. Appl. Phys.* **1993**, *74*, 3824–3826. [CrossRef]

85. Strössner, K.; Cardona, M.; Choyke, W. High pressure X-ray investigations on 3C-SiC. *Solid State Commun.* **1987**, *63*, 113–114. [CrossRef]

86. Prikhodko, M.; Miao, M.; Lambrecht, W.R. Pressure dependence of sound velocities in 3C-SiC and their relation to the high-pressure phase transition. *Phys. Rev. B* **2002**, *66*, 125201. [CrossRef]

87. Sarasamak, K.; Limpijumnong, S.; Lambrecht, W.R. Pressure-dependent elastic constants and sound velocities of wurtzite SiC, GaN, InN, ZnO, and CdSe, and their relation to the high-pressure phase transition: A first-principles study. *Phys. Rev. B* **2010**, *82*, 035201. [CrossRef]

88. Tang, M.; Yip, S. Atomistic simulation of thermomechanical properties of β-SiC. *Phys. Rev. B* **1995**, *52*, 15150. [CrossRef]

89. Xu-Dong, Z.; Shou-Xin, C.; Hai-Feng, S. Theoretical study of thermodynamics properties and bulk modulus of SiC under high pressure and temperature. *Chin. Phys. Lett.* **2014**, *31*, 016401.

90. Aleksandrov, I.; Goncharov, A.; Zisman, A.; Stishov, S. Diamond at high pressures: Raman scattering of light, equation of state, and high pressure scale. *Zhurnal Eksperimental'noi i Teoreticheskoi Fiziki* **1987**, *93*, 680–691.

91. Wang, Y.; Liu, Z.T.; Khare, S.V.; Collins, S.A.; Zhang, J.; Wang, L.; Zhao, Y. Thermal equation of state of silicon carbide. *Appl. Phys. Lett.* **2016**, *108*, 061906. [CrossRef]

92. Jakovenko, E.; Goncharov, A.; Stishov, S. SiC up to 35 GPa: Eos, phonon dispersion curves and sum rule. polytypes 6H and 15R. *Int. J. High Press. Res.* **1992**, *8*, 433–435. [CrossRef]

93. Becker, K. Eine röntgenographische Methode zur Bestimmung des Wärmeausdehnungskoeffizienten bei hohen Temperaturen. *Zeitschrift für Physik* **1926**, *40*, 37–41. [CrossRef]

94. Clark, D.; Knight, D. *Royal Aircraft Establishment*; Technical Report RAE-TR-65049; Royal Aircraft Establishment: Farnborough, England, 1965.

95. Hiroshige, S.; Takayoshi, I.; Masahiko, I. Annealing behavior of neutron irradiated β-SiC. *J. Nucl. Mater.* **1973**, *48*, 247–252. [CrossRef]
96. Kern, E.; Hamill, D.; Deem, H.; Sheets, H. Thermal properties of β-Silicon Carbide from 20 to 2000 °C. In *Silicon Carbide–1968*; Elsevier: University Park, PA, USA, 1969; pp. S25–S32.
97. Popper, P.; Mohyuddin, I. *The Preparation and Properties of Pyrolytic Silicon Carbide*; Special Ceramics; Academic Press: New York, NY, USA, 1965; Volume 45.
98. Price, R. *Structure and Properties of Pyrolytic Silicon Carbide*; Gulf General Atomic, Inc.: San Diego, CA, USA, 1969.
99. Taylor, A.; Jones, R. The crystal structure and thermal expansion of cubic and hexagonal silicon carbide. *Silicon Carbide* **1960**, 147–161.
100. Li, Z.; Bradt, R. Thermal expansion of the cubic (3C) polytype of SiC. *J. Mater. Sci.* **1986**, *21*, 4366–4368. [CrossRef]
101. Dorogokupets, P.; Dewaele, A. Equations of state of MgO, Au, Pt, NaCl-B1, and NaCl-B2: Internally consistent high-temperature pressure scales. *High Press. Res.* **2007**, *27*, 431–446. [CrossRef]
102. Fei, Y.; Ricolleau, A.; Frank, M.; Mibe, K.; Shen, G.; Prakapenka, V. Toward an internally consistent pressure scale. *Proc. Natl. Acad. Sci. USA* **2007**, *104*, 9182–9186. [CrossRef] [PubMed]
103. Yokoo, M.; Kawai, N.; Nakamura, K.G.; Kondo, K.-I.; Tange, Y.; Tsuchiya, T. Ultrahigh-pressure scales for gold and platinum at pressures up to 550 GPa. *Phys. Rev. B* **2009**, *80*, 104114. [CrossRef]
104. Franke, P.; Neuschütz, D.; Europe, S.G.T. *The Landolt-Börnstein Database*; Springer Materials: Heidelberg, Germany, 2004.
105. Gröbner, J.; Lukas, H.L.; Aldinger, F. Thermodynamic calculation of the ternary system Al-Si-C. *Calphad* **1996**, *20*, 247–254. [CrossRef]
106. Bhaumik, S.; Divakar, C.; Mohan, M.; Singh, A. A modified high-temperature cell (up to 3300 K) for use with a cubic press. *Rev. Sci. Instrum.* **1996**, *67*, 3679–3682. [CrossRef]
107. Hall, H.T. *High Temperature Studies*; Bringham Young University: Provo, UT, USA, 1956; p. 36.
108. Sokolov, P.S.; Mukhanov, V.A.; Chauveau, T.; Solozhenko, V.L. On melting of silicon carbide under pressure. *J. Superhard Mater.* **2012**, *34*, 339–341. [CrossRef]
109. Togaya, M.; Sugiyama, S. Melting Behavior of β-SiC at High Pressure. *Rev. High Press. Sci. Technol.* **1998**, *7*, 1037–1039. [CrossRef]
110. Ekimov, E.; Sadykov, R.; Gierlotka, S.; Presz, A.; Tatyanin, E.; Slesarev, V.; Kuzin, N. A high-pressure cell for high-temperature experiments in a toroid-type chamber. *Instrum. Exp. Tech.* **2004**, *47*, 276–278. [CrossRef]

crystals

MDPI

Article

High-Pressure Synthesis, Structure, and Magnetic Properties of Ge-Substituted Filled Skutterudite Compounds; $Ln_xCo_4Sb_{12-y}Ge_y$, Ln = La, Ce, Pr, and Nd

Hiroshi Fukuoka

Department of Applied Chemistry, Graduate School of Engineering, Hiroshima University, 1-4-1 Kagamiyama, Higashi-Hiroshima 739-8527, Japan; hfukuoka@hiroshima-u.ac.jp; Tel.: +81-82-424-7742

Academic Editor: Daniel Errandonea
Received: 18 October 2017; Accepted: 14 December 2017; Published: 15 December 2017

Abstract: A series of new Ge-substituted skutterudite compounds with the general composition of $Ln_xCo_4Sb_{12-y}Ge_y$, where Ln = La, Ce, Pr, and Nd, is prepared by high-pressure and high-temperature reactions at 7 GPa and 800 °C. They have a cubic unit cell and the lattice constant for each compound is 8.9504 (3), 8.94481 (6), 8.9458 (3), and 8.9509 (4) Å for the La, Ce, Pr, and Nd derivatives, respectively. Their chemical compositions, determined by electron prove microanalysis, are $La_{0.57}Co_4Sb_{10.1}Ge_{2.38}$, $Ce_{0.99}Co_4Sb_{9.65}Ge_{2.51}$, $Pr_{0.97}Co_4Sb_{9.52}Ge_{2.61}$, and $Nd_{0.87}Co_4Sb_{9.94}Ge_{2.28}$. Their structural parameters are refined by Rietveld analysis. The guest atom size does not affect the unit cell volume. The Co–Sb/Ge distance mainly determines the unit cell size as well as the size of guest atom site. The valence state of lanthanide ions is 3+.

Keywords: skutterudite; intermetallic compound; high-pressure; thermoelectric materials

---

1. Introduction

The filled skutterudite compounds have attracted much attention after the finding of the anomalous superconductivity of $PrOs_4Sb_{12}$ [1]. It has a cage structure composed of corner-sharing MX_6 octahedra and large guest atoms A, which are situated in the cages, where A, M, and X are electropositive elements (mainly alkaline earth and rare earth elements), transition metals (group 8 and 9), and electronegative elements (mainly group 15), respectively. A lot of isotypic compounds were synthesized and examined to expand the skutterudite chemistry.

The high-pressure and high-temperature reactions are very effective to prepare new skutterudite compounds, including superconductors [2–5], and a great many filled skutterudite compounds have been reported for the widespread combination of the elements by the high-pressure techniques. Surprisingly, there are unique derivatives IM_4Sb_{12}, M = Rh and Co, which have anionic guest atoms (iodine) in the cages [6–8]. Recently, bromine filled skutterudite has been reported [9].

The group 14 elements, however, are slightly difficult to introduce into the structure. Nolas et al. reported the first filled skutterudite compounds containing group 14 elements in the M site, $LnIr_4Sb_9Ge_3$, Ln = La, Nd, and Sm. In these compounds, the electrons from the guest trivalent cations compensate for the electron deficiency of the host network caused by the substitution of a group 14 element for the group 15 element [10–12]. We also prepared a series of Ge-substituted filled skutterudite compounds $LnRh_4Sb_9Ge_3$, Ln = La, Ce, Pr, and Nd, using high-pressure and high-temperature reactions [13]. They were stable at ambient pressure and the oxidation state of the guest atoms was 3+. Takizawa and Nolas also reported an interesting compound $Ge_{0.22}Co_4Sb_{11.4}Ge_{0.6}$, where germanium atoms were situated in the A site as well as in the X site [14].

In the present study, I used cobalt instead of Rh. Mori et al. also prepared a Ge-substituted skutterudite compound, which contains Yb in the *Ln* site, $Yb_yCo_4Sb_{11.5}Ge_{0.5}$ [15]. I performed high-pressure and high-temperature reactions in the *Ln*-Co-Sb-Ge system, *Ln* = La, Ce, Pr, and Nd, to obtain a new series of Ge-substituted filled skutterudite compounds.

2. Results and Discussion

Dark gray products were obtained after the high-pressure reactions. They were stable in air under ambient pressure. The XRD patterns of obtained samples are shown in Figure 1. The diffractions for each compound were indexed with a cubic unit cell having the systematic absence of a space group *I m-3*. The peak pattern indicated that the obtained compounds have the skutterudite structure. Small amounts of Sb (containing small amount of Ge) and monoclinic $Co(Sb,Ge)_2$ were also detected as shown in the figure.

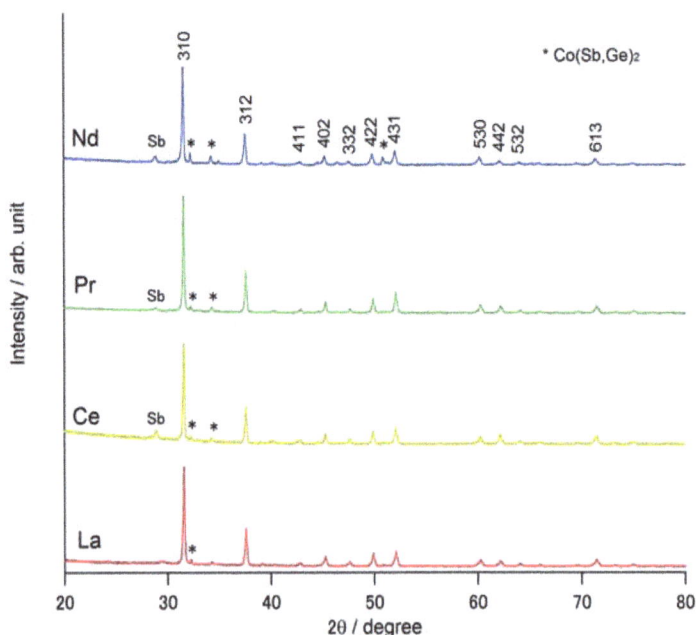

Figure 1. Powder XRD patterns of La, Ce, Pr, and Nd samples. The main phase is a Ge-substituted antimony skutterudite compound for each sample. Small diffraction peaks of Sb-Ge solid solutions and monoclinic $Co(Sb,Ge)_2$ are observed.

The presence of Ge in each compound was confirmed by chemical analysis. The chemical compositions for the samples determined by EPMA are summarized in Table 1. The relative number of atoms of each element is calculated according to an assumption that there is no defect in the Co site because the transition metal sites of the skutterudite structure are fully occupied in most skutterudite compounds. From these observations, the products of the high-pressure synthesis were identified as Ge-substituted filled skutterudite compounds.

I performed the Rietveld analysis for all compounds in order to confirm the site occupancy of the *Ln* sites and the Ge substitution for the Sb site. Rietveld structure analysis was performed using RIETAN-FP [16] on the powder XRD data collected with a D8 advance X-ray diffractometer (Bruker AXS, Karlsruhe, Germany) with CuKα radiation from 2θ range of 20° to 105°. The refined lattice constant for each compound is shown in Table 1. The results of the pattern fitting in Rietveld analysis are shown in Figure 2.

Table 1. Lattice constant and chemical composition of Ge-substituted skutterudite compounds.

Composition	$La_{0.57}Co_4Sb_{10.1}Ge_{2.38}$	$Ce_{0.99}Co_4Sb_{9.65}Ge_{2.51}$	$Pr_{0.97}Co_4Sb_{9.52}Ge_{2.61}$	$Nd_{0.87}Co_4Sb_{9.94}Ge_{2.28}$
Lattice constant (Å)	8.9504 (3)	8.94481 (6)	8.9458 (3)	8.9509 (4)

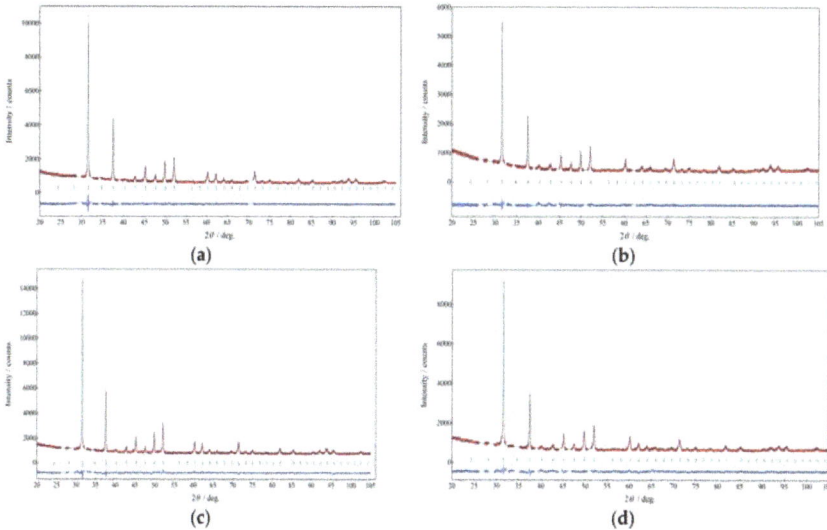

Figure 2. The XRD patterns and the results of pattern fitting; (**a**) $La_{0.57}Co_4Sb_{10.1}Ge_{2.38}$; (**b**) $Ce_{0.99}Co_4Sb_{9.65}Ge_{2.51}$; (**c**) $Pr_{0.97}Co_4Sb_{9.52}Ge_{2.61}$; (**d**) $Nd_{0.87}Co_4Sb_{9.94}Ge_{2.28}$. The observed data are shown as red points and the calculated fits and difference curves are shown as green and blue lines, respectively. Tick marks show the calculated diffraction positions.

All samples contain a very small amount of Sn-Ge solid solutions and monoclinic $Co(Sb,Ge)_2$ as shown in Figure 1. The diffraction peaks of those phases were excluded in Rietveld analysis so that they might not exert an adverse influence on the refinement.

The occupational parameters of *Ln* and Sb sites were refined as well as the atomic parameters and isotropic temperature factors for all sites. Split Pearson VII functions were used as profile functions. An overall isotropic temperature factor was refined only in the case of the La compound and a same isotropic temperature factor was applied to the Co and Sb/Ge sites only in the case of Ce compound due to the difficulty for convergence in the refinements.

The crystallographic data and some *R* factors are listed in Table 2. The atomic parameters and isotropic temperature factors for each compound are shown in Table 3. The refinement was well converged for each case. The *R* values are small and the *S* values are less than 1.3 for all compounds.

While the *Ln* sites of Ce, Pr, and Nd compounds were almost fully occupied, the La site occupancy turned out to be 0.7. I checked the possibility that some Ge atoms occupy the La site because some Ge atoms are situated in the guest site in a ternary skutterudite compound $Ge_{0.22}Co_4Sb_{11.4}Ge_{0.6}$ [14]. However, I could not get any meaningful results in all trials. I therefore concluded that the La site contained atom vacancies and did not contain any Ge atoms in the guest atom site. This site is fundamentally a site for cationic species in group 9–group 15 type skutterudite compounds. Therefore, if there is an electropositive element like La in the system, Ge appears not to be able to occupy the guest site.

The relatively larger isotropic temperature factors of the *Ln* sites may suggest an off-centered disorder of lanthanide ions. The analysis of it is, however, beyond the potential use of the present data.

The crystal structure of my compounds is illustrated in Figure 3. The refined occupational parameter of the *Ln* site for each compound is in good agreement with the composition determined by EPMA, when the standard deviations are taken into account. The M and X sites are occupied by cobalt and antimony atoms, respectively. Germanium atoms substituted randomly for some antimony atoms. The refined occupational parameters of the M site for the Ce and Pr compounds well reproduce the atomic ratios of Sb/Ge determined by EPMA, when the standard deviations are taken into account. Those parameters for La and Nd did not show good agreement with the Sb/Ge ratios determined by EPMA. This would be due to the fact that it is difficult to determine occupational parameters and temperature factors simultaneously. Especially in the case of the La compound, I applied an overall isotropic thermal parameter in the refinement, which might affect the poor estimation of the site occupancy of the La site. Even so, it can be clearly concluded that Ge atoms preferentially occupied the X site in the present systems.

Table 2. Crystallographic data and *R* indices of Rietveld analysis.

Formula	$La_{0.57}Co_4Sb_{10.1}Ge_{2.38}$	$Ce_{0.99}Co_4Sb_{9.65}Ge_{2.51}$	$Pr_{0.97}Co_4Sb_{9.52}Ge_{2.61}$	$Nd_{0.87}Co_4Sb_{9.94}Ge_{2.28}$
Space group	*I m-3* (204)	*I m-3* (204)	*I m-3* (204)	*I m-3* (204)
Lattice parameter $(a/Å)$	8.9504 (3)	8.94481 (6)	8.9458 (3)	8.9509 (4)
Unit cell volume $V/Å^3$	717.02 (4)	715.670 (9)	715.90 (4)	717.13 (6)
2θ range/degree	20–105	20–105	20–105	20–105
$R_{wp}/\%$	4.23	5.03	4.04	4.35
$R_p/\%$	3.33	3.94	3.16	3.44
$R_e/\%$	3.59	4.46	3.20	3.56
$R_B/\%$	3.86	7.15	7.17	6.40
$R_F/\%$	3.04	3.31	3.95	4.41
S	1.18	1.13	1.26	1.22

$R_{wp} = [\Sigma_i w_i\{y_i - I_i\}^2/\Sigma_i w_i y_i^2]^{1/2}$, $R_p = \Sigma_i |y_i - I_i|/\Sigma_i y_i$, $R_e = [(N - P)/\Sigma_i w_i y_i^2]^{1/2}$, $R_B = \Sigma_k |I_k('o') - I_k(c)|/\Sigma_k I_k('o')$, $R_F = \Sigma_k |[I_k('o')]^{1/2} - [I_k(c)]^{1/2}|/\Sigma_k [I_k('o')]^{1/2}$, $S = R_{wp}/R_e$; y_i: observed intensity, I_i: calculated intensity; w_i: weight; N: number of data; P: number of parameters; $I_k('o')$: estimated observed intensity of the k-th reflection; $I_k(c)$: calculated intensity of the k-th reflection.

Table 3. Structural parameters *ocp*, *n*, *x*, *y*, *z*, and $B/Å^2$ of the Ge-substituted cobalt antimony skutterudite compounds.

	ocp	*n**	*x*	*y*	*z*	$B/Å^2$
$La_{0.57}Co_4Sb_{10.1}Ge_{2.38}$						
La	0.704 (8)	1.41	0	0	0	overall
Co	1	8	0.25	0.25	0.25	0.03 (5)
Sb	0.72 (2)	17.28	0	0.3396 (2)	0.1599 (2)	
Ge	0.28	6.72	0	0.3396	0.1599	
$Ce_{0.99}Co_4Sb_{9.65}Ge_{2.51}$						
Ce	0.90 (3)	1.78	0	0	0	2.8 (5)
Co	1	8	0.25	0.25	0.25	0.05 (4)
Sb	0.85 (2)	20.3	0	0.3384 (5)	0.1611 (3)	0.05
Ge	0.15	3.7	0	0.3384	0.1611	0.05
$Pr_{0.97}Co_4Sb_{9.52}Ge_{2.61}$						
Pr	0.94 (2)	1.88	0	0	0	3.0 (4)
Co	1	8	0.25	0.25	0.25	0.6 (3)
Sb	0.77 (5)	18.48	0	0.3408 (2)	0.1581 (2)	0.5 (1)
Ge	0.23	5.52	0	0.3408	0.1581	0.5
$Nd_{0.87}Co_4Sb_{9.94}Ge_{2.28}$						
Nd	0.87 (3)	1.74	0	0	0	5.6 (5)
Co	1	8	0.25	0.25	0.25	0.2 (3)
Sb	0.63 (4)	15.2	0	0.3376 (2)	0.1572 (2)	0.3 (2)
Ge	0.37	8.8	0	0.3376	0.1572	0.3

* *n*: number of equivalent atoms per unit cell.

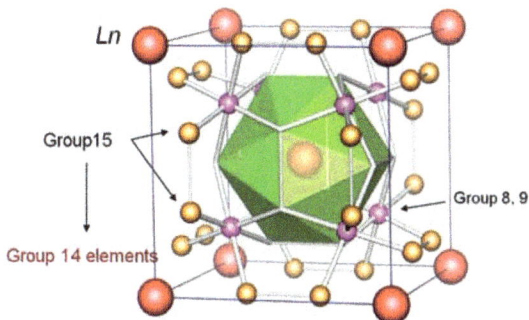

Figure 3. Crystal structure of the filled skutterudite compounds. In the present study, *Ln* sites are occupied by La, Ce, Pr, and Nd. The group 15 element is antimony, a part of which is substituted with germanium atoms. The transition metal sites are occupied by cobalt atoms.

The lattice constants of Ge-substituted compounds are smaller than that of $CoSb_3$ (9.039 Å [17]) because the radius of Ge is smaller than that of Sb. Figure 4 shows that the lattice parameter was not affected by the size of the guest ion, whether it was La, Ce, Pr, or Nd. They are almost constant. This means that the size of the unit cell is determined mainly by the covalent connection in the Co-(Sb,Ge) host network. Similar behavior was observed for the lattice constants of $LnRh_4Sb_9Ge_3$, *Ln* = La, Ce, Pr, and Nd [10]. The lattice constants for those Rh compounds were in a very narrow range from 9.112 to 9.107 Å.

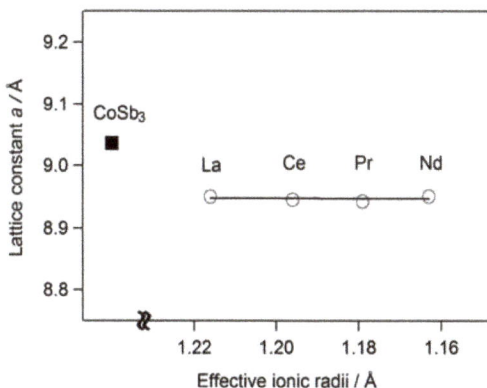

Figure 4. Lattice constant of the Ge-substituted cobalt skutterudite compounds.

The importance of the host network size is also proved by the fact that the lattice constants of the Ce and Pr compounds are slightly shorter than those of the La and Nd compounds. The former compounds contain a slightly larger amount of Ge in the X sites than the latter ones. The size of the host network is, thus, the principal parameter in these systems to determine the unit cell volume. This size effect of the host network also effectively explains the reason why it is difficult to prepare the analogs containing heavy rare earth elements, which have a smaller ionic radius than light ones. I tried to prepare a Lu analog by the same reaction condition but could not obtain it. Their small radius would not fit the host network size and such a system would become unstable. In contrast, the size of La^{3+} would be slightly too large for the cages. This indicates that difference of ionic size is why only the La site showed atomic vacancy.

The interatomic distances between host antimony/germanium atoms and guest *Ln* atoms are derived from the results of Rietveld analysis; they are 3.360, 3.353, 3.361, and 3.333 Å for the La, Ce, Pr, and Nd compounds, respectively. These values can be used as an indicator of the size of the A site. It is noteworthy that the interatomic distances of those compounds are very similar to that of $CoSb_3$ (3.352 Å [17]), even though the lattice constants of them remarkably decrease. Introduction of Ge atoms in the host network, therefore, contributes to the expansion of size of the A site, which can be correlated with the degree of f-electron localization in the lanthanide metal [18,19].

Magnetic susceptibility measurements revealed that the valence state of Ce and Nd was 3+. The temperature dependence of magnetic susceptibility of each compound is shown in Figure 5. The inset plots show Curie-Weiss behavior of the samples. Their susceptibilities obey a modified Curie-Weiss law, $\chi_{mol} = \chi_0 + C/(T - \theta)$, where χ_{mol} is a molar magnetic susceptibility, C is the Curie constant, θ is the Weiss temperature, and χ_0 is a paramagnetic term including a Van Vleck contribution. The data was approximated by the equation in the temperature range from 2 to 295 K for Ce and from 4 to 300 K for Nd. The obtained parameters in the fitting are summarized in Table 4. The effective moments calculated from their Curie constants are 2.12 μ_B and 3.12 μ_B for Ce and Nd, respectively. They are slightly smaller than the theoretical values of 2.54 μ_B and 3.62 μ_B for Ce^{3+} and Nd^{3+}. Although their valence states are basically 3+, we may have to consider mixed valence state or Kondo system especially for the Ce compound.

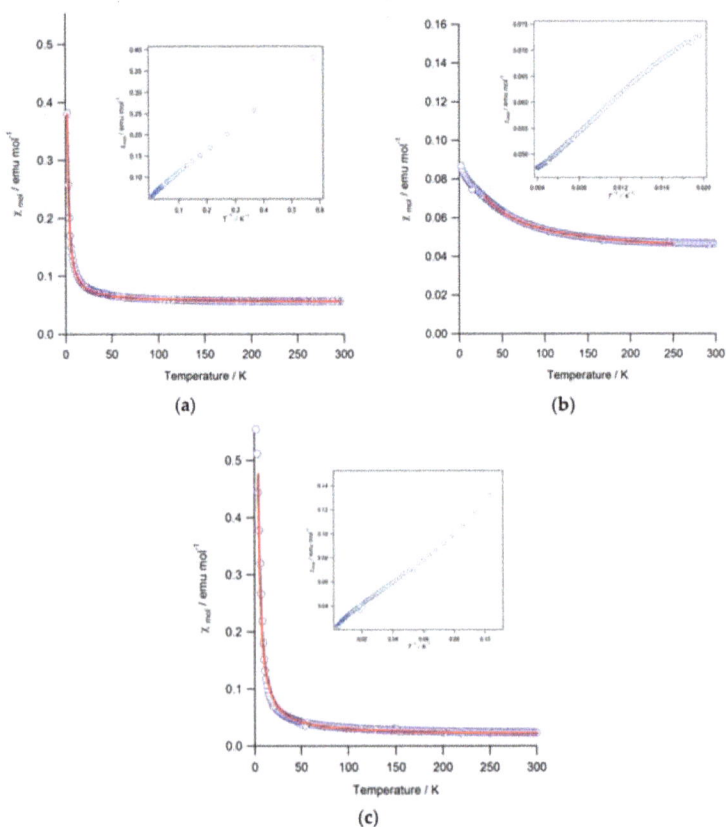

Figure 5. Temperature dependence of the magnetic susceptibility of (a) $Ce_{0.99}Co_4Sb_{9.65}Ge_{2.51}$; (b) $Pr_{0.97}Co_4Sb_{9.52}Ge_{2.61}$; and (c) $Nd_{0.87}Co_4Sb_{9.94}Ge_{2.28}$.

Table 4. Curie constant (*C*), Weiss temperature (*θ*), a paramagnetic term (χ_0) and the effective moment μ_{eff} (with theoretical values in parentheses) derived from regression calculations using a modified Curie-Weiss equation.

Compound	C (emu mol^{-1} K^{-1})	θ (K)	χ_0 (emu mol^{-1})	μ_{eff} (μ_B)	Temp. Range (K)
Ce$_{0.99}$Co$_4$Sb$_{9.65}$Ge$_{2.51}$	0.563 (3)	0.27 (1)	0.0544 (1)	2.12 (2.54)	2–295
Pr$_{0.97}$Co$_4$Sb$_{9.52}$Ge$_{2.61}$	1.78 (4)	−27 (1)	0.0401 (2)	3.77 (3.58)	30–250
Nd$_{0.87}$Co$_4$Sb$_{9.94}$Ge$_{2.28}$	1.22 (3)	1.33 (8)	0.0175 (7)	3.12 (3.62)	4–300

The data of the Pr compound did not obey the equation. I have, however, applied the equation to the data in a reduced temperature range from 30 to 250 K and obtained a well-fitting curve. The calculated effective moment of 3.77 μ_B is in good agreement with the theoretical value of 3.58 μ_B for Pr^{3+}. The valence state of Pr would be 3+ in the temperature range; however, further investigation should be necessary to determine the magnetic behavior of the Pr compound as well as the Ce compound.

The partial substitution of Ge atoms for Sb atoms induces a structural disorder in the host network. Such a disorder would have an effect of decreasing the lattice thermal conductivity [10–12]. It is desirable to evaluate the thermal conducting and thermoelectric properties for these compounds.

3. Materials and Methods

In the sample preparation I used rare earth elements (99.9%, Rare Metallic Co., Ltd., Tokyo, Japan), cobalt powder (99.9%, Aldrich Chem. Co., St. Louis, MO, USA), antimony powder (99.999%, Katayama Chemical, Osaka, Japan), and germanium (99.999%, Rare Metallic Co., Ltd., Tokyo, Japan). I first prepared digermanides *Ln*Ge$_2$ for La, Ce, Pr, and Nd with an argon-filled arc furnace because the elemental lanthanides are highly unstable and are easily oxidized in air. The mixtures of *Ln*Ge$_2$, Co, Sb, and Ge with a molar ratio of 1:4:9:1 (*Ln*:Co:Sb:Ge = 1:4:9:3) were reacted using a Kawai-type (6–8 type) high-pressure system according to the following process [20]. Each mixture was placed in an h-BN container with 2 mm in inner diameter and 4 mm in depth. The container covered with Ta foil, which was used as a heater, was put in a thermal-insulating pyrophyllite tube with 6 mm in diameter and 1 mm thick. A Pt/Pt-Rh thermo couple was used to control the reaction temperature during heating. This sample unit was placed in an octahedral MgO pressure medium, and was put at the center of eight tungsten carbide anvils. This reaction cell was pressed by a multi-anvil press. The samples were reacted at 7 GPa and 800 °C for 1 h, and was rapidly cooled down to room temperature. After the sample temperature became room temperature, the pressure was slowly released.

All products were characterized by powder X-ray diffraction (XRD) method and electron probe micro analysis (EPMA) (JEOL 733II, Tokyo, Japan). Rietveld structure analysis was performed using RIETAN-FP [16] on the powder XRD data collected with a D8 advance X-ray diffractometer (Bruker AXS, Karlsruhe, Germany) with CuKα radiation from 2θ range of 20° to 105°. The magnetic susceptibility of the samples were measured with a SQUID magnetometer (Quantum Design MPMS5s, San Diego, CA, USA) applying a magnetic field of 5000 Oe.

4. Conclusions

High-pressure and high-temperature reactions provide a new series of Ge-substituted filled skutterudite compounds, *Ln*$_x$Co$_4$Sb$_{12-y}$Ge$_y$, where *Ln* = La, Ce, Pr, and Nd. The guest site is fully occupied by lanthanide ions except for the La compound, whose occupancy *x* is 0.6. The germanium content, *y*, is different for each compound and is in the range from 2.3 to 2.6. The unit cell volume is not affected by the guest atom size. It is principally determined by the host network formed by the Co-Sb/Ge covalent bonds. The valence state of the guest lanthanide ions is basically 3+.

Acknowledgments: This work was supported by Grant-in-Aids for Scientific Research from the Ministry of Education, Science, and Culture of Japan, grant nos. 16037212, 16750174, 18750182, 18027010, 20550178, and 16K05724.

Conflicts of Interest: The author declares no conflict of interest.

References

1. Bauer, E.D.; Frederick, N.A.; Ho, P.-C.; Zapf, V.S.; Maple, M.B. Superconductivity and heavy fermion behavior in $PrOs_4Sb_{12}$. *Phys. Rev. B* **2002**, *65*, 1005061–1005064. [CrossRef]
2. Shirotani, I.; Shimaya, Y.; Kihou, K.; Sekine, C.; Yagi, T. Systematic high-pressure synthesis of ne wfilled skutterudites with heavy lanthanide, $LnFe_4P_{12}$ (*Ln* = heavy lanthanide, including Y). *J. Solid State Chem.* **2003**, *174*, 32–34. [CrossRef]
3. Sekine, C.; Uchiumi, T.; Shirotani, I.; Matsuhira, K.; Sakakibara, T.; Goto, T.; Yagi, T. Magnetic properties of the filled skutterudite-type structure compounds $GdRu_4P_{12}$ and $TbRu_4P_{12}$ synthesized under high pressure. *Phys. Rev. B* **2000**, *62*, 11581–11584. [CrossRef]
4. Kihou, K.; Shirotani, I.; Shimaya, Y.; Sekine, C.; Yagi, T. High-pressure synthesis, electrical and magnetic properties of new filled skutterudites $LnOs_4P_{12}$ (*Ln* = Eu, Gd, Tb, Dy, Ho, Y). *Mater. Res. Bull.* **2004**, *39*, 317–325. [CrossRef]
5. Shirotani, I.; Areseki, N.; Shimaya, Y.; Nakata, R.; Kihou, K.; Sekine, C.; Yagi, T. Electrical and magnetic properties of new filled skutterudites $LnFe_4P_{12}$ (*Ln* = Ho, Er, Tm and Yb) and YRu_4P_{12} with heavy lanthanide (including Y) prepared at high pressure. *J. Phys. Condens. Matter* **2005**, *17*, 4383–4391. [CrossRef]
6. Fukuoka, H.; Yamanaka, S. High-pressure synthesis, structure, and electrical property of iodine-filled skutterudite $I_{0.9}Rh_4Sb_{12}$—First anion-filled skutterudite. *Chem. Mater.* **2010**, *22*, 47–51. [CrossRef]
7. Li, X.; Xu, B.; Zhang, L.; Duan, F.; Yan, X.; Yang, J.; Tian, Y. Synthesis of iodine filled $CoSb_3$ with extremely low thermal conductivity. *J. Alloys Compd.* **2014**, *615*, 177–180. [CrossRef]
8. Zhang, L.; Xu, B.; Li, X.; Duan, F.; Yan, X.; Tian, Y. Iodine-filled $Fe_xCo_{4-x}Sb_{12}$ polycrystals: Synthesis, structure, and thermoelectric properties. *Mater. Lett.* **2015**, *139*, 249–251. [CrossRef]
9. Ortiz, B.R.; Crawford, C.M.; McKinney, R.W.; Parillab, P.A.; Toberer, E.S. Thermoelectric properties of bromine filled $CoSb_3$ skutterudite. *J. Mater. Chem. A* **2016**, *4*, 8444–8450. [CrossRef]
10. Nolas, G.S.; Slack, G.A.; Caillat, T.; Meisner, G.P. Raman scattering study of antimony-based skutterudites. *J. Appl. Phys.* **1996**, *79*, 2622–2626. [CrossRef]
11. Nolas, G.S.; Slack, G.A.; Morelli, D.T.; Tritt, T.M.; Ehrlich, A.C. The effect of rare-earth filling on the lattice thermal conductivity of skutterudites. *J. Appl. Phys.* **1996**, *79*, 4002–4008. [CrossRef]
12. Tritt, T.M.; Nolas, G.S.; Slack, G.A.; Ehrlich, A.C.; Gillespie, D.J.; Cohn, J.L. Low-temperature transport properties of the filled and unfilled $IrSb_3$ skutterudite system. *J. Appl. Phys.* **1996**, *79*, 8412–8418. [CrossRef]
13. Fukuoka, H.; Yamanaka, S. High-pressure synthesis and structure of new filled skutterudite compounds with Ge-substituted host network; $LnRh4Sb9Ge3$, *Ln* = La, Ce, Pr, and Nd. *J. Alloys Compd.* **2008**, *461*, 547–550. [CrossRef]
14. Nolas, G.S.; Yang, J.; Takizawa, H. Transport properties of germanium-filled $CoSb_3$. *Appl. Phys. Lett.* **2004**, *84*, 5210–5213. [CrossRef]
15. Mori, H.; Anno, H.; Matsubara, K. Effect of Yb filling on thermoelectric properties of Ge-substituted $CoSb_3$ skutterudites. *Mater. Trans.* **2005**, *46*, 1476–1480. [CrossRef]
16. Izumi, F.; Momma, K. Three-dimensional visualization in powder diffraction. *Solid State Phenom.* **2007**, *130*, 15–20. [CrossRef]
17. Schmidt, T.; Kliche, G.; Lutz, H.D. Structure refinement of skutterudite-type cobalt triantimonide, $CoSb_3$. *Acta Cryst. C* **1987**, *C43*, 1678–1679. [CrossRef]
18. Errandonea, D.; Boehler, R.; Schwager, B.; Mezouar, M. Structural studies of gadolinium at high pressure and temperature. *Phys. Rev. B* **2007**, *75*, 014103. [CrossRef]
19. Cunningham, N.C.; Qiu, W.; Hope, K.M.; Liermann, H.-P.; Vohra, Y.K. Symmetry lowering under high pressure: Structural evidence for f-shell delocalization in heavy rare earth metal terbium. *Phys. Rev. B* **2007**, *76*, 212101. [CrossRef]
20. Fukuoka, H. High-pressure and properties of new filled skutterudite compounds using a Kawai-type cell. *Rev. High Press. Sci. Technol.* **2006**, *16*, 329–335. [CrossRef]

MDPI

St. Alban-Anlage 66

4052 Basel

Switzerland

Tel. +41 61 683 77 34

Fax +41 61 302 89 18

www.mdpi.com

E-mail: books@mdpi.com

www.mdpi.com/books

www.ingramcontent.com/pod-product-compliance
Lightning Source LLC
Chambersburg PA
CBHW051714210326
41597CB00032B/5471